HETEROCYCLIC CHEMISTRY

HETEROCYCLIC CHEMISTRY

Fourth Edition

RAJ K. BANSAL, Ph. D.

ANSHAN LTD
6 Newlands Road
Tunbridge Wells, Kent.
TN4 9AT. UK

Co-published in the U.K. by

ANSHAN LTD., 6 Newlands Road, Tunbridge Wells, Kent TN4 9AT
In 2008

Tel-Fax: +44(0)1892557767
e-mail: info@anshan.co.uk
Web Site: www.anshan.co.uk

© 2008 by New Age International (P) Ltd., Publishers

ISBN: 978-1848-290-013

British Library Cataloguing in Publication Data
A Catalogue record for this book is available from the British Library

Dedicated affectionately to

Robin Bansal and Caroline Bansal

Preface

Heterocyclic chemistry is vastly expanding because of the enormous amount of research work being done in this area. Besides heterocyclic compounds have manifold applications in pharmacy, medicine, agriculture and allied fields. In this connection a great deal of changes have taken place since the third edition of this book was published. Two principal considerations prompted me to undertake a thorough revision to the book. First, my long experience in teaching coupled with the suggestions of the critics who pointed out the immediate need of improving certain sections of the book. Second, new and recent results of the past several years of research necessitated their inclusion.

The fourth edition of the book retains many of the features that have proved successful in earlier editions. The book still comprises of eleven comprehensive chapters and discusses the heterocyclic compounds ring by ring. New recent synthesis and methodology have been included. The emphasis remains on stereochemistry of reactions and their mechanisms. Much new material has been added and old information deleted. The art work has been redone to improve the pedagogical value of the illustrations. The names of all compounds under the diagrams are now written in boldface.

The book is meant primarily for a one year course in the study of heterocyclic chemistry at the graduate and the postgraduate level. However, a large number of students working in the area of biochemistry and medicine can also immensely benefit from it.

A set of new problems from the original literature are appended at the end of each chapter. The bibliography has been updated.

I am indebted to the readers of this book for constantly sending me their suggestions, comments and criticisms. I highly appreciate their efforts. I am particularly grateful to my former colleagues at the Indian Institute of Technology, Delhi for their advice and assistance in the task of improving and preparing this fresh edition.

Finally, I owe a sense of gratitude of my wife Mrs. K. Bansal for her pleasant cooperation and moral support during the period of revision.

Raj K. Bansal

Contents

Chapter 1

Nomenclature of Heterocyclic Compounds

A cyclic organic compound containing all carbon atoms in ring formation is referred to as a *carbocyclic compound.* If at least one atom other than carbon, forms a part of the ring system then it is designated as a *heterocyclic* compound.[1-15] Nitrogen, oxygen and sulfur are the most common heteroatoms but heterocyclic rings containing other hetero atoms are also widely known. An enormous number of heterocyclic compounds are known and this number is increasing rapidly. Accordingly the literature on the subject is very vast. Heterocyclic compounds may be classified into *aliphatic* and *aromatic.* The aliphatic heterocyclics are the cyclic analogues of amines, ethers, thioethers, amides, etc. Their properties are particularly influenced by the presence of strain in the ring. These compounds generally consist of small (3- and 4- membered) and common (5 to 7 membered) ring systems. The aromatic heterocyclic compounds, in contrast, are those which have a heteroatom in the ring and behave in a manner similar to benzene in some of their properties. Furthermore, these compounds also comply with the general rule proposed by Hückel. *This rule states that aromaticity is obtained in cyclic conjugated and planar systems containing* $(4n + 2)$ *π electrons.* The conjugated cyclic rings contain six π-electrons as in benzene, and this forms a conjugated molecular orbital system which is thermodynamically more stable than the non-cyclically conjugated system. This extra stabilization results in a diminished tendency of the molecule to react by addition but a larger tendency to react by substitution in which the aromatic ring remains intact.

A heterocyclic ring may comprise of three or more atoms which may be saturated or unsaturated. Also the ring may contain more than one hetero atom which may be similar or dissimilar.

The chemistry of heterocyclic compounds is as logical as that of aliphatic or aromatic compounds. Their study is of great interest both from the theoretical as well as practical standpoint. Heterocyclic compounds occur widely in nature and in a variety of non-naturally occurring compounds. A large number of heterocyclic compounds are essential to life. Various compounds such as alkaloids, antibiotics, essential amino acids, the vitamins, haemoglobin, the hormones and a large number of synthetic drugs and dyes contain heterocyclic ring systems. A knowledge of heterocyclic chemistry is useful in biosynthesis and in drug metabolism as well. Nucleic acids are important in biological processes of heredity and evolution.

There are a large number of synthetic heterocyclic compounds with additional important applications and many are valuable intermediates in synthesis.

1.1 NOMENCLATURE[14]

In heterocyclic chemistry there is a special name for each individual ring system and a trivial name for each compound. Trivial names convey little or no structural information but they are still widely used. The systematic name, in contrast, is designed so that one may deduce from it the structure of the compound. They tend to be long. However, a systematic nomenclature is still indispensable.

In recent years the International Union of Pure and Applied Chemistry (IUPAC) has made efforts to systematize the nomenclature of heterocyclic compounds.

According to this system single three-to-ten-membered rings are named by combining the appropriate *prefix* or *prefixes* (listed in Table 1.1) with a stem from Table 1.2.

Table 1.1 Prefix for Hetero Atoms

Hetero atom	Valence	Prefix
O	2	Oxa
N	3	Aza
S	2	Thia
Se	2	Selena
Te	2	Tellura
P	3	Phospha
As	3	Arsa
Si	4	Sila
Ge	4	Germa

Table 1.2 Common Name Endings for Heterocyclic Compounds

Ring size	Suffixes for fully unsaturated compounds		Suffixes for fully saturated compounds	
	With N	Without N	With N	Without N
3	-irine	-irene	-iridine	-irane
4	-ete	-ete	-etidine	-etane
5	-ole	-ole	-olidine	-olane
6	-ine	-in		-ane
7	-epine	-epin		-epane
8	-ocine		-ocin	-ocane

The endings in Table 1.2 also indicate the size of the ring and the state of hydrogenation with or without the presence of a nitrogen atom.

Accordingly some examples of compounds named on the basis of above two tables are cited below:

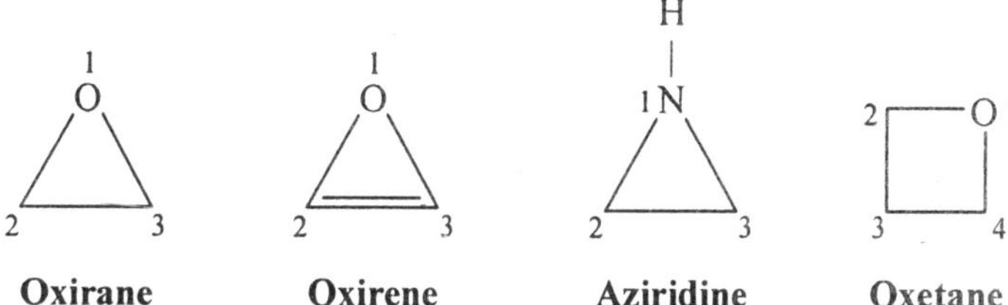

Oxirane   **Oxirene**   **Aziridine**   **Oxetane**

In all these examples the letter '*a*' has been omitted. However, in the following cases letter '*a*' has been retained.

Azacyclobutadiene   **2-Oxabicyclo [3.3.0] Octane**

Saturated or hydrogenated ring systems are named by varying the ending or by placing prefixes such as '*dihydro-*', '*tetrahydro-*', etc. The ending of the name will depend on the presence or absence of nitrogen (Table 1.2).

1,3,5-Triazine   **Tetrahydrofuran**

Two or more similar atoms contained in a ring are indicated by the prefixes '*di-*', '*tri*', etc. placed before the appropriate '*a*' term (Table 1.2).

If two or more different hetero atoms occur in the ring, then it is named by combining the prefixes in Table 1.1 with the ending in Table 1.2 in order of their preference, i.e. O, S and N. This is illustrated by the following examples:

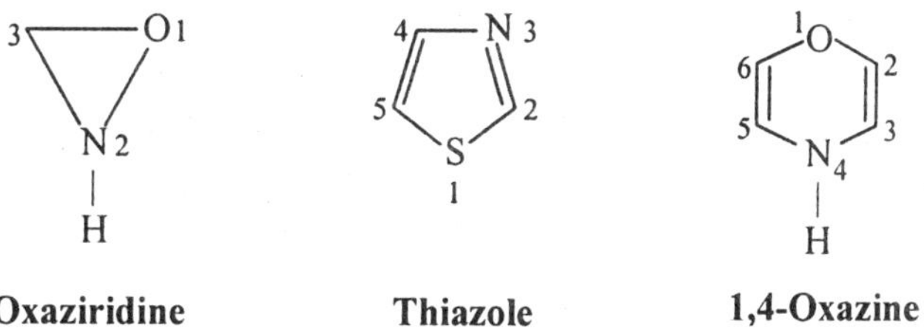

Oxaziridine   **Thiazole**   **1,4-Oxazine**

The position of a single hetero atom controls the numbering in a monocyclic compound but not necessarily in a bicyclic compound.

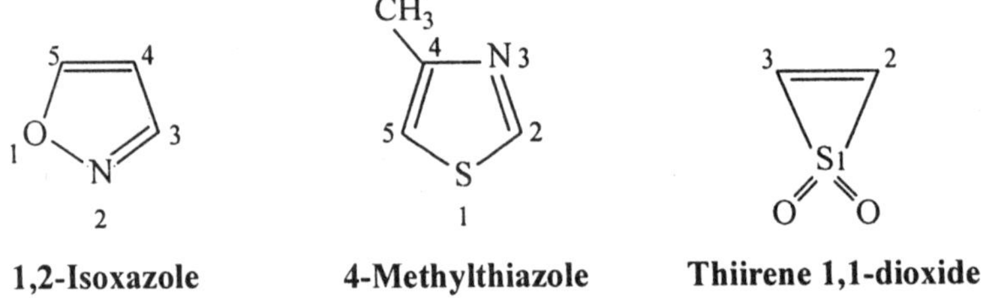

trans-2,4-Dimethylthietane   **2-Methylazete**   **3-Methylisoquinoline**

Numbering of the heterocyclic rings becomes essential when substituents are placed on the ring. Conventionally, the hetero atom is assigned position 1 and the substituents are then counted around the ring in a manner so as to give them the lowest possible numbers. While writing the name of the compound, the substituents are placed in an alphabetical order. In case the heterocyclic ring contains more than one hetero atom, the order of preference for numbering is O, S and N. The ring is numbered from the atom of preference in such a way so as to give the smallest possible number to the other hetero atoms in the ring. As a result the position of the substituent plays no part in determining how the ring is numbered in such compounds. The following examples illustrate this rule:

1,2-Isoxazole   **4-Methylthiazole**   **Thiirene 1,1-dioxide**

There are a large number of important ring systems which do not possess any systematic names rather non-systematic or common names are used for them. Some of such examples include the following:

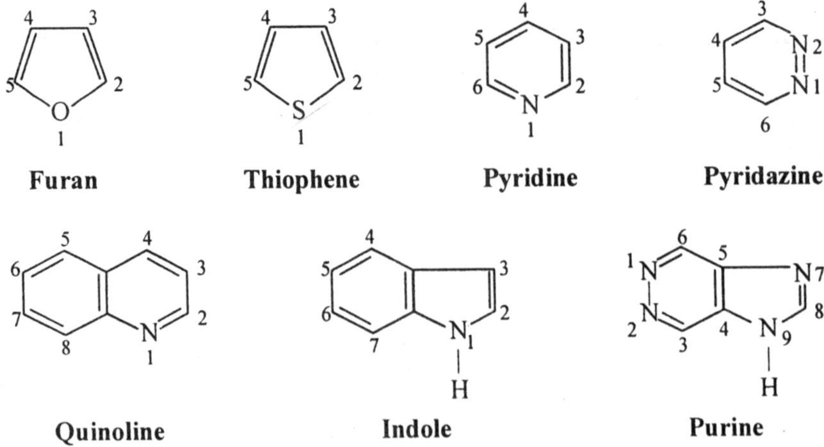

Furan   **Thiophene**   **Pyridine**   **Pyridazine**

Quinoline   **Indole**   **Purine**

Coumarin   **Acridine**

The numbering and nomenclature of heterocyclic rings become more complicated for condensed or fused ring systems, i.e., when a part of one ring is also a part of another ring. Such ring systems, however, are known by non-systematic or common names, such as indole, isatin, isoquinoline, etc., as indicated in the preceeding paragraph.

There is yet another system of nomenclature for fused rings that is commonly employed. According to this system, the side of the heterocyclic ring is labelled by the letters a, b, c, etc., starting from the atom numbered 1. Therefore side '*a*' being between atoms 1 and 2, side '*b*' between atoms 2 and 3, and so on as shown below for pyridine.

Pyridine

The name of the heterocyclic ring is chosen as the parent compound and the name of the fused ring is attached as a prefix. The prefix in such names has

Benzo [b] furan   **Benzo [b] pyridine**   **Benzo [c] thiophene**

the ending '*o*', i.e., benzo, naphtho and so on. The following examples explain this rule.

Benzo [d] thiepin   **Benzo [f] quinoline**

In a heterocyclic ring, other things being equal, numbering preferably commences at a saturated rather than at an unsaturated hetero atom, as depicted in the following examples:

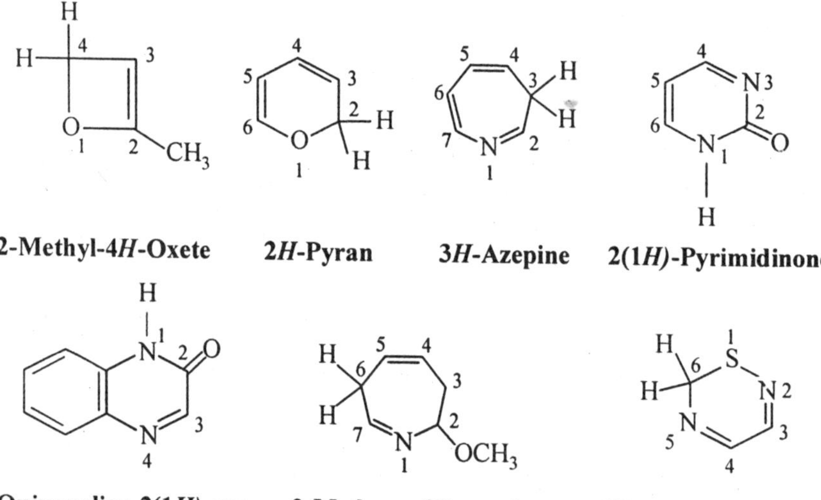

3-Ethyl-5-methylpyrazole   **1-Methylindazole**

In a heterocyclic ring with maximum unsaturation, if the double bonds can be arranged in more than one way, then their positions are specified by numbering those nitrogen or carbon atoms which are not multiply-bonded, i.e. bear an 'extra' hydrogen atom, by italic capital '1*H*' '2*H*' '3*H*', etc. The numerals indicate the position of these atoms having the extra hydrogen atom. The following examples illustrate this rule:

2-Methyl-4*H*-Oxete   **2*H*-Pyran**   **3*H*-Azepine**   **2(1*H*)-Pyrimidinone**

Quinoxaline-2(1*H*)-one   **2-Methoxy-6*H*-azepine**   **6*H*-1,2,5-Thiadiazine**

The position of the hydrogen atom in a partially saturated heterocyclic ring can be indicated by writing 1, 2-dihydro, etc. with the name of the compound. Alternatively, the position of the double bond can also be specified as Δ^1, Δ^2, Δ^3.,. etc., which indicates that 1 and 2; 2 and 3; 3 and 4 atoms respectively have a double bond.

5,5-Dimethyl-Δ^2-pyrroline N-oxide   **Δ^2-Pyrroline**

Δ^3-Tetrahydropyridine　　　**Δ^5-Dihydro-1,3,4-thiadiazine**　　　**Δ^2-Oxazoline**

A positively charged ring is denoted by the suffix "-ium".

Groups such as C = S and C = NH present in the ring are denoted by the suffixes "-thione" and "-imine".

Bicyclic bridged structures are quite common in heterocyclic chemistry. The nomenclature of such a structure consists of the prefix bicyclo, followed in square brackets the number of carbon atoms separating the bridge heads by the three possible routes in descending numerical order. This is followed by the alkane containing the same number of carbon as the whole bicyclic heterocyclic skeleton. The following examples illustrate the use of this rule.

3,6-Dioxabicyclo [3.1.0] octane　　　　**6-Hydroxyimino-7, 7-dimethyl-2-azabicyclo [2.2.1] heptan-5-one**

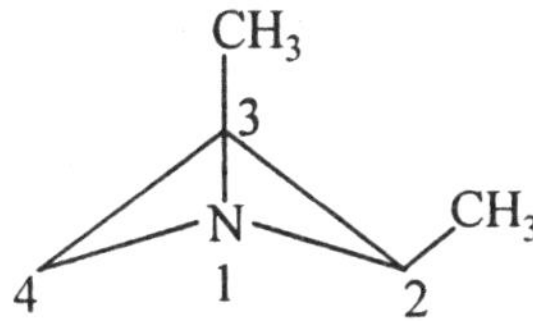

2,3-Dimethyl-1-azabicyclo [1.1.0] butane　　　　**6,6-Diphenyl-3-azabicyclo [3.1.0] hexane.**

REFERENCES

1. *"Advances in Heterocyclic Chemistry"*, Vols. 1 to 27, A. R. Katritzky and J. A. Boulton, (*Eds.*), Academic Press, New York (1963-1980).

2. *"The Chemistry of Heterocyclic Compounds"*, Vols. 1 to 29, A. Weissberger, (*Ed.*), Wiley Interscience, New York (1950 to 1975).

3. *"Physical Methods in Heterocyclic Chemistry"*, Vols. 1 to 5, A. R. Katritzky, (*Ed.*), Academic Press, New York (1963 to 1973).

4. *"Heterocyclic Chemistry"*, Vols. 1 to 9, R. C. Elderfield, (*Ed.*), Wiley, New York (1950 to 1967).

5. J. A. Joule and G. F. Smith, *Heterocyclic Chemistry,* Van Nostrand Reinhold Co., 2nd *ed.,* London (1978).

6. O. Büchardt, (*Ed.*), *Photochemistry of Heterocyclic Compounds,* John Wiley, New York (1976).

7. R. S. Chan, *Introduction to Chemical Nomenclature,* Butterworths, London (1974).

8. *Rodd's Chemistry of Carbon Compounds,* Vol. IV, Part A, B, S. Coffey, (*Ed.*), Elsevier, London (1973).

9. *C. R. C. Handbook of Chemistry and Physics,* R. C. Weast and M. J. Astle, (*Eds.*), C. R. C. Press, Inc., 63rd *ed.* Florida U. S. A. (1983).

10. J. H. Fletcher, O. C. Dermer and R. B. Fox, *Nomenclature of Organic Compounds, Principles and Practice,* American Chemical Society, Washington, D. C., *Adv. Chem. Ser.*

11. A. R. Katritzky, *Handbook of Heterocyclic Chemistry,* Pergamon Press, ' New York (1985).

12. G. R. Newkome and W. W. Pandler, *Contemporary Heterocyclic Chemistry,* John Wiley, New York (1982).

13. *"Comprehensive Heterocyclic Chemistry"*, Vol. 1-8, A. R. Katritzky.

14. A. D. McNaught in, *"Advances in Heterocyclic Chemistry"*, Vol. XX, A. R. Katritzky and A. J. Boulton (Eds.), Academic Press, New York (1976), pp. 175-319.

15. T. L. Gilchrist, *"Heterocyclic Chemistry"*, Pitman London (1985).

Three Membered Heterocyclic Compounds with One Hetero Atom

The three-membered heterocyclic compounds with one hetero atom have been known for a long time and are important from the synthetic and mechanistic point of view. They are formally derived from cyclopropane by replacing a carbon atom with a hetero atom. Such a change widely affects the physical and chemical properties of the resultant heterocyclic rings. This may, in part, be attributed to the compression of bond angles to a value around 116° between those expected of sp^3 (109.28°) and sp^2 (120°) hybridized carbon atoms. This compression of angles results in appreciable ring strain. Introduction of a double bond further enhances the ring strain. A double bond requires a 120° bond angle for sp^2 hybridized atoms and this inevitably increases the ring strain. The three-membered saturated heterocyclic compounds containing a nitrogen, oxygen or sulfur as the hetero atom are known as *aziridine, oxirane* and *thiirane* respectively. The unfavorable ring strain is not apparent in the synthesis of these ring systems as their formation is relatively easy because the two ends of a three atom intermediate can come closer for cyclization. Chemically they are more reactive than cyclopropane derivatives.

2.1. AZIRIDINES

Aziridines are the dihydro derivatives of parent azirines. Aziridine (1) is a saturated heterocyclic compound containing two carbons and one nitrogen atoms in a three-membered ring. Aziridines, as a class are of interest as biological alkylating[1] and anti-cancer agents.[2]

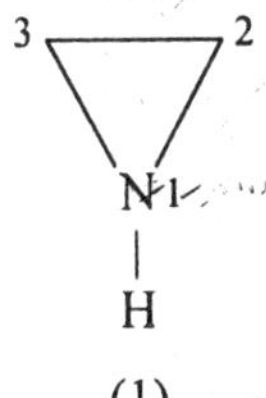

(1)

Besides, aziridine and its derivatives are produced commercially and are employed in plastic, coating and in the textile industries. The chemistry of these compounds has been extensively investigated in recent years.

Aziridine has often been called *azacyclopropane* or more commonly derivative of a parent alkene, *ethylenimine*. Its derivatives, for instance, compounds (2) and (3) are named N-methylpropylenimine, and cyclopentenimine, respectively. In the *Chemical Abstracts,* however, the name ethylenimine is used for the parent compound while its derivatives are indexed as aziridine, i.e. structure (2) is also called 1, 2-dimethylaziridine, and (3) 6-azabicyclo [3.1.0] hexane. The chemistry of these compounds has been reviewed.[3,4]

(2)   (3)

Aziridines and derivatives are potent pharmacological agents. The toxic effects of aziridine itself causes irritation of eyes, skin, and internal inflammation. Besides the ability of aziridine as an alkylating agent is of importance in industry and biology and this property has resulted in a number of industrial applications of aziridines. Certain antibiotics and anti-cancer agents possess the aziridine ring.

2.1.1. Physical and Spectroscopic Properties

Aziridine and its lower molecular weight derivatives are colorless liquids with a characteristic ammoniacal odor. Aziridine boils at 56°C and is miscible with water. It is strongly toxic to the skin and should be handled with care and contact with the skin be avoided. It is a weakly basic compound with a *p*Ka value of 7.98. A large number of alkylaziridines possess *p*Ka values in the range of 7.93-9.47. The reason for the weak basicity of aziridine and its derivatives is ascribed to the strain in the 3-membered ring compound. As distinct from this, the open-chain compound dimethylamine is a strong base with a *p*Ka of 10.87.

Microwave spectroscopy, which is used for the determination of bond distances and bond angles, has provided useful structural data for the aziridine ring system.

Aziridine ring is planar and rigid. The internal bond angles (C—C—N 59.6° and C—N—C 60.2°) in aziridine are very close to 60° compared to 113° for dimethylamine. The peripheral H—C—H bond angles are close to

118°. The C—C bond distance of 1.48Å is smaller than for the open-chain compound (C—C bond distance is of the order of 1.54Å) and cyclopropane. The ring strain from the combustion data has been estimated to be around 14 Kcal/mole which is apparently larger than that present in oxirane (13 Kcal/mole) or thiirane (9 Kcal/mole) but smaller than that present in cyclopropane (25 Kcal/mole). The existence of ring strain in aziridine has been further confirmed by infra-red spectroscopy, this is reflected in the increase in the C—H vibrational frequency from 1465 cm^{-1} to 1475 cm^{-1} and a decrease in the N—H vibrational frequency to 1441 cm^{-1} which is lower than that observed for secondary amines (1460 cm^{-1}). The calculated dipole moment value for aziridine is 2.09–2.40 D.

Simple aziridines absorb little light in the accessible region of the *u.v.* spectrum. However, there is an interaction between the substituents present on aziridines *via* the ring. This is reflected in the appearance of longer wavelength absorption maxima and greater molar absorptivities for the *trans* than the *cis* aziridines.[6]

2.1.2. Inversion in Aziridines

Examples of *cis-*and *trans-* aziridines are known. The two carbon atoms of aziridines are potentially chiral when unsymmetrically substituted ring should exist in four enantiomeric forms, *i.e.* two *cis* and two *trans.* If the carbon atoms bear the same substituents then the *cis* isomer is *meso* and optically inactive due to internal compensation of optical rotation.

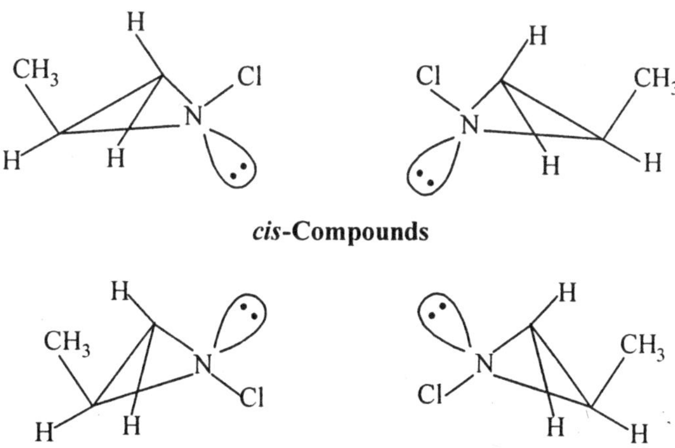

cis-Compounds

trans-Compounds

It is known that nitrogen compounds of the type $NR_1R_2R_3$ in which nitrogen carries three different groups is not superimposable on its mirror image, it is chiral, and should exist in two enantiomeric forms (4) and (5) each of which if separated from each other should exhibit optical activity. But such enantiomers have not been isolated for simple amines. The spectroscopic studies have shown that the activation energy between these two isomers is too small for resolution because of rapid nitrogen inversion as a result they are rapidly interconverted. The activation energy for interconversion of aziridines is higher than in acyclic amines, and Brois[7] in 1968 separated *syn* and *anti* enantiomers of R, S-2-methyl-1-chloroaziridine by *g.l.c.* He also asserted that they did not interconvert below 135°C.

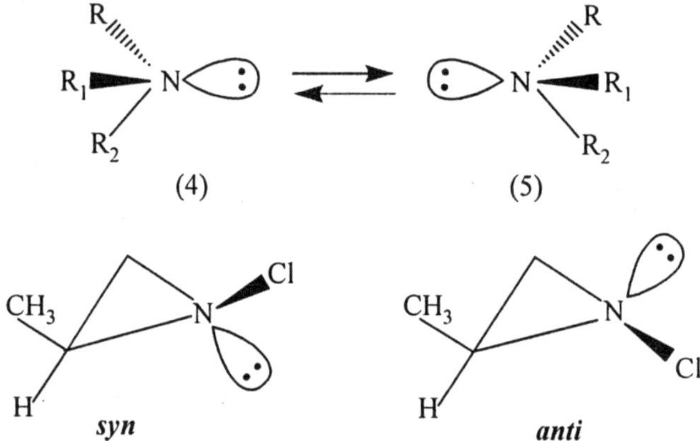

(4)   (5)

syn   *anti*

Optical resolution of aziridines by salt formation is not satisfactory because they are usually acid labile and tend to polymerize under these conditions. Generally an increase in the size and number of alkly groups on N—1, C—1, C—2 decreases the free energy between the invertomers and the introduction of an aryl group decreases this energy even further.[8]

2.1.3 Synthetic Methods

Aziridine and its derivatives have been prepared by a number of important methods.

1. *The Gabriel Ring Closure :* Aziridine was first obtained in 1888 by heating β-bromoethylamine in the presence of potassium hydroxide, equation (2.1), but the cyclic structure of this compound was confirmed after about a decade. The Gabriel ring closure follows a first order kinetics and the halogen atom is displaced intramolecularly by an S_N2 process, a particular

CH$_2$—CH$_2$ $\xrightarrow{\text{KOH, ethanol}}$ + KBr + H$_2$O   (2.1)

Aziridine

stereoisomeric form of the starting material reacts to yield a specific stereoisomeric form of the product. The cyclization to aziridine ring is stereospecific, *i.e.,* stereochemistry of the reactants are preserved in the products and that the ring closure involved a Walden inversion. Several variants of this synthesis have since been developed and form the subject of comprehensive reviews.[9] This synthesis, however, fails in the preparation of 2, 2, 3, 3-tetraalkylaziridines but a unique preparation of such aziridines involves a three step sequence of chloronitrosation of a tetraakylalkene in the first step followed by reduction ($SnCl_2$/HCl) of the nitrosochloride and cyclization in the presence of a base.[10] The process is demonstrated for the preparation of 2,2,3,3-tetramethylaziridine.

Optically active *erythro*-1, 2-diphenyl-2-bromoethylamine (6) gave optically active *trans*-2, 3-diphenylaziridine on cyclization in the presence of a base.

(6)

erythro

80% optically active
trans-2, 3-Diphenylaziridine

The optically active *threo*-compound (7) yielded optically inactive, *meso-cis*-2, 3-diphenlyaziridine.

(7)

threo

96% meso
cis-2, 3-Diphenylaziridine

2. *The Hassner Synthesis* : This method[11] involves a stereospecific addition of iodine isocyanate (INCO) to alkenes *via trans*-addition with the resultant formation of iodo isocyanate of the alkene. This is treated with methanol to form the carbamate which subsequently cyclizes in the presence of a base to yield a cyclic carbamate followed by its hydrolysis to an aziridine. The method is illustrated using styrene for the preparation of 2-phenylaziridine.

2-Phenylaziridine

3. *From Oxiranes* : Another method for the preparation of aziridine derivatives involves heating at 140°C of an oxirane with a substituted imino phosphorane under the strict exclusion of moisture and in the absence of a solvent. In this manner 1, 2-disubstituted aziridines are obtained in excellent yields[12], equation (2.2). A more recent method consists of treating the epoxide with sodium azide to form 2-azido alcohol.

This is then treated with a *tert.* phosphine to give an aziridine.[13] Styrene oxide, for instance, yields 2-azido-2-phenylethanol with NaN_3, this on treatment with triphenylphosphine leads to the corresponding aziridine. According to this procedure using *cis-* and *trans*-stilbene oxides, the

$$C_6H_5 \text{(epoxide)} + (C_6H_5)_3P = NCH_3 \xrightarrow{140°C} \text{1-Methyl-2-phenylaziridine} + (C_6H_5)_3P = O$$

1-Methyl-2-phenylaziridine

$$(2.2)$$

corresponding *threo*-and *erythro*-2-azido-1,2-diphenylethanols are obtained. Subsequent treatment with $(C_6H_5)_3P$ gives *cis*- and *trans*-2, 3-diphenylaziridines respectively thus resulting in a highly stereospecific formation of aziridines with retention of configuration.

A rather less general method of aziridine preparation involves the alkylation of an imine followed by deprotonation and ring closure with a base.[14]

1-*tert.* Butyl-2, 2-diphenyl-1-aziridine

4. *Nitrene Insertion* : Aziridines may be prepared by the direct addition of a nitrene into alkenes.[15] The nitrenes add to alkenes stereospecifically if singlet but non-stereospecifically when in the triplet state. The nitrenes can

Singlet   *cis*-Alkene   *cis*-Aziridine

be generated in a number of ways such as photolysis or thermolysis of ethoxycarbonyl azide or benzenesulfonyl azide, lead tetraacetate oxidation of N-aminophthalimide and by the action of triethyl phosphite on pentafluoronitrosobenzene. They yield of the desired product is rather low because of side-product formation. Alternatively, a pure product can be obtained by first permitting the azide to react with the alkene by a 1, 3-dipolar addition and the resulting Δ^2-1, 2, 3-triazoline is decomposed by *u.v.* irradiation.

Aziridination of alkenes with N-aminoheterocyclic compounds oxidized with lead tetracetate (LTA) is a familiar process. The intermediate has been assumed to be the corresponding N-nitrene. It has recently been observed that N-acetoxyaminoquinalozone is an effective aziridinating agent for alkenes.[16]

70%
syn-**Isomer**

The mechanism of this reaction has been proposed to proceed analogous to that for the epoxidation of alkenes by peracids but not a nitrene. This finding was supported by the reaction of this reagent with cyclohexenol which gave 70% of the *syn*-stereoisomer. A similar reaction of this alcohol with perbenzoic acid yield 90% of the *syn*-isomer.

5. *From Methylenes* : Like nitrenes, methylenes are also high energy species. Diazomethane transfers methylene to ternary iminium perchlorates and fluoroborates to form aziridinium salt. The first postulate[17] of the existence of an ethylenimonium or aziridinium compound (8) was postulated during the reaction of 1-β-chloroethylpiperidine hydrochloride with a limited amount of base.

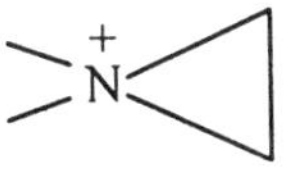

(8)

The perchlorate and fluoroborate salts are selected because their anions possess low order of nucleophilicity and are thus unable to open the very reactive positively charged ring. Leonard *et al*[18] prepared a large number of aziridinium salts in this manner.

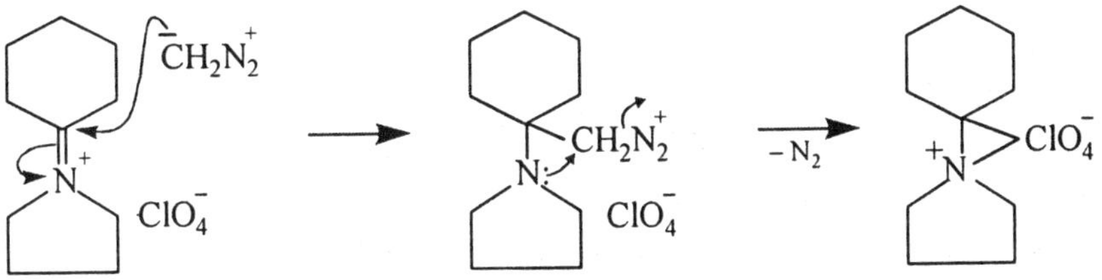

85%

6. *The Hoch-Campbell Method* : This method involves the use of a ketoxime and an excess of Grignard reagent. The reaction is found to be stereospecifc and regiospecific. The specificity is influenced by the type and concentration of the Grignard reagent and by geometric isomerism of the oxime (*syn* and *anti*). The 2-and 3-substituents in aziridine which arise from the ketone are invariably 'cis' oriented. An azirine (9) is probably involved as an intermediate in this reaction.[19] It is a general method for the preparation of 2, 2-disubstituted aziridines in which R is a hydrogen or any other alkyl group.

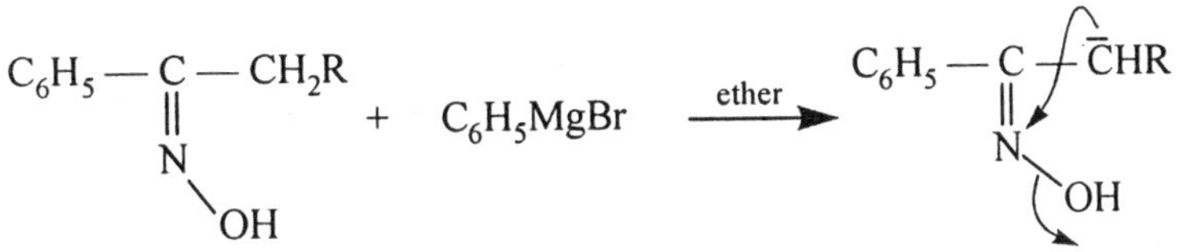

$$C_6H_5 - C - CH_2R + C_6H_5MgBr \xrightarrow{\text{ether}} C_6H_5 - C - CHR$$

R = –CH$_3$, –C$_6$H$_5$

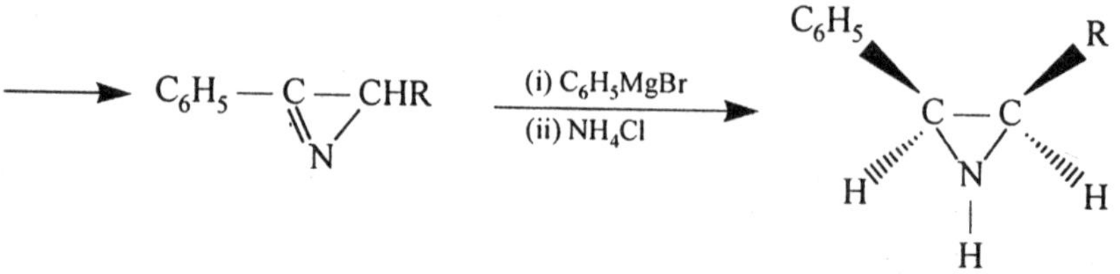

7. *From Ylides* : The ylide, diphenylsulfimide $Ph_2S^+—NH^-$ adds stereospecifically to E and Z conjugated alkenes which are electrophilic in

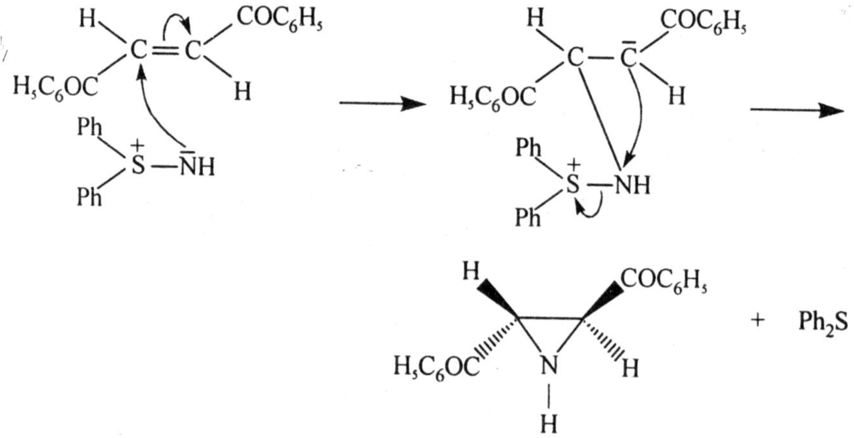

trans-2, 3-di-Benzoylaziridine

nature. First a Michael type addition takes place followed by collapse of the intermediate to aziridine.[20] This is depicted for the synthesis of 1, 2-dibenzoylaziridine. The reaction is stereospecific and where a chiral sulfinimine is used, the chirality is transferred to the aziridine with optical yields in excess of 25%.

8. *1, 3-Dipolar Cycloaddition* : High pressure induced 1, 3-dipolar cycloadditions of azides with electron-deficient olefins yields aziridines in high yields.[21] This is illustrated for the reaction of methyl acrylate with phenylazide. Some triazoline is also formed during this reaction.

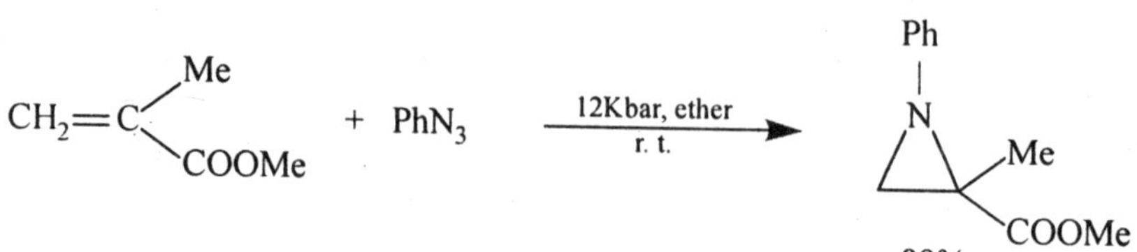

9. *Miscellaneous Methods* : Recently a number of additional methods have been developed for the aziridination of alkenes. Evans and coworkers[21a] have found that $Cu(acac)_2$ catalyst in the presence of (*N-* (*p*-toluenesulfonyl) imino) phenyliodinane (PhI = NTs) is highly effective as an aziridinating agent. The reagent functions as nitrene precursor.

2.1.3 Chemical Properties

One of the most important reactions of small ring compounds is the ring opening. These rings are strained and relief of strain makes ring-opening more facile. The aziridine ring is readily opened in the presence of several reagents to provide a wide variety of derivatives. The most readily cleaved bond is between carbon and nitrogen.

   1. ***Ring-Opening Reactions :*** The earlier methods for the preparation of ethylenimine included ring closure of a β-halogenated alkylamine in the presence of a base. This reaction could, however, be reversed in the presence of a mineral acid. Aziridine, thus yields β-chloroethylamine hydrochloride on treatment with HCl, equation (2.3) by C—N bond cleavage.

$$\overset{\text{HCl}}{\underset{\text{NaOH}}{\rightleftharpoons}} \quad ClCH_2CH_2NH_2 \cdot HCl \qquad (2.3)$$

2-Phenylaziridine similarly forms the corresponding salt, equation (2.4).

$$\xrightarrow{\text{HCl}} \quad C_6H_5\overset{|}{\underset{Cl}{C}}HCH_2NH_2 \cdot HCl \qquad (2.4)$$

The reaction is first order and the mechanism involves an initial protonation of the ring nitrogen atom followed by nucleophilic attack by the halide ion in a rate-limiting step.

$$+ H^+ \xrightarrow{\text{Fast}} \quad \xrightarrow{X^-, \text{slow}} H_2NCH_2CH_2X$$

$$\xrightarrow{\text{HCl}} XCH_2CH_2NH_2 \cdot HCl$$

After protonation of the amino group, the nucleophile attacks the carbon atom involved in the cleavage usually with inversion of configuration and retention of configuration at the carbon atom bearing the amino group.[22] The specificity of the reaction is very high since it is clearly S_N2.

Acid chloride similarly opens the ring to an acetamide derivative, *i.e.* $N - (\beta$-chloroethyl) acetamide, equation (2.5).

$$\text{aziridine} \quad + \quad CH_3COCl \longrightarrow ClCH_2CH_2NH\overset{\overset{\textstyle O}{\textstyle \|}}{C}CH_3 \qquad (2.5)$$

Ring opening of an aziridine is also affected by nucleophilic reagents like water, alcohols and amines. With water, aziridine yields β-hydroxyethylamine, equation (2.6).

$$\text{aziridine} \quad + \quad HOH \longrightarrow HOCH_2CH_2NH_2 \qquad (2.6)$$

Ring opening also takes place with sodiomalonic ester leading to 3-carbo-ethoxypyrrolidinone as the final product.[23]

$$\text{aziridine} \quad + \quad Na^+ \ {}^-CH(COOC_2H_5)_2 \longrightarrow$$

Protonated aziridines or aziridinium salts are exceptionally reactive towards nucleophilic attack because of the release of strain energy inherent in a small ring.[24] These reactions also proceed with extensive, if not complete inversion of configuration at the site of attack. When unsymmetrical aziridines are involved ring opening can occur in either of two directions. However, the nucleophile tends to attack the less hindered carbon atom with the result that ring opening in one direction is predominant. The ratio of the products is however, affected by the nature of solvent and reagents. Most aziridines which undergo direct nucleophilic ring opening carry electron-withdrawing groups at the nitrogen atom. Those which do not bear such groups require vigorous conditions.

$$\text{(aziridine with N—COOC}_2\text{H}_5) + ArNH_2 \longrightarrow H_5C_2OOCNCH_2CH_2NHAr$$

The aziridinium salts are easily attacked by nucleophiles to give stable ring-opened, equation (2.7), or ring-expanded products.[18] as shown for (10).

$$(2.7)$$

(10)

Aziridines, in additions are known to undergo different modes of ring opening under photochemical conditions. Initially a C—C bond is cleaved and the fragments then undergo a number of secondary reactions. Thus 1, 2, 3-triphenylaziridine on irradiation in alcohol yields N-benzylaniline and alkyl benzyl ether as the final products.[25] A competitive fragmentation to form N-benzaylaniline and phenylcarbene takes place and the latter is trapped by ROH to give alkyl benzyl ether.

$$\xrightarrow[\text{ROH}]{h\nu} \quad C_6H_5CH{=}NC_6H_5 + C_6H_5\ddot{C}H$$

$$\downarrow \text{ROH}$$

$$C_6H_5CH_2OR$$

Woodward and Hoffmann[26], in 1965 predicted that the thermal isomerization of cyclopropyl anion should be a conrotatory ring opening process while the photochemical ring opening should proceed by the disrotatory course. Aziridine which is isoelectronic with the cyclopropyl anion has been shown to open *via* the disrotatory process according to Huisgen *et al.*[27] The photochemical conversion of *trans*-dimethyl-1-(4-methoxyphenyl) aziridine-2, 3-dicarboxylate (11) in the presence of dimethylacetylene dicarboxylate gives *trans* tetramethyl-1-(4-methoxyphenyl)-3-pyrroline-2, 3, 4, 5-tetracarboxylate (12) according to a photolytic disrotatory ring cleavage.

Its formation has been interpreted in terms of an azomethine ylid intermediate (11a) which is intercepted by dimethylacetylenedicarboxylate

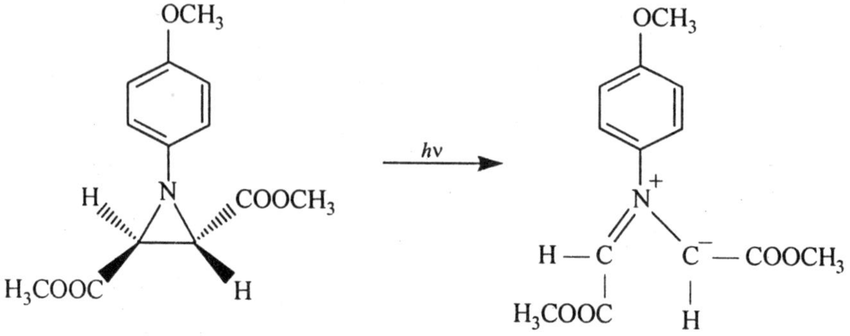

(11)   (11a)

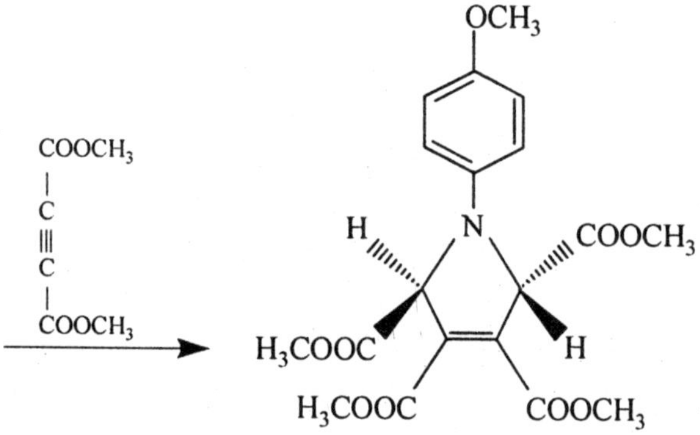

(12)

as the dipolarophile to give the heterocyclic compound.[28] Azomethine ylides are formed when aziridines undergo photolytic or thermolytic C—2, C—3 bond cleavage. The reaction is *stereospecific* as the stereochemistry of the groups is preserved in the end-products.

Trans-aziridine (11) undergoes thermal conrotatory ring opening to give a small equilibrium concentration of the corresponding *cis*-azomethine ylid intermediate while the *cis*-isomer gives the *trans*-ylid.

2. Thermal Reactions : Aziridines bearing unsaturated groups at the nitrogen atom rearrange under a variety of experimental conditions. For instance, N-acyl derivatives of aziridine frequently are converted to 2-substituted 2-oxazlines on distillation, equation (2.8).

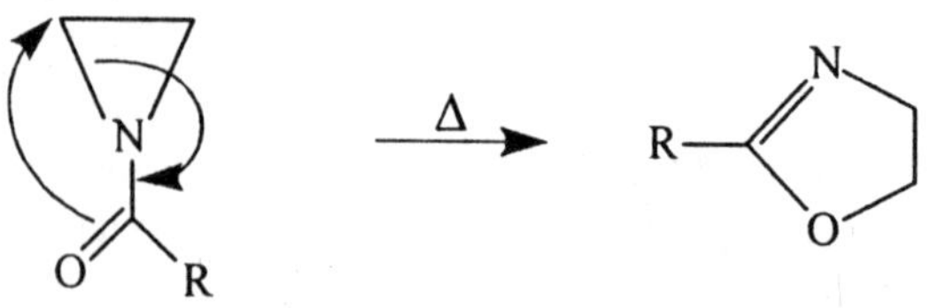

$$\Delta \qquad (2.8)$$

The rearrangement takes place by an intramolecular attack of oxygen on the ring carbon atom causing cleavage of the ring. The driving force for this reaction is obtained from the relief of ring strain which the opening of a small ring provides.

Iodine and thiocyanate ion are also effective catalysts for inducing the related opening and closure of the ring. Aziridines which contain unsaturation on the ring carbon also exhibit interesting and diverse thermal reactions. The pathway followed depends on the geometry and nature of the substituent. For instance, the compound (13) on heating affords the seven-membered heterocyclic ring.

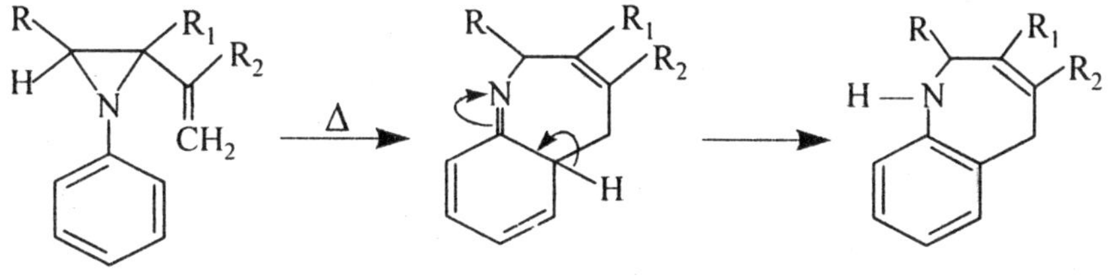

(13)

On the other hand, compound (14) isomer of structure (13) leads to 3-pyrrolines instead under the same conditions.

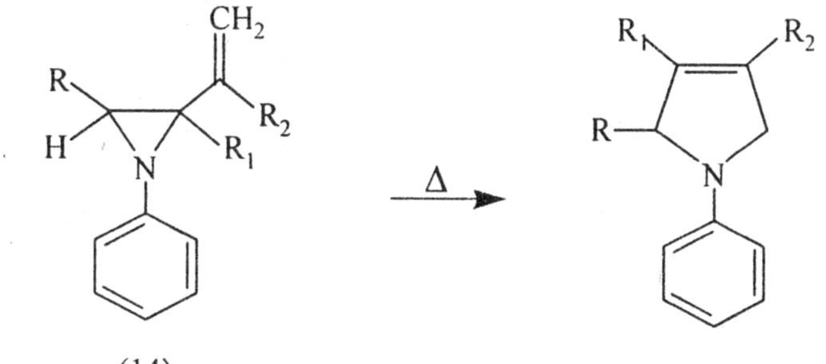

(14)

Thermolysis of an appropriate perester of aziridine yields an aziridinyl radical (15) which subsequently undergoes ring opening to give (16) as a dimerization product.[29]

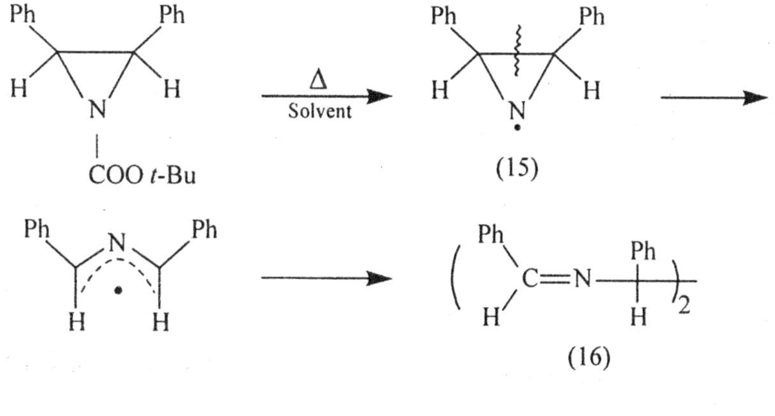

(15)

(16)

3. *The Friedel-Crafts Reaction* : Aziridine undergoes a facile Friedel-Crafts reaction with benzene in the presence of $AlCl_3$ resulting in the formation of β-phenylethylamine,[30] equation (2.9).

$$\text{(2.9)}$$

4. *Ozonolysis* : When N-*tert.* butylaziridine is treated with ozone in CH_2Cl_2, it results in the formation of N-*tert.* butylaziridine N-oxide[31], equation (2.10).

$$\text{(2.10)}$$

5. *Replacement of the H-atom* : The hydrogen on the nitrogen atom in aziridine is reactive and can be replaced by appropriate reagents as illustrated in the following reactions :

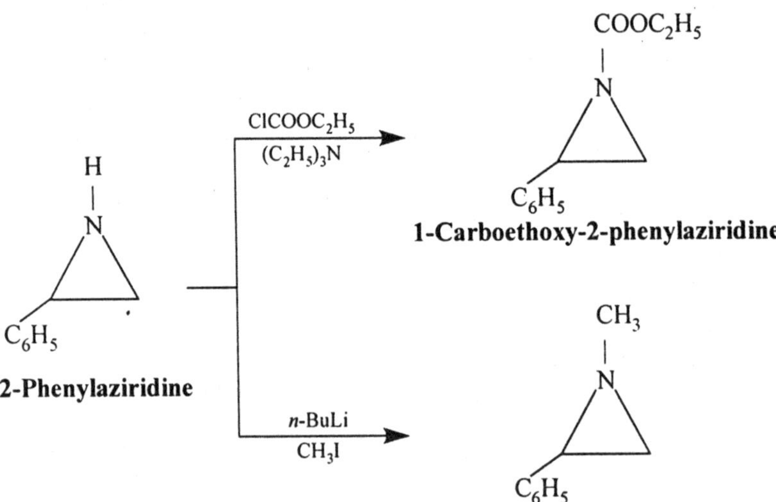

6. *Photochemical Reactions* : The photochemistry of aziridine derivatives has been extensively investigated. Irradiation may result in a number of expected photochemical reactions such as isomerization, rearrangement, fragmentation. Under photolytic conditions the following stereoisomeric urethanes undergo geometrical isomerism without any detectable fragmentation.

trans-Isomer   **cis-Isomer**

Photolytic behavior of the geometric isomers of 1-*t*-butyl-2-phenyl-3-benzoylaziridines has been investigated.[32] Irradiation of the *trans*-isomer in pentane gave 2, 5-diphenyloxazole and β-*t*-butylamino *trans*-benzalacetophenone

Formation of these products indicates both C—C and C—N bond cleavage in the *trans* isomer. Photolysis of the *cis*-isomer gave only the oxazole.

Photolysis of aziridine (17) affords a fused ring system (18) *via* an intramolecular 1, 3-dipolar cycloaddition reaction.[33] Conrotatory ring opening is obviously restricted by the ring system and in contrast the thermal process gives the rearranged product (19).

(19)

(17)

(18)

2.2 OXIRANES

The saturated heterocyclic three-membered ring containing one oxygen atom is known as *oxirane* (20). Oxirane and its derivatives have been known for a long time and have been used as key intermediates in the synthesis of a great variety of organic compounds. The oxirane ring has also been designated as *'epoxide'*, *'ethylene oxide'* or sometimes as 'α, β-*epoxyethane*' or 1, 2-*oxidoethane'*. The next member is propylene oxide (21).

(20)   (21)

Oxirane or ethylene oxide was first discovered in 1859 and is isomeric with acetaldehyde. It is an important industrial and research chemical.

It is used for the manufacture of 1, 2-ethanediol (ethylene glycol), emulsifiers, plastics and resins. Methyloxirane is used to make detergents and lubricants. They are also used in making 'epoxy' adhesives. These are cross-linked polyether thermoplastic made from a liquid resin. The resin is cross-linked with a hardner to the final resin. Epoxides also form a part of a large number of naturally occurring compounds. Such compounds have marked physiological properties showing carcinogenic, anticancer and antibiotic properties.

2.2.1 Physical and Spectroscopic Properties

Oxirane or ethylene oxide is a colorless liquid, b.p. 10.7°C, *i.e.,* it is a gas at room temperature. The addition of a methyl group as in propylene oxide (21) raises the boiling point considerably to 35°C whereas stilbene oxide is a solid. In contrast, to cyclopropane, it is polar and has a dipole moment of 1.88 D in benzene solution. Dimethyl ether, in comparison, has a lower value of 1.30 D. The higher value shows a considerable polarity of the C—O bond in oxirane. In other words, electron-donating groups decrease while electron-withdrawing groups increase the dipole moment. Oxirane and methyloxirane are miscible with water.

The structure of oxirane has been derived from microwave spectroscopy and the following structural data has been obtained, which further has been confirmed by electron diffraction studies. The C – C bond distance (1.472Å) is intermediate between that of a normal C – C bond distance (1.54Å) and that of a C = C bond distance (1.33Å). The H – C – H angle (116.5°) lies between the tetrahedral and the trigonal configurations. The O – C – C bond angle is of the order of 61.4° as compared to that of dimethyl ether (111.5°) and the C – O – C angle is 59.3°. Oxirane has been considered to have hybridization between sp^3 and sp^2 just as in the case of cyclopropane.[34] The strain energy in oxirane has been estimated to be 13 Kcal/mole. The strain is reflected by the C – H vibration frequency in the infrared which is 1500 cm^{-1} for aliphatic compounds. The J_{C-H} coupling constant increases in the order:[35] Oxirane < aziridine < thiirane. This value is still larger than cyclopropane. This increase seems to parallel the increase in the bond length of the ring C – C bond. The *n.m.r.* data also suggest the order of decreasing basicity as aziridine > oxirane > thiirane.

The literature contains sufficient data about mass spectral analysis of compounds containing an oxirane ring.

Aliphatic epoxides, undergo three principal types of fragmentation patterns. The first involves McLafferty rearrangement, and two types namely 'inside' and 'outside' rearrangement, however, the intensity of the ions favor an 'inside' rearrangement.

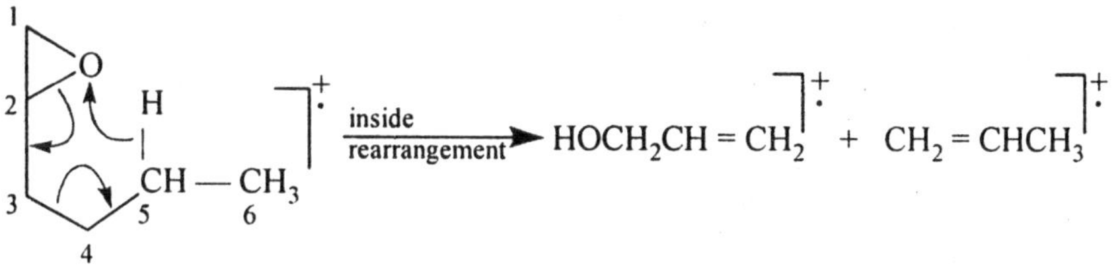
O
H
CH — CH3
1
2
3
4
5
6
inside
rearrangement
HOCH2CH = CH2
CH2 = CHCH3

It involves the fission of the bond connecting the oxygen atom to C – 2. In contrast the 'outside' rearrangement requires the cleavage of the connecting oxygen atom to C – 1. In the former the hydrogen transferred to oxygen from C – 5.

A unique behavior of epoxides is the transannular fission which involves the cleavage of C – C and C – O bonds of an oxirane ring. This fission may occur with or without the accompanying hydrogen transfer to oxygen. Unsymmetrically 2, 3-substituted chlorides lead to four possible ions upon electron induced fragmentation without any transfer of hydrogen. The cleavage of the two bonds connected to one carbon atom is an unfavorable process, as the ionized carbene species are not favored in electron impact induced fragmentation.

The third process is the single bond fission in which the larger alkyl group is lost. The mass spectra of some aromatic alkyl chlorides have been studied by high resolution and isotope labelling techniques. Various fragmentation processes such as α-cleavage, alkyl or aryl rearrangement

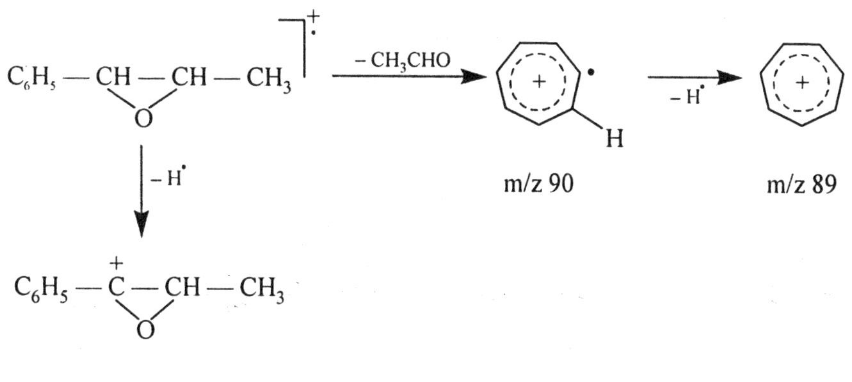
C6H5 — CH — CH — CH3
O
- CH3CHO
H
- H
m/z 90
m/z 89
- H
C6H5 — C — CH — CH3
O

and transannular bond rupture have been recognized. Transannular bond rupture without hydrogen rearrangement gives the most prominent ion m/Z 90, a ring expansion species which further loses one hydrogen to yield the peak at m/Z 89, the tropylium ion.[36]

To study the effects of functional groups on the cleavage of epoxy ring, several α,β-epoxyketones have been subjected to electron impact. The prominent cleavage in α,β-epoxyketones occurs between the carbonyl group and the oxraine ring with the retention of charge on the carbonyl moity. Generally the resulting acylium ion is responsible for the base peak.

$$\begin{bmatrix} \overset{R}{\underset{R}{>}}C \overset{O}{\overset{\diagup\diagdown}{—}} CH — \overset{O}{\overset{\|}{C}} — R_1 \end{bmatrix}^{+\cdot} \longrightarrow R_1C\equiv O^+ + \begin{bmatrix} \overset{R}{\underset{R}{>}}C \overset{O}{\overset{\diagup\diagdown}{—}} CH \end{bmatrix}^{\cdot}$$

2.2.2 Synthetic Methods

A large number of methods are available for the synthesis of compounds containing the oxirane ring. Several important ones are discussed here.

1. *From the Oxidation of Alkenes :* Probably the simplest method of preparing the parent compound oxirane is the direct oxidation of ethylene by air over a silver catalyst at elevated temperatures, equation (2.11).

$$H_2C{=}CH_2 \xrightarrow[\text{Ag}]{[O],\,\Delta} \overset{O}{\triangle} \qquad\qquad (2.11)$$

The epoxide is obtained in a moderate yield. Only half of the ethylene is converted into epoxide and the remaining into CO_2 and H_2O. In spite of the seeming disadvantage, this method is important industrially. Other alkenes, such as propylene and isobutylene are apparently converted into CO_2 and H_2O. The oxidation of an alkene to epoxide can be usefully accomplished with organic peracids, such as peracetic, perbenzoic or *m*-chloroperbenzoic acid in methylene chloride. But with *m*-chloroperbenzoic acid in sodium bicarbonate, excellent yields of epoxides have been obtained. The last acid is most suitable because it is stable and commercially available and also faster in reaction than the other two acids. The reaction is second order[37] and the mechanism[37] which was initially suggested by Bartlett[37] proceeds in a concerted manner[37] (equation 2.12). The peracid transfers an oxygen atom to the alkene followed by loss of carboxylic acid. The epoxidation takes a *cis*-course, i.e., a *cis*-alkene yields a *cis*-epoxide and a *trans*-alkene gives rise to a *trans*-epoxide. In other words, it is highly stereospecific. Although *cis*-addition of oxygen always occurs, a mixture of epoxides is possible if the olefin is not planar because the oxidant can attack from either side of the C—C double bond. Epoxidation of *cis*-4, 5-dimethylcyclohexene, for instance, results in 87 : 13 ratio of *cis* and *trans* epoxides. The most important factor affecting the direction of attack is steric and this depends not only on the stereochemistry of olefin but also on the peracid as well. Monoperphthalic acid in this regard requires more space than performic acid. A few cases which illustrate the steric effect in the olefins are the oxidation of 1-*t*-butyl-4-methylenecyclohexane (22), norbornene (23), 7, 7-dimethylnorbornene, (24), etc.

$$(2.12)$$

(22)   (23)   (24)

Strong directive effects by allylic hydroxy groups are observed. The hydroxy group assits *cis*-oxidation, i.e., the oxirane oxygen is *cis* with respect to the hydroxy group. Epoxidation of cyclohex-1-en-3-ol with perbenzoic acid in benzene gives stereoselectively a 91 : 9 ratio of (25) and (26). This has been attributed to hydrogen-bonding in the transition state between the peracid and the – OH group. The epoxidation of suitably substituted olefins always leads to a racemic mixture of oxiranes.

91%   9%
(25)   (26)

If, an optically active peracid for instance, (+) peroxycamphoric acid, is used than an asymmetric synthesis takes place leading to an optically active oxirane.[38]

In this reaction the rate is faster when alkyl groups are present on the alkene, but retarted if the alkene is conjugated with electron-withdrawing

groups, such as $-\overset{\overset{\textstyle O}{\|}}{C}-H$, $-C \equiv N$, etc. α, β-unsaturated aldehydes, ketones, esters, sulfones are difficult to epoxidize by peracids or metal catalyzed methods. The alternate route is the use of alkaline hydrogen peroxide[39a] which is non-stereospecific and of limited use, equation (2.13).

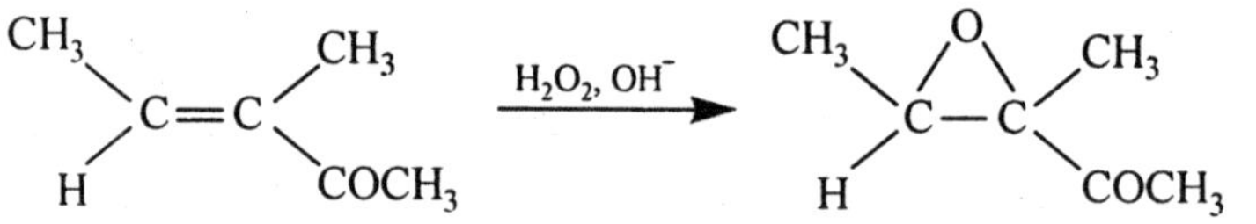

$$\text{(2.13)}$$

2-Acetoxy-2, 3-dimethyloxirane

This reaction has been catalyzed by polypeptides.[39b] Chalcone gives good yields of the corresponding epoxide under these conditions. A manganese complex as catalyst has also been reported.[39c] Phase transfer mediated epoxidation has also been achieved.[39d]

α, β-Unsaturated esters are not epoxidized in this manner, though many compounds are available by the Darzens condensation. In such cases anhydrous solution of t-butyl hydroperoxide in benzene, an organolithium compound in THF at $-80°C$ forms lithium t-butyl hydroperoxide and oxidation of α, β-unsaturated acid or sulfone results in excellent yield of epoxide derivative.[40]

$$CH_2 = CHCOOH + n\,BuLi \xrightarrow[\text{20°, 1hr}]{\text{Alc. } (CH_3)_3\,CCOOH} CH_2 - CHCOOC(CH_3)_3$$

Phenylsulfonyl oxirane has also been prepared in this way which is an important synthetic intermediate.[41]

This reagent is also suitable for highly hindered alkenes as in the following case.[42]

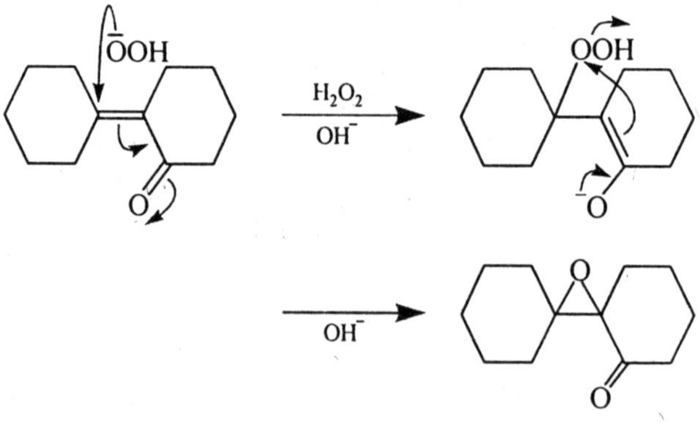

Surprisingly, the reaction of alkaline hydrogen peroxide with α, β-unsaturated nitriles leads to α, β-epoxyamides[43] rather than the anticipated α, β-unsaturated amides, according to the following mechanism. The

hydroperoxide ion first attacks the cyanide group. Hydrogen peroxide is a poor oxidising agent. Payne and coworkers[43] have used a mixture of benzonitrile and H_2O_2 in methanol which produces peroxy carboximidic acid which is an efficient expoxidizing agent.

Recently epoxidation of alkenes has been accomplished under milder conditions using sodium hypochlorite catalyzed by transition metal complexes.[44] Epoxides, under these conditions, are obtained in excellent yields. Cyclohexene epoxide, for instance, is obtained from cyclohexene in 70% yield, equation (2.14).

$$\text{C}_6\text{H}_{10} + \text{NaOCl} \xrightarrow[\text{CH}_2\text{Cl}_2,\ \text{H}_2\text{O, r. t.}]{\text{Mn (TPP) OAC, NH}_4^+\ \text{Cl}^-} \text{cyclohexene epoxide} \tag{2.14}$$

Cyclohexene epoxide

Certain epoxides have been obtained by photoxidation of alkenes as well in the presence of dyes[45] or triplet photosensitizers.[46] In the first case the results are consistent with the involvement of a singlet oxygen while in the second, the reaction is *stereoselective, i.e.,* a *trans*-epoxide is always obtained, the mechanism of the reaction, however, has not been fully understood.

The importance of asymmetric synthesis as a tool for obtaining enantioselective compounds has grown dramatically not only in synthetic organic, in medicinal and agricultural chemistry but also in the pharmaceutical industries. Several methodologies to obtain enantiomerically pure compounds are nowadays available.

1. Optical resolution of a racemate *via* separation of diastereoisomers.
2. Chiral chromatography separation.
3. Kinetic or bio-catalytic resolution.

Asymmetric metal catalyst represents one of the most active areas in organic chemistry and the number of chiral ligands for catalytic asymmetric transformation is growing rapidly.[47]

The first asymmetric epoxidation was performed by Herbert[38] though in a low optical activity. Asymmetric epoxidation of allyl alcohols by *t*-butyl hydroperoxide and dialkyl tartarates and transition metals was discovered by Sharpless in 1980.[48,49] The reaction involves the use of complexes formed by mixing titanium tetraalkoxide with diethyl esters of tartaric acid and *t*-butyl hydroperoxide (TBHP) in methylene chloride as the solvent. This reaction is able to oxidize asymmetrical prochiral substances to a product of predictable absolute configuration. Both enantiomeric tartaric acid esters are available. The steric bulk of the ester seems to be the deciding factor in the modification of catalyst function. One can predict the stereochemical outcome of this reaction while the asymmetric induction is generally high (79% ee) for most allyl alcohols. Thus for a given tartarate enantiomer the system delivers the epoxide oxygen to the same face of the carbon-carbon double bond regardless of the substituent pattern. If an allyl alcohol is drawn so that the hydroxyl group is at the lower right, oxygen is delivered at the bottom face in the presence of L-(+)-diethyl tartarate and from the top force in the presence of L-(−)-diethyl tartarate.

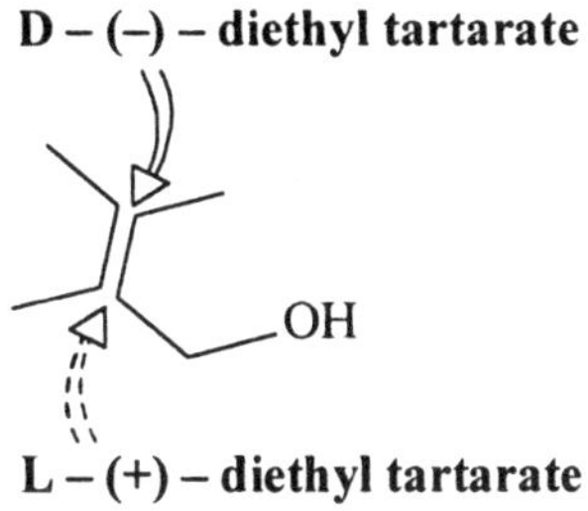

D – (–) – diethyl tartarate

L – (+) – diethyl tartarate

As a demonstration of this procedure alcohol (27) was converted to epimeric epoxides (28) and (29) in high yields.

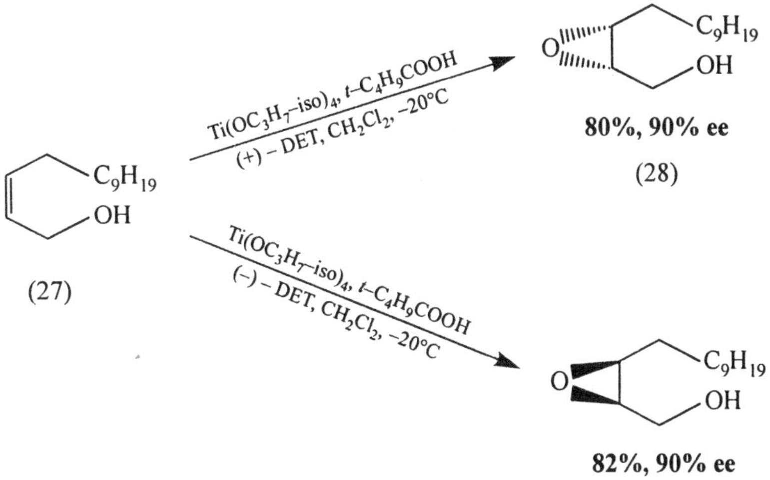

$Ti(OC_3H_7-iso)_4$, t-C_4H_9COOH, (+) – DET, CH_2Cl_2, –20°C → **80%, 90% ee** (28)

$Ti(OC_3H_7-iso)_4$, t-C_4H_9COOH, (–) – DET, CH_2Cl_2, –20°C → **82%, 90% ee** (29)

(27)

The following examples illustrate the same principle and its applications in organic synthesis.

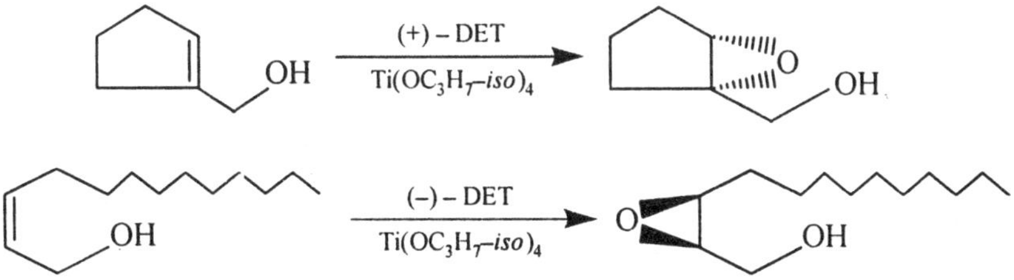

(+) – DET / $Ti(OC_3H_7-iso)_4$

(–) – DET / $Ti(OC_3H_7-iso)_4$ → **80%, 90% ee**

2. Ring Closure Methods : An excellent method for the introduction of an epoxide ring involves various ring closure procedures. A *trans*-chlorohydrin, for example, prepared by the addition of hypochlorous acid to an alkene, undergoes a bimolecular dehydrohalogenation in the presence of base with the resultant cyclization to an oxirane ring. Hypochlorous acid adds to an alkene in a *trans*-manner as depicted below in which ring closes in the presence of a base.

1-Methylcyclohexene epoxide

As in the case of an S_N2 displacement, there is an inversion of configuration at the carbon atom carrying the leaving group in the ring closure step. Cyclization, thus, occurs in a *trans* manner, accordingly, *threo-* and *erythro-*3-bromo-2-hydroxybutanes give *cis-* and *trans-*epoxides respectively.

threo-3-Bromo-2-hydroxybutane   *trans*-2-Expoxybutane

erythro-3-Bromo-2-hydroxybutane   *cis*-2-Expoxybutane

The rate of ring closure is favored by the presence of substituents at C–1, thus $HOC(CH_3)_2CH_2Cl$, cyclizes 11 times faster than $HOCH(CH_3)CH_2Cl$.

In a similar procedure an α-chlorohydrin is prepared from a ketone, bromochloromethane and Li metal in THF at low temperatures. The resulting chlorohydrin cyclizes to the corresponding epoxide.[50] This reaction is markedly facilitated by sonication.

$$PhCCH_3 + ClCH_2Br + Li \xrightarrow[\text{THF, }-20°C]{)))} $$

Similarly epoxides have been made first by making chlorosulfonate ester followed by subsequent ring closure. This process has been widely facilitated by phase transfer catalyst (PTC) without the necessity of isolating the ester. A one step synthesis gives the epoxide in excellent yields.[51] equation (2.15)

$$\begin{array}{c} \diagdown C - OH \\ | \\ \diagdown C - OH \diagup \end{array} \xrightarrow[\text{CH}_2\text{Cl}_2\text{, NaOH, }p\text{-toluenesulfonyl chloride}]{\text{Benzyltriethylammonium hydroxide (PTC)}} \quad (2.15)$$

A major synthetic pathway of ring closure methods is the Darzens condensation. This is the base-catalyzed reaction of a carbonyl compound with a halogenomethylene substance to yield a compound containing an oxirane ring. This reaction was first discovered by Erlenmeyer but later developed by Darzens. Benzaldehyde, for instance, on condensation with potassium *tert.* butoxide yields *tert.* butyl-β-phenylglycidate,[52] equation (2.16).

$$C_6H_5CHO + ClCH_2COOC(CH_3)_3 \xrightarrow[t\text{-BuOH}]{t\text{-BuO}^-K^+} C_6H_5CH - CHCOOC(CH_3)_3$$

$$(2.16)$$

An essential requirement for this reaction is that the halogenomethylene derivative possesses an activated α-hydrogen atom. The reaction seems to be general in its application though aliphatic aldehydes give poor yields because of side-product formation. A wide variety of base-solvent systems such as sodium amide, sodium hydroxide in dioxane, sodium hydride in ether, sodium ethoxide in toluene and phase transfer catalysts, have been used. It has been demonstrated[53] conclusively that this condensation proceeds *via* an enolate ion and not a carbene as intermediate. The reaction complies with a third order kinetics[54] being first order in each reactant.

The reaction is non-stereoselective as both *cis-* and *trans*-epoxides are obtained. The accepted mechanism is represented below: the enolate anion attacks the carbonyl component leading to the formation of *threo-* and *erythro*-halohydrins (30) and (31), which then cyclize to the *cis-* and *trans*-epoxides respectively. The diastereoisomer ratio can vary considerably depending on the structure of the reactants and the solvent.

cis-**Epoxide** (30)

trans-**Epoxide** (31)

Consideration of steric factors in predicting the stereochemistry of the products is important but sometimes thermodynamically less stable isomer is formed predominantly. This has been explained on the basis of maximum overlap of the carbonyl π-orbital of the ester and the developing p-orbital at the α-carbon atom in the transition state. Such an overlap is not the ony factor in determing the steric course of this reaction. Bachelor and Bansal[52] investigated the influence of the bulk of the group in the carbonyl moity and of the ester alkyl group on the *cis/trans* ratio on the Darzens synthesis of phenylglycidic esters. It was observed that the amount of the *cis*-epoxy ester increased with the increase in the size of the alkyl groups. This cannot be explained on the basis of steric effects alone. It was assumed that the configuration of the final product was determined in the aldolization step. The preferred configuration is the one leading to the *cis*-epoxide.

3. Methylene Insertion : A method that has proved quite useful in certain cases for epoxide formation is the addition of diazoalkanes, particularly diazomethane, to appropriate aldehydes. Diazomethane transfers a methylene group to the carbonyl group. According to this method simple aldehydes yield a mixture of products, acetaldehyde, for instance, gives a mixture of propylene oxide and acetone. Benzaldehyde, on the other hand, yields exclusively acetophenone by the following mechanism :

$$CH_3\overset{\overset{\displaystyle O}{\|}}{C}-H \xrightarrow[-N_2]{\overset{-}{C}H_2-\overset{+}{N}\equiv N} CH_3\overset{O}{CH}-CH_2 + CH_3\overset{\overset{\displaystyle O}{\|}}{C}CH_3$$

$$\qquad\qquad\qquad\qquad\qquad\qquad 28\% \qquad\qquad 28\%$$

$$C_6H_5\,CHO \xrightarrow{\overset{-}{C}H_2-\overset{+}{N}\equiv N} \qquad \xrightarrow{-N_2} C_6H_5\overset{\overset{\displaystyle O}{\|}}{C}CH_3$$

The methylene insertion from diazomethane though is a convenient procedure, it does not give good yield of epoxides. Corey and Chakowsky[55] discovered an ingenious application of dimethyloxosulfonium methylide $[(CH_3)_2\overset{\overset{\displaystyle O}{\|}}{S}{}^+ - CH_2^-]$ and dimethylsulfonium methylide $[(CH_3)_2\,S^+ - CH_2^-]$ as methylene transfer agents to the carbonyl group for oxirane formation. The formation of epoxide from an aldehyde or ketone is shown: The sulfur ylids are popular and useful reagents for the synthesis of oxirane as well as cyclopropane[56] There is, however, an important difference between the two reagents in their reaction towards α, β-unsaturated ketones such as benzalacetophenone. Dimethyloxosulfonium ylide gives exclusively a cyclopropane while dimethylsulfonium an epoxide.[57] In the first case, the sulfur ylide reacts preferentially by an initial Michael addition as shown below while in the second case, it is simply a transfer of the methylene group.

$$C_6H_5CH = CHCR \begin{cases} \xrightarrow{(CH_3)_2\overset{+}{S}-CH_2^-} C_6H_5-CH-CHCR + (CH_3)_2SO \\[3em] \xrightarrow{(CH_3)_2\overset{+}{S}-CH_2^-} C_6H_5-CH=CHC-CH_2 + (CH_3)_2S \end{cases}$$

where R = H or C_6H_5

The formation of cyclopropane derivative can be depicted in the following manner

$$C_6H_5CH = CH - CC_6H_5 \longrightarrow C_6H_5CH - CH = CC_6H_5$$

$$\longrightarrow C_6H_5CH - CHCC_6H_5 + (CH_3)_2\,SO$$

1-Benzoyl-2-phenylcyclopropane

Furthermore, dimethylsulfonium methylide attacks the carbonyl group from the more hindered side (i.e. axial) whereas the dimethyloxosulfonium derivative from the less hindered side (i.e. equatorial). Carbonyl compounds also react with sulfonium ylides to give epoxides.

$$Ph - \overset{O}{\overset{\|}{C}} - H \quad + \quad \overset{R}{\underset{R}{>}}S^- - \overset{+}{C}H\,C_6H_5 \longrightarrow$$

4. *Micellaneous Methods* : Certain hydroxy alkenes of type (32) undergo oxidation to ketone with simultaneous epoxidation of the methylene group in the presence of sodium hypochlorite.[59]

Catalytic oxygenation based on redox transition metal complexes $[Co(OAc)_2]$ and a multidentate ligand namely N, N'-bis (2-(4-imidazolyl) ethyl)-2, 6-pyridine carboxamide (BIPA) forms the basis for oxirane preparation.[60] The yields obtained though are not satisfactory by using this reagent.

2.2.3 Chemical Properties

Oxirane similar to aziridine undergoes facile ring opening reactions.

1. *Ring Opening Reactions* : Because of the strain present in the oxirane ring, it is highly labile and the ring can be readily cleaved by a variety of reagents. It has been used as an intermediate in organic synthesis for the preparation of many classes of compounds.[61] In solution, the reactions of oxirane involve both electrophilic as well as nucleophilic attack. The ring is opened in a *trans*-manner by water in the presence of a catalytic amount of acid to produce glycol, equation (2.17) and (2.18). The acid also neutralizes hydroxide ion and thus drives the reaction to the right by mass law effect.

$$\text{(epoxide)} \xrightarrow[]{H^+, H_2O} HOCH_2CH_2OH \qquad (2.17)$$

$$\text{(cyclohexene oxide)} \xrightarrow[]{H^+, H_2O} \text{(trans-diol)} \qquad (2.18)$$

trans-Cyclohexan-1, 2 diol

The mechanism involves an initial protonation of oxygen followed by attack of water. When the epoxide is not symmetrical, the opening of the oxirane

$$\triangle \xrightarrow[]{H^+, H_2O} \triangle \longrightarrow HOCH_2CH_2OH$$

ring can yield either of the two products or a mixture of two. Very often one direction of ring opening is very predominant. In the case of propylene oxide, a mixture of the following two products is obtained.

$$\xrightarrow[]{H^+, CH_3OH}$$

a → CH_3CHCH_2OH | OCH_3

2-Methoxypropanol (Minor)

b → $CH_3CHCH_2OCH_3$ | OH

1-Methoxy-2-propanol (Major)

It is easy to understand the ring opening in such unsymmetrical cases. As in the case of ordinary S_N2 substitution reaction, the rate of substitution follows the order 1° > 2° > 3°. In propylene oxide, therefore, the competition is between the substitution at 1° and at 2° carbon atoms. Since the reactivity is considerably greater at 1° carbon, the mode (b) of opening the ring takes place predominantly. This is the case with water, alcohols, amines, sodiomalonic ester and cyanoacetic ester.[62]

The nature of the substituent present on the oxirane ring has a powerful deciding effect on the direction of ring opening as is apparent from the reaction of styrene oxide.

$$C_6H_5 - CH - CH_2 \xrightarrow{HI} C_6H_5 - CH - CH_2 \xrightarrow{\bar{I}} C_6H_5CHCH_2OH$$
$$\underset{I}{}$$

The epoxide ring, in this case, opens in the direction[63] of the incipient benzylic carbocation to give 2-iodo-2-phenylethanol. On the contrary, if the carbocation becomes unstable due to the presence of electron-withdrawing groups, then the opening of the ring is reversed.

The mechanism of the reaction is different in neutral and in basic conditions. For instance, the regiospecificity in the acid- and base- catalyzed ring opening is demonstrated in the case of unsymmetrical (+)-2-methyloxirane.

(33)

(34)

In acid medium (+)–1, 2-dihydroxypropane (33) is formed with Walden inversion at the carbon atom in which the C – O bond is broken. In alkaline medium the (–) –1, 2-dihydroxypropane enantiomer (34) instead is obtained.

An interesting ring-opening reaction of aryl substituted epoxides takes place[64] in the presence of bitriflate. The rearranged products are aldehydes or Ketones depending on the nature of the substituents.
Probably a 1, 2-phenanium ion shift takes place in the famation of aldehydes and a 1, 2-hydride shift in ketones.

$$Ph\text{-}CH(H)\text{---}CH(Ph)\text{---}O \xrightarrow[\text{CH}_2\text{Cl}_2, \Delta]{0.1 \text{ mole \% of Bi (OTf)}_3, x\text{H}_2\text{O}} Ph_2CH\text{---}CHO$$

$$Ph\text{-}CH(H)\text{---}CH(CH_3)\text{---}O \xrightarrow[\text{CH}_2\text{Cl}_2, \Delta]{0.1 \text{ mole of Bi (OTf)}_3 . x\text{H}_2\text{O}} Ph\text{-}CH_2\text{---}CO\text{---}CH_3$$

2. Reaction with Organo-metallics : The reaction between oxirane and a Grignard reagent is extensively employed for the preparation of certain primary alcohols and lengthening of the C – C bond by two atoms. An intermediate (35) is initially formed which on subsequent hydrolysis yields the alcohol.

$$\triangle O \ + \ C_6H_5MgBr \xrightarrow{\text{ether}} C_6H_5CH_2CH_2\overset{\delta+}{}\overset{\delta-}{OMgBr}$$

$$(35)$$

$$\xrightarrow{H^+, H_2O} C_6H_5CH_2CH_2OH$$

In certain cases the reaction of epoxides with Grignard reagents gives rise to rearranged alcohols. Cyclohexene epoxide for instance, with ethyl-magnesium bromide results in some expected alcohol, *trans*-2-ethylcyclohexanol and in addition a cyclopentyl alcohol, a consequence of

trans-2-Ethylcyclohexanol

1-Cyclopentylpropan-1-ol

rearrangement[65]. A keto derivative[66] is obtained as a product *via* rearrangement of an epoxide in the presence of zinc bromide, a Lewis acid. If a good leaving group is present on the ring then α-bromo ketone is obtained,[67] equation (2.19) because of the presence of $MgBr_2$ in the Grignard reagent.

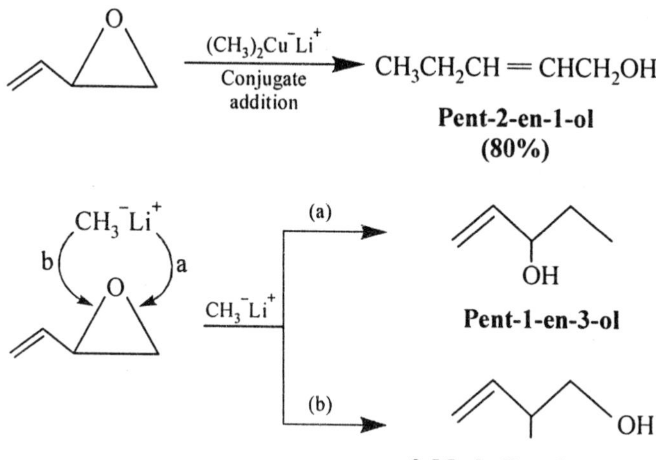

2-Bromocyclohexanone

The reaction of methyl metallated reagents with conjugated epoxides follows two distinct pathways, one by conjugated addition and the second by direct addition.[68] This is demonstrated for the reaction of 3, 4-epoxy-1-butene. The conjugated addition is similar to the role of Cu(I) in promoting the addition of Grignard reagent (here Li dimethylcuprate) to α, β-unsaturated carbonyl compounds.[69] On the contrary, the direct addition gives rise to a mixture of isomeric products by breaking either bond (a) or bond (b) by methyllithium.

3, 4-Epoxycyclohexene,[70] similarly reacts with these reagents to give a mixture of ring cleavage products.

3. *Reduction* : Metal hydrides reduce an epoxide ring to the corresponding alcohol, equation (2.20). Styrene oxide with $LiAlH_4$ yields α-phenylethanol. The hydride ion attacks C–3 with the simultaneous cleavage of the C – O bond.

4. *Thermal and Photochemical Reactions* : Aside, from the reactions in solution, the oxirane ring can undergo monomolecular thermal and photochemical reactions leading to various ring openings and fragmentation.

The glycidic esters which contain an epoxide ring rearrange on pyrolysis to aldehydes or ketones containing one carbon atom more than the starting carbonyl compound. Recently a series of glycidic esters was pyrolyzed by Bansal and coworkers,[71] and the corresponding aldehydes (ketones) were obtained which are, otherwise, accessible with great difficulty conventionally. The mechanism of pyrolysis proceeds in the following steps. The loss of gaseous isobutylene is followed by the liberation of carbon dioxide and the resulting enol tautomerizes to an aldehyde or ketone depending on the groups R_1 and R_2. This method provides a viable procedure for the preparation of aldehydes and ketones.

(A glycidic ester)

Certain appropriately substituted epoxides undergo a similar *cope rearrangement* to aldehydes. For instance, *cis*-1-ethynyl-2-vinyloxirane thermally isomerizes to *cis*-1-carboxyaldehyde-2-diethylcyclopropane.[72] A recent study by Kwaski and coworkers[73] has revealed the following facts about the photochemical or thermal decomposition of an oxirane :

- (i) Whatever the conditions (thermal or photochemical), the major reaction is the formation of two radical fragmentation $CH_3^{\cdot}$ and $CHO^{\cdot}$.
- (ii) Photochemically, formation of the above fragments is favored toward red shift of the irradiation wavelength while that of atomic oxygen and ethylene increases at short wavelength.
- (iii) Presence of sensitizers increases the amount of stable acetaldehyde. The common postulated intermediate in the two cases is a "hot" acetaldehyde molecule.

The activation energy for the formation of CH_3CHO is 57 Kcal/mole. It appears, that photochemically and thermally, the reaction pathway always involves $C - O$ bond cleavage as a primary process.

For substituted oxiranes, a dichotomy of mechanism arises since both $C - O$ and $C - C$ bonds can be cleaved depending on the nature of the substituents.[74] Alkyl substituents have been found to favor $C - O$ bond cleavage both thermally and photochemically and require an activation energy of the order of 52 Kcal/mole. The $C - C$ bond rupture requires relatively a much lower (5 – 7 Kcal/mole) energy of activation. Aryl substituents favor $C - C$ bond breakage to yield 1, 3-dipolar species as primary products.[75] These dipolar intermediates which have been characterized at low temperatures, generally afford a carbene and a carbonyl compound[76]. Photolysis of styrene oxide in solution yields a mixture of products.[77]

$$\text{(styrene oxide, } C_6H_5\text{)} \xrightarrow{h\nu} C_6H_5\overset{O}{\overset{\|}{C}}CH_3 \ + \ C_6H_5CH_2CH_2OH \ + \ C_6H_5\underset{\underset{OH}{|}}{C}HCH_3$$

Formation of ring contraction products is a common feature in oxirane chemistry.[78]

$$\xrightarrow{h\nu} \qquad \longrightarrow$$

5. *Miscellaneous Reactions* : Polymeric ethers are obtained by the action of antimony pentafluoride on oxiranes. Oxirane itself for instance, is converted to its trimeric form 1, 4, 7-trioxycyclononane (36) in SbF_5[79], equation (2.21).

$$\text{(oxirane)} \xrightarrow{SbF_5,\ HF,\ 20°C} \text{(36)} \qquad\qquad (2.21)$$

(36)

Extrusion of oxygen from oxiranes takes place on reaction with triphenylphosphine. The *cis*-oxirane yields mainly the *trans*- olefin. Triphenyl phosphine selenide also deoxygenates oxiranes in the presence of trifluoroacetic acid and produces olefins with retention of configuration.

The formation of the intermediate episelenide and loss of selenium proceed with inversion.[80] Deoxygenation also takes place with selenium and tellurium reagents.[81,82] Reactions of oxiranes with selenophenoxide ion constitutes a single step method for the preparation of allyl alcohols.[83]

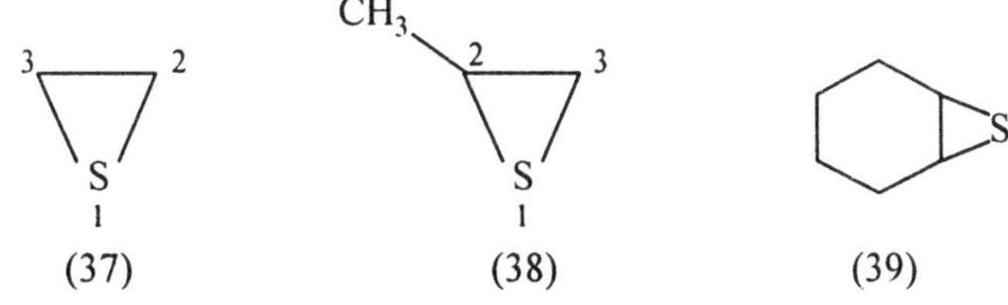

2.3 THIIRANES

Thiirane (37) is the name applied for a three-membered ring comprising of one sulfur atom. It has been known for a long time.[85] Thiirane has also been designated as *ethylene sulfide, episulfide or thiacyclopropane*. The higher homologs of thiirane have been called by the name alkene sulfides, for instance, propylene sulfide (38), cyclohexene sulfide (34), etc. Thiiranes have not been found to occur naturally.

(37)   (38)   (39)

Synthetic compounds, however, find considerable applications in the textile industry. Ethylene sulfide has been employed as an intermediate in polymeric compositions. Addition of small amounts of dithioglycidol (3-mercaptopropene sulfide) to the polymerization of acrylonitrile yields polymers of increased thermal stability. Aryloxypropene sulfides have been proposed as light and heat stabilizers for polyvinyl chloride (PVC) as well as copolymer of PVC. Certain thiiranes are used in consmetics. Fluorinated thiiranes may be employed as refrigerants and as fire extinguishers. Dithioglycidyl thiophosphates have a strong insecticidal effect whereas S-acyl derivatives of dithioglycidol inhibit tuberculosis.

2.3.1 Physical and Spectroscopic Properties

Thiirane is a colorless liquid, b.p., 55-56°C. It is sparingly soluble in water and the solubility is still less in organic solvents. It possesses a dipole moment of 1.66 D which is higher than that of dimethyl sulfide (1.40 D) but lower than that of oxirane (1.88 D). The difference in dipole moment indicates that the polarity of the C – S bond is smaller than the C – O bond. The structure of thiirane is similar to that of oxirane except that there can also exists S– oxides and S, S– dioxides. The structural[86] information of the ring system has been derived from microwave spectroscopy. The C – C bond distance (1.429 Å) is intermediate between ethane (1.54 Å) and ethylene (1.33 Å) C – C bond distances. This suggests partial double bond character of the C – C bond in thiirane. The C – S bond length (1.819 Å) is almost of the same order as in dimethyl sulfide (1.810 Å). The H – C – H angle is 116.0° while the C – S – C bond angle is 48.4°. The strain energy has been calculated to be 9 Kcal/mole which is also reflected in the C – H vibration frequency in the infra-red and is higher (1475 cm^{-1}) than the normal C – H frequency of 1465 cm^{-1}.

Among the three-membered heterocyclic rings, thiirane has the lowest strain energy and also shows the lowest electron density at the hetero atom.[86]

2.3.2 Synthetic Methods

Thiirane and its derivatives have been prepared by the following significant methods :

1. *From 2-Mercaptoethanol :* Probably the most useful method of preparing thiirane is from 2-mercaptoethanol by treating it with phosgene in ethyl acetate and pyridine to give monothioethylene carbonate which on decarboxylation leads to thiirane.[87]

2. *Ring Closure Reactions :* Just as epoxides are obtained from 2-haloalcohols (chlorohydrins) and alkali, thiiranes have also been prepared by treatment of 2-halosulfides (or 2-halomercaptans) with alkali. 2-Halosulfides are prepared from 2-hydroxysulfide which in turn can be made from 1, 2-halohydrin.

Maximum yields are obtained by maintaining the *pH* between 7.5 to 9.5. The reagent for this purpose is sodium bicarbonate or sodium acetate. Thus cyclohexene sulfide is obtained by the reaction of sodium bicarbonate on *trans*-2-chlorocyclohexanethiol.

3. *From Epoxides* : Thiirane derivatives are often prepared by the action of a wide variety of reagents such as thiocyanate ion, thiourea, thioamides, triphenyl phosphene sulfide, etc. on epoxides.[85,88,89] The first synthesis of thiirane from epoxide and inorganic thiocyanate was reported in a patent specification. The *trans*-or *cis*-stereochemistry of oxirane is preserved and proceeds in the following four steps: Optically active (R, R)-oxiranes should give rise to optically active (S, S)-thiiranes.

The entire reaction is accompanied by Walden inversion at both the carbon atoms of the ring.[90]

Thiirane is also obtained by the action of thiourea on epoxides.[91] The epoxide ring is first opened by thiourea in the manner shown above which then cyclizes to thiirane, releasing urea.

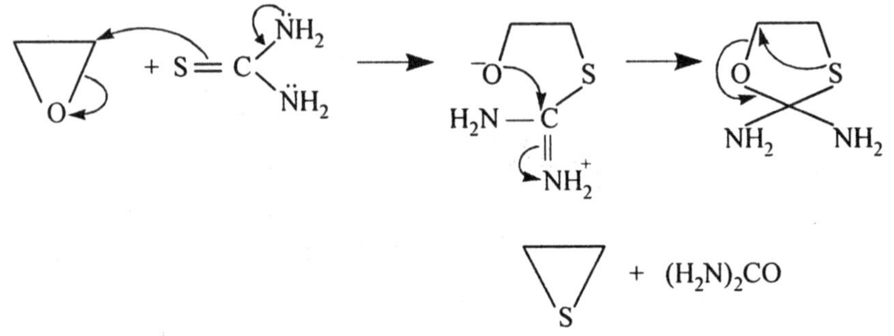

Epoxides are also cleaved by thiocyanic acid to thioethanols which then cyclize to thiiranes in the presence of a base.[92]

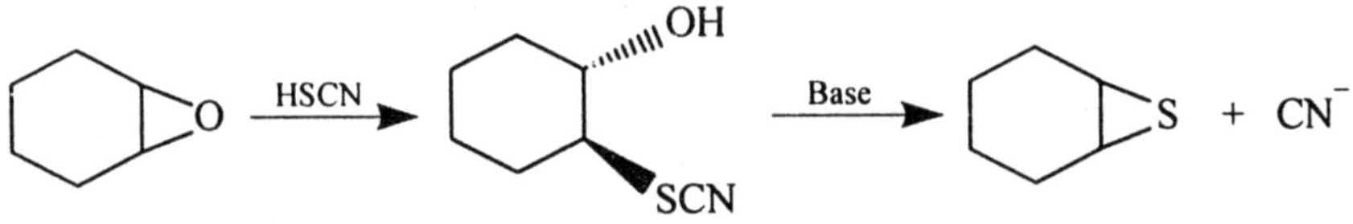

Thiiranes are also obtained by the action of triphenylphosphine sulfide on epoxides,[93] in the presence of trifluoroacetic acid by a replacement of oxygen by sulfur atom.

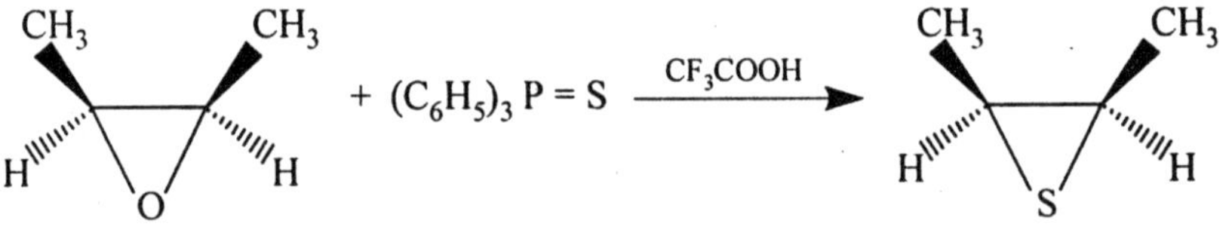

2, 3-Dimethylthiirane

$$+ (C_6H_5)_3 P = O$$

4. *From Cyclic Ethylene Carbonates :* Reaction[94] between a cyclic carbonate of 1, 2-diol (1, 3-dioxolone) and an alkali thiocyanate at a high temperature yields the corresponding thiirane according to the following five steps:

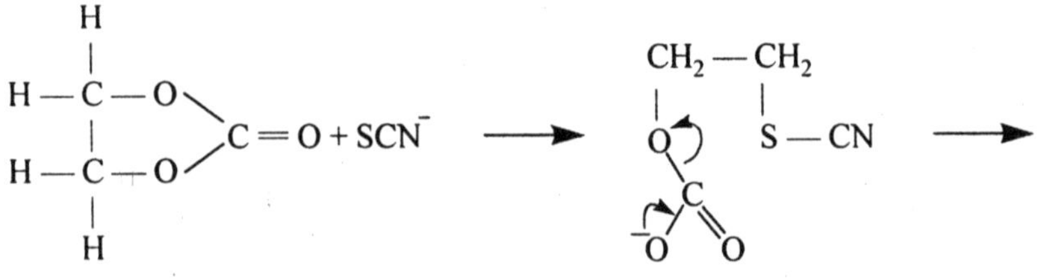

5. *From Aldehydes and Ketones* : Thiirane derivatives have been prepared
from 2-(thiomethyl)-2-oxazoline and an aldehyde or ketone in the presence of
n-butyllithium in the following manner.[95]

Lithioxazoline (40) adds to a ketone and rearrangement of the intermediate
followed by fragmentation produces the thiirane derivative.

6. From Diazoalkanes : The reaction between diazoalkanes and thioketones has been extensively studied and is useful for alkene synthesis.[96] The intermediate formed is denitrogenated to give thiirane.

7. From Alkene Episulfidation : An alkene episulfidation[97] by nitrogen sulfide (N_2S) generated *in situ* from 5-aryloxy-2, 3, 4-thiatriazole forms a three-membered zwitterionic species.

This on denitrogenation yields a thiirane derivative. The reaction is stereospecific.

2.3.3 Chemical Properties

Thiirane is relatively unstable than oxirane and undergoes a multitude of reactions.

1. Ring Opening Reaction : All reactions of thiiranes involve ring opening and since the electron density at the sulfur atom is lower than that at the oxygen atom in oxiranes, in some cases they are thus less reactive towards electrophilic reagents. Their reactivity towards nucleophilic reagents seems similar to or a little greater than that of oxiranes. Thiirane ring is readily cleaved by a large number of nucleophilic and electrophilic reagents, to form mercaptans as oxiranes are obtained from alcohols. Some of these reactions are summarized given below:

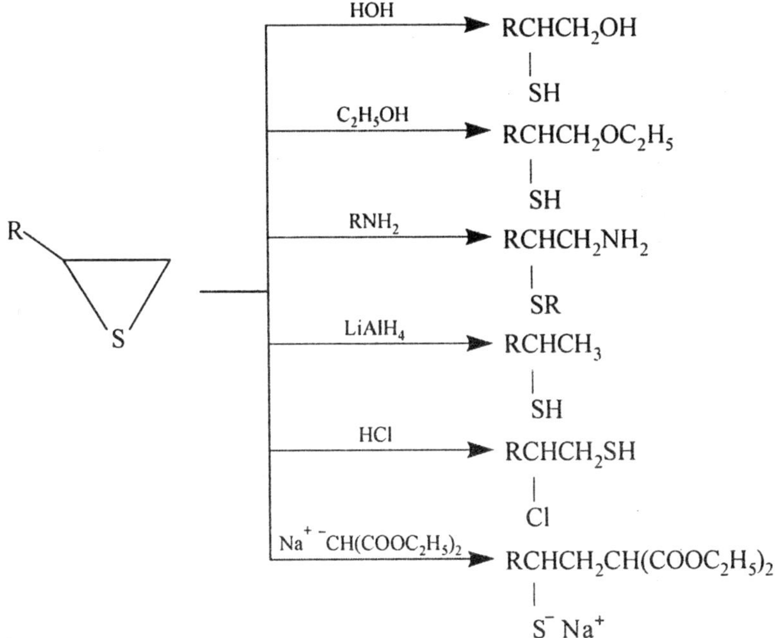

A nucleophile first attacks at an unsubstituted carbon atom to open the ring in an unsymmetrical episulfide. In the case of acidic reagents in contrast, the ring opening depends on the ability of the group attached to the ring carbon atom to stabilize the incipient carbocation.

If excess conc. hydrochloric acid is added to ethylene sulfide, then both the monomer and the dimer of thiirane may be isolated.

$$\text{(thiirane)} + HCl \longrightarrow HSCH_2CH_2Cl \longrightarrow HSCH_2CH_2SCH_2CH_2Cl$$

The reaction of acyl chlorides with episulfides, as also in the case of epoxides, takes place by electrophilic attack of the reagent on the sulfur atom.

$$CH_3\text{-thiirane} + CH_3COCl \longrightarrow \longrightarrow CH_3CHCH_2SCOCH_3$$

2. Desulfurization : Certain reagents function in a manner as to remove the sulfur atom of the sulfides leading to alkene formation and a sulfur-phosphorus derivative. There is considerable current interest in the desulfurization of sulfur compounds. These reactions can be of substantial preparative importance. A variety of reagents have been employed for this purpose, for instance, superoxide ion (O^-_2)[99] alkaline oxidation $(O_2, t\text{-}BuO^-)$,[100] transition metal complexes[101], butyllithium[102] and trialkyl phosphorus compounds.[103] The reaction proceeds with complete stereospecificity. This is demonstrated by the accompanying examples.

trans-**2-Butene episulfide**   *trans*-**2-Butene**

cis **2-Butene episulfide**   *cis* **2-Butene**

3. Oxidation : Thiirane is oxidized with sodium metaperiodate in aqueous ethanol forming ethylene sulfoxide,[104] equation (2.24)

$$\text{(thiirane)} + NaIO_4 \xrightarrow[20\text{-}25°C]{\text{Aq ethanol}} \text{(ethylene sulfoxide)} + NaIO_3 \qquad (2.24)$$

This compound undergoes dethionylation[77] at 100°C to yield ethylene and sulfur monoxide. The ring is also opened under acidic conditions in different solvents to give different products.[105]

$$\text{(ethylene sulfoxide)} \xrightarrow{H^+, CH_3OH} CH_3OCH_2CH_2SSCH_2CH_2OCH_3$$

with the S bearing a double-bonded O.

A thiol sulfonate, 97%

$$\xrightarrow{H^+, C_2H_5OH} C_2H_5SCH_2CH_2SSC_2H_5$$

A disulfide, 40%

Thiirane 1-oxides are also desulfurized stereospecifically by treatment with phenyllithium

4. Formation of Peroxysulfenic Acid : Photolysis of thiirane in methanol/ methylene blue as sensitizer and light forms a peroxysulfenic acid[106]. For instance, 7-thiabicyclo [4.1.0] heptane (42) under these conditions gives (43). The resulting acid oxidizes olefins to epoxides. Norbornene on oxidation with

(43) provides 37% yield of *exo*-norbornene epoxide.

(42)

(43)

5. *Thermal and Photochemical Reactions* : Pyrolysis of thiiranes results in the fission of C – S bond of the ring with the formation of an alkene while a sulfur atom is split off. This is particularly true if thiirane is substituted by more than two aromatic nuclei. Styrene sulfide can be distilled under vacuum but stilbene sulfide decomposes to 1, 2-diphenylethylene and sulfur, equation (2.25).

$$C_6H_5CH - CHC_6H_5 \xrightarrow{\Delta} C_6H_5CH = CHC_6H_5 + S \qquad (2.25)$$

The C – C bond in thiiranes is cleaved thermally in a few cases. Ultraviolet light generally gives rise to the formation of an alkene from episulfide just as thermal reactions and other reagents do, equation (2.26).

$$\xrightarrow{h\nu} H_2C = CH_2 + S \qquad (2.26)$$

Similarly 4,4,6, 6-tetramethyl-1-thiaspiro [2.3] hexan-5-one yields[107] 3, 3, 5, 5, -tetramethylene cyclobutanone on photolysis (2537 Å) in methanol, equation (2.27).

$$\xrightarrow[CH_3OH]{2537°C} \qquad O = \quad = CH_2 + S \quad (2.27)$$

2.4   AZIRINES, OXIRENES AND THIIRENES

The unsaturated three-membered heterocyclic compounds containing one hetero atom and one double bond namely, *azirines* (44) and (45), *oxirene* (46), and *thiirene* (47) belong to the $4n\pi$-electron ring system which defy

Hückel aromaticity rule. According to Breslow's postulate[108] that in some members of the $4n\pi$ series cyclic delocalization of π electrons leads to

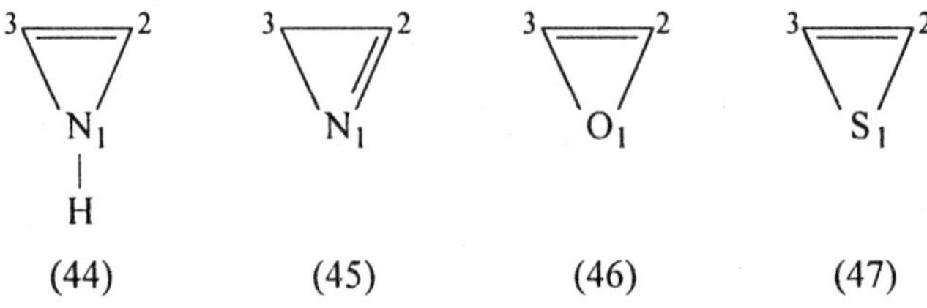

(44)   (45)   (46)   (47)

strong destabilization of the compound, in contrast to the stabilization characteristics of aromaticity. These ring systems exist only as unstable intermediates and thus may be anti-aromatic, i.e., electronic conjugation is destabilizing. Indeed, appropriate orbital calculations confirm this prediction[109] for 2-azirine, oxirene and thiirene which are all computed to be thermo-dynamically less stable than their acyclic isomers.[110,111] Azirine is the name used to denote the azacyclopropene system. There are two possible isomeric azirine structures, namely the enemine, 2-azirine (44) and the imine, 1-azirine (45). Ring Index and the Chemical Abstracts have designated them as 1*H*-azirine and 2*H*-azirine. The general chemistry of these two ring systems, particularly that of 1-azirine is well documented.

Oxirenes, the family of unsaturated epoxides, are as yet unknown as isolable compounds although they have been known as possible intermediates in the peroxidation of acetylenes[112] and in the Wolff rearrangement of α-diazoketones.[113-116]

This provides an indirect though a compelling evidence for their transient existence as short-lived intermediates.

Thiirene (47) is isoelectronic with cyclobutadiene and also belongs to the class of $4n\pi$ electron anti-aromatic heterocycles. It is not surprising, therefore, that thiirenes have only been observed as transient intermediates.[117-119] 1, 2, 3-Thiadiazole (48) for instance, on photolysis[119] yields a thioketene *via* a thiirene which implicates the latter as a precursor.

2.4.1 Physical and Spectroscopic Properties of Azirines

Azirines are generally vile-smelling compounds and highly irritant to skin. They readily polymerize unless stored under cold and in an inert atmosphere. The instability of azirines is not only due to the larger bond-angle strain inherent in an unsaturated three-membered ring, but also due to the overlap of the nitrogen lone-pair electrons with the $C = C$ bond, a situation which gives rise to a destabilizing anti-aromatic $4n\pi$ electron system. A measure of this destabilization has been provided by *ab initio* calculations[120]; which indicate that 1-azirine (2*H* azirine) is 40 Kcal/mole more stable than the 2-isomer[121], 2-azirine (1*H* azirine). The azirinyl cation (49) has extensive π-delocalization and can be obtained as stable aromatic species.

(49)

The *n.m.r.* spectra of 1-azirines are consistent with the strained nature of this compound.[122] The $^{13}C - H$ coupling constant for the methylene hydrogens in 2-phenyl-1-azirine is 178 Hz which indicates 36% *S*-character of the exocyclic carbon orbitals at C—3. Due to this large degree of *S*-character in the orbital containing the electron ion-pair on the nitrogen atom, 1-azirines are non-basic. Thus 2-phenyl-3-methyl-1-azirine is insoluble even in 10% hydrochloric acid. Azirines substituted at the 2-position are most stable and are easily prepared. The dipole moment values for 2-azirine and 1-azirine are 2.51 D and 2.56 D respectively.

2.4.2 Synthetic Methods

The synthesis of 1-azirines has spurred considerable research in these heterocyclic systems as a result 1-azirine system is well known. Todate, however, there seems to be no authentic examples of 2-azirine, though several attempts have been made for their synthesis.[123]

1. *Neber Rearrangement* : Neber proposed 1-azirine as an intermediate in the base-catalyzed rearrangement of oxime *p*-toluenesulfonyl derivative to α-amino ketone. This reaction is familiarly known as the *Neber rearrangement.*[124]

$$RCH_2-\underset{\underset{OH}{\overset{|}{N}}}{\overset{\|}{C}}-R_1 \quad \xrightarrow[\text{TsCl}]{^-OC_2H_5} \quad R\bar{C}H_2-\underset{\underset{OTs}{\overset{|}{N}}}{\overset{\|}{C}}-R_1 \quad \xrightarrow[-OTs^-]{} \quad RCH-\underset{N}{C}=C-R_1$$

$$\xrightarrow{H^+, H_2O} \quad RCH-\underset{\underset{NH_2}{|}}{\overset{\overset{O}{\|}}{C}}-R_1$$

The occurrence of 1-azirine structure has been proved by its isolation and its conversion under the experimental conditions to easily identifiable products, by Baumgarten and coworkers.[125, 126] This reaction is probably successful because of the presence of strong electron-withdrawing groups which enhance the acidity of the α-hydrogen atom and thus the ring formation is facilitated. The Neber rearrangement however, does not generally constitute a useful method for 1-azirine synthesis because of poor yields. Synthesis of optically active 3-amino-1-azirines have been realized using the modified Neber rearrangement.[126]

2. *From Vinyl Azides* : The first general synthesis of 1-azirines was developed by Smollinsky[127] and he employed the vapor phase pyrolysis of vinyl azides. Vinyl azides prepared by the addition of iodine azide to an alkene,[127] are converted to 1-azirines under three different conditions. First procedure involves thermolysis of vinyl azides in 50-60% yield.

$$H_2C=CHR \xrightarrow{IN_3} H_2C-CHR \xrightarrow[-HI]{\text{Pot. }t\text{-butoxide}} CH_2=CR$$

with the central carbons bearing I and N_3, and the product bearing N_3.

$$\xrightarrow{\Delta, -N_2}$$

Photolysis of vinyl azides under nitrogen atmosphere at $-30°C$ also permits isolation[128] of 1-azirine. This reaction probably proceeds by nitrene insertion into the $C = C$ bond[129], equation (2.28). Of the two, the first method is the most general one.[130]

$$C_6H_5 \underset{H}{\overset{}{>}}C = C \underset{H}{\overset{N_3}{<}} \xrightarrow[-N_2]{h\nu} C_6H_5 - CH - CH \qquad (2.28)$$

3-phenyl-1-azirine

A third procedure,[131] developed recently, involves milder conditions for the preparation of 1-azirine also from vinyl azides. This procedure consists of refluxing a toluene solution of an azide in the presence of catalytic amount of a tertiary amine, as shown for β-styryl azide, equation (2.29).

$$C_6H_5 \underset{H}{\overset{}{>}}C = C \underset{CH_3}{\overset{N_3}{<}} \xrightarrow[\text{reflux, } -N_2]{\ddot{N}(CH_3)_3, \text{ Toluene}} C_6H_5 - CH - C - CH_3 \qquad (2.29)$$

2-Methyl-3-phenyl-azirine

3. From α-Bromo Ketoxime : Hassner[132] reported the preparation of 1-azirine using α-bromo ketoximes, the reaction consists of five steps. All the steps, however, could be accomplished in the same reaction vessel without isolating the intermediate. The reaction sequence is shown using α-bromopicolone (50). The hydroxy group of the α-bromo ketoxime is first protected with a ketal

(50)

(51)

(52)   **2-*t*-Butyl-1-azirine**

followed by treatment with triphenylphosphine to form (51). This is then deprotected to regenerate the phosphonium salt. This salt is converted to oxazophospholine (52) which is thermolyzed to 2-*t*-butylaziridine.

4. *From Pyrolysis or Photolysis of Isoxazoles :* Nishiwaki *et al.*[133] observed that pyrolysis of 5-alkoxy substituted isoxazoles resulted in the formation of isolable 1-azirines equation (2.30). The photolysis,[134] on the other hand, of 3, 5-diphenylisoxazole produces 2-phenyl-3-benzoylazirine in 82% yield equation (2.31). This reaction has been found to be wavelength dependent as it is reversed at 3000 Å.

3-Benzoyl-1-phenyl-1-azirine

2.4.3 Chemical Properties

1-Azirine undergoes several types of reactions and is an extremely useful reagent for the preparation of different types of well known and novel heterocyclics.

1. *Reaction with Acids :* Azirines are non-basic and thus insoluble in dilute acids, however, protonation of 1-azirine followed by ring opening takes place in strong acid solutions.[135] The reaction with chloroacetic acid or benzoic acid opens the ring which subsequently undergoes rearrangement to give (53).

(53)

Reaction of 2-phenyl-1-azirine with benzoic acid results in compound (54).

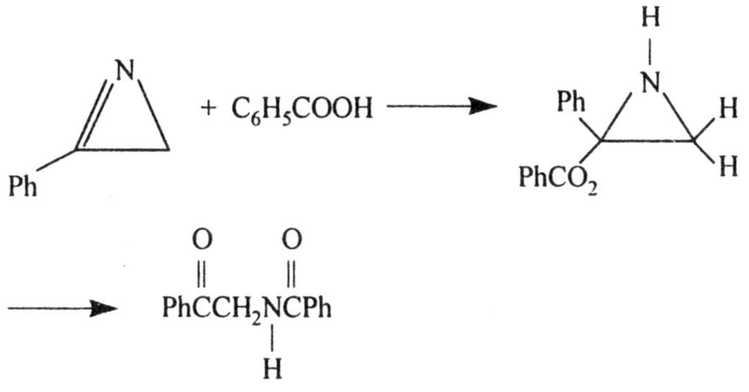

$$PhCCH_2NCPh$$

(54)

The reaction of 2-phenyl-1-azirine with acid chlorides or acid anhydrides in the presence of triethylamine gives the oxazole (55) directly.

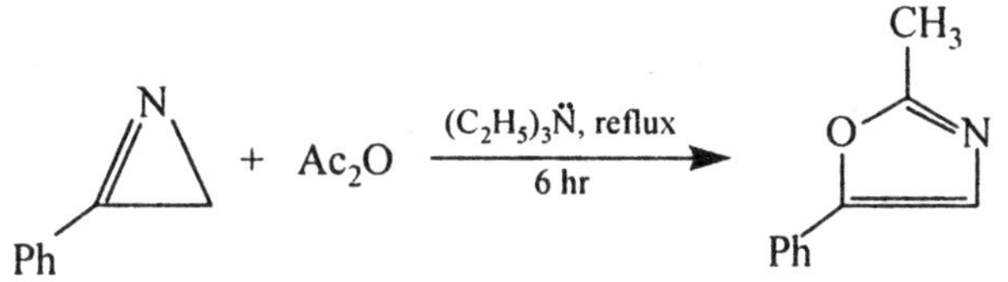

(55)

2. Formation of Complexes : 1-Azirines form stable complexes[136] with dichloro *bis* (benzonitrile) palladium (II).

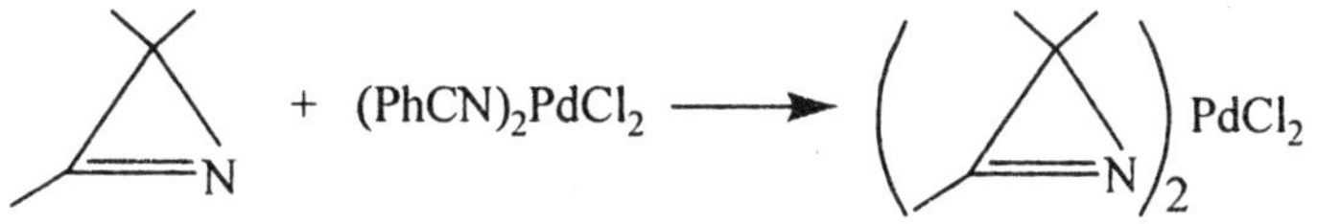

3. Ring Opening Reactions : 1-Azirine ring can be opened either thermally or photochemically and the products obtained in the two cases are different. The thermal reaction of 1-azirines can involve either $C - N$[137] or $C - C$[138-139] bond cleavage depending on the location of the substituents. It has been suggested[140] that the presence of a stabilizing group such as phenyl at $C - 2$ tends to induce the formation of products formed from initial $(C - C)$ bond cleavage, whereas its location at $C - 3$ results in products formed by initial rupture of $C - N$ bond. The ultimate products formed are due to fragmentation *via* a carbene and a nitrile molecule.[141] Thus 2-phenyl-3, 3-dimethyl-2*H*-azirine yields phenyl cyanide and dimethyl carbene, equation (2.32).

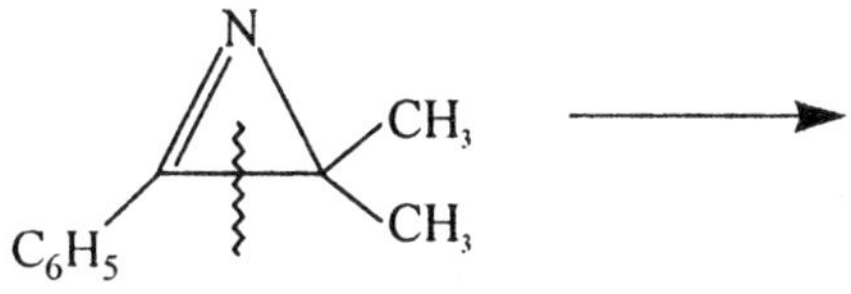

$$C_6H_5C \equiv N + CH_3\ddot{C}CH_3 \quad (2.32)$$

Formation of indole has also been observed in certain cases[142], thus 2-methyl-3-phenyl-2*H*-azirine rearranges to 2-methylindole *via* a $C - N$ bond cleavage.

2-Methylindole

The photochemical mode of ring opening, on the other hand, involves a C – C bond cleavage resulting in the formation of a 1, 3-dipolar nitrile ylide.[142, 143]

(56)   (57)

CH$_2$ = CHCOOCH$_3$
Cycloaddition

(58)

Cyclization
(arrows)

aromatization

(59)

which enters into inter- and intra-1, 3-dipolar cycloaddition with a wide variety of dipolarophiles to give a 5-membered ring. In the absence of a dipolarophile, the intermediate undergoes intramolecular ring closure. For example, 3-(*O*-vinylphenyl)-2, 2-dimethyl-2*H*-azirine (56), on photolysis, gives rise to the ylide (57) which cycloadds to methyl acrylate to give (58) and to 1-N-isopropylidene-endene-3-amine (59) in its absence.

These studies have been theoretically confirmed.[144]

Mukai and Sukawa[145] have shown that azirine is an intermediate in the photoconversion of 4-phenyl-2, 3-oxazobicyclo [3.2.0] hepta-3, 6-diene to 2-phenyl-1, 3-oxazepine.

4. The Diels-Alder Reaction : The Diels-Alder cycloaddition reaction which has proven to be extensively useful for the formation of carbocyclic ring compounds, has been found to be equally effective for the formation of heterocycles using 1-azirines as the dienophiles. 1-Azirines contain a reactive double bond and participate as a 2π-component in these type of reactions:[146] Thus, 1, 3-diphenylisobenzofuran cycloadds to 3-methyl-2-phenylazirine to form the adduct (60). The reactivity of the azirine double bond is also reflected in the facile addition of ethyl bromoacetate[147] under the Reformatsky reaction.

(60)

5. Reduction : A number of azirines have been reduced by lithium aluminum hydride to aziridines, equation (2.33).

$$(2.33)$$

2-Methyl-3-phenylaziridine

The reaction proceeds in high yields and is stereospecific. It can be a useful aziridine synthesis. The approach of the hydride ion occurs exclusively from the side opposite to the group at position-3 to give *cis*-aziridines.

Grignard reagents are also known to react with 1-azirines to give aziridines. Furthermore, the attack of the Grignard reagent occurs stereospecifically from the less sterically hindered side of azirine. This is analogous to the stereospecific reduction of azirines to aziridines by lithium aluminum hydride.

2.4.4 Oxirenes

Their exists a significant amount of data which indicate that the parent oxirene[148] and its alkyl derivatives are at best, of low stability.[149] Earlier claims of oxirene preparation by oxidation of alkynes proved incorrect.[150] However, the demonstration of the involvement in α-diazo ketone decomposition of an oxirene species opened up the possibility of its existence in other reactions as well. Cormier[151] has shown that oxirene (61) could be generated from the diazo ketones (62) *via* the intermediate carbenes. Oxidation of alkynes to oxirenes formed enzymatically in the living cells has been advanced.[152]

Oxirene derivatives have also been involved in elimination, photolytic and retrocycloaddition reactions by other workers but alternative explanations are always available.[152, 153]

REFERENCES

1. A. Dureault, I. Transchepain and J. C. Depezay, *J. Org. Chem.*, **54**, 5324 (1989).
2. O. C. Dermer and G. E. Hamm, *Ethylenimine and Other Aziridines*, Academic Press, New York (1969).
3. P. E. Fanta in, *Heterocyclic Compounds*, A. Weissberger, (*Ed.*), Part I, Interscience, New York (1964), p. 524.
4. H. C. Fruton in, *Heterocyclic Compounds*, E. C. Elderfield, (*Ed.*), Vol. I, John Wiley, New York (1950), p. 61.
5. C. E. O'Rourke, L. B. Clapp and J. O. Edwards, *J. Am. Chem. Soc.*, **78**, 2159 (1956).
6. N. H. Crowwell, R. E. Bambury and J. L. Adelfang, *J. Am. Chem. Soc.*, **82**, 4241 (1960).
7. S. J. Brois. *J. Am. Chem. Soc.*, **90**, 506 (1968).
8. H. Kessler and D. Leibfritz, *Tet. Letters*, 4297 (1970).
9. W. McCoull and F. A. Davis, Synthesis, 1347 (2000); N. H. Cromwell, *Rec. Chem., Progr.* (Kresge-Hooker Sci Lib) **19**, 215 (1958) ; Also see P. L. Levins and Z. B. Papanstassion, *J. Am, Chem. Soc.*, **87**, 826 (1965).
10. G. L. Closs and S. J. Brois, *J. Am. Chem. Soc.*, **82**, 6068 (1960).
11. A. Hassner and C. Heathcock, *Tetrahedron*, **20**, 1037 (1964).
12. R. Appel and M. Halstenberg, *Chem. Ber.*, **109**, 814 (1976).
13. Y. Ittah, Y. Sasson, I. Shahak, S. Tsaroom and J. Blum, *J. Org., Chem.*, **43**, 4271 (1978).
14. J. A. Deyrup and W. A. Szabo, *J. Org. Chem.*, **40**, 2048 (1975).
15. W. Lwowski and I. Mattingly, *Jr. J. Am. Chem. Soc.*, **87**, 1947 (1965).
16. R. S. Atkinson and B. J. Kelley, *Chem. Comm.*, 1362 (1987); R. S. Atkinson and B. J. Kelley, *ibid.*, 624 (1988).
17. W. A. Skinner, A. P. Martinez, H. F. Cram, L. Goodman and B. R. Baker *J. Org. Chem.*, *16*, 148 (1961).
18. D. R. Crist and N. J. Leonard, *Angew. Chem.*, **81**, 953 (1969); N. J. Leonard and K. Jann, *J. Am. Chem. Soc.*, **84**, 4806 (1962); *ibid.*, **82**, 6418 (1960); N. J. Leonard, *Rec. Chem. Progr.* (Kresge-Hooker Sci Lib), **26**, 211 (1965); N. J. Leonard and L. E. Brady, *J. Org. Chem.*, **30**, 817 (1965).
19. A. Laurent and A. Muller, *Tet. Letters*, 759 (1969).
20. N. Furukawa, T. Yoshimura, M. Ohotsu, T. Akasaka and S. Oae., Tetrahedron, **36**, 37 (1980).
21. G. T. Anderson, J. R. Henry and S. M. Weinreb, *J. Org. Chem.*, **50**, 6946 (1991).

21a. D. A. Evans, M. M. Faul and M. T. Bilodeau, J. Org. Chem., **56**, 6744 (1991).

21b. G. T. Anderson, J. R. Henry and S. M. Weinreb, *Ibid.* **56**, 6946 (1991).

22. R. Ghirardeli and H. J. Lucas, *J. Am. Chem. Soc.,* **72,** 734 (1975).

23. H. Stamp, *Angew, Chem.,* **74**, 694 (1962).

24. J. D. Cox, *Tetrahedron,* **19**, 1175 (1963).

25. H. Nozaki, S. Fujita and R. Noyori, *Tetrahedron,* **24**, 2193 (1968).

26. R. B. Woodward and H. R. Hoffman, *J. Am. Chem. Soc.,* **87**, 395 (1965).

27. R. Huisgen, W. Scheer and H. Huber, *J. Am. Chem. Soc.,* **89**, 1753 (1967); Also see H. Durewell, *Aust. J. Chem.,* **30**, 1367 (1977).

28. V. Bhatt and M. V. George, *Tet. Letters,* 4133 (1977).

29. S. Sustmann, R. Sustmann and C. Ruchardt, *Chem. Ber.,* **108**, 1527 (1975).

30. J. E. Earley, C. E. O'Rouree, L. B. Clapp, J. O. Edwards and B. C. Lawes, *J. Am. Chem. Soc.,* **80**, 3458 (1958).

31. A. P. Kozikowski, H. Ishida and K. Isobe, *J. Org. Chem.,* **44**, 2788 (1979).

32. A. Padwa. and W. Eisenhardt, *J. Am. Chem. Soc.,* **93**, 1400 (1971).

33. M. Klaus and H. Prinzbach, *Angew. Chem. Int. Edin (Engl),* **10**, 273 (1971).

34. N. H. Cromwell and G. V. Hudson. *J. Am. Chem. Soc.,* **75**, 872 (1953).

35. C. A. Kingsbury, D. L. Durham and R. Hutton, *J. Org. Chem.,* **43**, 4696 (1978).

36. Q. N. Porter and J. Baldas, *"Mass Spectrometry of Heterocyclic Compounds"*, 2nd ed., John Wiley, New York (1985).

37a. F. Fringuelli, R. Germani, F. Pizzo nd G. Savelli *Chem. Comm.,* 1427 (1989).

37b. P. D. Bartlett, *Rec. Chem. Progr.,* **11**, 47 (1950).

37c. For details see, R. K. Bansal, Organic Reaction Mechanisms, 3rd. *ed.,* Tata McGraw Hill, New Delhi.

38. R. C. Ewins, H. B. Henbest and M. A. Mckervey, *Chem. Comm.,* (1967); Also see E. J. Corey, S. Shibata and R. K. Bakhshi, *J. Org. Chem.,* **53**, 2861 (1988).

39a. H. E. Zimmermann L. Singer and B. S. Thyagrajan, *J. Am. Chem. Soc.,* **81**, 108 (1959).

39b. P. A. Bentley, *Chem. Comm.,* 1616 (2001).

39c. J. Brinksman, *Chem. Comm.,* 537 (2000).

39d. B. Lygo and P. G. Wainwright, *Tetrahedron,* **55**, 6289 (1999).

40. C. Clark P. Hermans. O. M. Cahn, C. Moore, H. C. Taljaad and G. Van Vuuren, *Chem. Comm.,* **1378** (1986).

41. M. Ashwell and R. F. W. Jackson. *ibid.,* 645 (1988).

42. J. Reese, *Chem. Ber.* **75.**, 384 (1942).

43. G. B. Payne, *Tetrahedron*, **18**, 763 (1962); Y. Ogatta and Y. Sawaki, *ibid.*, **20**, 2065 (1964); G. B. Pyane and F. H. Wilhams, *J. Org. Chem.*, **26**, 659 (1961); *ibid.*, **26**, 663 (1961).

44. E. Guilmet and B. Meunier, *Tet. letters*, 4449 (1980).

45. W. R. Adams, *Oxidation*, **2**, 65 (1971); K. Gollnick, *Adv. Photochemistry*, **6**, 1 (1968).

46. N. Shimizu and P. D. Bartlett, *J. Am. Chem. Soc.*, **98**, 4193 (1976).

47. I. Ojima (Ed.) Catalytic Asymmetric synthesis, VCH Publishers, N. Y. (2000); C. Bolm and K. Muniz, *Chem. Soc., Rev.*, **28**, 51 (1999); C. Bonini and G. Righi, *Tetrahedron*, **58**, 4981 (2002).

48. T. Katsuki and K. B. Sharpless, *J. Am. Chem. Soc.*, **102**, 5974 (1980).

49. For details see B. E. Rossiter in "*Asymmetric Synthesis*," Vol. 5, J. D. Morrison (*Ed.*), Academic New York (1985). Chap 7. T. Katsuki in Comprehensive Asymmetric Synthesis, E. N. Jacobsen, A. Pfaltz and H. Yamamoto (Eds.), Springer-Verlag, Heidelberg (1999), Vol. II, pp. 621-648.

50. C. Einhorn, C. Allavena and J. L. Luche, *Chem, Comm.*, 337 (1988).

51. W. Szeja, *Synthesis*, 984 (1985).

52. F. W. Bachelor and R. K. Bansal, *J. Org. Chem.*, **34**, 3600 (1969); *Ind. J. Chem.*, **15B**, 574 (1977).

53. H. E. Zimmerman and L. Ahramjian, *J. Am. Chem. Soc.*, **82**, 5459 (1960).

54. R. K. Bansal, and K. Sethi, Bull, *Chem. Soc. Japan*, **53**, 1197 (1980).

55. E. J. Corey and M. Chakowsky, *J. Am. Chem. Soc.*, **84**, 867, 3782 (1962); also see Y. Lin, S. A. Lang, C. M. Seifert, S. G. Child, G. O. Morton and P. E. Fabio, *J. Org. Chem.*, **44**, 4701 (1979).

56. R. S. Bly, C. M. BuBose and G. B. Konizer, *J. Org. Chem.*, **33**, 596 (1956).

57. E. L. Eliel and D. L. Delmonte, *J. Org. Chem.*, **21**, 596 (1956),

58. V. K. Aggarwal et. al, *J. Am. Chem. Soc.*, **124**, 5747 (2002); V. K. Aggarwal *et. al, Chem. Rev*, **97**, 2341 (1997).

59. A. Foucaud and E. le Rouille, Synthesis, 787 (1990), A. Foucaud and M. Bakouetila, *Synthesis*, 854 (1987).

60. T. Hirao, T. Moriuchi, S. Mikami, I. Ikeda and Y. Ohshiro, *Tet. Letters*, 1031 (1993).

61. For a review, see, J. G. Smith, *Synthesis*, 629 (1984), T. J. Mason, *Heterocyclic Chem.*, **3**, 1 (1982), J. Rebek, *Heterocycles* **15**, 517 (1981), K. Mori *et al.*, *Tetrahedron*, **38**, 3705 (1982), W. L. Nelson and T. R. Burks, *J. Org. Chem.*, **43**, 3641 (1978); A. S. Rao *et al.*, Synthesis, 142 (1984).

62. C. A. Stewart and C. A. Wanderwerf, *J. Am. Chem. Soc.*, **76**, 1259 (1954).

63. A. Orekhoff and M. Tiffeneau, *Bull. Chem. Soc. France*, **146**, 697 (1908).

64. R. S. Mohan et. al, *Tetrahedron Letters,* **42**, 8129 (2001).

65. S. Winstein and R. S. Hendersen, in, *"Heterocyclic Compounds"*, R. C. Elderfield, (*Ed.*), Vol. I. John Wiley, New York (1950); Chapt. 1.

66. R. L. Settine and C. M. McDaniell. *J. Org. Chem.,* **32**, 2910 (1967).

67. F. de Reinach-hirtzbach and T. Durst, *Tet. letters*, 3647 (1971).

68. R. W. Herr and C. R. Johnson, *J. Am. Chem. Soc.,* **92**, 4979 (1970).

69. H. O. House, W. L. Respess and G. M. Whittesides *J. Org. Chem.,* **31**, 3128 (1966).

70. D. M. Wieland and C. R. John, *J. Am. Chem. Soc.,* **93**, 3047 (1971).

71. R. K. Bansal, K. Sethi and S. K. Jain, *Ind, J. Chem.,* **20B**, 121 (1981).

72. N. Manisse and J. Chuche, *J. Am. Chem. Soc.,* **99**, 1272 (1977).

73. M. Kawaski *et al., J. Chem. Phys.,* **59**, 2076, 5321 (1973); Also see, S. Braslavsky and J. Helcklen, *Chem. Rev.* **77**, 4473 (1977).

74. M. C. Flowers, R. M. Parker and M. A. Volsey, *J. Chem. Soc.,* (B), **239** (1970); G. A. Lee, *J. Org. Chem.,* **41**, 2656 (1976).

75. R. Huisgen, Angew. *Chem. Int. Edin* (Engl.), **16**, 572 (1977).

76. G. W. Griffin *ibid.,* **10**, 537 (1971).

77. W. Reusch *J. Am. Chem. Soc.,* **85**, 3894 (1963).

78. C. K. Johnson, B. Dominy and J. Dale and K. Dassvatn, *Chem. Comm.,* 295 (1976);

79. D. L. J. Clive and C. V. Denyer, *Chem. Comm.,* **253** (1973).

80. H. Paulsen, F. R. Heilker, J. Feldmann, and K. Heyns, *Synthesis*, 636 (1980).

81. D. L. J. Clive and S. M. Menchen, *J. Org. Chem.,* **45**, 2347 (1980).

82. K. B. Sharpless and R. F. Lauer, *J. Am. Chem., Soc.,* **95**, 2697 (1973).

83. B. M. Trost and L. S. Melvin, Jr., *Sulfur Ylides-Emerging Synthetic Intermediates,"* Academic Press, New York (1975).

84. For a review see, M. Sanders, *Chem. Rev.,* **66**, 297 (1966).

85. G. L. Gunningham *et al., J. Chem. Phys.,* **19**, 676 (1951).

86. D. D. Reynolds, *J. Am. Chem. Soc.,* **79**, 4951 (1957).

87. T. H. Chan and J. R. F. Finkenbine, J. *Am. Chem. Soc.,* **94**, 2821 (1972).

88. L. Lautenschlaeger and N. V. S. Schwartz, *J. Org. Chem.,* **34**, 3991 (1969).

89. C. C. Price and P. F. Kirk, *J. Am. Chem. Soc.,* **75**, 2396 (1953).

90. G. L. Braz, *J. Gen, Chem.,* (USSR), **21**, 757 (1951).

91. E. E. Van Tamelen, *J. Am. Chem. Soc.,* **73**, 3444 (1951).

92. T. H. Chan W. E. Childrens and P. S. Furth, *J. Am. Chem. Soc.,* **94**, 288 (1972), also see M. J. Slih and C. H., Robinson, *J. Org. Chem.,* **53**, 5947 (1988).

93. S. Searles, H. R. Hays and E. F. Lutz, *J. Org. Chem.,* **27**, 2832 (1962).

94. A. I. Meyers and M. E. Ford, *Tet. Letters*, 2861 (1975).

95. C. E. Diebert, *J. Org. Chem.,* **35**, 1504 (1970).

96. Z. Yu and Y. Wu, *J. Org. Chem.,* **68**, 6049 (2003). For a recent review

on thiiranes and its derivatives See, W. Ando, N. Choi and N. Tokitoh, in "*Comprehensive Heterocyclic Chemistry* II", A. R. Katritzky, C. W. Rees and A. Padwa (Eds.), Pergmon Press, U. K. (1996), Vol. I A, p. 173; W. Chow and D. N. Harpp, *Sulfur Rep.*, **15**, 1 (1993).

97. E. Block, A. J. Yencha, M. Aslam, V. Eswarakrishnan, J. Luo and A. Sano, *J. Am. Chem. Soc.*, **110**, 4748 (1988).

98. G. Crank and M. I. H. Makin, *Chem, Comm.*, 53 (1984).

99. Y. H. Kim, H. J. Kim and G. H. Yon, *ibid.*, 1064 (1984).

100. J. A. Gladysz, S. Togaslin, J. A. Gladysz, J. G. Fulchin and M. Hasegamma, *J. Org. Chem.*, **45**, 3044 (1980).

101. T. Kato, *Chem. Comm.*, 127 (1981); B. M. Trost and S. D. Ziman, *J. Org. Chem.*, **38**, 932 (1973).

102. D. N. Harpp, *J. Org. Chem.* **45**, 5185 (1980). J. I. G. Gadogan and R. K. Mackie, *Chem. Soc. Rev.*, **3**, 87 (1974).

103. G. E. Haztzell and J. N. Page, *J. Am. Chem. Soc.*, **86**, 2616 (1966).

104. K. Kondo, A. Negishi and I. Ojima, *J. Am. Chem. Soc.*, **94**, 5786 (1972).

105. T. Akasaka, M. Kako, H. Sonobe and W. Ando, *J. Am. Chem. Soc.*, **110**, 494 (1988).

106. J. G. Pacific and C. Diebert, *J. Am. Chem. Soc.*, **91**, 4595 (19609).

107. R. Breslow, *Acc. Chem. Res.*, **6**, 393 (1973).

108. D. T. Clark, Theor. *Chem. Acta.*, **15**, 225 (1969).

109. G. P. Strausz, R. K. Gosvi, F. Bernardi, P. G. Mezey, J. D., Goddard and I. G. Csizmadia, *Chem. Phys. Lett.*, **53**, 211 (1978).

110. An *ab initio* study of the infrared spectra of these compounds, See P. Carsky, B. A. Mess and L. J. Schaad *J. Am. Chem. Soc.*, **105**, 396 (1983).

111. P. W. Concannon and *J. Ciabattoni*, *ibid.*, **95**, 3284 (1973).

112. H. E. Harvey and S. J. Heath, *Trans. Faraday Soc.*, **68**, 512 (1972).

113. For a review see, H. Meierand K. F. Zeller, Angew, *Chem. Int. Edin.* (Engl.), **14**, 32 (1975).

114. O. P. Strausz, *et al.*, *J. Am. Chem. Soc.*, **95**, 124 (1973).

115. S. A. Maltin and P. G. Sammes, *J. Chem. Soc. Perkin Trans.*, **1**, 2623 (1972).

116. G. N. Schrauzer and H. Kisch., *J. Am. Chem. Soc.*, **96**, 6768, (21974); *ibid.*, **95**, 2501 (1973).

117. J. Fenvick, G. Frater, K. Ogi and O. P. Strausz, *ibid.*, **89**, 4806 (1976), *ibid.*, **90**, 7360 (1968).

118. A. Krantz and *J. Laureni, ibid.*, **96**, 6769 (1974) ***ibid.***, **103**, 486 (1981).

119. M. J. S. Dewar and C. A. Ramsden, *Chem.* 688 (1973).

120. J. Pittman, J. r. A. Kress, J. B. Palievson P. Walton and L. D. Kispert *J. Org. Chem.* **39**, 373 (1974).

121. F. W. Fowler and A. Hassner. *J. Am. Chem. Soc.*, **90**, 2868 (1968).

122. For a review see, F. W. Fowler, Heterocyclic *Chem.*, **13**, 45 (1971).

123. For details see R. K. Bansal, *Organic Reaction Mechanisms*, 3rd, ed., Tata McGraw Hill, New Delhi, (1986). Chap. 8.

124. H. E. Baumgarten J. E. Dirks, T. M. Peterson and D. C. Wolf. *J. Am. Chem. Soc.*, **82**, 4422 (1960).

125. I. Piskunova, A. V. Eremeev, A. F. Mishnev and I. A. *Vosekalna, Tetrahedron*, **49**, 4671 (1993).

126. G. Smolinsky, *J. Org. Chem.*, **27**, 3557 (1962); *ibid.*, **83**, 4483 (1961); A. Padwa *et al.*, *J. Org. Chem. Soc.*, **82**, 442 (1960).

127. K. Isomura, S. Kobayashi and H. Tamiguchi, *Tet, letters*, 3499 (1968).

128. F. P. Woerner, *Angew. Chem.*, **80**, 119 (1968).

129. A. G. Hartmann; D. A. Robertson and B. K. Gillard, *J. Org. Chem.*, **37**, 322 (1972); F. P. Woerner and H. Reimlinger, *Chem. Ber.*, **103**, 1908 (1970).

130. M. Komatsu, S. Ichijima, Y. Olshiro and T. Agama, *J. Org. Chem.* **38**, 4341 (1973).

131. A. Hassner and V. Alexanian, *J. Org. Chem.*, **44**, 3861 (1969).

132. T. Nishiwaki, T. Kitimura and A. Nalzano, *Tetrahedron*, **26**, 453 (1970).

133. E. F. Ullman and B. Singh, *J. Am. Chem. Soc.*, **88**, 1844 (1966).

134. N. J. Leonard and B. Zwanenburg, *J. Am. Chem. Soc.*, **89**, 4456 (1967).

135. A. Hassner, C. A. Bunnell and K. Haltinanger, *J. Org. Chem.*, **43**, 57 (1978).

136. N. S. Narashimhan, H. Heimgarten, H. J., Jansen and H. Schmid, *Helv. Chin.. Acta*, **56**, 1351 (1973); T. Nishikawa, *Chem. Comm.*, 565 (1972).

137. L. A. Wendling and R. G. Bergman, *J. Am. Chem. Soc.*, **96**, 308 (1974).

138. W. Seiber, P. Gilgen, S. Chalou, H. J. Hanson and H. Schmid S. Kobayashi and H. Taniguchi, *Helv. Chim. Acta.* **56**, 1679 (1973).

139. A. Padwa, P. Gilgen, S. Chalou, H. J., Hansen and H. Schmid *Acc. Chem. Res.*, **9**, 371 (1976); *Angew. Chem. Int. Edn.* (Engl.), **15**, 123 (1976).

140. K. Isomura S. Kobayashi and H. Taniguchi., *Chem. Comm.*, 1252 (1980); *Tet. letters*, 3499 (1968).

141. A. Padwa and Per J. H. Carlsen, *J. Am. Chem. Soc.*, **99**, 1514 (1977).

142. A. Padwa and N. Kamigatta, *ibid.*, **99**, 1871 (1977); A. Padwa and A. ku *ibid.*, **100**, 2181 (1978).

143. B. Bigot, A. Serin and A. Devaguet, *J. Am. Chem. Soc.*, **99**, 1514 (1977).

144. T. Mukai and H. Sukawa, *Tet. letters*, 2283 (1973).

145. P. B. Kryczka and A. Laurent, *Tet. letters*, 31 (1977).

146. G. Smolinsky, *J. Am. Chem. Soc.*, 4483 (1961).

147. For a review see, E. G. Lewars, *Chem. Rev.* **83**, 519 (1983).

148. A. Greenberg and J. F. Liebman, *Strained Organic Molecules*, Academic Press, New York (1978).

149. H. Schubach and V. Franzen, *Ann, Chem.*, **557**, 60 (1952); T. F. Rutledge, *Acetylenes and Allenes*, Reinhold, New York (1969), S. A. Matlin and P. G. Sammers, *J. Chem. Soc., Perkin Trans.*, 2851 (1973).

150. R. A. Cormier, *Tet. letters,* 2021 (1980).
151. P. R. O. de Montellano and K. L. Kunze, *J. Am. Chem. Soc.,* **102,** 7373 (1980).
152. H. Hart. J. B. Jiang and M. Sasaoka, *J. Org. Chem.,* **42,** 3840 (1977).

REVIEW PROBLEMS

2. 1 Predict the major product of the following reactions :

(a) [structure: *trans*-stilbene oxide derivative with H, C_6H_5 substituents] $\xrightarrow{\text{(i) NaN}_3 \quad \text{(ii) (C}_6\text{H}_5)_3\text{P}}$

(b) [cyclopentadiene] + $C_6H_5-\overset{\overset{\text{O}}{\|}}{C}-$[2H-azirine ring with H and CH_3] $\xrightarrow[\text{reflux}]{\text{Toluene}}$

(c) $BrCH_2\underset{\underset{Br}{|}}{CH}CN \xrightarrow{\text{Liq. NH}_3, \text{(C}_6\text{H}_5)_3\text{N}}$

(d) [octahydronaphthalene diene] $\xrightarrow{\substack{\text{(i) NOCl} \\ \text{(ii) SnCl}_2 \\ \text{(iii) OH}^-}}$

(e) [2,2-dimethylthiirane with CH_3, CH_3 and S] $\xrightarrow{\text{HBr}}$

(f) [cyclohexane spiro epoxide with $-CH_2CH_3$ and O] $\xrightarrow{\text{LiAlH}_4}$

(g) $\underset{CH_3}{\overset{CH_3}{>}}C{=}O + Br\underset{\underset{R}{|}}{CH}\overset{\overset{O}{\|}}{C}SC(CH_3)_3 \xrightarrow[\text{DMF}]{t\text{-BuO}^-\text{K}^+}$

(h) [cyclohexene oxide] $\begin{cases} \xrightarrow{\text{CH}_3\text{Li}} \\ \xrightarrow{\text{(CH}_3)_2\text{CuLi}} \end{cases}$

(*i*) [bicyclic N-methyl iminium structure] $\xrightarrow[\text{(ii) CH}_2\text{N}_2]{\text{(i) HClO}_4}$

(*j*) [methylenecyclohexane] $\xrightarrow[\text{H}_2\text{O}_2]{\text{C}_6\text{H}_5\text{CN}}$

(*k*) [2,3-dimethylthiirane structure] $\xrightarrow{(\text{C}_6\text{H}_5)_3\text{P}}$

(*l*) [cyclohexane with CH$_3$, H, OH, Cl substituents] $\xrightarrow{\text{OH}^-}$

(*m*) [bicyclic diester structure with COOC$_2$H$_5$ groups] $\xrightarrow{\text{C}_6\text{H}_5\text{N}_3}$

(*n*) [bicyclic episulfide structure, S] $\xrightarrow[\text{aq., ethanol}]{\text{NaIO}_4}$

(*o*) [7-oxabicyclo structure] $\xrightarrow{\text{Ch}_3\overset{+}{\text{S}}\overset{-}{\text{CH}_2}}$ (with O double bond above S)

(*p*) $\text{H}_2\text{C}=\overset{\displaystyle |}{\underset{\displaystyle \text{Br}}{\text{C}}}-\overset{\displaystyle |}{\underset{\displaystyle \text{NHC}_3\text{H}_7}{\text{CH}_2}}$ $\xrightarrow{\text{NaNH}_2}$

(*q*) [benzene ring with two COCl groups] + [aziridine, N–H] $\xrightarrow[\text{Benzene}]{(\text{C}_2\text{H}_5)_3\ddot{\text{N}}}$

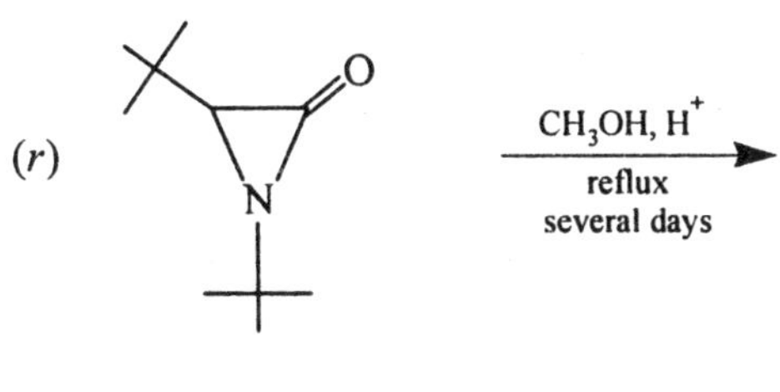

(r) $\xrightarrow[\substack{\text{reflux} \\ \text{several days}}]{CH_3OH,\ H^+}$

(s) $\xrightarrow{Br^-}$

(t) $\xrightarrow[THF,\ -40^\circ C]{(CH_3)_3S^+I^-,\ n\text{-BuLi}}$

(u) $\xrightarrow{h\nu}$

(v) $+\ C_6H_5\overset{O}{\overset{\|}{C}}OOH\ \xrightarrow[CH_2Cl_2]{Reflux}$

(w) $+\ ClCH_2COOC_2H_5\ \xrightarrow[C_2H_5OH]{NaOC_2H_5}$

(x) $R-N\diagup\ \diagdown S\ \xrightarrow[n\text{-Pentane}]{Ph_3P}$

(y) $\xrightarrow{C_6H_5Li}$

2. 2 Suggest a reasonable mechanism for each of the following reactions:

(a)

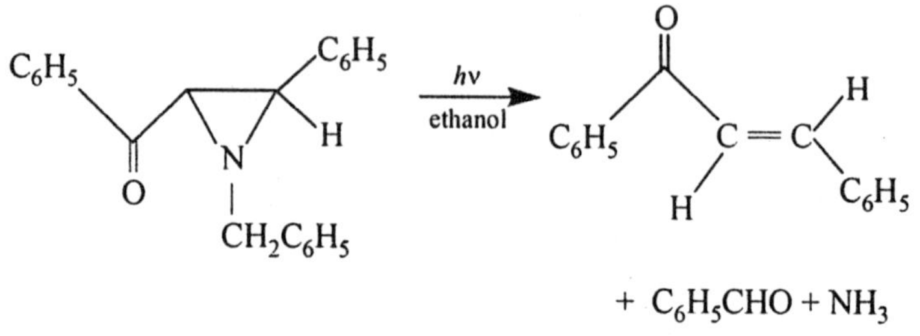

(b)

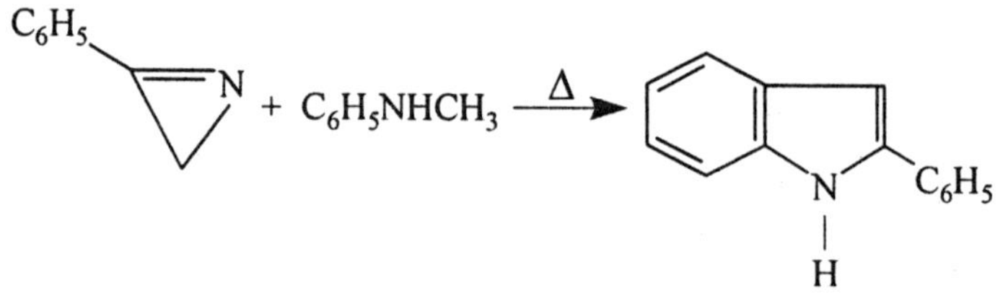

(c) $CH_2 = CHCN \xrightarrow{\text{H}_2\text{O}_2, \text{ NaOH}}$

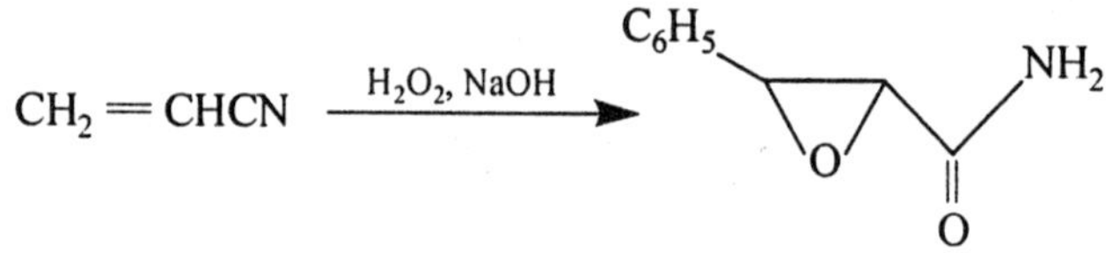

(d)

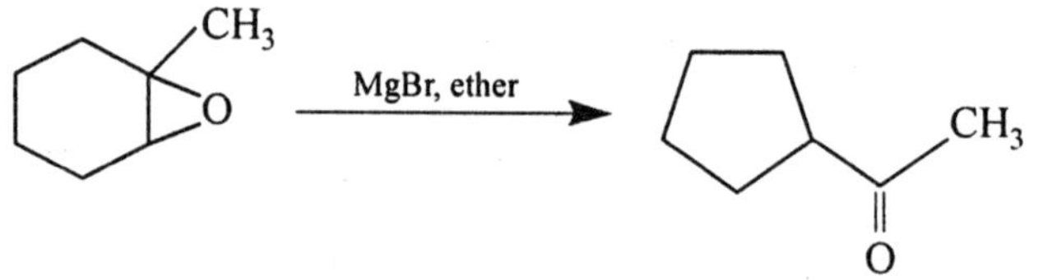

(e) $(C_6H_5)_2CN_2 + (Ar)_2 C = S \xrightarrow{\Delta} (C_6H_5)_2C = C(Ar)_2 + S$

(f)

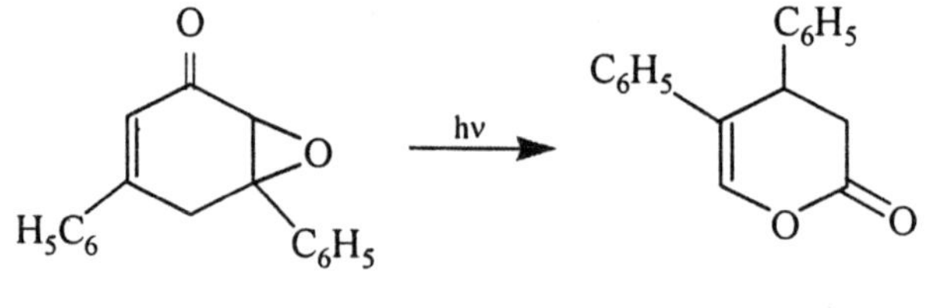

(g)

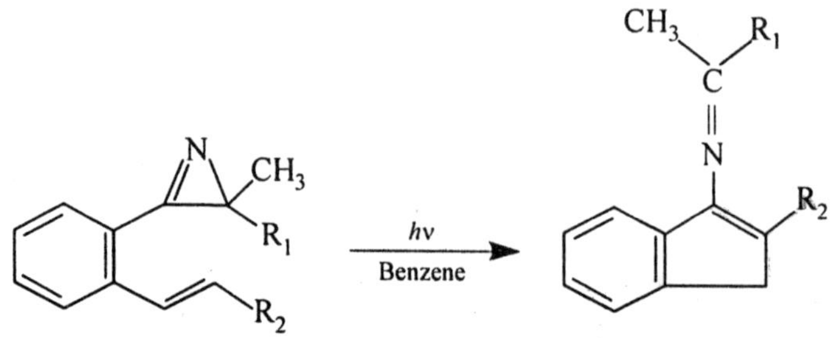

(*h*) 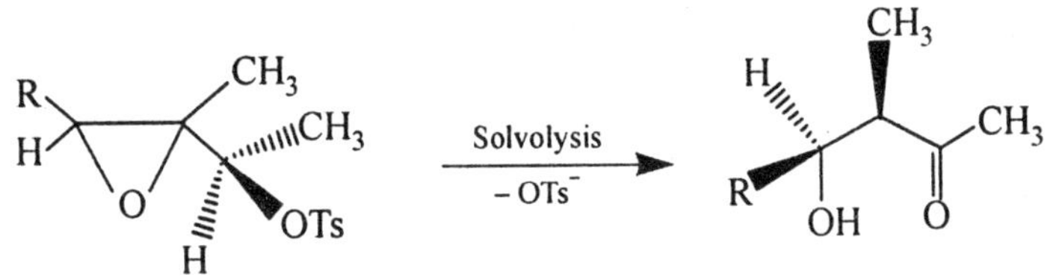

(*i*) 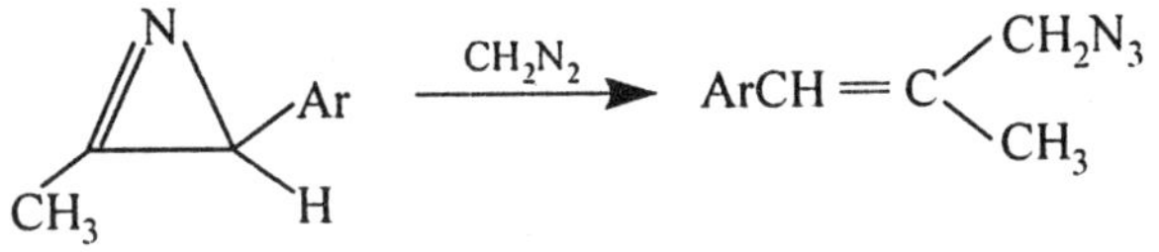

(*j*) 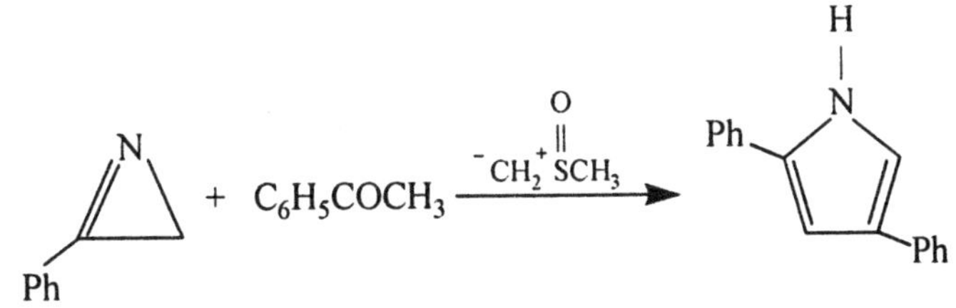

(*k*) 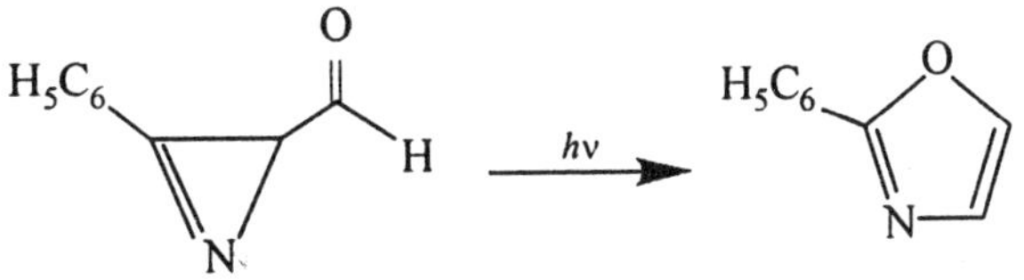

2. 3 Offer a suitable explanation for the following observations:

(a) α-Lactams do not undergo resonance.

(b) Reaction of propylene oxide with thiophenol in base yields about 90% of the product. In acid solution, on the other hand, a poor yield of the product is obtained. What do you conclude about the attacking species?

(c) Aziridine is relatively basic than methyl- or dimethyl-amine.

(d) Oxirane has a higher dipole moment than thiirane.

(e) Electron-donating groups increase the rate of epoxidation of an alkene with perbenzoic acid.

(f) 2-Azirines are unstable.

2. 4 Account for the formation of the product in the following reaction:

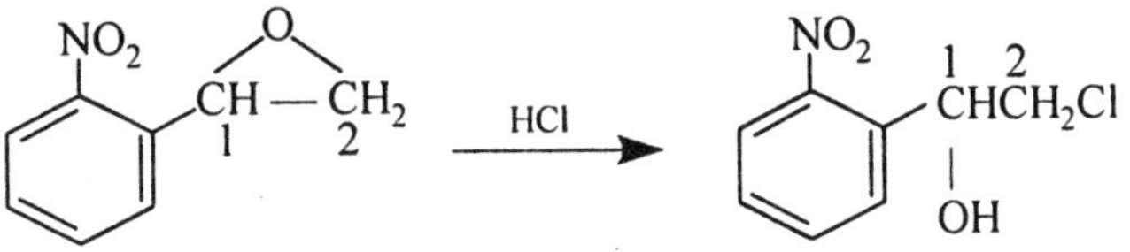

2. 5 Discuss the ring opening reactions of oxirane with organometallics.

2. 6 Elucidate the mechanism of the Darzens glycidic ester condensation. How is this reaction affected by steric and electronic factors?

2. 7 Suggest a method for the conversion of oxiranes into thiiranes.

2. 8 Discuss the ring opening reactions of thiirane by electrophiles and nucleophiles.

2. 9 Why are 2-azirine, oxirene and thiirene called anti-aromatic? Explain.

2. 10 List the reagents employed for the desulfurization of a thiirane. Give appropriate examples.

2. 11 The *n. m. r.* spectrum of 1-ethylaziridine at room temperature shows the triplet quartet of the ethyl group and two signals of equal peak area. When the temperature is raised to 120°C the latter two signals merge with a single signal. Explain.

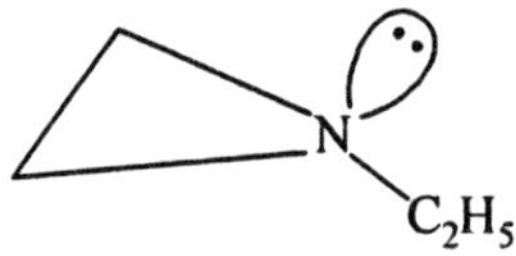

Three Membered Heterocyclic Compounds with Two Hetero Atoms

The three-membered heterocyclic compounds with one hetero atom possess high strain energy but their preparations could be carried out under milder conditions because of facile ring closure as compared to the closure of open-chains to larger rings. It is thus not surprising that the three-membered rings with two hetero atoms could also be prepared smoothly.[1] In the past several years, a large number of such compounds, indeed, have been synthesized. These compounds offer very interesting reactions and in addition, are important from theoretical standpoint. They are highly reactive and display some unusual properties.

The three-membered heterocyclic rings containing two hetero atoms have been prepared by a variety of experimental techniques and many of them are well investigated classes of compounds. Several diaziridines and N-substituted diaziridines are known but the substituent does not seem to markedly effect the reactivity of the ring system. Among the diazirines only the $3H$ - diazirines (N = N) are known while 1H - diazirines with a C—N double bond are unknown. Diazirines bearing one or two alkyl substituents at the carbon atom have been obtained. In the oxaziridine ring positions 1, 2 and 3 are assigned to oxygen, nitrogen and carbon respectively. Oxaziridinones are not known. The nitrogen atom in these rings can be substituted by alkyl or aryl groups and the substituents are known to influence the chemical behavior.

NMR studies of oxaziridine and diaziridine reveal a marked degree of configurational stability of the pyramidal nitrogen in contrast to the nitrogen of amines which inverts rapidly. There are, however, inversion barriers to saturated three-membered rings with two hetero atoms which offer a chance to isolate invertomers.

Simple diaziridines and oxaziridines do absorb in the near *u.v.* As a matter of fact lack of absorption was one of the arguments which assisted in distinguishing between a cyclic three-membered ring structure and unsaturated open-chain structure like nitrones and hydrazones. Oxaziridine, however, can undergo conjugation to a certain degree with unsaturated groups. Oxazirines, in contrast, exhibit characteristic *u.v.* spectra in the gas phase and absorption between 280-350 *nm* consists of a series of intense bands.

3.1 DIAZIRIDINES

Diaziridine (1) is a saturated compound containing two nitrogen atoms in a three-membered ring. The first authentic synthesis of diaziridine was achieved by Schmitz[2] in 1959.

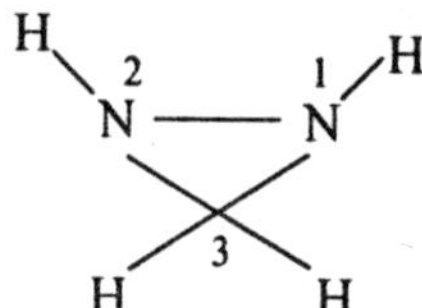

3.1.1 Physical and Spectroscopic Properties

Diaziridines and its derivatives are weak bases and form salts with mineral acids. The solubility in aqueous mineral acids, however, decreases with the increase in the chain length of alkyl substituents. Diaziridines, especially with N-substitution crystallize well from solution in organic solvents while those that are N,N-disubstituted do not crystallize but are miscible with all organic solvents. One mole of diaziridine on acid hydrolysis decomposes into one mole of carbonyl compound and one mole of hydrazine. This shows the presence of an N—N bond in diaziridine. Evidence for the presence of ring structure is obtained by the mutual interconversion of diaziridine and diazirine. The replacement of a CH_3 group in diaziridine raises the barrier to nitrogen inversion considerably and as a result nitrogen inversion is slow.

The infra-red spectroscopy provides a great deal of information about the small membered rings because of strain present in them. Small rings display high C – H absorption frequencies (between 3080 and 3000 cm^{-1}) for the ring C – H bonds. In *n.m.r.* the C – H absorption in diaziridine takes place around 1.35 ppm.

3.1.2 Synthetic Methods

The synthesis of the title compound has been achieved by the following methods.

1. *From Hydroxylamine-O-Sulfonic Acid and Ketone:* A ketone when treated with hydroxylamine-O-sulfonic acid and ammonia gives rise to a diaziridine. Cyclohexanone, for instance yields 3,3-pentamethylene-diaziridine[3], equation (3.1). If an amine[4] is employed then the corresponding 1-substituted diaziridine is obtained (equation 3.2).

$$(3.1)$$

$$(3.2)$$

2. *From Schiff's Bases:* Diaziridines have also been conveniently prepared from Schiff's bases. A Schiff's base adds onto a molecule of chloroamine, an aminating reagent and then the ring closure of the terminal addition product in the presence of a base[4] takes place to give diaziridine.

3,3–Pentamethylene-1–methyldiaziridine

3. *From Diazirines:* Substituted diaziridines can also be prepared by the addition of a Grignard reagent to the reactive double bond of a diazirine, as illustrated in the following example (equation 3.3).[5]

$$(3.3)$$

4. *Photoisomerization:* Diazepine (2) converts rapidly even in sunlight, to its diaziridine isomer (3).

(2)   (3)

3.1.3 Chemical Properties

Diaziridines display chemical reactions in which substitution at the nitrogen atom or fission of C – N bond takes place:

1. **Reduction:** Schmitz and coworkers[6] investigated the effect of reducing agents and found that different reducing agents give rise to different products. Under catalytic reduction, a diaziridine takes two moles of hydrogen resulting in the formation of two molecules of amines (equation 3.4). Lithium aluminum hydride reacts only if one of the N-atoms is unsubstituted and the products formed in this case are different than those obtained from catalytic hydrogenation, equation (3.5).

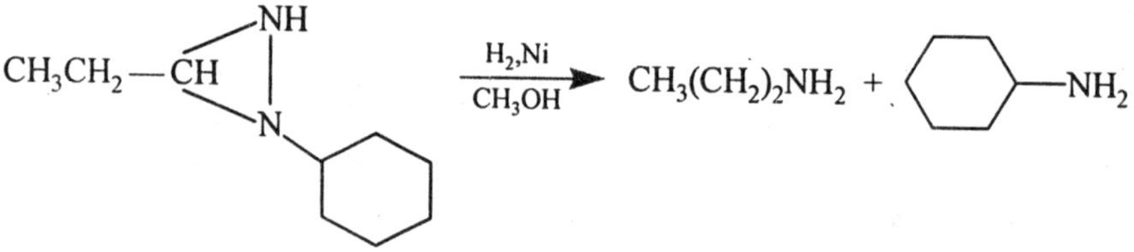

$$(3.4)$$

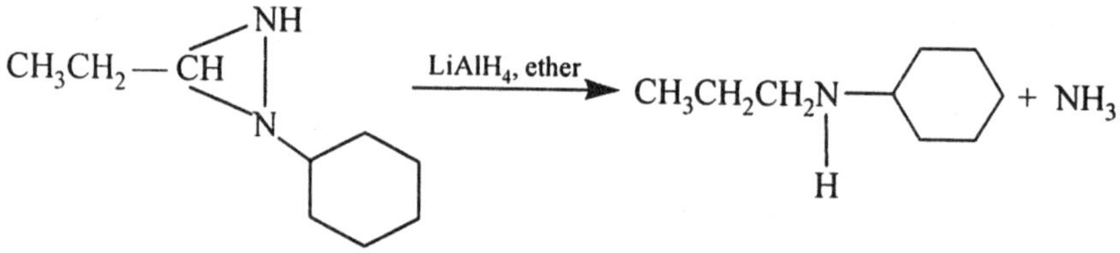

$$(3.5)$$

2. **Hydrolysis:** Fission of the diaziridine ring also takes place under hydrolytic conditions, for instance, 1-ethyl-3–methyldiaziridine yields acetaldehyde and ethylhydrazine, equation (3.6).

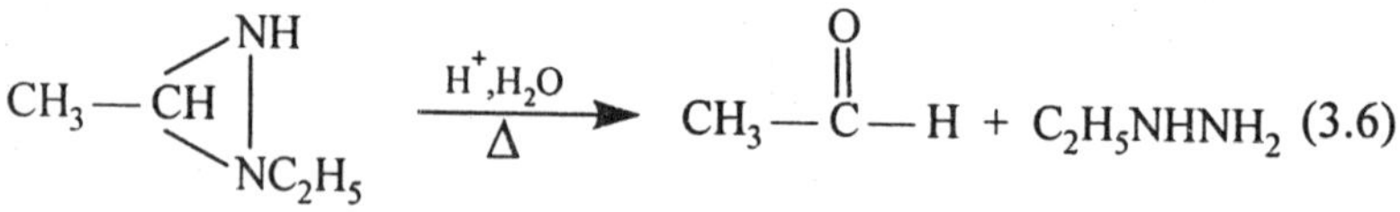

$$(3.6)$$

3. **Oxidation:** Diaziridines unsubstituted on both the nitrogen atoms tend to decompose at 125°C by a redox reaction resulting in one mole of a diazirine, a ketone and ammonia from (4). The reaction takes place even below 60°C when copper salts are present as catalysts.

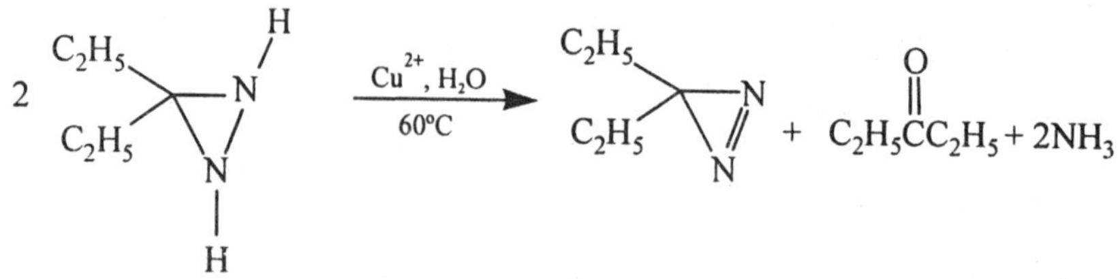

$$(4)$$

4. *Diaziridinones:* Diaziridinones similarly undergo facile ring opening in aqueous hydrochloric acid and subsequently decarboxylate to a hydrazine derivative.[7]

Diaziridinones lack an amide-type of delocalization of nitrogen lone-pair of electrons and are thus sluggish towards nucleophilic attack.

All diaziridines which contain at least one N – H bond undergo certain reactions typical of a secondary amine[8], the condensation with chloral is a representative example (equation 3.7).

$$(3.7)$$

Diaziridines react with activated acetylenes in a manner that the ring is no longer intact in the final product.[9]

3.2   DIAZIRINES

Diazirine is the name given to the class of compounds containing two nitrogen atoms in an unsaturated three-membered ring. Two isomeric diazirines (5) and (6) are possible, but (6) has anti-aromatic character and is yet unknown. Calculations show it to be 27 Kcal/mole less stable than its tautomer (5). Diazirine structure was first suggested for diazomethane (7) as it was the only valence bond structure which could be written for this compound. The linear structure (7) was, however, assigned to diazomethane on the basis of spectroscopic analysis and was confirmed[10] unequivocally in 1959.

Diazirine is calculated to be 23.2 Kcal/mole more stable than cyclopropene. Ab initio calculations on the diazirinyl anion point that the anion may be stable in solution.[11]

$$CH_2 - \overset{+}{N} \equiv N \longleftrightarrow CH_2 = \overset{+}{N} = \overset{-}{N}$$

(5)   (6)   (7)

3.2.1 Physical and Spectroscopic Properties

The parent diazirine is a gas while most of the diazirines are colorless liquids. They are insoluble in water but are completely miscible with organic solvents. They are also thermally more stable than the diazo compounds. Diazirines have no basic character and thus are not attacked by strong mineral acids. They are even unaffected by bases. Their *i.r.*[12] and *n.m.r.*[13] spectra have been reported.

3.2.2 Synthetic Methods

Diazirines may be prepared by the following methods:

1.   *From Oxidation of Diaziridines:* Studies on diaziridines have indicated that they possess considerable reducing properties and they can be readily oxidized by a number of common oxidizing agents[1], i.e. Ag_2O, HgO and $KMnO_4$ to the corresponding diazirines.

3,3 – Pentamethylenediazirine

Though the yields with silver oxide are good but this reagent interferes with certain functional groups particularly the amino group. In view of this, the use of solid iodine in one equivalent of triethylamine has been recommended.[14]

2.   *From t-Alkylazomethine and Dihaloamine:* A superior method for the synthesis of the parent heterocycle involves the addition of dichloro- or difluoro-amine to a formaldehyde imine (*t*-butylazomethine) in carbon tetrachloride.[12]

3-Halogenodiazirines are easily obtainable by halogenation of alkyl- or aryl-amidines in aqueous dimethylsulfoxide.[13]

This reaction is of interest in that the cyclization of nitroenamine (8) results in an unstable diazirine which spontaneously rearranges to diazirinium cation (9). Attack by the chloride ion yields 3-chlorodiazirine.

3.   A modified procedure of diazirine synthesis involves the following steps starting from *m*-bromophenol. It was first converted to the acetal followed by its reaction with *n*-butyllithium and methyltrifluoro acetate to give a ketone. Its reaction with hydroxylamine, formation of aziridine and finally oxidation with silver oxide yields (10).

3.2.3 Chemical Properties

Despite their highly strained structures, diazirines are unreactive. They show no tendency to react with acids and alkalies at room temperature, and with other electrophilic reagents. The diazirines, however, react with Grignard reagents at 0°C to give 1-substituted disziridines. Other organometallic reagents also add in the same manner.[16]

1. *Reaction with Nucleophiles and Electrophiles:* Diazirines show sensitivity to the attack of nucleophiles and electrophiles when they attack on a functional group outside the ring. Aryl bromodiazirine is attacked by an azide ion to give substitution product which subsequently converts to benzonitrile in 90% yield.[17]

Similarly attack by an acid followed by ring opening proceeds by initial protonation of the functional group in hydroxydiazirine and methylvinylazirine. The product in both cases is a ketone.

2. *Photochemical and Thermal Decomposition:* The principal interest in diazirine chemistry concerns their controlled thermolytic and photolytic decomposition which provide a source of carbenes. These can be trapped by a variety of alkenes.[18] The formation of the final products is a function of the molecular structure. Diazirine itself yields singlet methylene for instance, 2-cyclopropyl-3-chlorodiazirine,[19] on photolysis yields, cyclopropylchlorocarbene which can be trapped by an alkene or can alternatively undergo a ring expansion of yield 1-chlorocyclobutene. The thermolysis of

3-phenyl-3-*n*-butyldiazirine[19] similarly yields a carbene which by a hydride shift yields a mixture of *cis*- and *trans*-1-phenyl-1-pentene plus 50% of valerophenone and in addition 1-phenyldiazopentene was also isolated as an intermediate.

$$C_6H_5,CH_3(CH_5)_3 \overset{C}{\underset{}{}}\!\!\begin{array}{c}N\\ \parallel\\ N\end{array} \xrightarrow{\Delta} C_6H_5,CH_3(CH_5)_3\overset{}{C}=\overset{+}{N}=\overset{-}{N} \xrightarrow{-N_2}$$

$$C_6H_5,CH_3(CH_5)_3\overset{}{C}: \xrightarrow[\text{Shift}]{H^-} C_6H_5CH=CHCH_2CH_2CH_3$$

In view of these observations Liu and coworkers[20] in a reinvestigation of the thermolytic decomposition of 3-methyl-3-phenyldiazirine have suggested the following two pathways for its decomposition, one involving the initial

$$C_6H_5,CH_3\overset{C}{\underset{}{}}\!\!\begin{array}{c}N\\ \parallel\\ N\end{array} \xrightarrow{\Delta}$$

$$C_6H_5,CH_3\overset{}{C}=\overset{+}{N}=\overset{-}{N}$$

Hexane → $C_6H_5,CH_3\overset{}{C}=N-N=\overset{}{C},C_6H_5,CH_3$ (11)

Gas phase, $-N_2$ → $C_6H_5,CH_3\overset{}{C}:$

$$\downarrow$$

$$C_6H_5-CH=CH_2$$

$$\downarrow\ C_6H_5,CH_3^-\overset{}{C}:$$

$$\begin{array}{c}H \qquad\qquad CH_3\\ \triangle\\ C_6H_5 \qquad\qquad C_6H_5\end{array}$$

***cis*-1,2-Diphenyl-1-methylcyclopropane**

formation of phenylmethyldiazine (11) and a second involving a direct decomposition to the corresponding carbene. The first process is favored in a non-polar solvent while the second in the gas phase.

Another interesting example in which the formation of the product depends on the starting substrate is exemplified[21] as follows in which a contraction of the ring takes place.

Cyclopentyl N-(methyl, phenyl) Ketone

3.3. OXAZIRIDINES

Oxaziridine or oxazirane (12) has one atom each of O, N and C in a three-membered ring. It was first prepared by Emmons[22] in 1957. The corresponding unsaturated oxazirene (13) is yet unknown.

(12)   (13)

3.3.1 Physical and Spectroscopic Properties

Oxaziridines are colorless liquids and can be distilled below 100°C. They possess a characteristic unpleasant smell. They have low basicity compared to amines and do not form salts with acids. Oxaziridines have a relatively weak N – O bond. They possess a configurationaly stable nitrogen at ordinary temperature. For N-alkyl nitrogen the inversion barrier is of the order of 24 - 31 Kcal/mole. Oxaziridines do not absorb in the *u.v.* region while in *i.r.* they display a band near 1400 cm^{-1} which is considered characteristic of oxaziridines.[23] Thus *u.v.* and *i.r.* spectra eliminate structures with a C – N double bond. The nuclear magnetic resonance spectra[22] of substituted oxaziridines are also in agreement with the three-membered ring formula.

Some oxaziridines have been shown to occur as intermediates in numerous investigations of photo rearrangement of aromatic amine oxides.[24,25] and in the photochemical Beckmann rearrangement.[26] The reaction has been shown to proceed by intramolecular oxygen transfer through ^{18}O labelled oximes.[27-29]

3.3.2 Synthetic Methods

The formation of oxaziridines occurs surprisingly quite smoothy with ease comparable to that of three-membered ring containing one hetero atom.

1. *From Carbonyl Compounds:* Two approaches have been used for synthesizing oxaziridines from carbonyl compounds. In the first, a carbonyl compound is converted to an imine (Schiff's base) by treating with an amine which on peroxidation yields the corresponding oxaziridine.[30]

2-Pentyl-3-phenyloxaziridine

This reaction is quite a general one as imines of widely varying structures can be prepared. The chief limitation, however, is the extremely labile nature of some oxaziridines. The mechanism of this reaction is similar to the epoxidation of alkenes. Recently,[31] monoperoxycamphoric acid[32] has also been found to effect the same reaction.

2-*t*-Butyl-3-methyloxaziridine

Ozonolysis of cyclohexylidene cyclohexane in the presence of ammonia also produces an oxaziridine derivative.[33]

Oxidation of sulfonimines, prepared from amination of carbonyl compounds, on oxidation with buffered potassium peroxymonosulfate (oxone) in place of *m*-chloroperbenzoic acid yields a racemic mixture of oxaziridines.[34]

$$RSO_2NH_2 + O{=}CHAr \longrightarrow RSO_2N{=}CHAr \xrightarrow{\text{Oxone}}$$

(R, R)   +   (S, S)

80-95%

Optically active (+) or (-) - (camphorylsulfonyl) oxaziridine is available from the corresponding optically active sulfonimine, on reaction with oxone.

(−) $\xrightarrow{\text{Oxone}}$ (+)

80%

In this case oxidation can take place only from the end face of C–N bond due to steric blocking of the *exo* face and as a result a single oxaziridine is obtained.

The second approach is based on the consideration that hydroxylamine-O-sulfonic acid or chloramines react with aldehydes or ketones in alkaline solution to yield oxaziridines.[35]

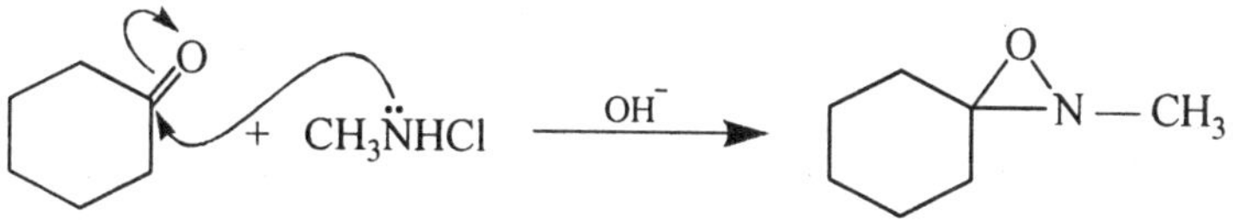

The yields in this method are generally poor as oxaziridines are unstable in alkaline solution.

2.   From α - Lactams: An α-lactam on oxidation with a peracid also yields an oxaziridine; for instance, 1,3-di-*t*-butylaziridinone, a stable α-lactam gives with *m*-chloroperbenzoic acid, 2,3-di-*t*-butyloxaziridine[36] *via* the rearrangement shown below:

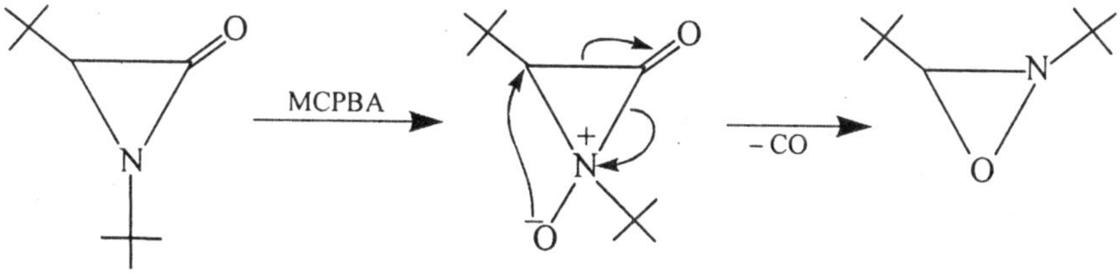

3.   *Photolysis of Nitrones:* Another important method consists of the photolytic rearrangement of nitrones to oxaziridines.[36] This is exemplified in the conversion of phenyl *t*-butylnitrone to 2-*t*-butyl-3-phenyloxaziridine.[37] The *t*-butyl nitrones of benzaldehyde and 4-nitrobenzaldehyde yield about 90% of oxaziridine on *u.v.* irradiation as depicted for (14).

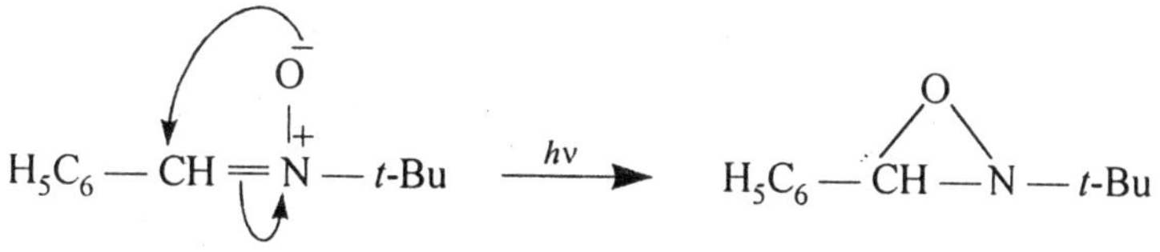

(14)   **2-*t*-Butyl-3-phenyloxaziridine**

3.3.3 Chemical Properties

Oxaziridines are active oxygen compounds comparable in many ways to organic peroxides. They can be assayed iodometrically with potassium iodide in acetic acid, according to the following stoichiometry.

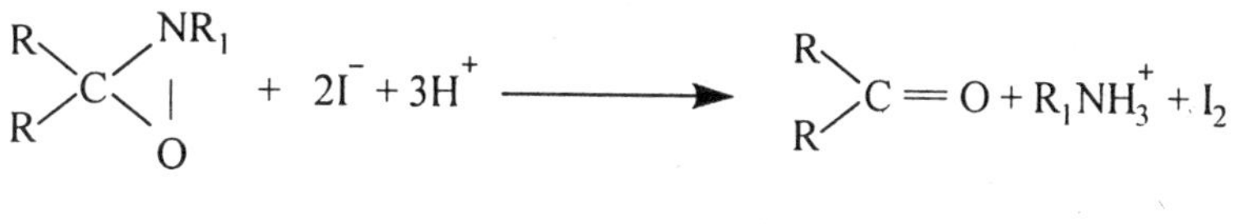

1.   *Ring Opening Reactions:* These heterocycles have high energy and as a result without exception involve the opening of the ring at C – O or N – O bond by a large number of reagents such as acids, bases, reducing agents and thermally or photochemically.

Acid hydrolysis of an oxaziridine takes place with N – O bond fission. A phenyl substituted oxaziridine cleaves at the 3-position under acidic conditions to give an aldehyde and hydroxylamine,[22] according to the following mechanism:

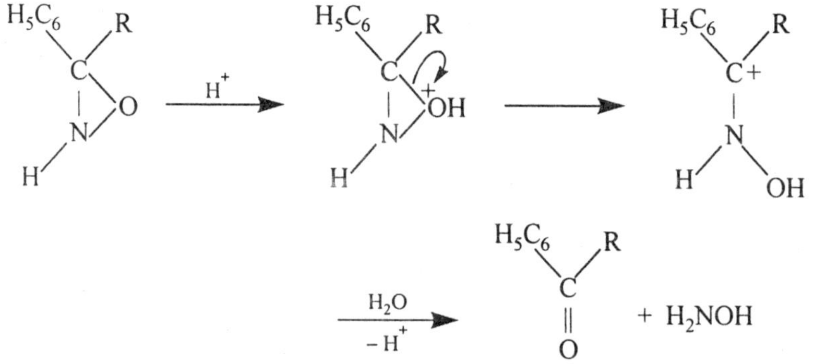

$$\xrightarrow{\text{H}_2\text{O}, \; -\text{H}^+} \quad + \; \text{H}_2\text{NOH}$$

If, on the other hand, N-substituent is also present, then the course of the reaction is changed and instead an aldehyde and ammonia are formed. A migration of a hydride ion to the nitrogen atom takes place in this case.

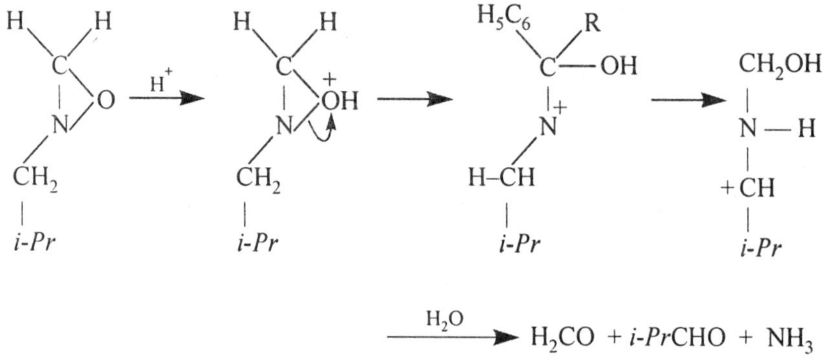

$$\xrightarrow{\text{H}_2\text{O}} \; \text{H}_2\text{CO} + i\text{-}Pr\text{CHO} + \text{NH}_3$$

If however, no hydrogen is available than an alkyl group shifts to nitrogen to liberate an amine.[38, 39] The oxaziridine ring is rather sluggish towards attack by basic reagent at ordinary temperature. Suitably substituted[40-42] oxaziridines however, decompose in the presence of a base by N – O bond fission analogous to acid fission. This reaction presumably involves the abstraction of a proton leading to the formation of a carbanion followed by its fragmentation to a mixture of aldehyde and aldimine. The later being unstable hydrolyses to aldehyde and ammonia.

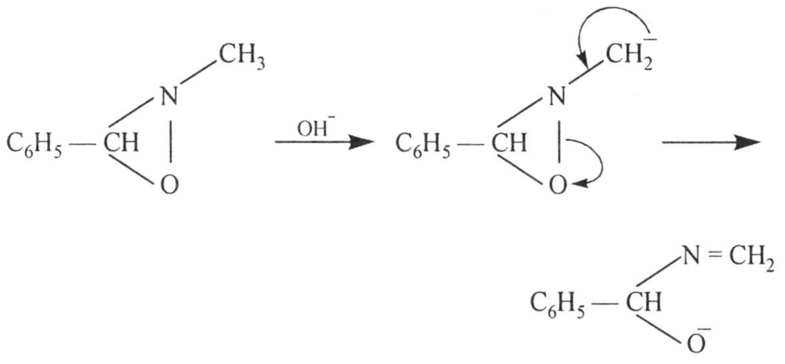

$$\xrightarrow{H_2O} C_6H_5CHO + HN = CH_2 \xrightarrow{H_2O} HCHO + NH_3 + C_6H_5CHO$$

Oxaziridines, in addition, undergo reductive fission with lithium aluminum hydride (equation 3.8). This reaction also offers a method to distinguish oxaziridines from the isomeric nitrones.

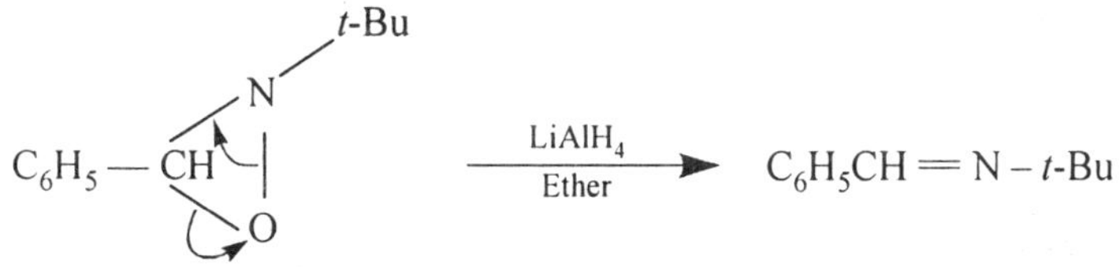

$$\xrightarrow[\text{Ether}]{\text{LiAlH}_4} C_6H_5CH = N - t\text{-Bu} \quad (3.8)$$

Whereas 2-*t*-butyl- 3-phenyloxaziridine affords benzylidine-4-butylamine, the reaction of the isomeric nitrone results in N-benzyl-N-*t*-butyl hydroxylamine, equation (3.9).

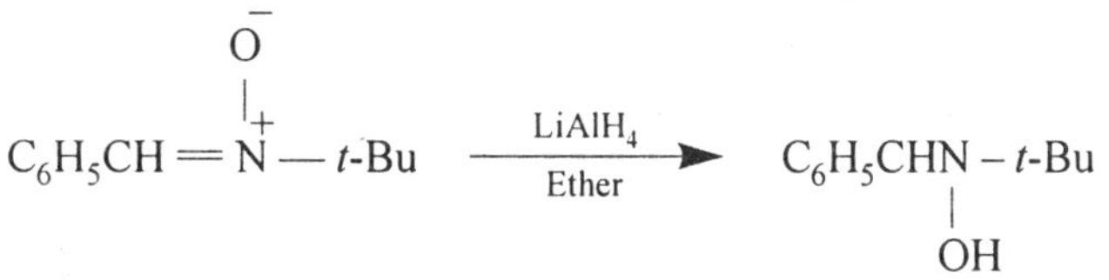

$$C_6H_5CH = \overset{\overset{\bar{O}}{|}}{\underset{+}{N}} - t\text{-Bu} \quad \xrightarrow[\text{Ether}]{\text{LiAlH}_4} \quad \underset{\underset{OH}{|}}{C_6H_5CHN - t\text{-Bu}} \quad (3.9)$$

2. *Reaction with Nucleophiles:* Oxaziridines react with almost every nucleophile and they must be handled in inert solvent such as ether and toluene. Attack by a nucleophile takes place at the nitrogen atom of the three - membered ring with simultaneous N – O bond cleavage.

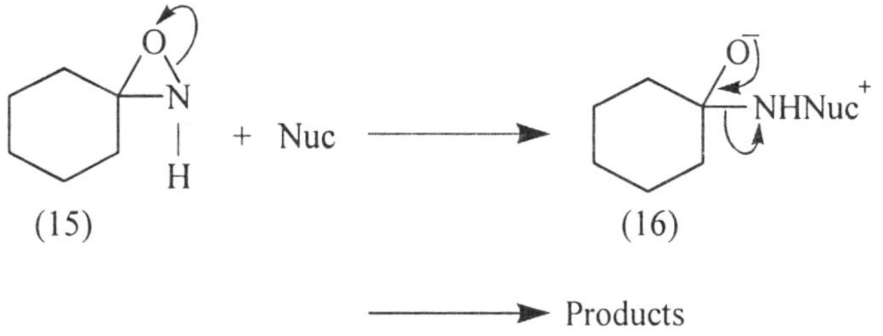

(15)   + Nuc $\longrightarrow$ (16)

$$\longrightarrow \text{Products}$$

For instance[44], 1-oxa-2-azaspiro [2.5] octane (15) reacts with a nucleophile with simultaneous opening of the ring. The intermediate (16) can undergo a number of interesting reactions.

A secondary amine also reacts with (15) at room temperature to give cyclohexane hydrazone. Its hydrolysis yields N, N-dialkylhydrazine and constitutes an efficient method for hydrazine preparation.

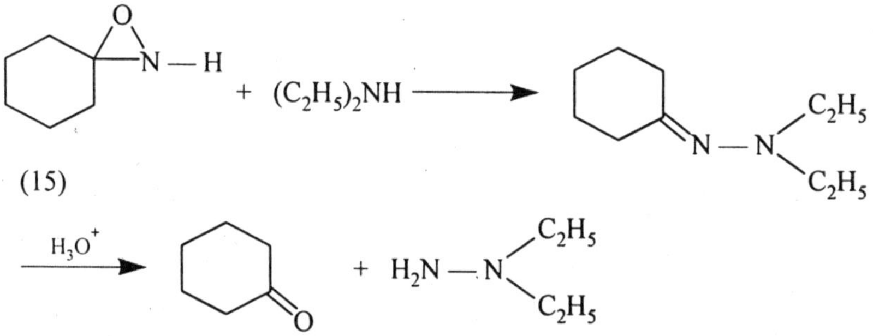

Triethylamine reacts with (15), to yield hexamethylenetriamine.[45]

3.   *Thermal and Photochemical Reactions:* Thermally, oxaziridine ring is easily decomposed to give products depending on the presence of substitution at the 3-position.

The C-aryl oxaziridines are isomerized to nitrones by C – O bond fission at a high temperature, equation (3.10).

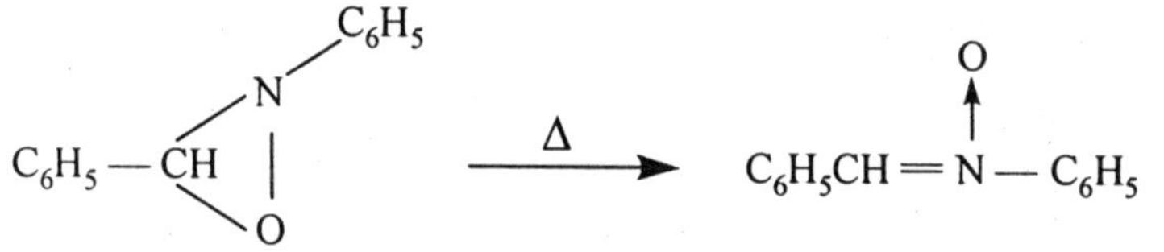

$$C_6H_5CH = N - C_6H_5 \quad (3.10)$$

The C-alkyl oxaziridines, on the other hand, are converted to amides by N – O bond rupture, equation (3.11).

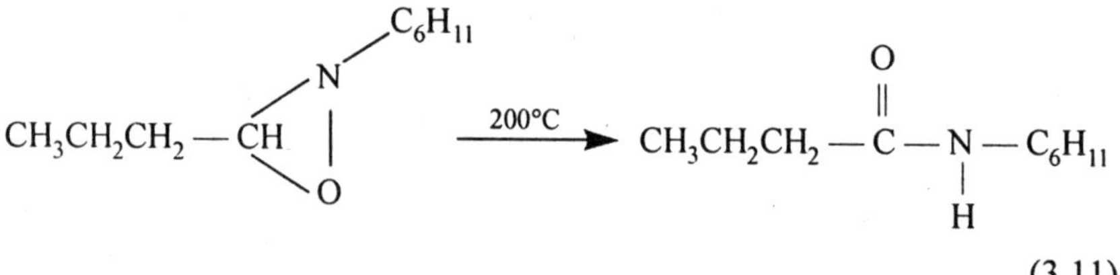

$$(3.11)$$

In contrast, oxaziridines derived from cyclic ketones give products due to ring expansion, equation (3.12).

$$\text{(3.12)}$$

Oxaziridines rearrange photochemically to amides[46] with a simultaneous migration of a group from carbon to the nitrogen atom, equation (3.13).

$$\text{(3.13)}$$

This conversion in analogous to the photochemically induced Beckmann rearrangement.[47]

Photolysis of bicyclic oxaziridines result in ring contraction[48] or ring expansion[49] products. A free radical mechanism has been postulated by Black and Watson[43] for such reactions as shown below for the conversion of (17) to (18).

(17)

(18)

4. *Epoxidation of Alkenes:* Oxaziridines have assumed some importance as useful reagents for the epoxidation of alkenes. For instance, 2-benzenesulfonyl-3-aryloxaziridine reacts with an alkene at room temperature to give a 25% yield of the corresponding epoxide.[47] The reaction is stereospecific and transfer of oxygen takes place during the reaction.[50-51]

Chiral oxaziridines provide syntheticallly useful reagents for the asymmetric transfer of an oxygen atom to a variety of substances.

REFERENCES

1. For a review see, E. Schmitz in, "*Advances in Heterocyclic Chemistry*", A. R. Katritzky and A. J. Boulton, (*Eds*), Academic Press, New York (1963), p. 83.
2. E. Schmitz; *Angew. Chem.,* **71,** 127 (1959).
3. H. J. Abendroth and G. Henrich, *Angew. Chem.* **71,** 283 (1959).
4. E. Schmitz and R. Ohme and D. R. Schmidt, *Chem. Ber.* **95,** 2714 (1962).
5. E. Schmitz and R. Ohme, *Chem. Ber.,* **94.** 2166 (1961).
6. E. Schmitz and D. Habisch, *Chem. Ber.,* **95,** 680 (1962).
7. C. A. Renner F. D. Greene, *J. Org. Chem.,* **41,** 2813 (1976); F. D. Greene, J. C. stowell and W. R. Bergmark, *ibid.,* **34,** 2254 (1969).
8. E. Schmitz and R. Ohme, *Chem. Ber.,* **95,** 795 (1962).
9. H. W. Heine, T. R. Hoye, P. G. Williard and R. C. Hoye, *J. Org. Chem.,* **38,** 2984 (1973).
10. J. A. Pople, *Tropics in Current Chem.,* **40,** 1 (1973).
11. R. L. Kroeker, S. M. Bachrach and S. K. Kass, *J. Org. Chem.,* **56,** 4062 (1991).
12. W. H. Graham, *J. Am. Chem. Soc.,* **84,** 1063 (1962).
13. W. H. Graham *ibid.,* **87,** 4396 (1965).
14. R. F. R. Church and M. J. Weiss, *J. Org. Chem.,* **35,** 2465 (1970).
15. J. E. Baldwin, A. J. Pratt and M. G. Moloney, *Tetrahedron,* **43,** 2565 (1987) also see J. E. Baldwin, C. D. Jesudason, M. G. Moloney, D. R. Morgon and A. J. Pratt, *Tetrahedron,* **47,** 5603 (1993).
16. M. T. H. Liu, *Chem. Soc. Rev.,* **11,** 127 (1982).
17. X. Creary, A. F. Sky and G. Phillips, *J. Org. Chem.,* **55,** 2005 (1990); D. P. Cox, R. A. Moss and J. Terpinski, *J. Am. Chem. Soc.,* **105,** 6513 (1983).
18. R. A. Moss and M. E. Fantina, *J. Am. Chem. Soc.,* **100,** 6788 (1978).
19. B. M. Jennings and M. T. H. Liu, *J. Am. Chem. Soc.,* **98,** 6416 (1976).
20. M. T. H. Liu and K. Ramakrishnan, *J. Org. Chem.,* **42,** 3450 (1977).
21. R. Livingstone, in "*Rodd's Chemistry of Carbon Compounds*", 2nd *ed.,* Vol. IVa. (*Ed.*), S. Coffey, Elsevier, London (1973).
22. W. D. Emmons, *J. Am. Chem. Soc.,* **78,** 6208 (1956); *ibid.,* **79,** 5739 (1957).

23. O. Büchardt, *Chem. Ber.*, **70,** 231 (1970).

24. F. Bellamy and J. Streith, *J. Heterocycl. Chem.*, **4.** 1391 (1976).

25. Y. Ogata, *J. Org. Chem.*, **47,** 3684 (1984).

26. T. Sasaki, *J. Chem. Soc.*, 1239 (1970).

27. T. Oine and T. Mukai, *Tet. Letters,* 157 (1969).

28. H. Izawa, P. de Mayo and T. Tabata, *Canad. J. Chem.*, **47,** 51 (1969).

29. J. S. Splitter and M. Calvin, *J. Org. Chem.*, **30,** 3427 (1965); D. H. Aue and D. Thomas, *J. Org. Chem.*, **39,** 3855 (1974).

30. W. H. Pirkle and P. L. Rinaldi, *J. Org. Chem.*, **42,** 2080 (1977).

31. N. A. Milas and A. C. McAlevey, *J. Am. Chem. Soc.*, **42,** 2182 (1977).

32. J. S. Belew and J. T. pearson, *Chem, and Ind.*, 1246 (1959).

33. M. Schulz, D. Becker and A. Rieche, *Angew. Chem.*, **77,** 548 (1965).

34. For a review see, F. A. Davis and A. C. Sheppard, *Tetrahedron,* **45,** 5703 (1989).

35. Y. Hata and M. Watanabe, *J. Am. Chem. Soc.*, **101,** 1323 (1979); Y. Hata and M. Watanabe, *Tet. letters,* 3827 (1972).

36. J. S. Splitter and M. Calvin, *J. Org. Chem.*, **23,** 651 (1958).

37. J. F. Garvey and J. A. Hashmall, *J. Org. Chem.*, **43,** 2380 (1978).

38. A. P. Butler and B. C. Challis, *J. Chem. Soc.*, (B), 778 (1971).

39. F. A. Davis *et al.*, *Chem. Comm.*, 25 (1977); D. R. Boyd., *ibid.*, 162 (1976).

40. K. Suda, F. Hina and C. Yijima, *J. Org. Chem.*, **51,** 4232 (1986).

41. L. Henn, D. M. B. Hickey C. J. Moody and C. W. Rees *J. Chem. Soc.*, *Perkin Trans.* **1,** 2189 (1984).

42. W. H. Rastetter, W. R. Wagner and M. A. Findies, *J. Org. Chem.*, **47,** 419 (1982).

43. O. Oliveros, M. Riviere, J. P. Malrieu and Ch. Teichteil, *J. Am. Chem. Soc.*, **19,** 318 (1979); M. L. Druclinger *et al.*, *J. Heterocyclic Chem.*, 13, 1001 (1976); F. A. Davis and U. K. Nadir, *Tet. letters,* 1721 (1977); E. Meyer and G. W. Griffin, *Angew, Chem. Int. Edin* (Engl), **6,** 634 (1967).

44. For a comprehensive review see, S. Andreae and E. Schmitz, *Synthesis,* 327 (1991).

45. Y. Hata and M. Watanabe, *J. Org. Chem.*, 46, 610 (1981).

46. H. Suginome and F. Yagihashi, *J. Chem. Soc.*, *Perkin Trans.*, **1,** 2488 (1977).

47. D. St. C. Black and K. G. Watson, *Aust. J. Chem.*, **26,** 2505 (1973).

48. E. Desherces, H. Riviere, J. Parello and A. Lattes, *Tet. letters,* 851 (1975).

49. F. A. Davis, N. F. Abdul, S. B. Awad and M. E. Harakal, *Tet. Letters,* 917 (1981); F. A. Davis, M. E. Harakal and S. B. Awad, *J. Am. Chem. Soc.*, **105,** 3123 (1983); R. D. Bach and G. J. Wolber, *ibid.*, **106,** 1410, 1410 (1984).

50. R. D. Bach. B. A. Coddens, J. J. W. McDonall and H. B. Schlegel, *J. Org. Chem.*, **55,** 3325 (1990).

51. F. A. Davis and M. C. Weismiller, *ibid.*, **55,** 3715 (1990).

REVIEW PROBLEMS

3. 1 Predict the major product of the following reactions:

(a) $C_6H_5CH = NC(CH_3)_3 \xrightarrow{O_3}$

(b) [cyclic structure with NH and NH bridged by C=O] $\xrightarrow[\text{(ii) Base}]{\text{(i) } t\text{-BuOCl}}$

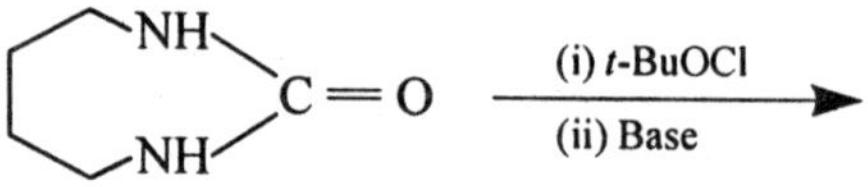

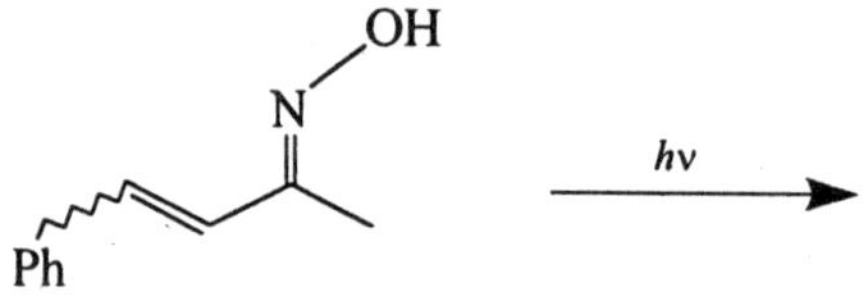

(c) [structure with N–OH oxime, Ph and CH] $\xrightarrow{h\nu}$

(d) [dihydroisoquinoline structure] $\xrightarrow{CH_3NHCl}$

(e) [cyclohexane spiro diazirine, N=N] $\xrightarrow{\Delta}$

(f) [cyclohexane with NH and N–CH$_3$] $+ HC \equiv CCOOC_2H_5 \longrightarrow$

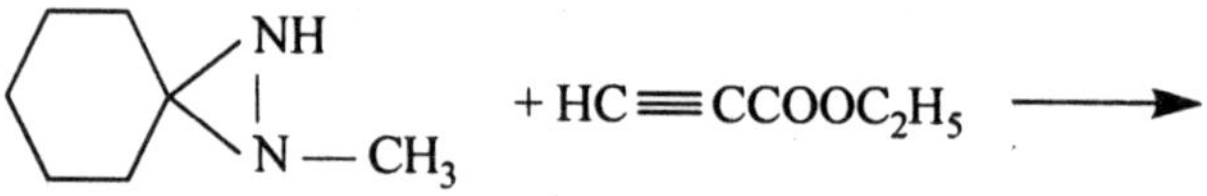

(g) $H_2C = N - t\text{-Octyl} \xrightarrow{HNCl_2}$

(h) [structure with CH$_3$, N–C, O, Ph, H oxaziridine] $\xrightarrow{(CH_3)_2NH}$

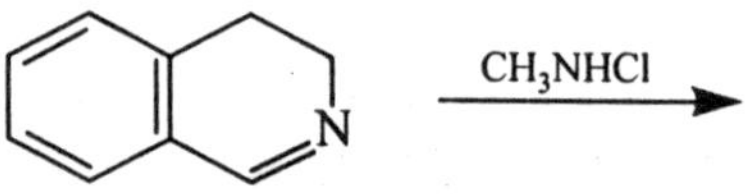

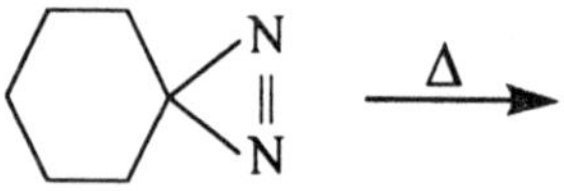

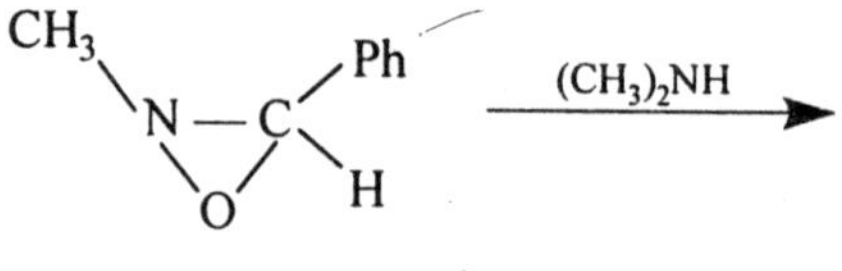

(i) [2-methyl-1-tetralone structure] $\xrightarrow[\text{(ii)}]{\text{(i) Base}}$ [camphorsultam oxaziridine reagent, R, H, SO$_2$, N, O]

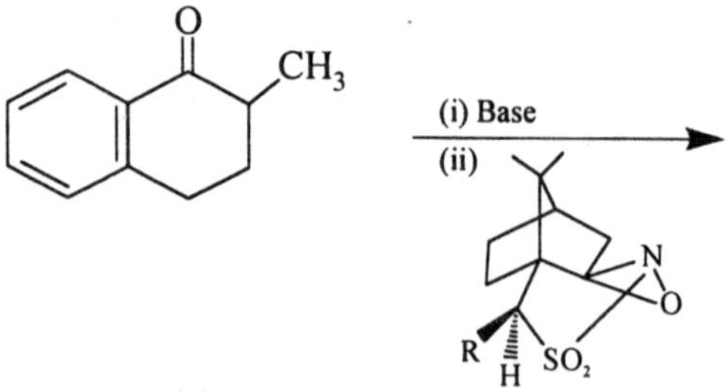

3. 2 Postulate a suitable mechanism for each of the following reactions:

(a)

(b) $CH_3 - C(=NH)(NH_2) \xrightarrow{NaOCl}$

(c)

(d) $\xrightarrow{H_2O,\ H^+}$

(e) $\xrightarrow{hv}$

3. 3 Discuss three different modes of ring openings of oxaziridines by acids.

3. 4 List different methods for the epoxidation of an alkene. Give one example of each.

3. 5 Show the formation and reactions of a carbene generated from a diazirine with suitable examples.

3. 6 Explain the occurrence of an oxaziridine as intermediate in the Beckmann rearrangement.

3. 7 Which is more basic, an oxaziridine or diaziridine? Justify your answer.

3. 8. Write an account of the thermolytic ring opening of diazirines.

3. 9. Suggest a method with appropriate examples to distinguish between an oxaziridine and a nitrone.

Four Membered Heterocyclic Compounds with One Hetero Atom

The four-membered heterocyclic ring compounds are heterocyclic analogues of cyclobutane and N, O and S containing heterocyclics are known as *azetidine, oxetane* and *thietane* respectively. These are relatively less strained than the three-membered heterocyclic rings, but are most difficult to prepare by direct intramolecular cyclization procedures. This difficulty, in part, arises due to the change in the position of the atoms undergoing reactions. As a result, direct cyclization occurs only if the atoms combining are present in some appropriate orientation. The preference for ring closure, understandably is because of their presence in close proximity. But as the chain length increases so does the total number of conformations which do not allow easy ring closure.

4.1 AZETINES AND AZETIDINES

Azacyclobutadiene is known formally as *azete* (1), the partially unsaturated ring is called *azetine* or more appropriately 1-*azetine* (2) and the isomeric structure (3) as 2-*azetine*. Azetidine (4) is the name given to the completely saturated four-membered nitrogen containing compound, it is also designated as *trimethyleneimine*. Azete is anti-aromatic and unstable. The benzazetes such as phenylbenzazete (5) are more stable and have been prepared.

$$\text{(1)} \qquad \text{(2)} \qquad \text{(3)} \qquad \text{(4)} \qquad \text{(5)}$$

Not very many compounds are known to occur in nature which contain the azetidine ring structure. Synthetic azetidine derivatives have also not yet yielded any useful results for pharmacological evaluation. However, L-azetidine-2-carboxylic acid naturally occuring antimetabolite of proline has been isolated from *Liliaceae*. Its 3-isomer has been prepared in the laboratory.

4.1.1 Physical and Spectroscopic Properties

Alkyl-and aryl-azetines appear to be more stable thermally than alkoxy-1-azetines and alkoxy-2 azetines. The latter have the tendency to polymerize.[1,2] *I.r.*[1,3,4] and *n.m.r.*[1,4,5] for 1-azetines are available.

Azetidine is a colorless liquid, b.p. 61°C and has an ammonical or ammonia-like odor. Azetidine and lower members are soluble in water and fume in air. It is considerably a stornger base (*p*Ka 11.25) than aziridine (*p*Ka 7.98). The next member N-methylazetidine possesses a *p*Ka of 10.40. Azetidine and its N-alkyl derivatives behave in many respects as typical secondary and tertiary amines respectively. 1, 2-and 1, 3-Diazitidines are known and can be prepared by photochemical or thermal addition of dialky azidocarboxylates to olefins.

Azetidine has a non-planar structure. It has three carbon atoms all of which if properly substituted can be chiral centers, and two in turn can be subject to *cis-trans* isomerism. A small barrier of 1.26 Kcal/mole has been estimated for ring flipping in aziridine compared to 1.44 Kcal/mole for cyclobutane. A substituent at nitrogen can be either axial or equatorial and rapid inversion at nitrogen normally occurs. *N.m.r.* studies also support a configurationally mobile non-planar structure for azetidine in which nitrogen inversion takes place.

4.1.2 Synthetic Methods

Following are the general synthetic routes for 1-azetines and azetidines.

1.   Pyrolysis of Cyclopropyl Azides: A useful route to 1-azetines involves pyrolysis of cyclopropyl azides.[35]

2.   Photocycloaddition: Contrary to earlier reports, 1-azetines are also obtained, although in low yields by [2 + 2] cycloaddition of alkenes and aryl cyanides.[4,8] A representative example of this method is the preparation of

2-phenyl-3,3,4,4-tetramethyl-1-azetine from benzonitrile and 2,3-dimethyl but-2-ene. 2-Azetines are unknown but seem to have been obtained by the thermal [2 + 2] cycloaddition of N-sulfonylimines ($PhCH = NSO_2Ar$) and ketene N,N-acetals [$CH_2 = C (NR_2)_2$].

2-Phenyl-3,3,4,4-tetramethyl-1-azetine

3. Cyclization Methods: Azetidine and its derivatives have been prepared by a number of methods which include intramolecular cyclization and cycloaddition. The heterocyclic rings are formed by intramolecular ring closure from their acyclic counterparts. The activation energy of ring closure depends largely on ring strain in the cyclic product and the probability of the two ends coming into close proximity. These two factors make the three-membered rings relatively easy to form. Although three-membered rings possess a higher ring strain, the probability of the two ends of the chain to attain the right conformation is very high. This later factor makes the four-membered ring to form rather less easily.[10]

γ-Haloalkylamines in the presence of a base have been employed to prepare azetidine and its derivatives by intramolecular cyclization. The parent compound, azetidine is obtained from γ-bromopropylamine and base.

$$Br - CH_2CH_2CH_2 - NH_2 \xrightarrow[\Delta]{OH^-} \quad + \quad Br^- + H_2O \quad (4.1)$$

A poor yield of azetidine is obtained in this method, as the rate of ring closure is very slow. Best results are obtained when the halogen is primary; secondary halogens undergo competing reactions while tertiary do not react. When substituents are present on the chain, the ring closure is facilitated[10], for instance, 1, 3, 3-trimethylazetidine is obtained in 80% yield from 3-bromo-2, 2-dimethyl-N-methylpropane (6) (equation 4.2). This is attributed to the Thorpe-Ingold effect.

Cyclization takes place by nucleophilic displacement of the halo group at one end of the chain. The process is thus reverse to the known ring-opening reactions which can be effected easily in the 3-and 4-membered ring compounds.

$$(4.2)$$

(6)

Sulfate esters of γ-amino alcohols have also been used in place of γ-haloamines to affect the cyclization.[11]

$$^-OSO_2CH_2CH_2CH_2NH_2^+ R \xrightarrow{NaOH} \square\!\!-N\!\!-R + Na_2SO_4 + H_2O$$

4. **From Isoxazole Derivatives:** An alternative method for the preparation of azetidines involves the ring opening and closure of isoxazoles.[12] Thus, the 3, 5-dimethylisoxazole ring is first opened by treatment with sodium in *n*-pentanol and subsequently reacted with tosyl chloride and pyridine. Cyclization is then accomplished in the presence of a strong base to give (7). The final product, 2, 4-dimethylazetidine is obtained by reducing (Na/pentanol) this intermediate product.

$$\text{(i) Na, } n\text{-Pentanol}$$
$$\text{(ii) TsCl, Pyridine}$$

$$CH_3CHCH_2 - CH - CH_3$$
$$|\qquad\qquad |$$
$$HNTs\qquad OTs$$

$$\xrightarrow[-\,OTs^-]{NaOC_2H_5}$$

$$CH_3 \text{—} \square \text{—} CH_3 \qquad \xrightarrow{Na, \, n\text{-Pentanol}}$$
$$\underset{Ts}{N}$$

(7)

$$CH_3 \text{—} \square \text{—} CH_3$$
$$\underset{H}{N}$$

2, 4-Dimethylazetidine

5. **From Aziridines :** Another unequivocal synthesis of azetidines involves the reaction between substituted aziridine[13] such as 1, 2-diphenyl-3, 3-dicarbomethoxyaziridine with sulfurane. The reaction is carried out in benzene at room temperature to give excellent yields of 3-benzoyl-2, 2-dimethoxycarbonyl-1, 4-diphenylazetidine.

6. *From γ-Lactones:* Cromwell and Rodebaugh[14] have converted a γ-lactone to azetidine-2-carboxylic acid successfully by the following sequence of reactions in 54% overall yield.

Methyl α, γ-dibromobutyrate was obtained by bromination of γ-butyrolactone in the presence of PBr_3 followed by treatment with a slight excess of methyl alcohol saturated with dry HBr gas. The dibromo compound was refluxed with benzhydrylamine in acetonitrile to yield the 1, 2-disubstituted azitidine. It was hydrolyzed to the corresponding acid. Finally hydrogenolysis at a pressure of 45 psi over Pd-C catalyst gave azetidine-2-carboxylic acid.

7. By Intramolecular Participation: An interesting example of oxime group participation in intramolecular reactions is exhibited in the following compound which forms azetine N-oxide.[15]

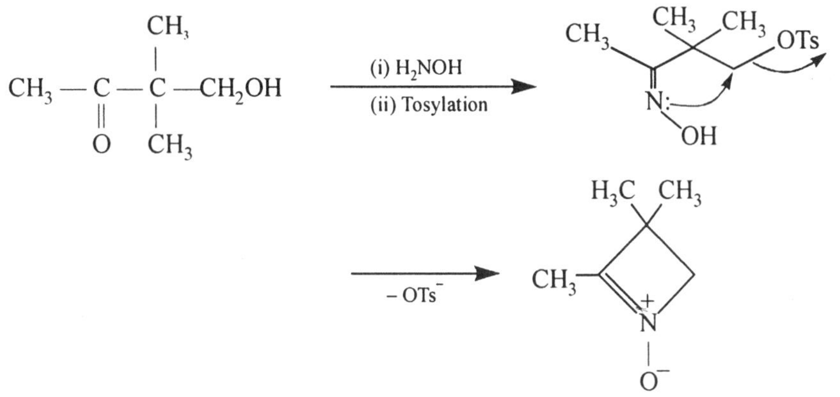

8. Reaction of Amines with Halogenoalkyl Oxiranes: Amines react with epichlorohydrin to give 1-alkylamino-3-chloro-2-alkanols. Presence of a bulky group at the nitrogen atom suppresses side reactions. The intermediate (8) cyclizes on heating to 50°C. This method provides 3-hydroxyazetidines.

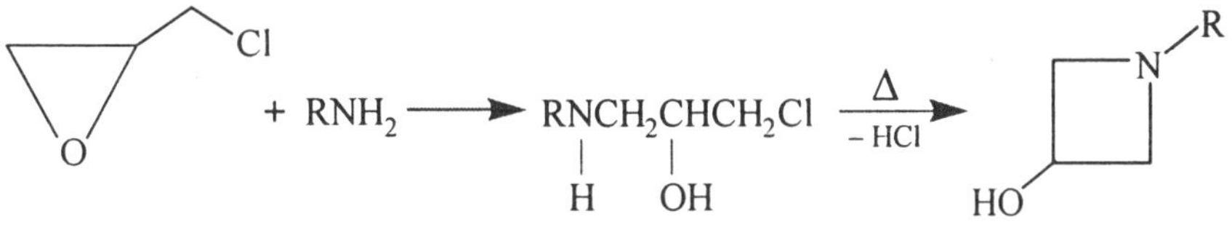

$$+ RNH_2 \longrightarrow RNCH_2CHCH_2Cl \xrightarrow[-HCl]{\Delta}$$

(8)

Using this procedure 1,3,3-trinitroazetidine[9] has been synthesized recently[16] in 35% yield.

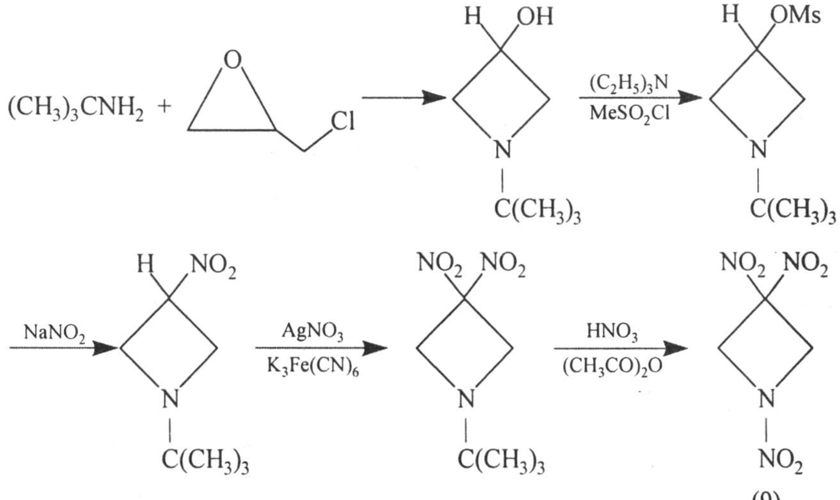

(9)

1-*t*-Butyl-3-azitidol on reaction with methanesulfonyl chloride gives the corresponding mesylate. The mesylate group is displaced by sodium nitrite in the following step. Potassium ferricyanide and silver nitrate yields the dinitro derivative. Finally the *t*-butyl group is removed on nitration with acetyl nitrate at 0°C.

4.1.3 Chemical Properties

The four-membered heterocyclic compounds display several of the reactions characteristic of their lower homologs but with a lesser degree of reactivity because of decrease in ring strain.

1.   Ring Opening Reactions: As expected the azetidine ring is cleaved more sluggishly than aziridine. The cleavage is considerably accelerated in the presence of acid catalysts. Opening of azetidine in the presence of hydrochloric acid results in 3-chloropropylamine hydrochloride[17], equation (4.3). Ring cleavage occurs by the attack of a nucleophile on the protonated azetidine.

$$\text{azetidine} \xrightarrow{\text{2HCl, } \Delta} \text{ClCH}_2\text{CH}_2\text{CH}_2\text{NH}_2\text{HCl} \qquad (4.3)$$

2.   Formation of Azitidine Derivatives : Azetidine behaves like a secondary aliphatic amine and thus it reacts with reagents such as carbon disulfide to form a salt, with nitrous acid to give N-nitrosoazetidine and N-hydroxymethylazetidine with formaldehyde.

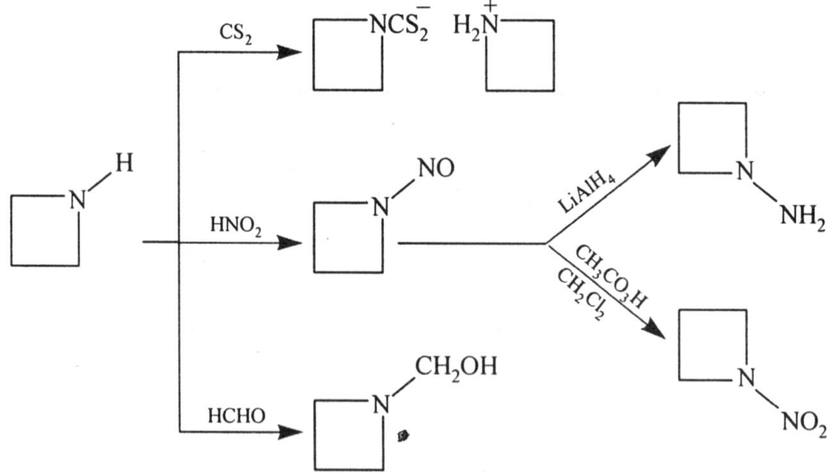

The quaternization of N-alkylazetidine with an alkyl halide is a neat reaction, equation (4.4).

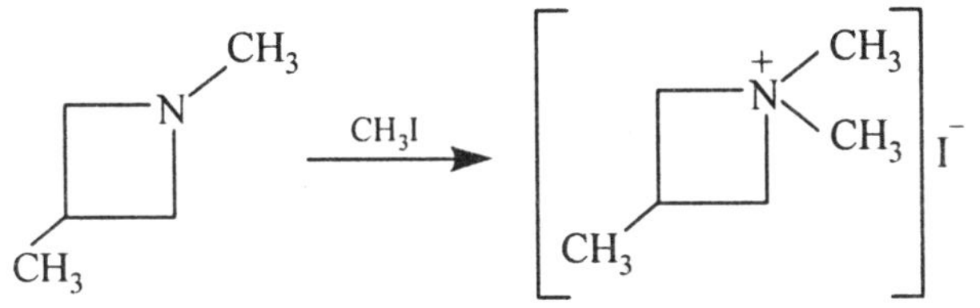

$$(4.4)$$

3.   *Photochemical Reaction:* The photochemical reactions of azetidines have not have been well investigated and much needs to be done in this regard. One representative system that has been studied by Padwa and coworkers[18] is azetidinyl ketone (10) which on irradiation rearranges to diarylpyrroles (11) and (12).

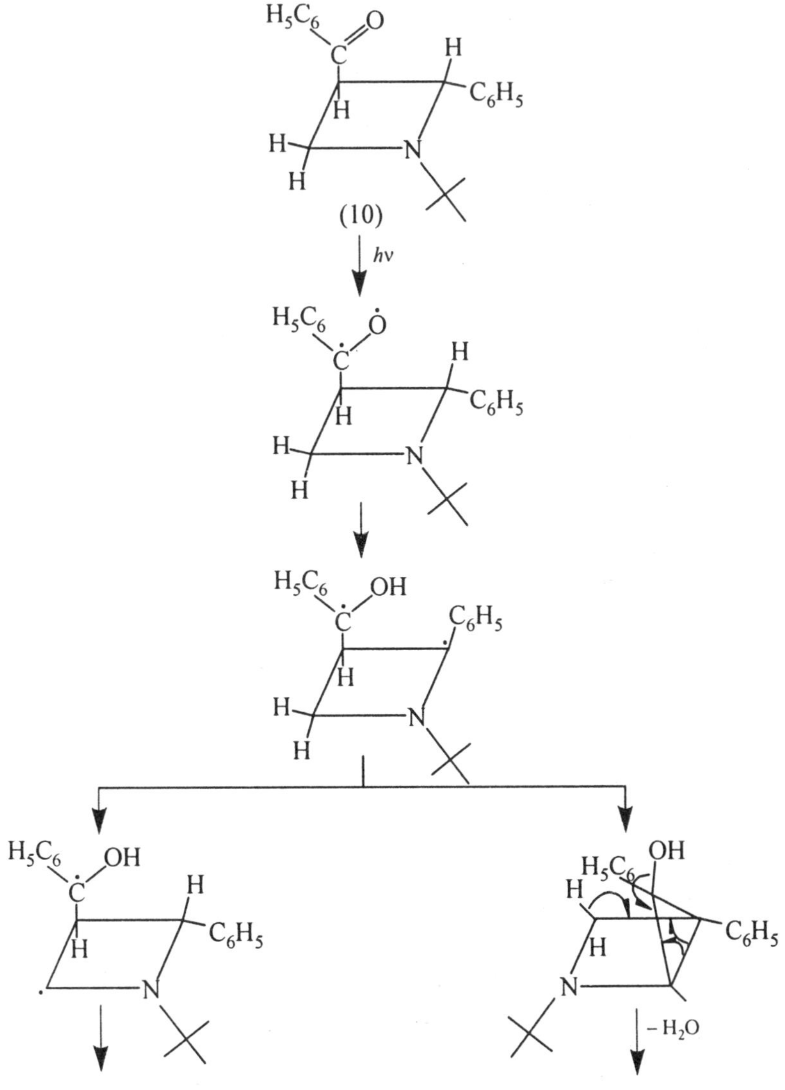

(10)

hv

$-H_2O$

(12)

(11)

A mechanism consistent with this rearrangement has been formulated. It involves an intramolecular hydrogen shift and subsequent generation of a spin-unpaired 1, 3-biradical intermediate. Spin-inversion followed by ring closure yields a transient azabicyclo [2.1.0.] pentane. This readily loses a molecule of water to give disubstituted pyrroles.

4.2 OXETANES

Four-membered cyclic ethers are known as *oxetanes*. Oxetane (13) is a saturated compound the corresponding unsaturated compound is known as *oxete* or 2-*oxetene* (14). Oxetane has been named as *oxacyclobutane* with the oxygen atom assigned position-1 or alternatively 1, 3-*trimethylene oxide* or simply *trimethylene* oxide. Oxetanes are known compounds while the unsaturated derivatives are unknown.

(13)   (14)

The four-membered ring is not common in nature but it occurs in some physiologically active compounds. An interesting compound is the bicyclic oxetane thromboxane A_2, TXA_2 (15). It is a naturally occuring prostaglandin and is a potent aggregatory agent for platelets in human blood as well as constriction of vascular and bronchial smooth muscles. Synthetically obtained oxetanocin-A (16) is a potent antiviral activity agent against herpes simplex virus I and II and humancytomegalo virus.

(15)   (16)

4.2.1 Physical and Spectroscopic Properties

Oxetane is a colorless liquid, b.p. 47-80°C and is miscible with water and in most organic solvents. It can donate electrons much more easily than the three-membered heterocyclic compound, oxirane. The structure of oxetane has been deduced from microwave and Raman spectroscopy. The C – O – C bond angle is of the order of 94.5° and the C – C bond length is 1.54 Å which is greater than the C – O bond length of 1.46 Å. It follows that the molecule is not a perfect square. The classical Baeyer-strain theory predicts that the four-membered ring compounds be less strained than the three-membered ring compounds. The three-membered rings, however, are considered planar but the four-membered rings have been shown to be puckered. Cyclobutane, for instance, is puckered. Puckering helps to reduce internal bond angles and thus causes added ring strain but this is more than offset by the relief of non-bonded (steric) interactions between the eclipsed neighbouring hydrogen atom which the planar model imposes. Four-membered parent heterocyclic rings such as oxetane and thietane have been shown, on the contrary, to be planar and this is thought to be due to reduction in the number of non-bonded interactions when a bivalent hetero atom replaced a methylene group in the ring. Presence of substitutents on the ring causes the ring to be non-planar, for instance, 2-methyloxetane and 3-methyloxetane are puckered. The calculated value of dipole moment of oxetane 2.01D is higher than either diethyl ether (1.22D) or dimethyl ether (1.31D). This high value indicates more electron density on the oxygen atom in oxetane than in acyclic aliphatic ethers.

Oxetane absorbs only in the vaccum *u.v.* as discussed earlier. Cross-ring cleavage to form alkane and carbonyl fragments is the predominant type of fragmentation of oxetanes. In *m.s.* of oxetane it has been observed that the relative abundance of ethylene radical is several times greater than that of the formaldehyde radical ion at an ionizing potential of 70 ev. In contrast, 2, 2-dimethyloxetane fragments much more to the radical ion of acetone than to isobutylene cation.

4.2.2 Synthetic Methods

Oxetane and its derivatives have been prepared by a number of general methods.

1.   Cyclization Reactions: The commonly employed and a direct route for the preparation of oxetanes is the reaction of a halohydrin in the presence of a base.[20] Intramolecular cyclization takes place with the concurrent loss of HBr as exemplified in equation (4.5). This method gives fairly good yields of oxetanes but the ring closure is slow. In contrast to oxirane where substituents enhance the rate of cyclization, in oxetane formation the cyclization is also markedly influenced by the presence of substituents but in addition, depends

$$(C_2H_5)_2CCHCH_2Br \xrightarrow{\ OH^-\ } \quad + \ Br^- \ + \ H_2O \quad (4.5)$$

2, 2-Diethyloxetane

on their location as well. It has been observed[21] that the yield of oxetane is increased by alkyl substituents located on the carbinol carbon structure (17) while decreased if present on the halogen bearing carbon atom structure (18).

    (17)   (19)   (18)

1-Oxaspiro [3, 5] nonane

This is illustrated in the accompanying reaction for the formation of (19). In the latter case an *E2* elimination takes place which decreases the yield of oxetane. In place of halogen, a trimethylamino acetate or tosylate group can be used as the leaving group, equations (4.6) and (4.7).

$$\xrightarrow[-\,N(CH_3)_3]{K^+\textit{t}\text{-BuO}^-} \qquad\qquad (4.6)$$

$$(4.7)$$

2. *Photochemical Methods:* A more general and viable method for preparing oxetanes by photocycloaddition of ketones to alkenes has been described. It was demonstrated at the turn of the century that oxetanes are available by photoaddition of ketones to alkenes and was known as the *Paterno-Buchi reaction.*[22, 23] The addition of benzophenone to 2-methyl-propene produces a 9 : 1 mixture of 3, 3-dimethy-2, 2-diphenyloxetane and 2, 2-dimethyl-4, 4-diphenyloxetane.

Mechanistically a ketone triplet is the chemically active species which adds onto a ground state alkene[24], but cases are known in which a ketone singlet attacks the alkene molecule[25] or the reverse of this latter process. According to another view[26] it has been claimed that the reaction my take place by a charge-transfer excitation complexes of both the reactants.

To explain the regioselectivity of the photoaddition it was proposed that the reaction of electron-rich olefins and excited ketones involves an interaction of the electron-deficient carbonyl lone-pair orbital with the electron rich π olefin orbitals to produce a diradical intermediate (20) which is thermodynamically more stable than (21). The radicals can subsequently cyclize to lead to the final product.

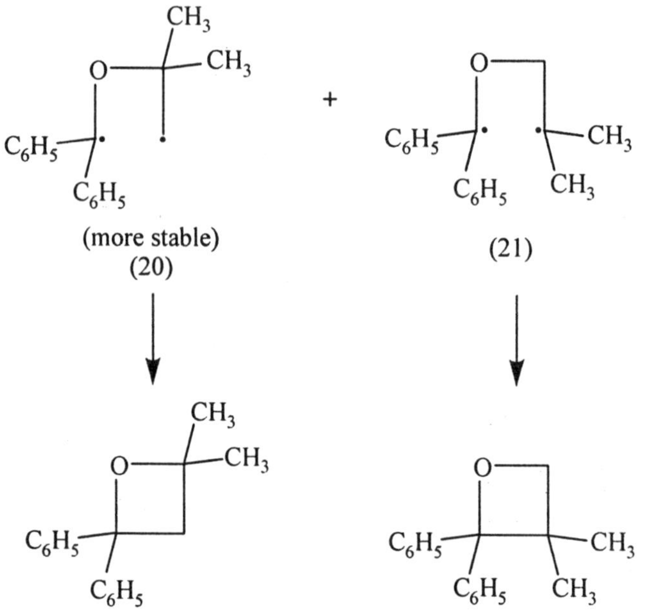

(more stable)
(20)

(21)

Norbornene cycloadds to benzophenone to give predominantly the *exo* oxetane, because approach from the less hindered side is preferred.

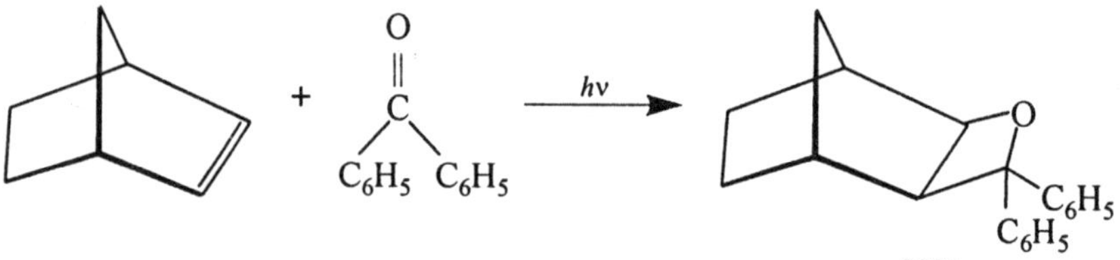

80%

In case the olefin contains easily abstractable hydrogen atoms (H-donor then pinacol formation also takes place in a cometitive reaction.[27-29] Higher yields of oxetanes, however, can be affected provided substituted acyclic or cyclic olefins are irradiated in the presence of a carbonyl compound.[30]

Ketones, similarly, add to allenes on irradiation to produce oxetanes.[31]

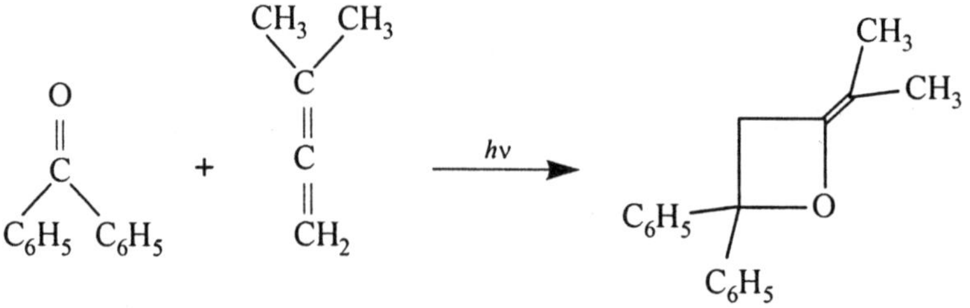

2-Isopropylidine-4, 4-diphenyloxetane

3. ***From Alkoxyacetylenes:*** An attempt has been made to prepare oxete. Midleton[32], in order to achieve this, treated trifluoromethyl ketone with ethoxyacetylene in the presence of a Lewis acid at $-78°C$ which presumably yielded an oxete derivative only as a transient intermediate and immediately rearranged to an ester at room temperature.

**1-Carboethoxy-2,
2-di(trifluoromethyl) ethylene**

4. Vinyloxetane has been obtained from *trans* epoxy allyl ether in the presence of a strong base according to the following sequence.[33]

5. Dimethyloxosulfonium methylide transfers a methylene to a ketone to produce an oxirane. A similar transfer to an oxirane yields an oxetane.[34]

$$\longrightarrow \qquad \underset{O}{\Box}\!\!<\!\!\begin{array}{c}R\\R\end{array} \quad + \quad (CH_3)_2SO$$

4.2.3 Chemical Properties

Oxetane displays chemical properties similar to that of oxirane in many respects.

1. ***Ring Opening Reactions:*** In contrast to oxirane, the oxetane ring is cleaved only under vigorous conditions (usually at a high temperature) by nucleophiles. With hydroxide ion, for instance, oxetane ring is cleaved 10^3 times more slowly than oxirane.[35] Other reagents such as amines[36], thiophenol[37] and β-naphthol open the ring on prolonged heating at about 150°C.

Reaction of oxirane with Grignard reagents and organolithiums is a well-known method of preparing certain primary alcohol and of lengthening the carbon chain by two carbon atoms. Oxetane ring is similarly cleaved by Grignard reagents[39] although higher reaction temperatures are required. Oxetane with phenylmagnesium bromide yields 3-phenylpropanol-1, equation (4.8).

$$\text{(4.8)}$$

If position-2 carries a substituent then attack takes place at the less substituted C – 4 atom.

This reaction is less satisfactory with substituted oxetanes because of decreased rate of reaction and formation of by-products. The organoaluminum compounds react[40] similarly with oxetanes.

2, 2-Dimethyloxetane is also cleaved by lithium 2, 4-di *t*-butylphenylide (LDBB) to provide primary lithium *tert* oxyanium ion, however the same reaction in the presence of trialkylaluminum and LDBB gives exclusively 3-lithio-3, 3-dimethyltriethylaluminum propoxide.[38] The former on trapping with (E) –2-methyl-2-butenal affords (+) ocimenoly oxide while the latter gives a diol though in a poor yield. Thus this reductive cleavage provides a regiocontrol reaction.

Oxetanes are known to undergo a number of acid-catalyzed reactions which may be employed for the synthesis of various organic compounds.[41] Though the structure and hybridization of oxirane and oxetane are fundamentally different, the reactivity of the two ring systems is identical under acidic conditions. The lesser degree of ring strain can be compensated for by the greater electron donor capability of the ring oxygen atom is oxetane.[41] Several reactions of ring opening of 2-methyloxetane are depicted below:

It is noticed that two products are generally obtained, their formation may be explained on the formation of an oxonium ion intermediate which does not open to a free carbocation rather is attacked both at C – 2 and C – 4.

If a free stable carbocation is possible then the reaction proceeds to give a single product. This is the case with 2-phenyloxetane which forms 3-chloro-3-phenylpropan-1-ol. The course of the reaction is altered because of the formation of a stable benzyl cation. In contrast, in the Friedel-Crafts reaction[42] with 2-methyloxetane or 2-phenyloxetane in benzene, a single product 3-phenylbutanol is obtained. In addition oxetanes react with carbenes[43] and also can be converted to trimethylene carbonate in a 1 : 1 cycloaddition reaction with CO_2 in the presence of tetraphenylantimony iodide.[44]

2.   *Photochemical and Thermal Reactions:* Photochemistry of oxetanes has not been much investigated. However, oxetanes bearing chromophoric substituents tend to undergo interesting photochemical transformations.[45] Oxetane derivative (22) available easily from benzophenone and 1, 1-dimethylallene undergoes photorearrangement to the isomeric cyclobutane derivative (22a), the reaction is sensitized by triphenylene:

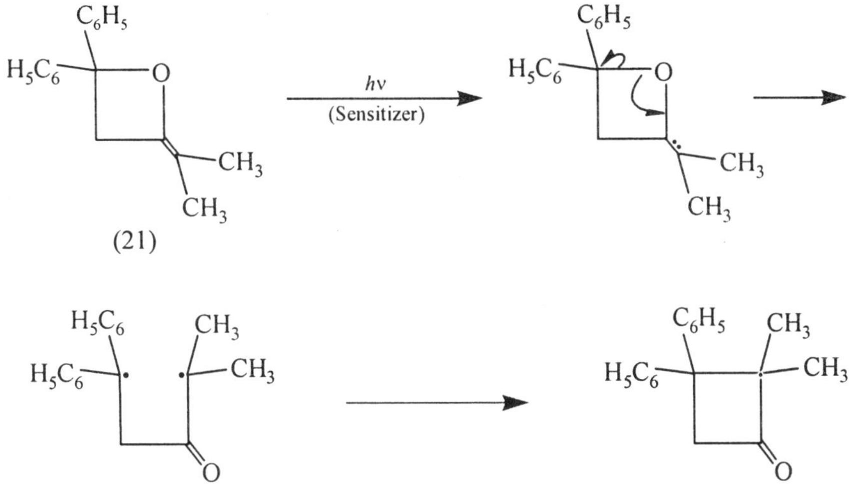

Pyrolysis of *trans*-3-*n*-propyl-2-phenyloxetane at 250 – 300°C proceeds in a stereospecific manner and produces almost exclusively *trans*-1-phenylpent-1-ene

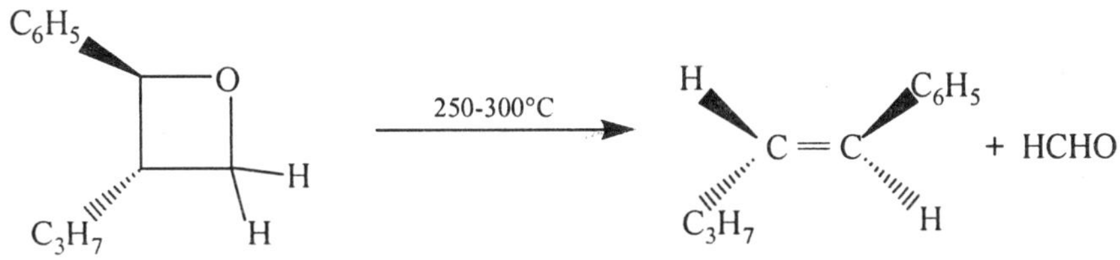

The reaction proceeds by an intermediate briadical and when phenyl is replaced by methyl, the reaction follows a non-stereospecific course, *i.e.* produces a mixture of both *cis*- and *trans*-alkenes.

4.3 THIETANES

Thietane (23) is a four-membered heterocyclic compound containing one sulfur atom in the ring. The corresponding unsaturated compound is called *thiete* (24) or *thietene*. Other names given to this ring system are *trimethylene sulfide, 1, 3-propylene sulfide,* or *thiacyclobutane.*

4.3.1 Physical and Spectroscopic Properties

Thietane is a colorless liquid, b.p., 94.9°C and m.p., − 73.2°C. The molecular dimensions have been obtained from electron-diffraction studies. The C − S bond distance of 1.85 Å is larger than the C − C bond length of 1.55 Å. The C − C − C (78 ± 1°) and C − S − C (97 ± 5°) bond angles differ considerably from those in non-cyclic molecules. For instance, the value of C − C − C bond angle is 109.28° in alkanes while the C − S − C bond angle in dimethylsulfide is 105°. In addition to the maximum at 215-220 *nm* which is typical of cyclic sulfides, the ultraviolet spectrum of cyclic sulfides of different ring sizes shows that the weak absorption maximum (275 *nm*) of oxetane is observed at larger wavelength than that of ethylene sulfide and tetra- and penta-methylene sulfides. This is ascribed[46] to the higher electron density at the sulfur atom in thietane. The order of electron density at the sulfur atom of cyclic sulfides of different ring sizes has been found to decrease in the following order:

$$4 > 5 > 7 > 3$$

Thietane is also planar similar to oxetane but puckering of the ring varies with substitution. Its dipole moment in benzene solution has been calculated to be 1.78 D which also shows some charge polarization, towards sulfur.

4.3.2 Synthetic Methods

Thietane has been obtained by the following general methods:

1. **_From 1, 3-Dihaloalkanes:_** The earliest and the common method[47, 48] for the preparation of thietane involved the reaction of 1, 3-dihaloalkanes with sodium or potassium sulfide, equation (4.9). This reaction is generally carried out in alcohol / water mixture at 70° but the yields are seldom above 50%.

$$\text{(4.9)}$$

3,3-Dimethylthietane

The principle reason for the low yield is the formation of polymeric sulfides. Lancaster and Smith[49] repeated the preparation of thietane from 1, 3-dihaloalkanes with an aqueous solution of sodium sulfide containing hexadecyltriethylammonium chloride as a phase transfer catalyst[50] and obtained thietane in a yield of 70% (equation 4.10).

$$\text{(4.10)}$$

Alternatively, the dihaloalkanes can be treated first with thiourea and the thiurium salt is subsequently decomposed[51] in the presence of a base as exemplified in the following reaction.

$$ClCH_2CH_2CH_2Br + (H_2N)_2 C = S \longrightarrow [ClCH_2CH_2CH_2SC(NH)_2)_2]^+ Br^-$$

$$\xrightarrow{OH^-} ClCH_2CH_2CH_2SH \xrightarrow[-HCl]{}$$

The 3-chloromercaptan[21] so obtained can readily be converted to thietane by alkali, however, the preparation of this halo compound is rather intricate.

In alkaline medium 2-hydroxy-3-chloropropyl mercaptan is converted at 50-60°C to 3-hydroxythietane (3-thietanol) in approximately 80% yield, equation (4.11).

$$\xrightarrow[-Cl^-]{OH^-}$$ (4.11)

2.   *From Chloromethylthiirane* : Alkaline ring opening of chloromethylthiirane to 3-chloropropane sulfide and its subsequent cyclization[52] yields 3-thietanol. Epichlorohydrin[53] in the presence of barium hydroxide also gives 3-thiethanol.

3.   *From the Decomposition of Cyclic Carbonates:* A highly stereospecific synthesis for the preparation of 2, 3-disubstituted thietane involves the decomposition of cyclic carbonates (24) of 1, 3-diols (1, 3-dioxan-2-one) in the presence of thiocyanate ion[54] the 4R-4-methyl derivative is converted to 2S-2-methylthietane.

(24)

The first step[55] in this reaction is assumed to be a nucleophilic attack of the thiocyanate ion at C-6 with the resultant opening of the ring. This view is in agreement with the observation that the reactivity decreases upon substitution of dioxane at position-6 and also that a nucleophile does not react. In the subsequent step loss of CO_2 takes place and ring closure and opening reaction produce thietane. Thietane so produced is simultaneously distilled off in a yield ranging from 30-65%. Thietanes have also been obtained by photochemical methods.[56]

4. Thietane-3-one is readily obtained by the hydrolysis[57] of acetal (25). The ketone is reduced with $NaBH_4$ to 3-thietanol which can be desulfurized in the usual manner with Raney nickel to give isopropyl alcohol.

4.3.3 Chemical Reactions

1. *Ring Opening Reactions:* Thietane molecule is less reactive than thiirane but more so than the inert thiophene. This is explained to be due to lower strain in the thietane molecule than in thiirane. The thietane ring is opened rather slowly and that too under drastic conditions. With ammonia it forms 3-aminopropanethiol, equation (4.12).

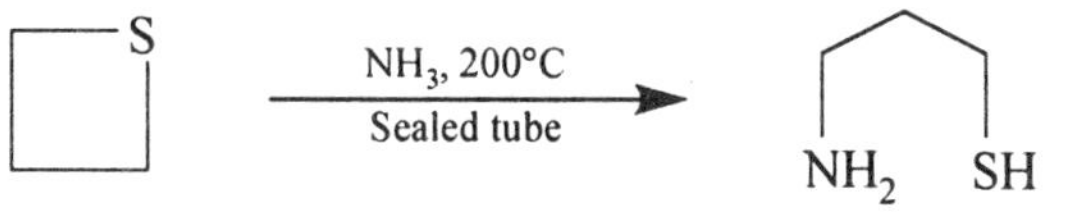

$$(4.12)$$

Thietane reacts similarly with acidic reagent acetyl chloride in the presence of $SnCl_4$ in benzene to form 3-chloropropylthioacetate.

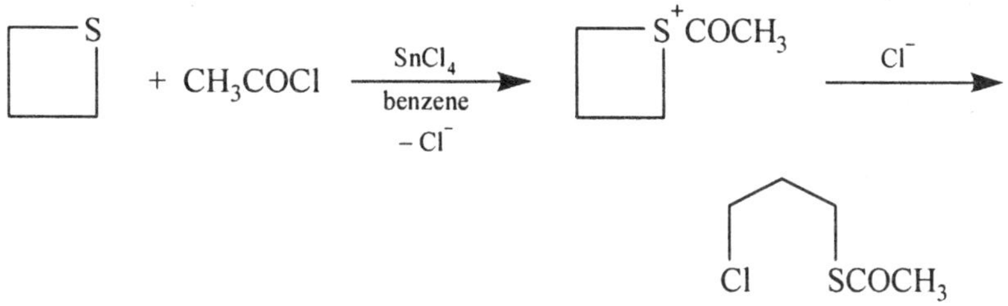

In 2-methylthietane, the $S - C_1$ bond is attacked preferentially by acetyl chloride.

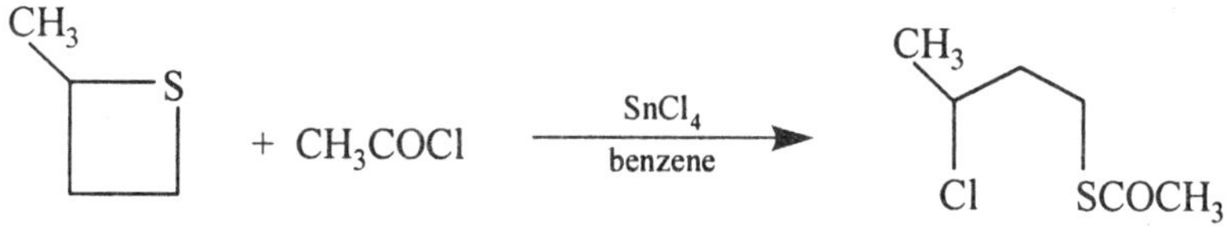

Thietane enters into reaction with halogens such as chlorine or bromine in a manner similar to thiirane with the fission of the ring to produce 3-chloropropyl disulfide[58].

2.   *Formation of Quaternary Salts:* Methyl iodide reacts with thietane to give a quaternary thietanium salt but the ring is immediately cleaved. In addition other polyhalides are also formed.[59, 60]

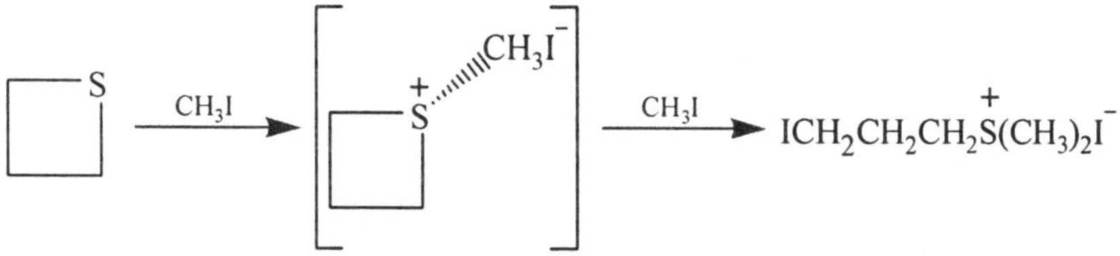

The quaternary thietanium salt decomposes in the presence of *n*-butyllithium to produce cyclopropane.[60]

Desulfurization of thietane and thiete derivatives takes place upon treatment with Raney nickel, as illustrated by the following example:

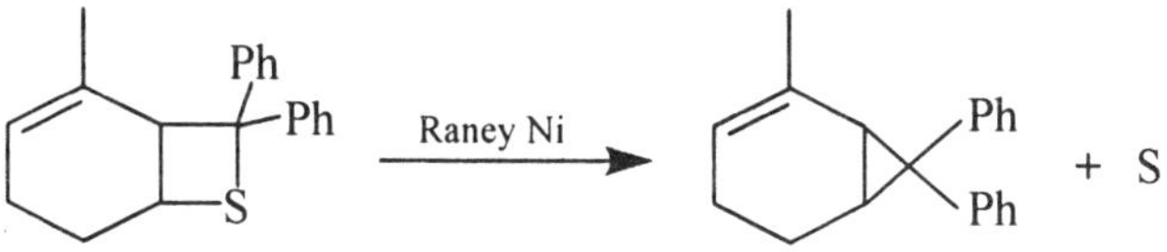

5. *Oxidation:* Thietane is easily oxidized by hydrogen peroxide to produce successively a 1-oxide (26) and eventually the sulfone. (27)

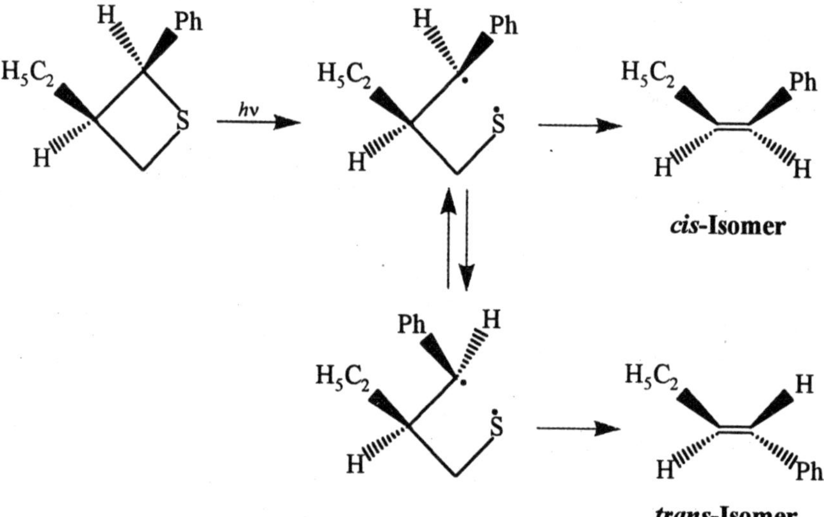

                                     (26)   (27)

6. *Photochemical Reactions: cis-* and *trans-*3 -Ethyl-2-phenylthietane on photolysis yield the corresponding *cis-* and *trans-*alkenes *via* a biradical intermediate[61] with *trans-*predominating.

4.4 CARBONYL DERIVATIVES

The 2-carbonyl derivatives of the four-membered heterocyclic compounds containing one hetero atom in the ring are known as 2-*azetidinone* (28), 2-*oxetanone* (29) and 2-*thietanone* (30) respectively. All these rings possess a common property that the ring system can be easily cleaved by various types of reagents.

4.4.1 2-Azetidinones[62]

The carbonyl derivative of azetidine is designated as 2-azetidinone (28) or more commonly known as β-lactam. Azitidinone derivatives occupy a central place among medically important compounds due to their diverse and interesting antibiotic properties. This ring system has been known since 1907 but the investigation of their chemistry laid dormant till 1943. In that year it was found that the important penicillin and cephalosporin series of antibiotics contain the β-lactam ring. Since then these compounds have been studied extensively and a variety of synthetic methods have been developed for their preparation.

1. Physical and Spectroscopic Properties

2-Azetidinone is a colorless solid, m.p., 73 – 74°C. It is highly soluble in ethanol and chloroform. The physical state of other β-lactams varies widely with the degree and nature of the substituents. The structural data obtained from X-ray studies using sodium benzylpenicillin shows that all the four bond angles and bond distances have different values. One highly characteristic and extremely useful property of these compounds is the infra-red absorption spectrum which provides a relative confirmation of the presence of this four-membered ring system. The carbonyl stretching frequency in an acyclic amide usually has a value of about 1665 cm^{-1} (6μ) is shifted for azetidinone in the range of 1755 – 1735 cm^{-1} (5.7 – 5.7 μ) in monocyclic lactams. This implies that the carbonyl group in β-lactam behaves like an ester carbonyl. In the fused ring thiazolidine β-lactam, an additional hypsochromic shift to about 1785 cm^{-1} (5.6 μ) occurs because of additional constraint in the

$$-\overset{\displaystyle O}{\overset{\displaystyle \|}{C}} - N$$

linkage. This also accounts for the higher reactivity of the carbonyl group in the ring.

2. Synthetic Methods

A great deal of work[63] has been done in connection with the synthesis of β-lactams since their discovery.

1. *From Direct Cyclization of Amino Acids:* A commonly employed method is the cyclization of free amino acids using acyl chloride, phosphorus trichloride or thionyl chloride as illustrated below.[64] No cyclization of the amino acid takes place on heating rather it splits into an amine and an acid.

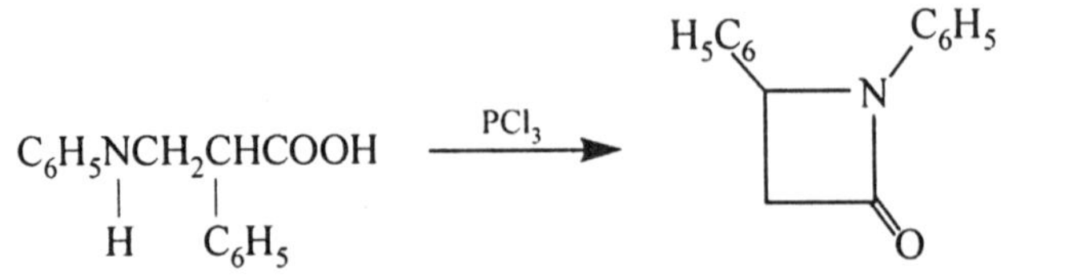

1,4-Diphenyl β-lactam

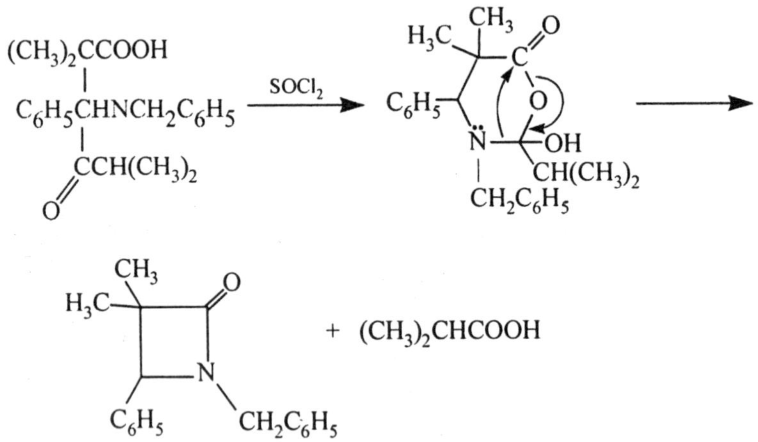

1-Benzyl-3, 3-dimethyl-4-phenyl-β-lactam

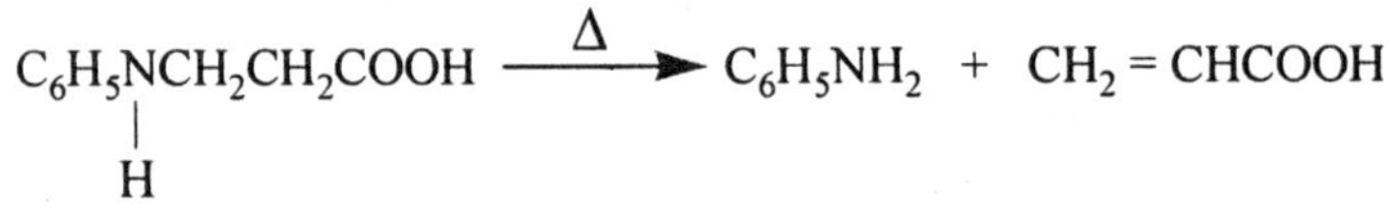

Ring closure between α, β-amino ester and Grignard reagent is generally used to produce a β-lactam. The parent compound[64] according to this method is obtained in a yield of about 50–75%. Further reaction of the Grignard reagent with the carbonyl group is prevented because of the hindered nature

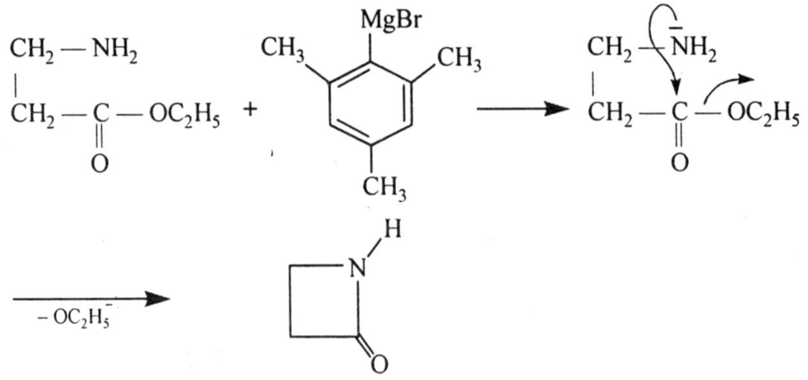

2-Azetidinone

of mesityl magnesium bromide. Another one stage[64] cyclization to a β-lactam derivative utilizes N-phenyl diethyl acetamido malonate in the presence of a base, equation (4.13).

$$(4.13)$$

2.   *Direct Combination Methods:* An older technique involves the direct combination of two appropriately substituted components. For instance, ketenes condense with imines to form β-lactams.[65] Chlorosulfonyl isocyanate adds

1,4.4-Triphenyl-β-lactam

similarly with a number of alkenes to lead to the corresponding 4,4-dimethyl N-chlorosulfonyl-β-lactam.[66] The chlorosulfonyl group can easily be removed by treatment with thiophenol in pyridine.[67]

4, 4-Dimethyl-2-azetidinone

3.   *From Substituted Azetidines:* N-substituted azetidine-2-carboxylic acid (31) can be converted into azetidinone by the following sequence involving a dicarbanion intermediate.[68] The azetidine carboxylic acid is treated with lithium diisopropylamide (LDA) and the resultant dicarbanion is decarboxylated oxidatively.

4.   *From Aziridines:* Because of the ready availability of aziridines, a one-pot synthesis has been developed for the preparation of azetidinones.[68] The reaction involves treatment of an aziridine derivative with lithium iodide followed by treating the reaction mixture with nickel tetra carbonyl and finally addition of solid iodine. 1-Benzyl-2-methylaziridine under these conditions gives 1-benzyl-4-methyl-2-azitidinone in 50% yield. It is noticed that the less substituted C – N bond is carbonylated.

5.   *Cyclization of β-Amino Acids:* β-Amino acids cyclize cleanly in the presence of diphenylphosphoric chloride to β-lactams. N-benzyl-3-aminobutyric acid leads to N-benzyl-2-azitidinone[69] in 72% yield. This reaction is solvent-dependent as the yield of the product decreases by altering the solvent, THF (68%), CH_2Cl_2 (61%).

6.   *Insertion of Carbenes:* Generation and intramolecular insertion of carbenes in a C – C bond in appropriate substrates results in azetidinone formation. N,

N-diethyldiazoacetamide, for instance on photolysis yields[70] 1-ethyl-4-methyl-2-azetidinone in a yield of 57%.

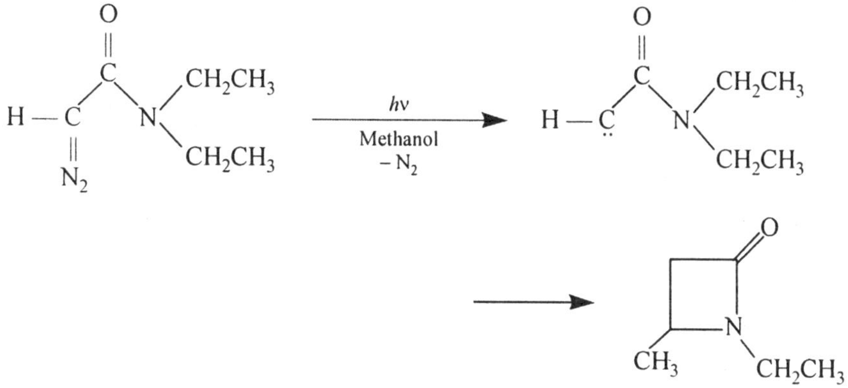

3. Chemical Reactions

β-Lactams exhibit a number of interesting reactions.

1. *Ring Opening Reactions:* Three-and four-membered ring lactams are considerably more reactive than the acyclic amides or even the larger ring lactams. A β-lactam is not as much strained as an α-lactam and also not nearly as reactive as 2-oxetanone. The occurrence of the resonance form (32a) of β-lactam makes the carbonyl carbon of β-lactam more electrophilic than the acyclic amide, and after the addition of a nucleophile, ring strain facilitates

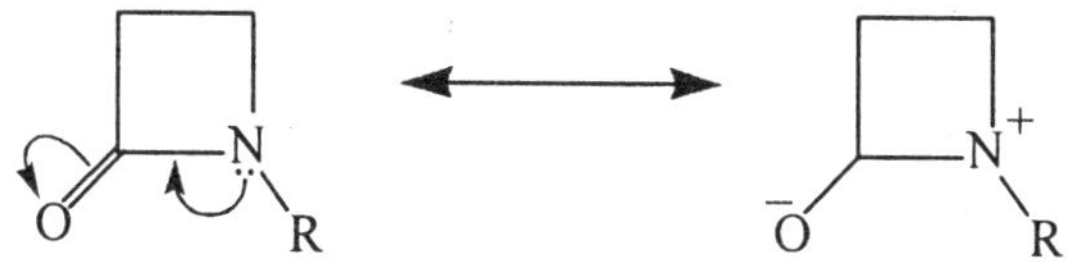

(32)   (32a)

the opening of the ring.[71] The final result is the formation of β-amino acids as shown in the following reaction; β-aminopropionic acid is obtained from β-lactam in the presence of a base, equation (4.14).

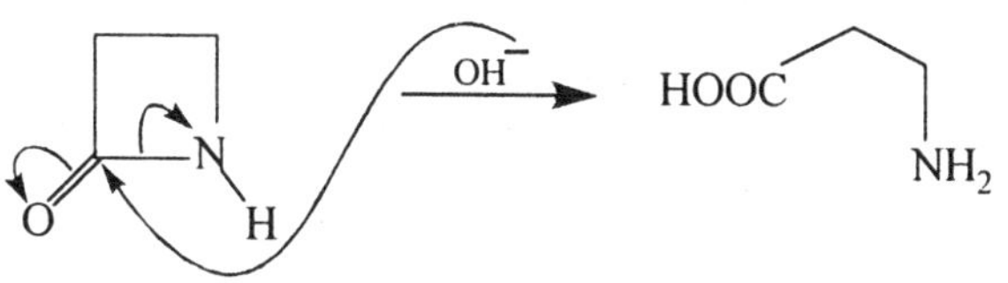

(4.14)

An amine similarly opens the ring but the reaction requires a high temperature, equation (4.15).

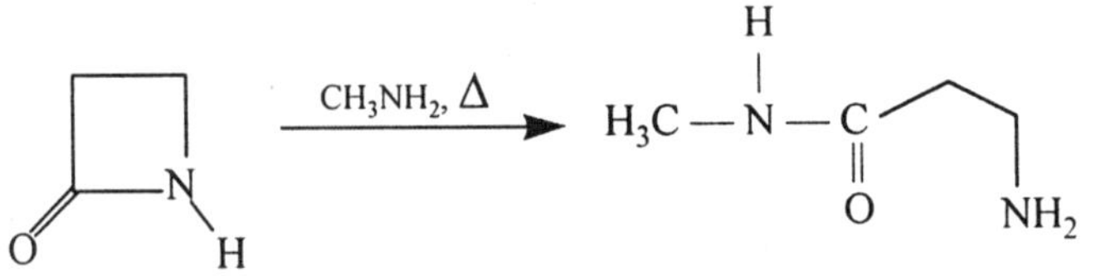

$$(4.15)$$

The β-lactam ring is also susceptible to the action of alcoholic HCl, equation (4.16).

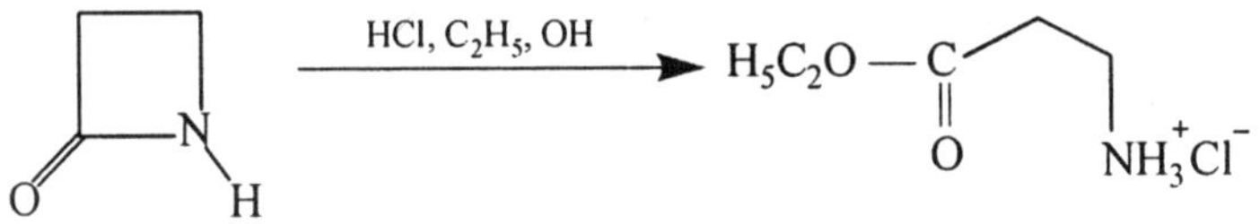

$$(4.16)$$

Analogues of penam derived by replacing sulfur atom by oxygen oxepenam structure, interestingly reacts with benzylamine to give an alkene derivative.[72]

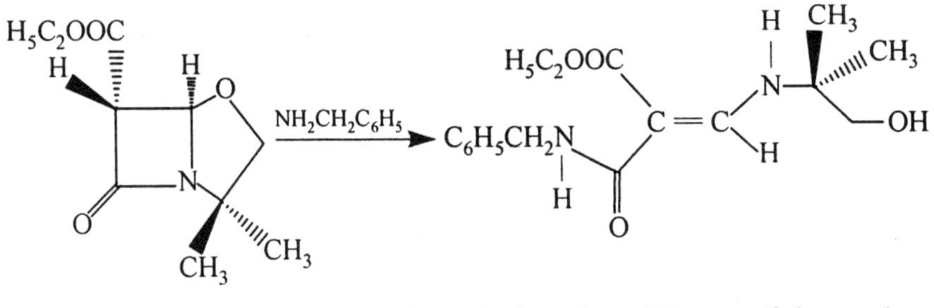

2.   *Reduction:* The reduction of N-substituted azetidinones (β-lactams) to azetidines is an important synthetic method. The ring is, however, cleaved with $LiAlH_4$ leading to 3-N-methylamine alcohol.[73] Mechanistically an attack by the hydride ion occurs initially.

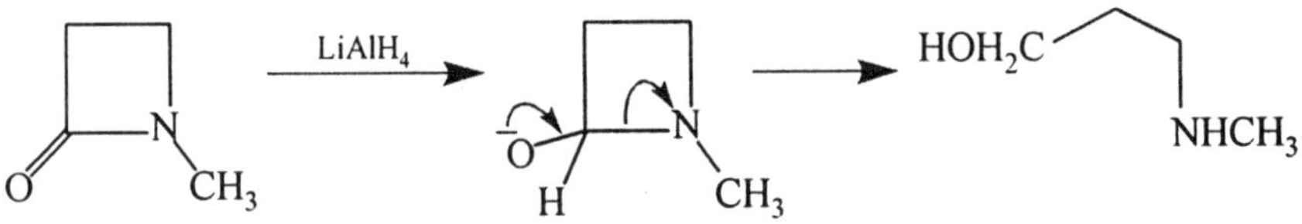

3.   *Reaction with Phosphorus Pentasulfide:* β-Lactams are converted[74] to thiolactams on reacting with P_2S_5 in benzene at 25°C, equation (4.17).

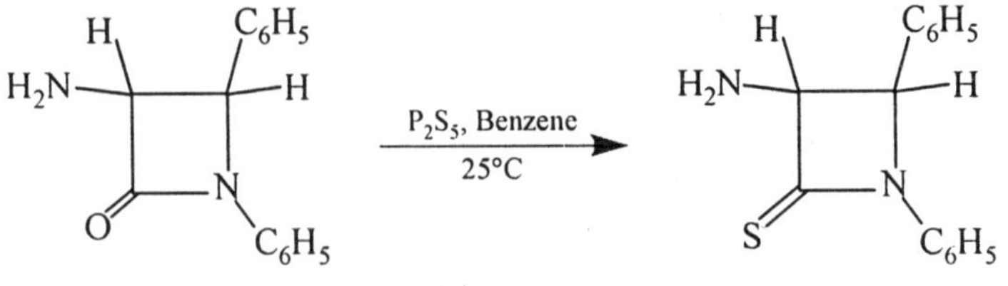

$$(4.17)$$

4. *Photochemical Reactions:* β-Lactams undergo interesting photochemical reactions. A bicyclic product (32), for instance, is obtained on photolysis of

(32)

N-phenyl-β-lactam.[75] Thietane undergoes homolytic ring opening[76] by attack of the alkoxy and other free radicals at the sulfur atom as shown here, equation (4.18).

$$ROSCH_2CH_2CH_2R_1$$

(4.18)

4. Naturally Occurring and Biological Active Compounds

The most important antibiotics containing the β-lactam ring are the penicillin (33) and cephalosporin (34). The name *antibiotics* is used for those substances that are produced by microorganisms and inhibit the growth of other microbes. The different penicillins have been produced by various strains of *Penicillin* molds. The basic structures commonly encountered in β-lactam antibiotics are the penam (35) and cepham (36). It is thought that the high reactivity of the β-lactam is essential to the antibiotic activity of these compounds. There is

(33)

(34)

a constant need for β-lactam antibiotics to combat bacteria which have built up resistance against the traditional penicillins. Resistance to penicillins and the relatedd cephalosporin is mainly caused by the formation of enzymes capable of opening the β-lactam ring common to these antibiotics.

(35)   (36)

Penicillin was first reported by Fleming in 1929 but he did not study its potential chemotherapic activity. In 1940, however, a group of workers isolated it in homogeneous form from a culture of the molds *penicillium rotatum* and demonstrated its powerful effect against bacteria.[77-80] A great deal of work was done in the following years and modern large scale fermentation methods have enabled penicillin production to reach a level of thousand tons per year. Penicillin and related penams have proven to be of enormous value for the treatment of bacterial infections. Research chemists have made structural modifications on the penicillins in their search for materials with better therapeutic properties. This has resulted in broader or more selective specta, increased potency and increased stability to lactamases. Some of the important penicillins used in medical practice are the following:

Penicillin G

(37)

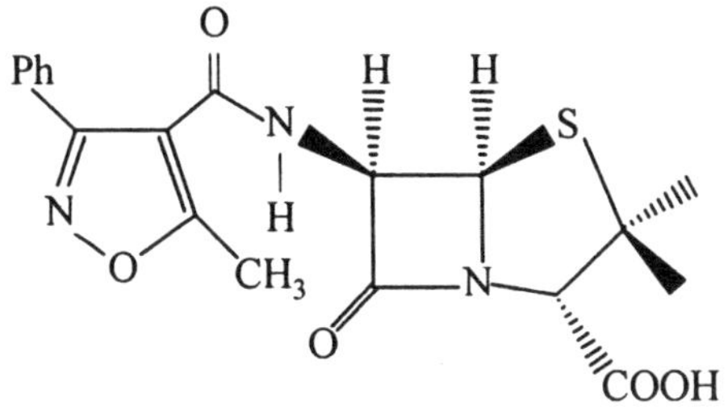

Oxacillin

(38)

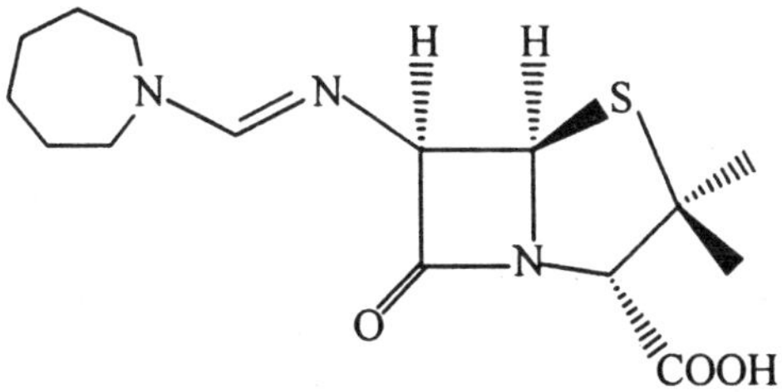

Mecillinam

(39)

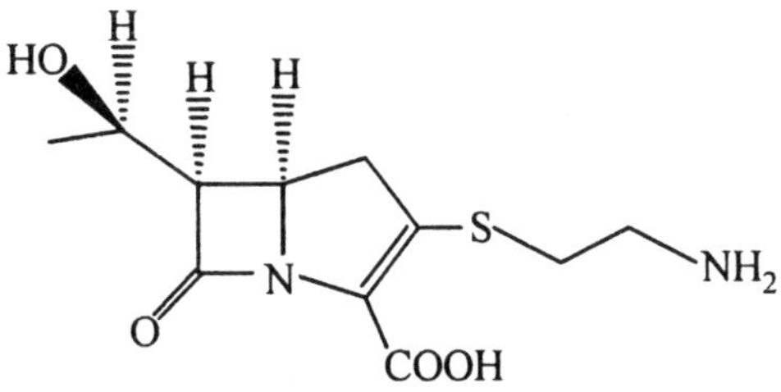

Theienamycin

(40)

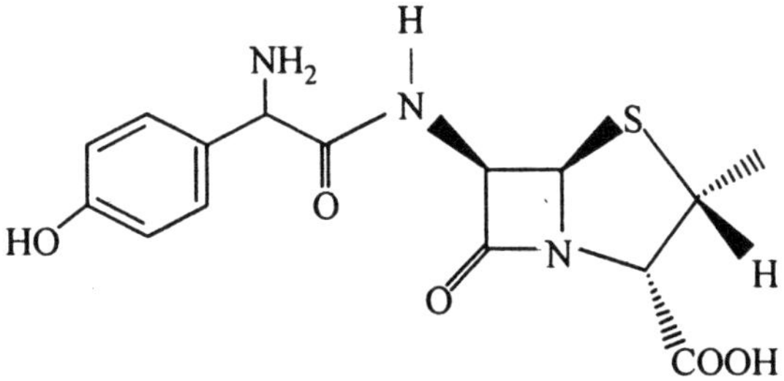

Amoxcillin

(41)

The accepted structure of penicillin was first proposed on the basis of degradation work but Mrs. D. Crowfoot Hodgkin provided a proof for its

structure by X-ray crystallographic methods. The first total synthesis of a
penicillin was achieved in 1957 by Sheehan and coworkers.[81] D-penicillamine
was condensed with an aldehyde to yield a thiazolidine from which the
phthaloyl group was removed by hydrogenolysis. Subsequent acylation with
phenoxyacetyl chloride and cleavage of the *t*-butyl ester with anhydrous
hydrogen chloride gave the penicillinoic acid. Cyclization of the acid under
mild conditions in the presence of dicyclohexylcarbodiimide gives the desired
penicillin (G).

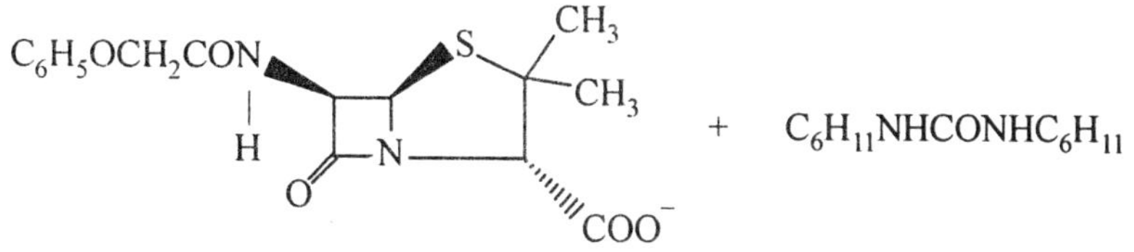

$$C_6H_5OCH_2CON \quad\cdots\quad + \quad C_6H_{11}NHCONHC_6H_{11}$$

Penicillin G

Cephalosporin: Because of the constant need and search for more potent antibiotics the related cephalosporins are found to be clinically more useful against pathogenic bacteria. They also belong to the class of β-lactam antibiotics. Cephalosporin C (42) a product of *Cephalosporin acremonium* was first isolated in 1955, and its structure was established subsequently[82] by these workers. The cephalosporin group of antibiotics are costlier to produce than the penicillins and yet have a much wider spectrum of activity than the penicillins. A total synthesis of cephalosporin (43) was first reported by Woodward and cowrkers.[83]

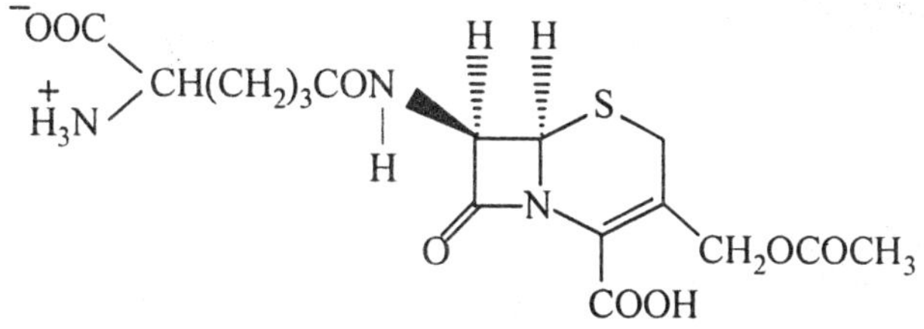

Cephalosporin C
(42)

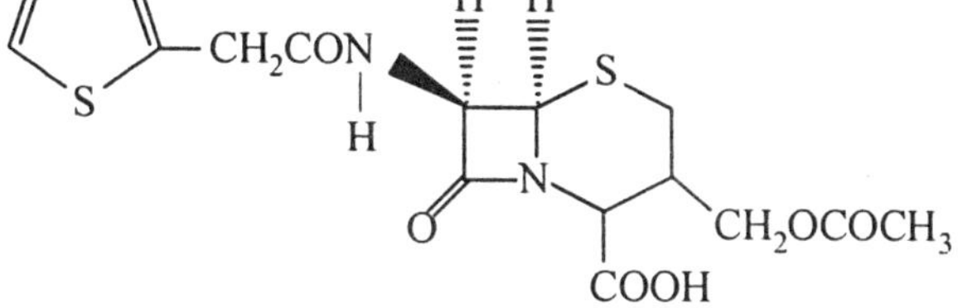

Cephalosporin
(43)

Some additional chemically useful cephalosporins (44–46) are the following:

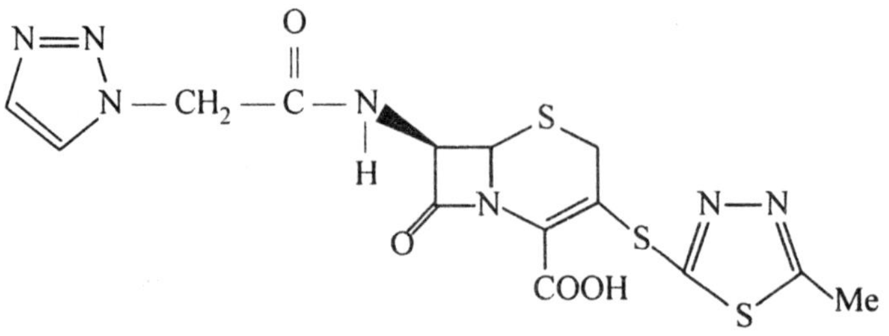

Cefazoline
(44)

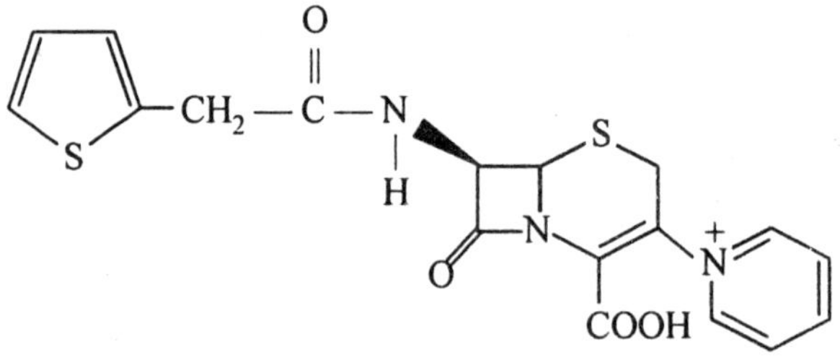

Cephaloridine
(45)

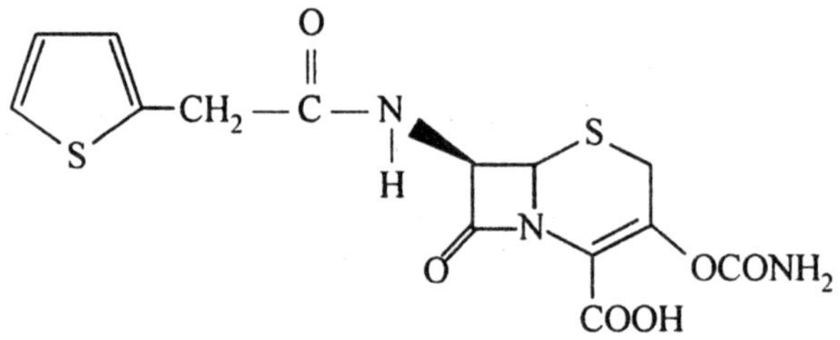

Cefoxitin
(46)

Considerable efforts are being expended in attempting the production of cepham (36) type of antibiotics from the more readily available penicillin by semi-synthetic processes.[84-86] One such scheme is due to Morin and coworkers[87] starting from methyl ester of phenoxymethyl penicillin sulfoxide (47).

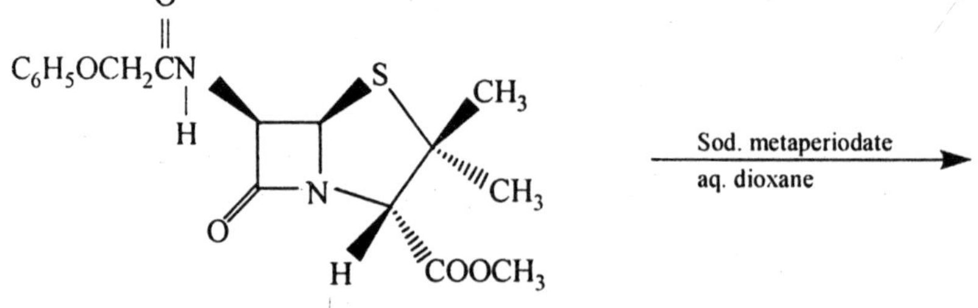

$$\xrightarrow[\text{aq. dioxane}]{\text{Sod. metaperiodate}}$$

(47)

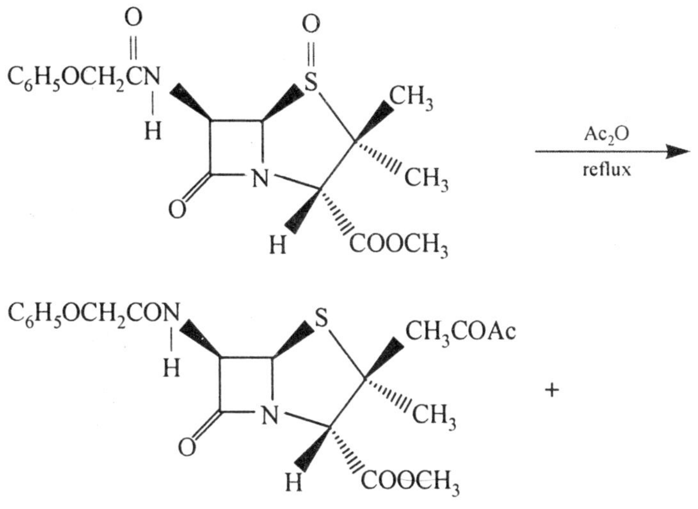

$$C_6H_5OCH_2\overset{\overset{\textstyle O}{\|}}{C}N\cdots \quad S=O \quad CH_3 \quad CH_3 \quad COOCH_3 \quad \xrightarrow[\text{reflux}]{Ac_2O}$$

$$C_6H_5OCH_2CON\cdots \quad S \quad CH_3COAc \quad CH_3 \quad COOCH_3 \quad +$$

(48)

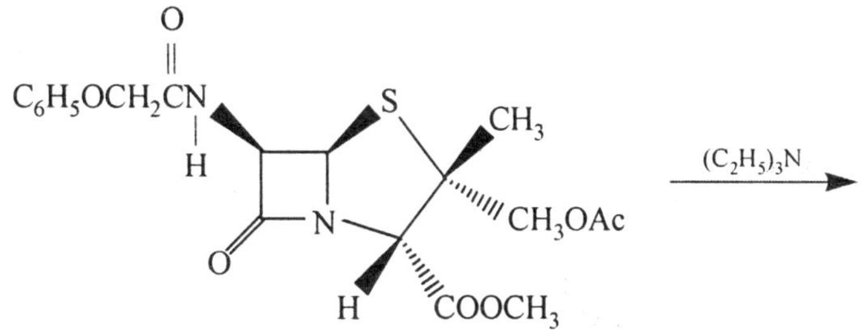

$$C_6H_5OCH_2\overset{\overset{\textstyle O}{\|}}{C}N\cdots \quad S \quad CH_3 \quad CH_3OAc \quad COOCH_3 \quad \xrightarrow{(C_2H_5)_3N}$$

(48a)

$$C_6H_5OCH_2\overset{\overset{\textstyle O}{\|}}{C}N\cdots \quad S \quad OAc \quad COOCH_3$$

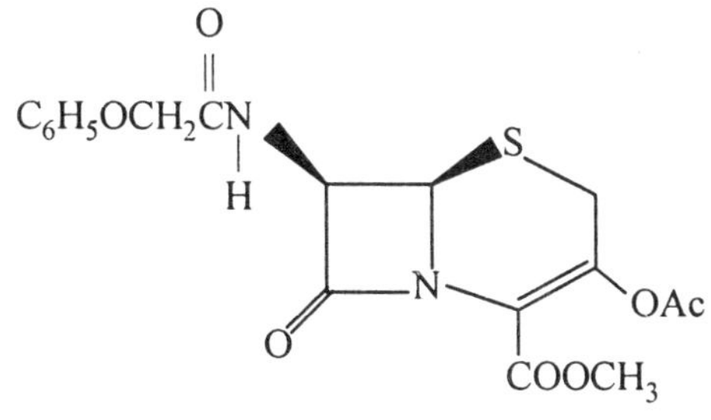

Oxidation of (47) and subsequent treatment of the product with acetic anhydride yields two products (48) and (48a). The latter on reacting with triethylamine gives the required ceph-3-em structure.

4.4.2 Oxetanone

The most important functional derivatives of oxetane are the oxetan-2-ones. The name β-lactone (49) is more universally in use than 2-oxetanone or oxetan-2-one. Its derivatives are generally named as substituted derivatives of β-propiolactone. The β-lactones are internal esters which are strained

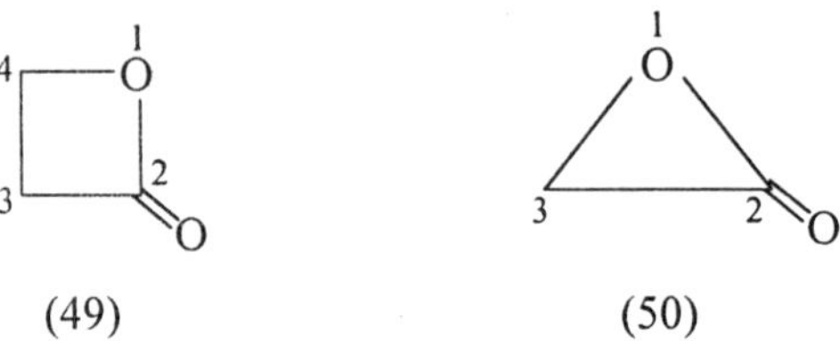

molecules but less than the α-lactones (50). The simplest β-lactone or β-propiolactone was first obtained in 1916. It possesses carcinogenic properties. β-Lactone inhibits the growth of certain micro-organisms and is thus employed as an external disinfectant. Its aqueous solution is also bactericidal. It is available commercially and is thus important synthetic chemical for polymers and other useful compounds.

1. Physical and Spectroscopic Properties

2-Oxetanone is a colorless liquid, b.p., 50°/10 mm (m.p. 33.5°). It is a planar molecule and its structure has been found by electron-diffraction measurements.

Five-and four-membered lactones are strained relative to acyclic esters and a shift of the carbonyl stretching absorption to a higher wave number is expected. In the infrared the carbonyl stretching frequency occurs at 1800 cm^{-1} which is higher than in acyclic esters. This is ascribed to the deformation of bond angles or the presence of strain in the ring and is illustrated for methyl acetate, γ-lactone and β-lactone.

CH$_3$COCH$_3$		
in CCl$_4$ 1741 cm^{-1}	1775 cm^{-1}	1841 cm^{-1}

2. Synthetic Methods

Because of the mobility of the β-lactone rings not very many methods are available for their preparations.

1. *Cyclization Methods:* A hydroxy acid contains both hydroxy and carboxyl groups, therefore, it can undergo intramolecular esterification to yield a cyclic ester, a lactone. A direct lactonization of a β-hydroxy acid is not generally a

useful method because it is an equilibrium process and also because of the formation of α, β-unsaturated acid due to elimination. Therefore, only when the lactone has five-or six-membered ring then there is a substantial amount of lactone present under equilibrium conditions. Instead treatment of the β-halo acid with one equivalent of base (Na_2CO_3 or Ag_2O) under controlled

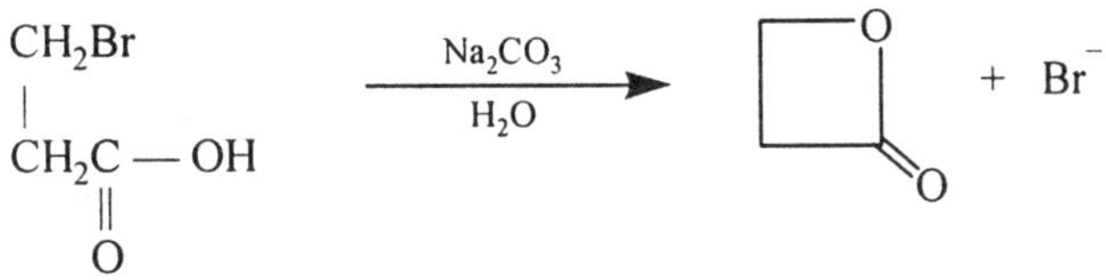

$$\underset{\underset{O}{\overset{\|}{\underset{}{}}}{\overset{CH_2Br}{\underset{|}{CH_2C-OH}}} \quad \xrightarrow[\text{H}_2\text{O}]{\text{Na}_2\text{CO}_3} \quad \text{(}\beta\text{-lactone)} \quad + \ Br^- \qquad (4.19)$$

conditions affects the ring closure, equation (4.19). The reaction is carried out at room temperature.[88] If appropriate substituents are present as in 1-bromo-1-*p*-bromobenzoylcyclohexan-2-carboxylic acid (51) than the reaction follows a stereospecific ring closure with inversion of configuration at the carbon carrying the halogen atom.[89] On the other hand, no ring formation takes place

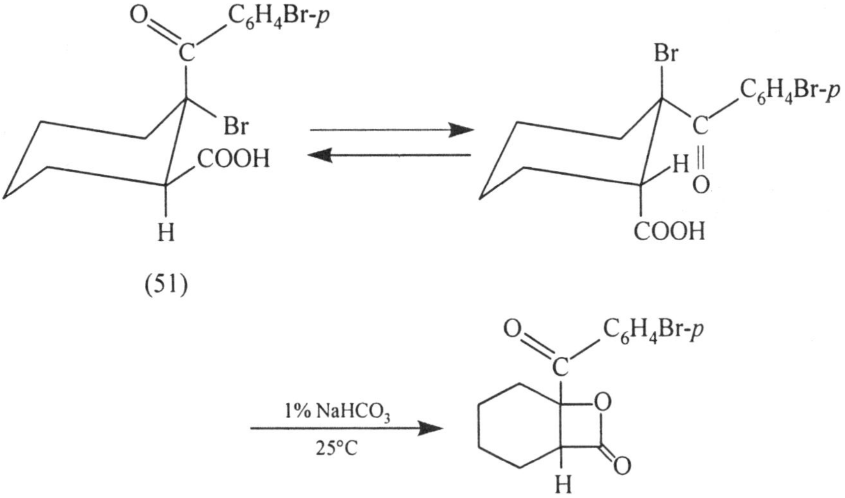

(51)

in (52), because of steric reasons the carboxyl group cannot attack from the backside and instead only normal solvolysis takes place.

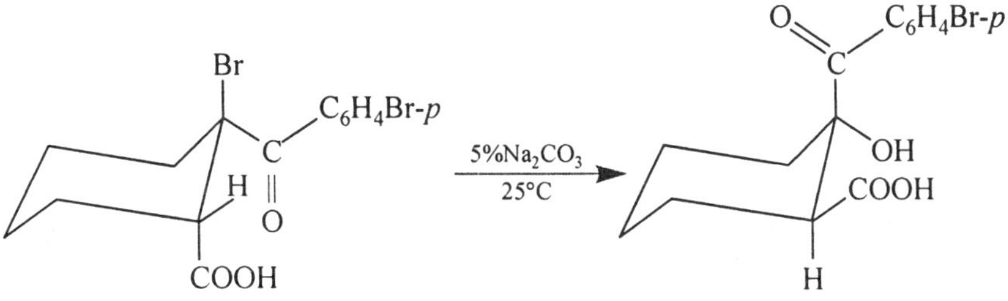

(52)

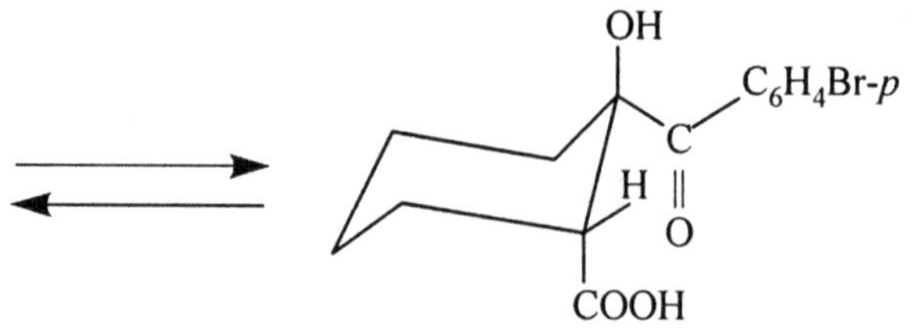

2.   *Direct Combination Methods:* Ketenes react with carbonyl compounds in the presence of catalysts like $ZnCl_2$, BF_3 $(OC_2H_5)_2$, boric acid to give good yields of β-lactones *via* cycloaddition. β-Lactone is commercially obtained in this manner (equations 4.20 and 4.21).

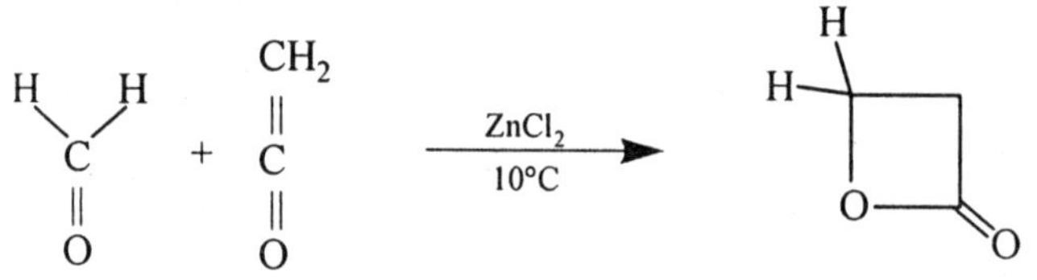

$$(4.20)$$

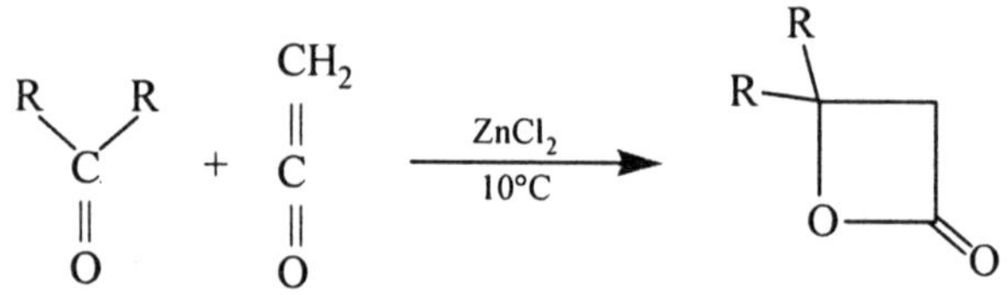

$$(4.21)$$

3.   *From 3-Bromobutyric Acid:* R – (+) – 3-methyloxetanone has been prepared from 3-bromobutyric acid.[90] The racemic, acid obtained from hydrobromination of crotonic acid, was resolved with R – (+) -α-(1-naphthyl) ethylamine which afforded S – (+) – 3-bromobutyric acid in 90% ee.

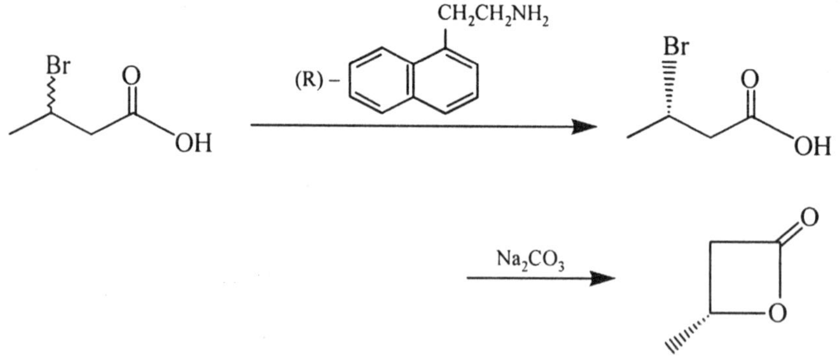

Cyclization of the acid with aqueous sodium carbonate in chloroform yielded the acid in 72% yield. The oxetanone is used as a starting material for the synthesis of R – (+)-citronellol and R – (+)-pulegone.

3.   Chemical Reactions

1.   *Ring Opening Reactions:* β-Lactams always result in ring opening in their reactions due to the presence of strain. The β-lactones in contrast, are generally stable at room temperature under similar conditions but are easily cleaved when heated. The β-lactone derivatives are obtained by cycloaddition of ketones to ketenes but dissociate on heating above 20°C. Thus 4, 4-dimethyl-3, 3-diphenyloxetanone obtained from acetone and diphenyl ketene decomposes to 2-Methyl-3,3-diphenylpropene. This ring-opening

2-Methyl-3, 3-diphenylpropene

reaction is highly stereospecific[91] and proceeds with retention of configuration, (equation 4.22) as in the following example:

$$\text{(4.22)}$$

1-(4-Chlorophenyl) propene

Oxetanone polymerizes slowly on standing at room temperature. This process can be accelerated by heating or by the addition of acids or bases. The resulting polymer has a linear structure.

Lactones being cyclic esters can hydrolyze to the corresponding open-chain drivatives, depending on the reaction conditions (*i.e. p*H). In neutral or slightly acidic medium a bimolecular alkyl-oxygen fission B_{AL}–2 (mode a) occurs whereas in strong acid solutions (or basic medium) the reaction occurs by a bimolecular acyl-oxygen B_{Ac}–2 (mode b) mechanism. The position of cleavage was revealed by the hydrolysis of 3-methyloxetanone by ^{18}O labelled water.[92]

Nucleophiles give products of ring opening of B_{AL}–2 mechanism.[96] Alcoholysis follows the same pathway as hydrolysis.

The generalization of ring-opening by phenols and amines in oxetanone however, does not hold good. Phenol is an acidic substance and thus it reacts slowly with a β-lactone to give rise to β-phenoxypropionic acid (mode a) but in the presence of catalytic amount of acid an ester is formed, a consequence of acyl-oxygen fission (mode b). This departure of the ring opening is explained[93] in terms of the acidity of phenol. A low acid concentration in methanol still reflects the catalyzed attack of the solvent on β-lactone (mode a) but in phenol it follows the limiting acid catalyzed mechanism as is observed in very strong acidic medium. In the presence of ammonia or amine, in contrast, a relative proportion of the two products is obtained as a result of opening of the ring in both directions.

2. *Photochemical Reactions:* Recently it has been demonstrated by Corey and Streith[94] that irradiation of α-pyrone [(pyrex filter) > 291 *nm*] under matrix isolation conditions at $-20°C$ results in the formation of β-lactone which on continued irradiation affords cyclobutadiene and CO_2 gas is evolved.

Phthaloyl peroxide on photolysis yields a benzyne *via* benzopropiolactone.[95]

REFERENCES

1. G. E. Test et al., *Heterocyclic Chem.,* **4,** 619 (1967).
2. D. Bohrmann, *Ann.* **725,** 124 (1969).
3. G. Szeiemies, U. Siefken and R. Rinck, *Angew. Chem. Int. Edn.*(Engl), **12,** 161 (1973)
4. N. C. Yang *et al., Chem. Comm.,* 729 (1976).
5. A. B. Levy and A. Hassner, *J. Am. Chem. Soc.,* **93,** 2051 (1971).
6. Q. N. Porter and J. Baldas, *Mass Spectrometry of Heterocyclic Compounds,* 2nd ed., John Wiley, New York (1985).
7. For a review see, N. H. Cromwell and B. Phillips, *Chem. Rev.,* **79,** 331 (1979).
8. T. H. Koch, R. H. Higgins and H. F. Schuster, *Tet. letters,* 431 (1977).
9. F. Effenberger and R. Maier, *Angew. Chem. Int. Edn.*(Engl.), **5,** 416 (1966).
10. H. W. Weine, A. D. Miller, W. H. Barton and R. W. Greiner, *J. Am. Chem. Soc.,* **75,** 4778 (1973).

11. S. Searless. *Jr.* M. Tamares, F. Block and L. Aquaterman, *J. Am. Chem. Soc.*, **78,** 4917 (1956).

12. J. P. Freeman, D. G. Pucci and G. Binsch, *J. Org. Chem.*, **37,** 1984 (1972).

13. M. Vaultier, R. D. Bougot, D. Danion, J. Hamelin and R. Carrie, *J. Org. Chem.*, **40,** 2990 (1975).

14. N. H. Cromwell and R. M. Rodebaugh, *J. Heterocyclic Chem.*, **6,** 435 (1969); R. F. C. Brown B. T. Bunstan and S. S. Ternhell., *Tet. letters,* 3111 (1984).

15. D. St. C. Black, *Tet. letters.* 4283 (1974).

16. T. G. Archibald, R. Gilardi, K. Baum and C. George, *J. Org. Chem.*, **55,** 2920 (1990).

17. S. Searles and G. F. Butler, *J. Am. Chem. Soc.*, **76,** 56 (1954).

18. A. Padwa, R. Grunber and L. Hamilton, *J. Am. Chem. Soc.*, **89,** 3077 (1967); A. Padwa, *ibid.,* **90,** 4456 (1968); **92,** 100 (1970).

19. I. C. Cotteril and S. M. Roberts, *J. Chem. Soc.*, **4,** *Perkin Trans.* I, 2585 (1992),; Also see G. W. J. Fleet *et al., Tet. letters,* 1674 (1991).

20. S. Searles, *Jr.* R. G. Nickerson and W. K. Witsiepe., *J. Org. Chem.*, **24,** 1839 (1959).

21. A. Rosowsky and D. S. Tarbell, *J. Org. Chem.*, **26,** 2255 (1961).

22. E. Pasterno and G. Chieffi, *Gazz. Chim. Ital.*, **39,** 342 (1909).

23. For a comprehensive review see D. R. Arnold, *Adv. Photochem.*, **6,** 301 (1968).

24. N. J. Turro and P. A. Wriede, *J. Am. Chem. Soc.*, **92,** 320 (1970).

25. N. C. Yang and R. L. Loeschen, *Tet. letters,* **257** (1968).

26. S. Farid and S. E. Shealer, *Chem. Comm.*, 296 (1973).

27. S. Farid., J. C. Sorty and J. L. R. Williams., *ibid.,* 711 (1972).

28. J. S. Bradshaw, *J. Org. Chem.*, **31,** 237 (1966).

29. For a review see A. Mustafa, *Chem. Rev.*, **60,** 181 (1947).

30. N. C. Yang, M. Nussim, M. J. Jorgenson and S. Murov, *Tet. letters,* 3657 (1964).

31. H. Gotthardt, R. Steinmetz and G. S. Hammond, *J. Org. Chem.*, **33,** 2774 (1968); D. R. Arnold and A. H. Glick, *Chem. Comm.*, 813 (1966).

32. W. J. Middleton, *J. Org. Chem.*, **30,** 1307 (1965).

33. M. Muskatirovic and Z. D. Todic, *J. Chem. Soc., Perkin II,* 1701 (1975).

34. I. K. Okuma, Y. Tanaka, S. Kaji and H. Ohta, *J. Org. Chem.*, **48,** 5133 (1983).

35. J. G. Pritchard and F. A. Long, *J. Am. Chem. Soc.*, **80,** 4162 (1960); M. Yamaguchi and I. Hirao, *Tet. letter,* 4549 (1984).

36. S. Searles, *Jr.,* and V. P. Gregory, *ibid.,* **76,** 2789 (1954).

37. S. Searles, *Jr., et al., ibid.* **73,** 4515 (1953).

37a. S. Searles, *Jr.,* and C. F. Butler, *ibid.,* **76,** 56 (1954).

38. B. Mudryk and T. Cohen, *J. Org. Chem.*, **56,** 5760 (1991).

39. S. Searles, *Jr., ibid.,* **73,** 4515 (1953).

40. B. M. Miller, *J. Organometallic Chem.,* **14,** 253 (1968).

41. S. Searles, *Jr., in The Chemistry of Heterocyclic Compounds,* Vol. 10, Part 2, A Weissberger, *(Ed.), Interscience,* New York (1964), Chap. 9. M. Yamaguchi *et al., Tetrahedron,* **40,** 4261 (1984); T. Suzuki, H. Saimoto, H. Saimoto, H. Tomoka, K. Oshima and H. Nozaki, *Tet. letters,* 3597 (1982); P. F. Hudrlik and C. N. Wan. *J. Org. Chem.,* **40,** 2963 (1975).

42. M. Segi, M. Takebe, S. Masuda, T. Nakajima and S. Suga, *Bull. Chem. Soc. Japan,* **55,** 167 (1982).

43. K. Friedrich, U. Jansen and W. Kirmse, *Tet. letters,* 193 (1985); *ibid.,* 193 (1985).

44. M. Yamaguchi, K. Ishibato, I. Hirao, A. Baba, H. Kashiwagi and H. Matsnda, *Tet. letters,* 1159 (1984); A. Baba, *Tet. letters,* 1323 (1985).

45. H. Gottehardt, R. Steinmetz and G. Hammond, *J. Org. Chem.,* **33,** 2774 (1968); D. R. Arnold and A. H. Glick, *Chem. Comm.,* 813 (1966).

46. R. E. Davis, *J. Org. Chem.,* **33,** 1380 (1958).

47. M. Sander, *Chem. Rev.,* **66,** 341 (1966).

48. Y. Etienne, R. Sandles and H. C. Lumbroso, in *Heterocyclic Chemistry of Three and Four-membered Rings,* part II, *Interscience,* London (1969).

49. M. Lancaster and D. J. H. Smith, *Synthesis,* 582 (1982).

50. E. V. Dehmlow, *Angew. Chem. Int. Edn.* (Engl.), **13,** 170 (1974); J. Docks, *Synthesis,* 441 (1973).

51. F. G. Bordwell and B. M. Pitt, *J. Am. Chem. Soc.,* **77,** 572 (1955).

52. E. P. Adams *et al., J. Chem. Soc.,* 2605 (1960).

53. D. C. Dittmer and M. C. Christy, *J. Org. Chem.,* **26,** 1324 (1961).

54. S. Searles, H. R. Hays and E. F. Lutz, *J. Org. Chem.,* **27,** 2832 (1962).

55. J. H. Stewart and C. H. Burnside *J. Am. Chem. Soc.,* **75,** 243 (1953).

56. J. D. Coyle and P. A. Rapley, *Tet. letters,* 2247 (1984).

57. S. Searles, H. R. Hays and E. F. Lutz, *J. Org. Chem.,* **27,** 2828 (1962).

58. H. Stewart and J. E. Burnside, *J. Am. Chem. Soc.,* **75,** 243 (1953).

59. D. C. Palmer and E. C. Taylor, *J. Org. Chem.,* **51,** 846 (1986).

60. B. M. Trost, W. L. Schinski and I. B. Mantz, *J. Am. Chem. Soc.,* **91,** 4320 (1969).

61. D. R. Rice and R. D. Stier, *Chem. Comm.,* 166 (1973).

62. For a review see, Y. Etiene and N. Fischer, *Heterocyclic Compounds,* Vol. XIX, Part 2, A. Weissberger, *(Ed.),* Interscience, New York, (1964), pp. 729-880.

63. For a review see, A. K. Mukherjee and R. C. Srivastava, *Synthesis,* 327 (1973); G. S. Singh, *Tetrahedron,* **59,** 7631, (2003); I. Ojima and F. Delage, *Chem. Soc. Rev.,* **26,** 377 (1997).

64. S. Searles and R. E. Wann, *Chem. and Ind.,* 2097 (1964).

65. J. C. Sheehan and A. K. Bose, *J. Am. Chem. Soc.,* **73,** 1761 (1951); B.

G. Chatterjee *et al., J. Org. Chem.,* **30**, 4101 (1969), D. H. Hua and A. Verma, *Tet. letters,* 547 (1985).

66. R. Garf, *Ann. Chem.,* **661**, 111 (1963); H. Hoffman and H. J. Diehr, *Tet. letters,* 1855 (1963).

67. H. H. Wasserman and B. H. Lipschutz, *Tet. letters,* 4613 (1976).

68. W. Chamchaang and A. R. Pinhas, *J. Org. Chem.,* **55**, 2943 (1990).

69. S. Kim, P. H. Lee. and T. A. Lee, *Chem. Comm.,* 1242 (1988).

70. R. R. Rando, *J. Am. Chem. Soc.,* **92**, 6707 (1970).

71. A. K. Murkherjee and A. K. Singh, *Synthesis,* 547 (1978).

72. B. T. Golding and D. R. Hall, *Chem, Comm.,* 293 (1978).

73. E. Testa, G. Piffieri and L. Fonlanetle, *Ann. Chem.,* **696,** 108 (1966).

74. K. R. Henery-Logan, H. P. Knoepfel and J. Rodricks, *J. Heterocyclic Chem.,* **5**, 433 (1965).

75. M. Fischer and A. Mattheus, *Chem. Rev.,* **69,** 342 (1969).

76. D. E. Applequist and L. F. Machenzie, *J. Org. Chem.,* **41**, 2262 (1976).

77. H. T. Clark, J. R. Johnson and R. Robinson, *The Chemistry of Penicillins,* Princeton University Press, New York, (1949).

78. E. P. Abraham, *Q. Rev.,* **21,** 231 (1967).

79. D. H. R. Barton, *Pure Appl, Chem.,* **33**, 1(1973); C.A. **78**, 7143 (1973).

80. M. Schiozaki, *Synthesis,* 691 (1990); S. Hamessian, D. Sesilets and Y. L. Bennani, *J. Org. Chem.,* **55**, 3098 (1988).

81. J. C. Scheehan and K. R. Logan, *J. Am. Chem. Soc.,* **81**, 5838 (1959); *ibid.,* **84**, 2983 (1962).

82. E. P. Abraham and G. C. Newton, *Biochem. J.,* **79**, 377 (1961).

83. R. B. Woodward *et al., J. Am. Chem. Soc.,* **88**, 852 (1966).

84. For a review see, P. G. Sammes, *Chem. Rev.,* **76**, 113 (1976).

85. P. A. Lemke, D. R. Brannor and E. H. Flynn, *Cephalosporin and Penicillium,* Academic Press, New York (1972).

86. R. P. G. Copper, L. D. Hatfield and D. O. Spry, *Chem. Rev.,* **73**, 32 (1973).

87. R. B. Monn, B. G. Jackson, R. A. Mueller, E. R. Lavagnino, W. B. Scanion and S. L. Andrews, *J. Am. Chem. Soc.,* **91**, 1401 (1969).

88. H. E. Zaugg, *ibid.,* **72**, 2998 (1950).

89. P. D. Bartlett and P. N. Rylander, *J. Am. Chem. Soc.,* **73**, 4275 (1951).

90. T. Sato, T. Kawara, A. Nishizava and T. Fujisawa, *Tet. letters* 3377 (1980); Also see J. R. Shelton, D. E. Agostini, and J. B. Lambo, *J. Polym. Sci,* A-1, **9**, 2789 (1971).

91. D. S. Noyce and E. H. Banitt, *J. Org. Chem.,* **31**, 4043 (1966).

92. R. K. Bansal, *Organic Reaction Mechanism,* Tata McGraw-Hill 3rd. ed., New Delhi (1998).

93. H. E. Zaugg, *ibid.,* **72**, 3001 (1950).

94. E. J. Corey and J. Streith, *J. Am. Chem. Soc.,* **86**, 950 (1964).

95. O. L. Chapman *et al., ibid.,* **95**, 4061 (1973).

REVIEW PROBLEMS

4.1 Predict the products of the following reactions:

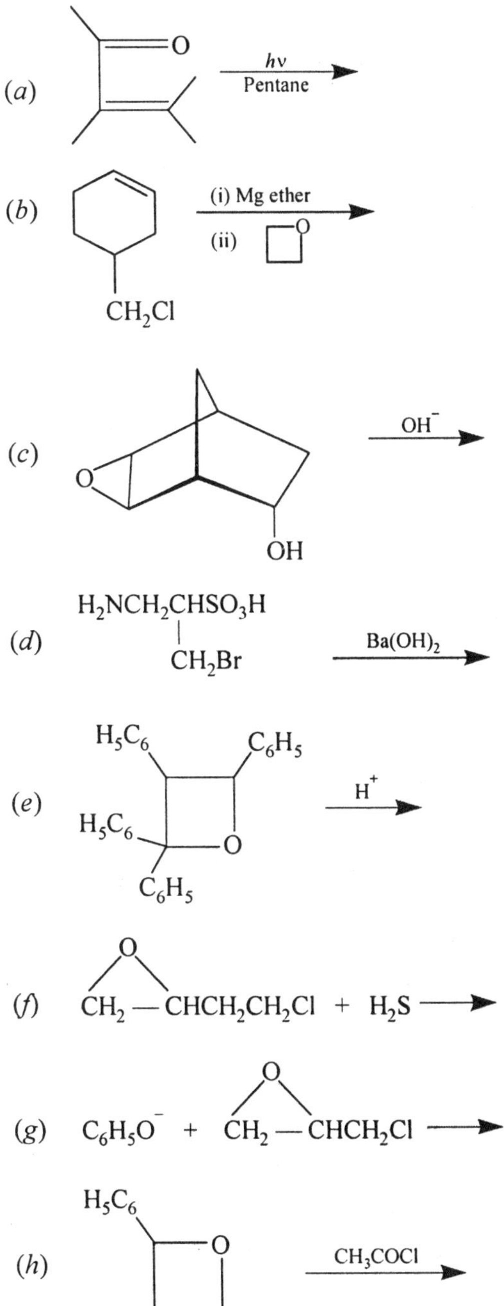

(i) O_2N—C_6H_4—Br + [azetidine (NH)] $\xrightarrow{\text{50°C, Sealed tube}}$

(j) [3-oxo-1-oxaspiro[3.5]... oxetanone spirocyclohexane] $\xrightarrow{\text{NaBH}_4}$

(k) $m\text{-}C_6H_5CH_2\overset{\overset{\displaystyle O}{\|}}{C}HN$—[penam, 2,2-dimethyl thiazolidine/β-lactam] $\xrightarrow{\text{HN(C}_2\text{H}_5)_2,\ \text{CH}_3\text{OH}}$

(l) [epoxide: H, ϕ, Cl, H substituents] $\xrightarrow[\text{H}_2\text{S}]{\text{Ba(OH)}_2}$

(m) [2-methyl-4-(benzylamino)butanoic acid: CH_3, $HOOC$, $NHCH_2C_6H_5$] $\xrightarrow[\text{(C}_2\text{H}_5)_3\text{N, CH}_3\text{CN}]{\overset{\overset{\displaystyle O}{\|}}{\text{Ph}_2\text{PCl}}}$

(n) [1,2-dihydronaphthalene] $\xrightarrow[\text{ether}]{\text{ClO}_2\text{SNCO}}$

(o) [2-thia-spiro thietane cyclohexane] + I_2 $\xrightarrow{\text{CCl}_4}$

(p) [2-phenyloxetane, C_6H_5] + CH_2MgBr $\xrightarrow{\text{ether}}$

(q) [2-phenyloxetane, C_6H_5] + CH_3COCl $\longrightarrow$

(r) $\underset{H_3C}{\overset{HO}{\diagdown}}CHC \overset{\overset{O}{\parallel}\ \overset{O}{\parallel}}{-}COCOC_2H_5 \quad \xrightarrow{OH^-}$

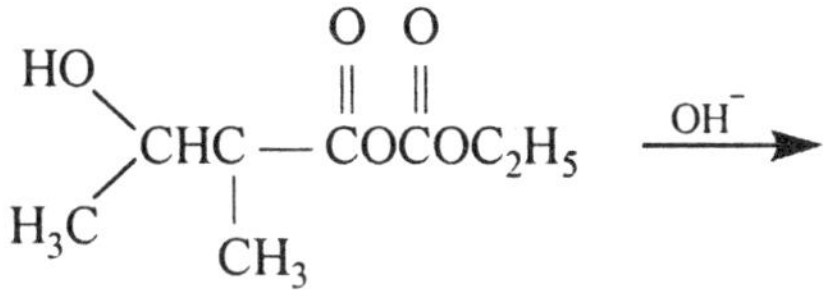

with CH_3 substituent

(s) $NCH_2C(CH_3)_2CH_2Cl \quad \xrightarrow{Base}$

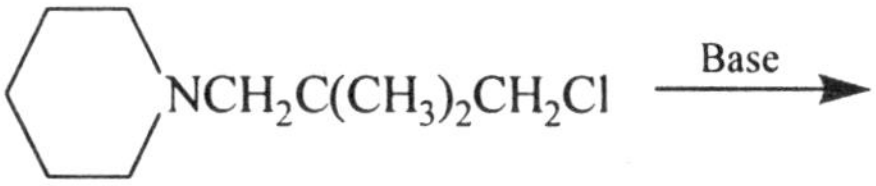

(t) $+ \ BrCH_2COOC_2H_5 \quad \xrightarrow{Zn}$

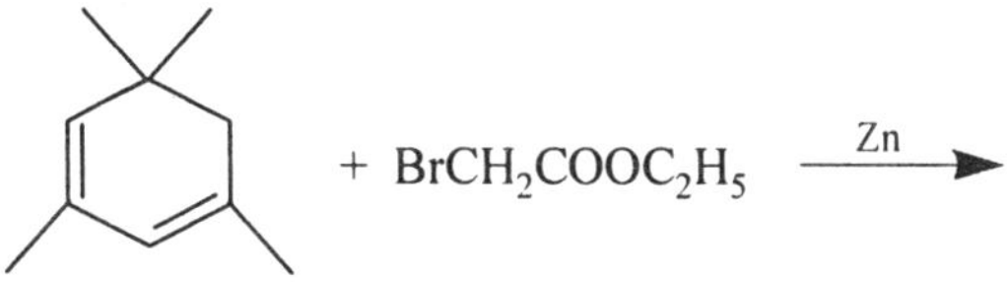

(u) $O \diagup\diagdown S \ + \ H_2O_2 \longrightarrow$

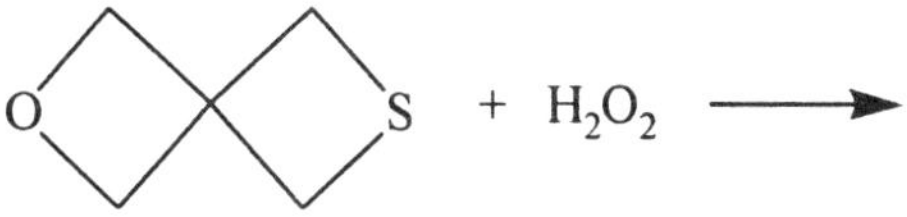

(v) $\xrightarrow[\text{(ii) Ni(CO)Cl}]{\text{(i) LiI, THF}}$

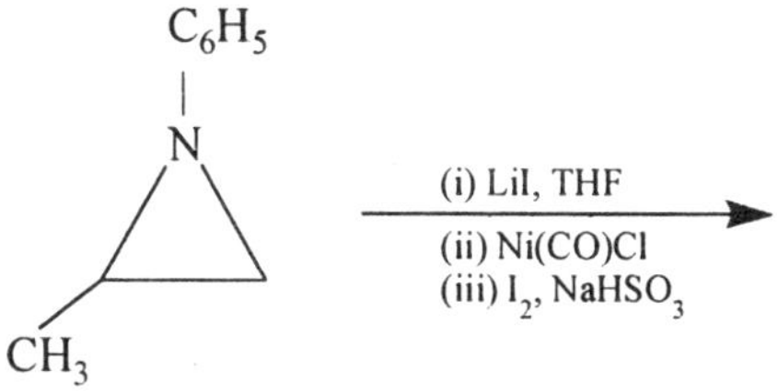

(iii) I_2, NaHSO$_3$

(w) $\xrightarrow{CH_2NH_2}$

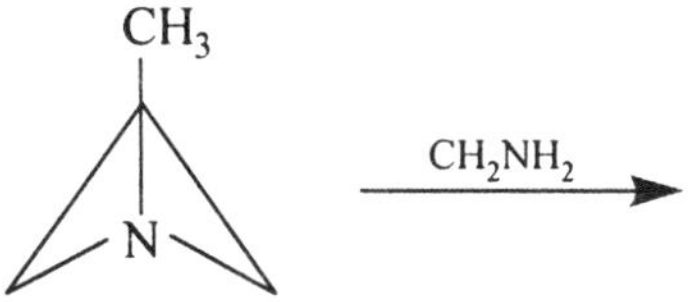

4.2 Write a probable mechanism for each of the following reactions:

(a) $\xrightarrow[\text{2hr}]{\text{Aq HCl, 0°C}}$ with $-CH_2OH$ and Cl

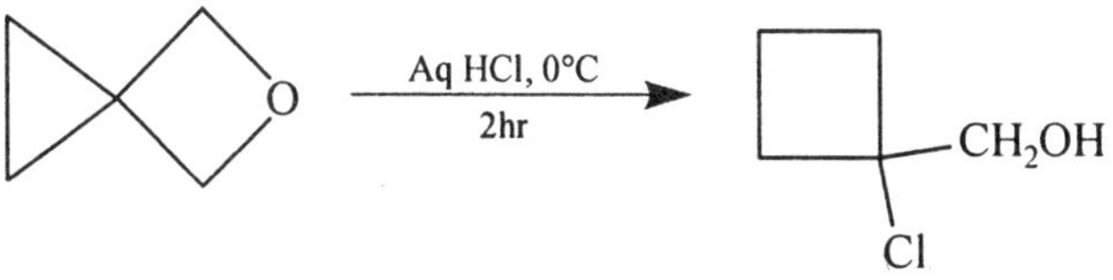

(b) $\xrightarrow[\substack{\text{Non protic} \\ \text{solvent}}]{hv}$ with CH_3

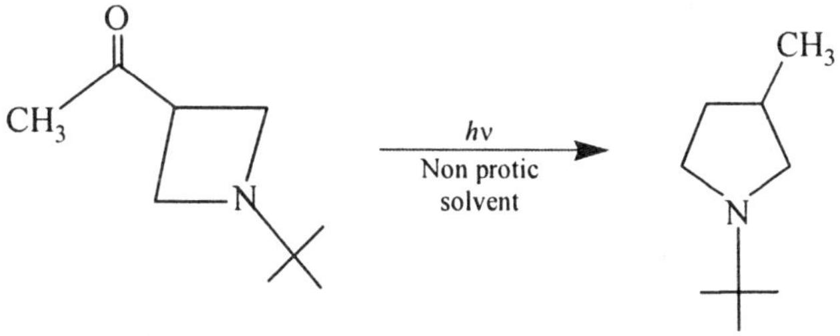

(c) $\xrightarrow[\text{(ii) H}^+]{\text{(i) NaBH}_4}$

(d) $\xrightarrow{\text{OH}^-}$

(e) $\xrightarrow[\Delta]{\text{Polar solvent}}$

(f) $\xrightarrow[\text{Acetone}]{h\nu}$

(g) $\xrightarrow[\text{Ethanol}]{h\nu}$

(h) $\xrightarrow{\text{OH}^-}$

(i) $H_2NCH_2CCOOH \xrightarrow{\text{HNO}_2}$

4.3 Offer explanation for the following observations:

(*a*) Ring opening of oxetane by nucleophiles is slower than that of oxirane.

(*b*) A three-membered ring is easier to form than a four-membered one, although the former has a larger ring strain.

4.4 Write the product and discuss the mechanism of the reaction:

$$\text{structure} \xrightarrow[\text{10°C}]{\text{1\% BF}_3,\ \text{CH}_3\text{COOH}} \ ?$$

4.5 Which ring is cleaved more readily, an aziridine or azetidine? Explain with examples.

4.6 Discuss the ring opening reactions of 2-methyloxetane.

4.7 Illustrate by an example how an isoxazole derivative can be used to prepare azetidine.

4.8 Write a short note on the Paterno-Büchi reaction.

4.9 What does a higher dipole moment value of oxetane (2.01 D) than thietane (1.47 D) imply? What are the consequences of this difference?

4.10 Discuss the formation of the product when 2-phenythietane and 2-nitrophenylthietane are treated with acetyl chloride.

4.11 Suggest a synthesis for penicillin G.

4.12 Discuss the ring opening reactions of β-lactam by various reagents.

4.13. Suggest a synthesis for 2-phenylcyclohexanone starting from phthaloyl peroxide.

4.14 Which is more easily hydrolyzed – an amide linkage in a β-lactam or in an acyclic amide? Why?

4.15 Effect the following conversion:

Five Membered Heterocyclic Compounds with One Hetero Atom

The parent five-membered heterocyclic rings namely *pyrrole* (1), *furan* (2) and *thiophene* (3) are formally derived from benzene by replacing the two – CH groups with one hetero atom, i.e., N, O or S respectively, Pyrrole and thiophene occur naturally whereas furan arises from the decomposition of sugars. The structures of these heterocycles would suggest that they have highly

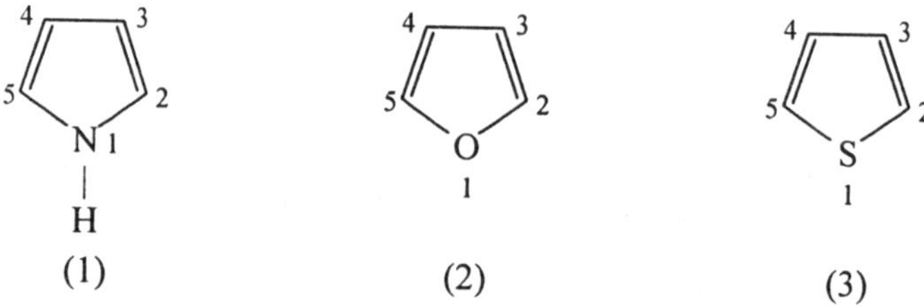

reactive diene character analogous to cyclopentadiene. Rather these compounds give reactions which suggest that they possess considerable aromatic character like benzene. These molecules are characterized by a high degree of reactivity towards substitution by electrophilic reagents rather than addition and they also show the effect of a ring current in their *n.m.r.* spectra. From the molecular orbital standpoint these molecules are described as consisting of planar pentagon with sp^2 hybridized carbon atoms. Each ring atom has one electron remaining in the p_z orbital while each hetero atom contributes two such p- electrons to the aromatic sextet.

The aromatic character of these heterocyclic rings may also be expressed using resonance structures which demonstrate that a pair of electrons from the hetero atom is delocalized into the ring as represented below:

(4)   (4a)   (4b)

(4c)   (4d)

These heterocycles are thus endowed with considerable aromatic character and possess high resonance energies.

	N	N (H)	O	S
36	23	21-2	15-8	29

Resonance energy (Kcal/mole)

The resonance energy is a very difficult property to measure for simple heterocyclic compounds and for this reason there is a wide range of values published for each of them in the literature. Nevertheless, the aromaticity falls in the order, thiophene > pyrrole > furan, an order which parallels the calculated resonance energies of these compounds.

Since the electronegativities of the hetero atom is in the order oxygen > nitrogen > sulfur, resonance structures (4a)-(4d) are less important in the case of furan relative to pyrrole and thiophene. Oxygen atom is reluctant to release its electrons and as a consequence, furan is the least aromatic of the three heterocycles.

The compounds of the general formula (4) are considered to be π-excessive, i.e., rich in electrons because the five sp^2 hybridized atoms sustain a 6π-electron system. In other words, there are six electrons in five orbitals.

Additional evidence to support the delocalized structure in these rings is derived from the dipole moments of these heterocycles compared to their non-aromatic counterparts. In the saturated compounds the dipole points towards the hetero atom. As a result the net dipole moment of furan and thiophene is reduced. Recent *n.m.r.* studies also indicate that the hetero atom is at the negative end of the dipole in furan and thiophene since the dipole moments of these compounds are smaller than their respective tetrahydro derivatives, forms

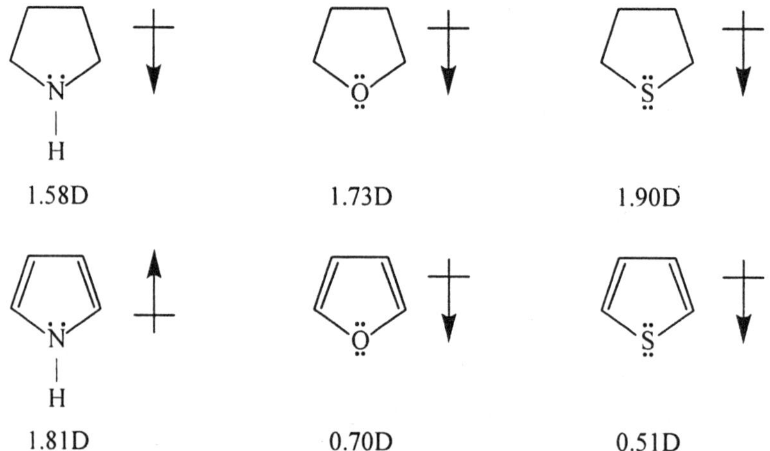

(4a) to (4d) must counteract the inductive effect of the hetero atom to some extent. In pyrrole, on the other hand, the dipole always points towards the ring away from the nitrogen atom whereas in the saturated analogue, pyrrolidine it points towards the nitrogen.[1] This evidence indicates that the partial delocalization of the lone-pair of electrons on nitrogen in pyrrole lies over the ring system as a whole. Because of this the electron-pair on the nitrogen atom is not available for donation and pyrrole loses its basic character, a property usually associated with organic amines. Furan reacts violently with acids while in thiophene the extra electron-pair can become coordinated with acidic reagents in certain cases without the destruction of the aromatic character of the ring. All these five-membered heterocyclics behave differently towards the familiar Diels-Alder reaction.

The mass spectrometric analysis of the three five-membered heterocycles exhibits interesting fragmentation patterns. In pyrrole the uneven valence and even atomic weight of the principal isotope ^{14}N produces a molecular ion of uneven mass unless, of course, nitrogen carries a substituent.

Similarly oxygen has only one principal naturally occurring isotope. Sulfur, on the other hand, has a natural isotope distribution ^{32}S/^{34}S of 25:1 and thus ensures two molecular ions for thiophene, two mass units apart of appropriate intensity ratio. The principal fragmentation patterns for the three heterocyclics are depicted below:

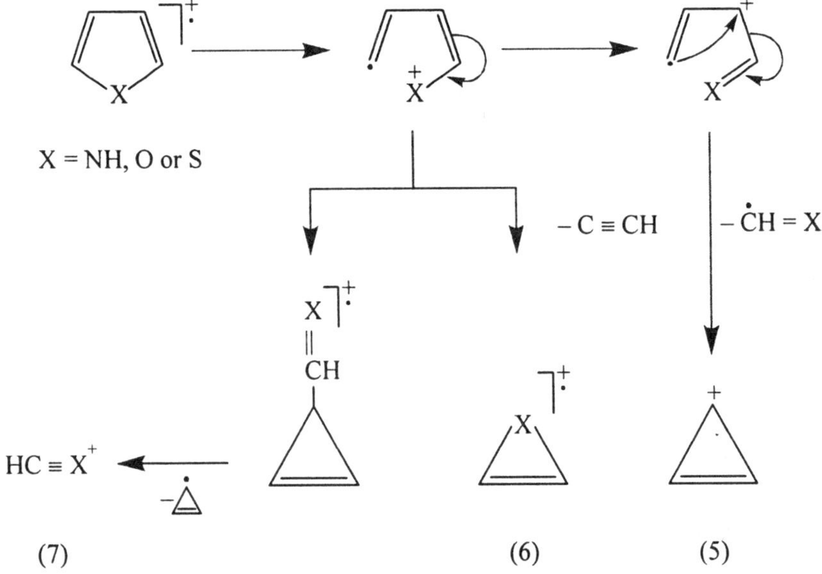

It has been noticed that the molecular ions of pyrrole and thiophene are the base peaks in their respective spectra, whereas the molecular ion of furan is strongest peak (70%) after the cyclopropenyl cation (5) which rather constitutes the base peak. The same ion, i.e. the cyclopropenyl cation is also an important feature of the spectrum of pyrrole but less so in the spectrum of thiophene. An additional fragment which is of significance for pyrrole and thiophene but absent for furan is (6) formed by the loss of acetylene from the molecular ion. Another ion (7) is much less abundant with furan than for pyrrole and thiphene.

Similarly mass spectrometric spectra are obtained for 2- and 3-alkyl derivatives of these heterocyclic compounds. Besides modest contributions from ions corresponding to (5) to (7) above, a principal fragmentation pathway is initiated by β-cleavage of the alkyl substituent.

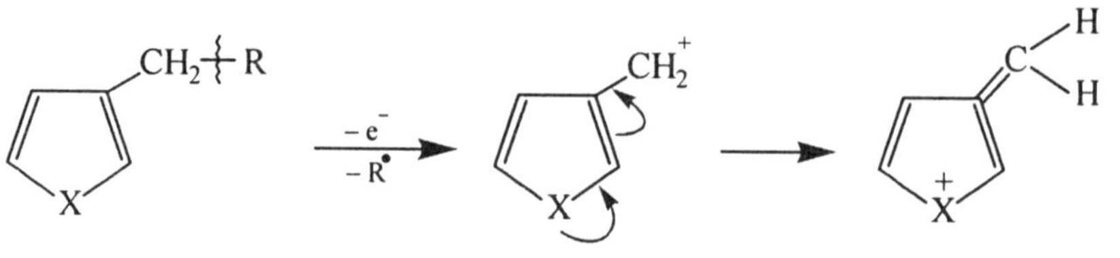

(9)

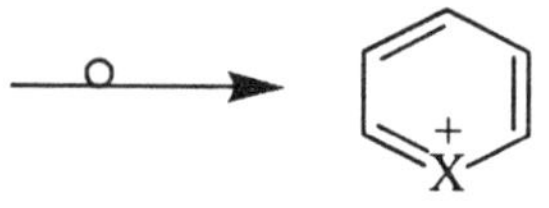

(10)

It is believed that the ions (8) and (9) which result after β-cleavage tend to rearrange to a common ion (10) which is generally the base peak.

5.1 TAUTOMERISM IN FIVE MEMBERED HETEROCYCLICS

Tautomerism is defined as a phenomenon in which two or more isomeric structures exist in dynamic equilibrium with each other, i.e. the energy barrier between them is small. Tautomers readily interconvert but isomers (isomerism) do so much less readily. Thus the following two forms of carboxylic acid are tautomers (11) whereas 1-butene and 2-butene are isomers which can be separated.

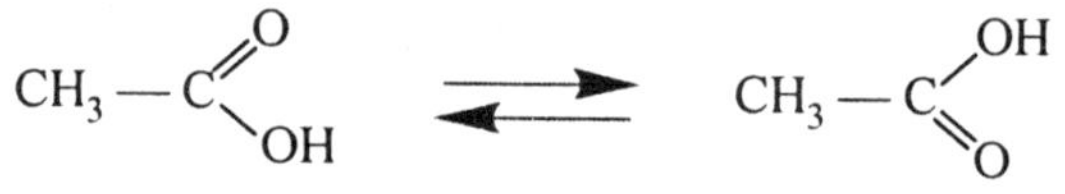

(11)

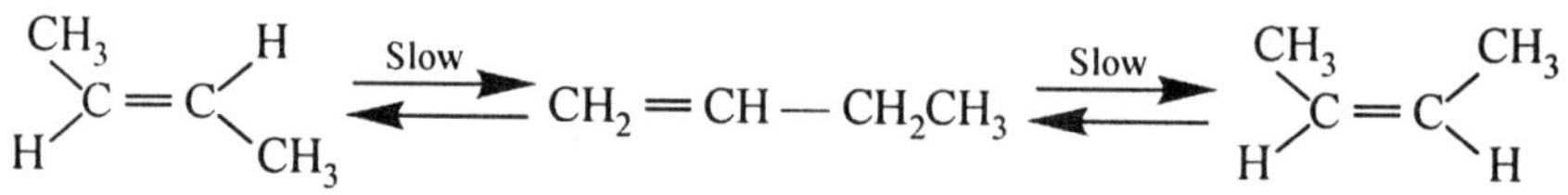

trans   *cis*

In other words, the latter molecules exhibit isomerism. We will be concerned mainly with *prototropic tautomerism* where a proton moves from one position to another. A methyl group can also migrate from one site to another but the activation energy for such a migration is very high. In heterocyclic chemistry two types of tautomerism have been recognized namely annular and side-chain. In the former, the atom or atomic group is exchanged between the ring carbon - or hetero-atom while in the latter, the exchange takes place between a ring and a side-chain atom. Reactivity and reaction mechanism in heterocyclic chemistry are better rationalized if the tautomeric structures of the compounds are known.

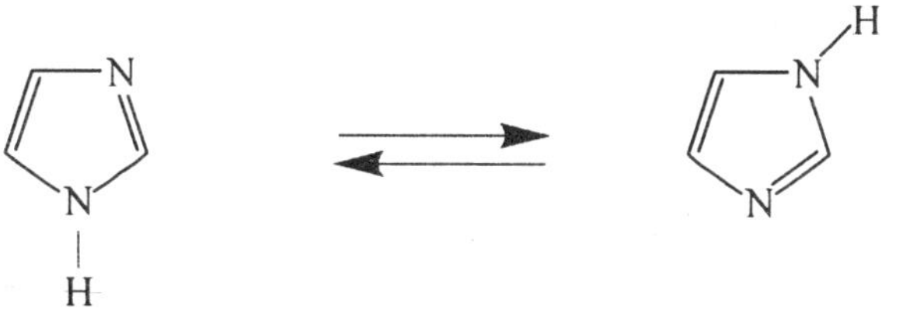

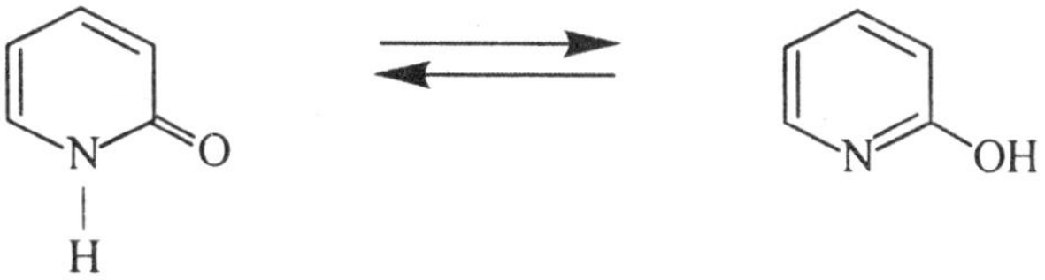

For structural reasons tautomerism not involving a functional group is permitted in pyrroles only but not in furans and thiophenes. For pyrrole (12), the pyrrolenine forms such as (12a) and (12b) are highly destabilized in annular

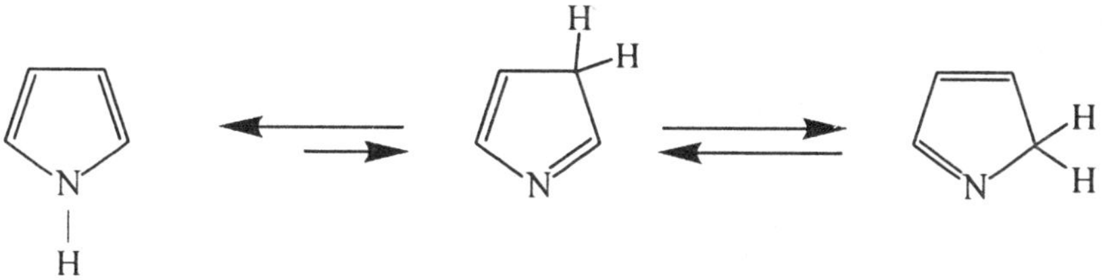

(12)   (12a)   (12b)

tautomerism and thus highly unfavorable. In these hetero aromatic rings the tautomeric equilibria almost always involve at least one non-aromatic tautomer. Extensive investigations have demonstrated that 2-hydroxypyrrole (13) exist as Δ^3 - pyrrolin - 2- one (13b) in ring chain tautomerism. The two tautomeric forms (13a) and (13b), have been prepared and their structures confirmed by *n.m.r* analysis. It has further been shown that (13a) and (13b) are in

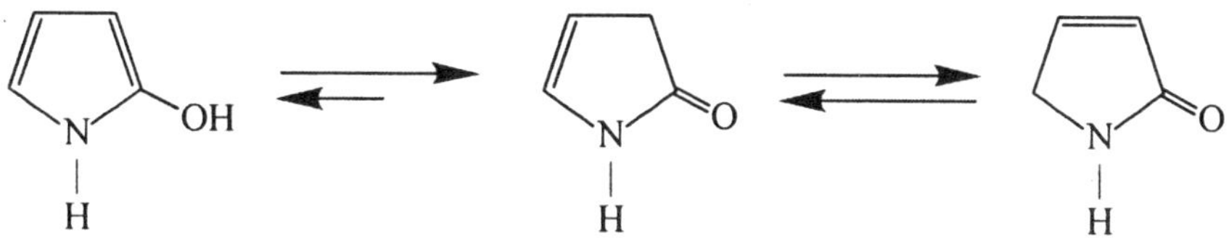

(13)   (13a)   (13b)

equilibrium at room temperature in a polar solvent. In 2 - hydroxyfuran[14] similarly the 2 - oxo form and usually the Δ^3-oxo (14a) form is more stable than the Δ^4 - oxo form (14b) because of conjugation of double bond with the carbonyl group in (14a).

(14)   (14a)   (14b)

2-Hydroxythiophene also involves a complex pattern of tautomerism with Δ^3 - and Δ^4 - thiolen - 2 - ones.

2 - Aminopyrrole as well as 2 - aminothiophene exist in the amino forms. Same is the case with 2-pyrrolethiols, 2 - mercaptofuran and 2 - mercaptothiophene.

5.2 PYRROLES

Pyrrole occurs in bone oil, coal tar and in products derived from proteins. It is also one of the most ubiquitous throughout the plant as well as animal kingdom because of its involvement as a sub-unit of haem, the chlorophyll, vitamin B_{12} and some bile pigments. Pyrrole itself was first obtained by Runge in 1834. Subsequently Anderson obtained pyrrole in 1857 from coal tar and was later synthesized by heating the ammonium salt of mucic acid. With *p*-dimethylaminobenzaldehyde, pyrrole gives an intense red color, this is referred to as *Ehrlich test* and is regarded as characteristic of pyrroles.

In naming pyrrole and its derivatives, the nitrogen atom is assigned position-1. The position of the substituent may be specified either in Arabic numerals or in Greek letters.

(11)   (15)   (16)

The radical (16) is named as 3-pyrryl. Pyrrole may tautomerize to 2*H*-(17) and 3*H*-(18) pyrroles and derivatives of both these structures are known.

(17)   (18)

2*H*-Pyrrolenine   **3*H*-Pyrrolenine**

The partially saturated dihydropyrroles are called pyrrolines, three (19-21) of these namely Δ^1-, Δ^2- and Δ^3- pyrrolines are possible. The fully saturated tetrahydropyrrole is designated as pyrrolidine (22).

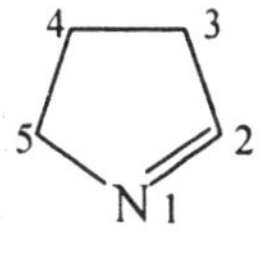

(19)	(20)	(21)	(22)
Δ^1-pyrroline	Δ^2-pyrroline	Δ^3-pyrroline	Pyrrolidine
·or	or	or	
3,4-Dihydro-*2H*-pyrroline	4,5-Dihydro-*2H*-pyrroline	2,5-Dihydro-*2H*-pyrroline	

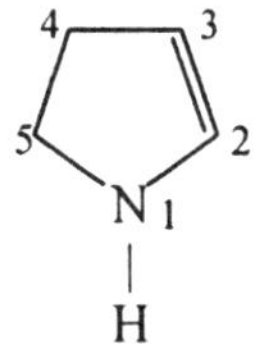

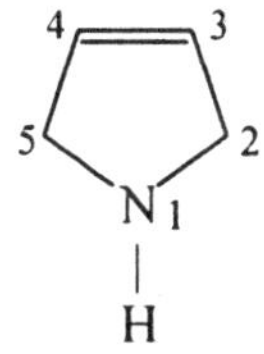

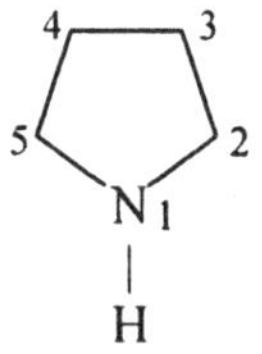

5.2.1 Physical and Spectroscopic Properties

Pyrrole is a colorless liquid, b.p., 130°C. It has odor resembling that of chloroform. The boiling point of pyrrole is higher than that of furan (32°C) and thiophene (84°C). It seems likely that pyrrole forms intermolecular H-bond and the various models (23-25) have been suggested.[2]

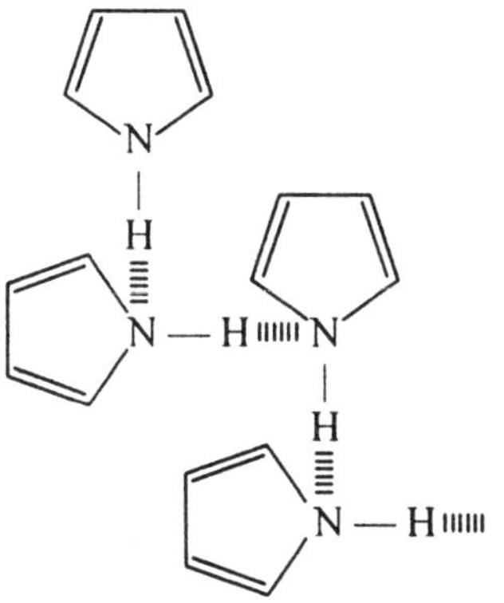

(23)	(24)	(25)

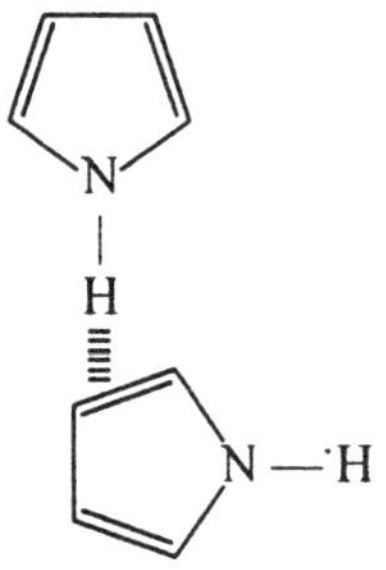

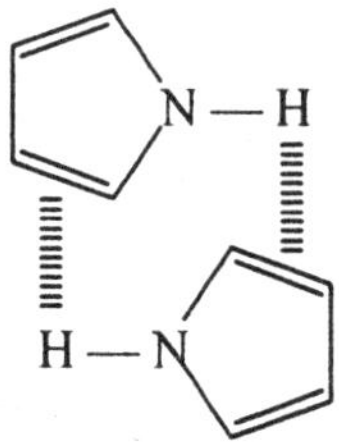

There is also a clear evidence from *i.r.* spectra and from other physical techniques that association occurs between pyrrole molecules due to H-bonding of the N-H group of one pyrrole with the π-electron system of another (models 24 and 25).

Pyrrole slowly turns brown on exposure to air. It is slightly soluble in water but freely soluble in most organic solvents. Pyrrole itself is completely planar and the molecular dimensions are obtained from electron diffraction studies.[3] These data are in accordance with the aromatic character of pyrrole.

The 2, 3- and 4, 5- bond distances are about 1.371 Å which are greater than that of C—C bond in ethylene (1.34Å) but smaller than the C—C bond distance of 1.54Å in saturated compounds. The 3,4- bond distance is of the order of 1.42Å. This value is close to the C—C bond distance of 1.39 Å in benzene. In short, the bond distances are between C=C and C—C bond lengths, and thus supports a resonance hybrid structure for pyrrole. However, the conjugation in pyrrole is less perfect because the bond angles are less than 120° as in benzene. The resonance energy is approximately 21 Kcal/mole which is also considerably less than that of benzene.

The five sp^2 hybridized C atoms sustain a 6π-electron system. From the molecular orbital standpoint pyrrole is considered as consisting of planar pentagon with sp^2 hybridized carbon atoms. Each of the four carbon atoms has one electron remaining in a p_z orbital. The nitrogen atom has two electrons in the p orbital. These p orbitals overlap to give a total of six electrons in the π orbital. These p orbitals overlap to give a total of six electrons in the π system (aromatic sextet) as for benzene. This closed shell provides stability to the ring.

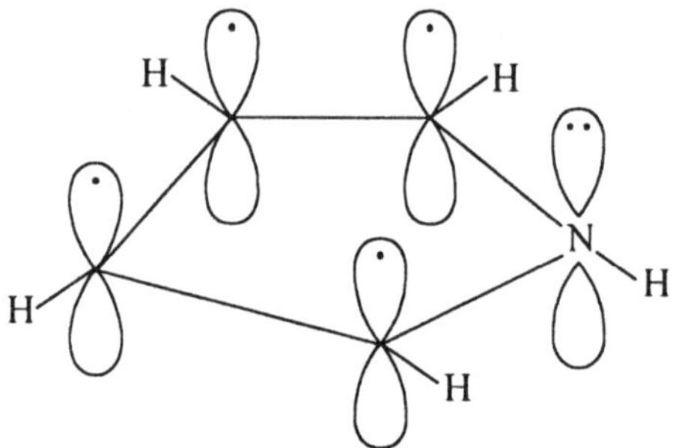

Since the electron pair is involved in the aromatic sextet, it is not available for protonation. It is consequently a very weak base, pKa 0.4. It has *pseudo* acidic property and forms salts with strong bases. The alkyl substituted pyrroles have greater basic strengths. The fully saturated pyrrole, i.e. pyrrolidine is highly basic (pKa 11.1).

The dipole moment of pyrrole has been found to be 1.81 D in non-polar solvents such as cyclohexane or benzene. 1-Methylpyrrole has a value of 1.92 D.

Electron density calculations by the valence bond and molecular orbital methods[4] confirm that the π-electron density is higher at C_2 and C_5 than at C_3 and C_4. This implies that electrophilic attack in pyrrole takes place preferentially at the former two positions. This deduction is in accordance with the several expected observations as we will see later.

Pyrrole derivatives exhibit chirality due to restricted rotation about the pivot C—N bond. 1-Phenylpyrrole derivatives bearing bulky groups in the *ortho* position can be resolved into enantiomers. For instance, 1- (2-carboxyphenyl) - 2, 5-dimethylpyrrole-3-carboxylic acid has been resolved into enantiomeric forms (26) and (26a).

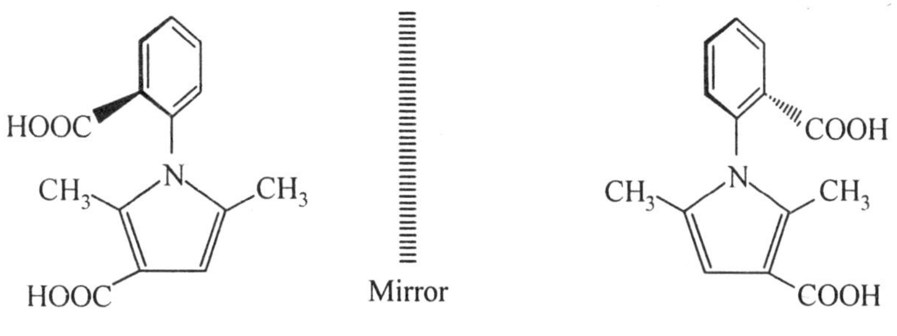

(26)   Mirror   (26a)

5.2.2 Synthetic Methods

The synthesis of the pyrrole ring system has been achieved in a variety of ways.

1.   *From Furans:* Commercially pyrrole is obtained by fractional distillation of coal tar or bone oil. Alternatively it can be obtained from furan by passing it over ammonia and steam and heated (400°C) in the presence of aluminum oxide catalyst. A primary amine may be employed to prepare 1-substituted pyrroles; equation (5.1).[5]

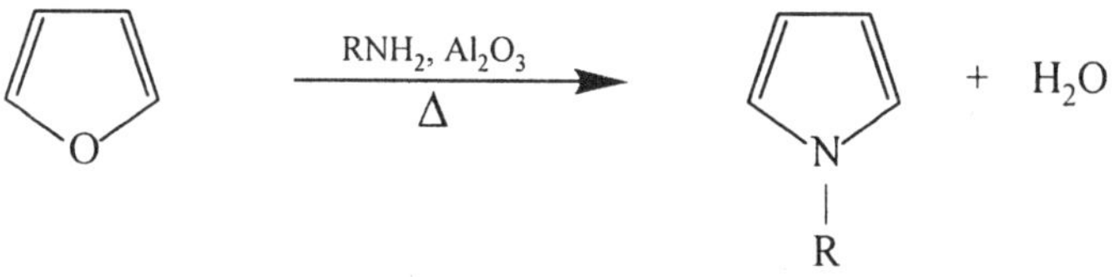

$$\text{furan} \xrightarrow[\Delta]{RNH_2, Al_2O_3} \text{pyrrole} + H_2O \qquad (5.1)$$

2.   *From Ammonium Mucate:* A classical method to prepare pyrrole consists of heating ammonium mucate. This method was first described by Schwartz in 1860 but was improved later. The ammonium salt dissociates into the free acid (27) which dehydrates followed by decarboxylation and cyclization in ammonia yields the parent compound. This method was applied[6] for the preparation of N-substituted pyrroles but the yields were found to be disappointing.

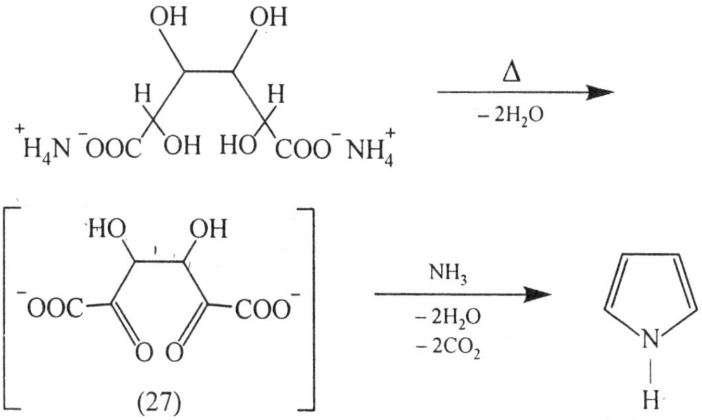

3. ***Ring Closure Methods:*** A number of methods starting from appropriate components and resulting in ring closure to pyrrole or its derivatives are known.

(***i***) ***The Knorr Synthesis:*** This constitutes the most important and widely used method of pyrrole synthesis. Originally it involved the condensation of an α-amino ketone or an ester with another dicarbonyl compound carrying an active methylene group in the the presence of acetic acid. 3, 5-Dicarboethoxy-2, 4-dimethylpyrrole was obtained by this method from α-aminoacetoacetic ester and acetoacetic ester as depicted. The precise mechanism of the reaction is not fully understood, however, it seems that the dicarbonyl component first reacts with the amino group of the second component and the pyrrole ring is formed in subsequent steps. The α-amino ketone has a tendency to react by itself, therefore, its methylene group should be sufficiently acidic.[7]

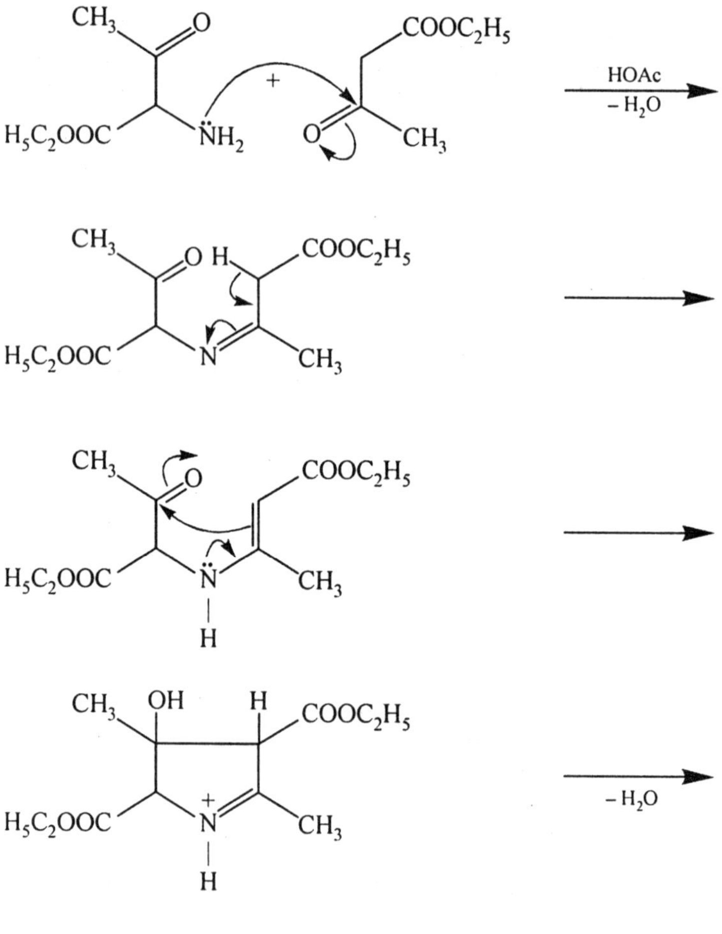

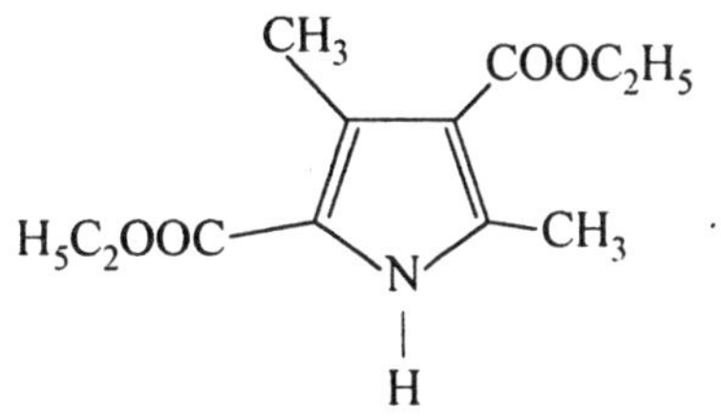

(ii) ***The Paal-Knorr Synthesis:*** This is the most general method[8] and it involves the condensation of a 1, 4-diketone with ammonia or a primary amine. Pyrrole itself is formed from succinaldehyde and ammonia whereas 2, 5-dimethylpyrrole is obtained from acetylacetone and ammonia (used as ammonium sulfate). The yields of pyrroles in general, are very good. This method is widely applicable for the preparation of substituted pyrroles.[9, 10]

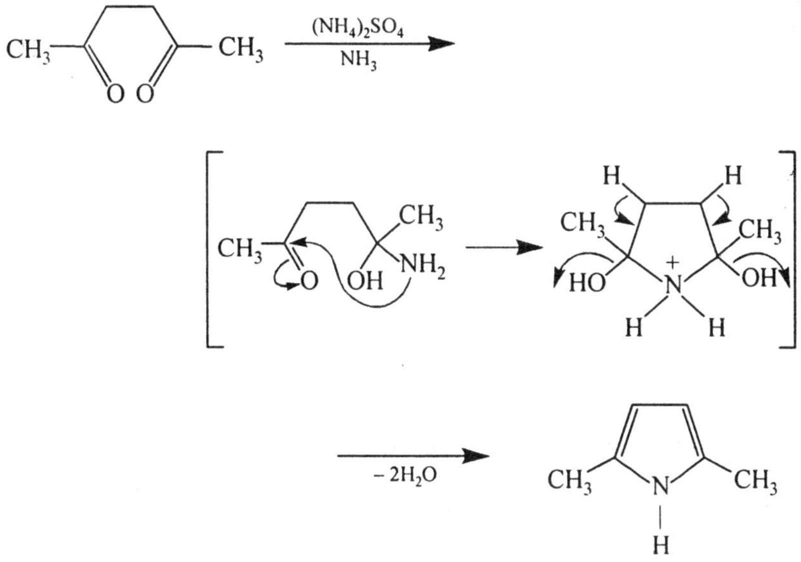

(iii) ***The Hantzsch Synthesis:*** In this method cyclization to a pyrrole takes place by reacting an α-haloketone or aldehyde with a β-keto ester (or β-chloroketone) in the presence of a nitrogen containing base such as ammonia or an amine. The base functions both as a reactant and as a catalyst. The yields of pyrroles are moderate to good.

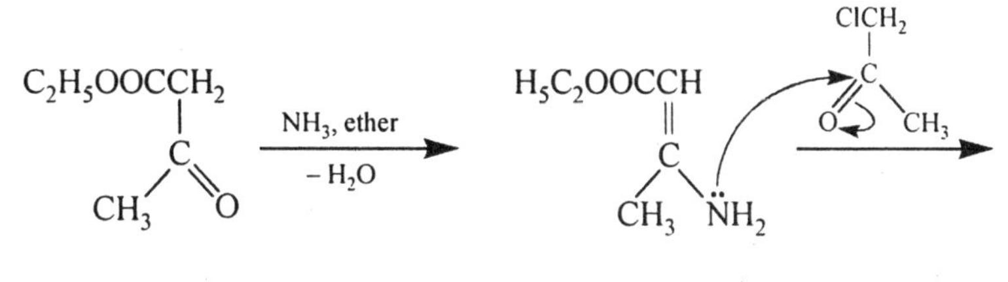

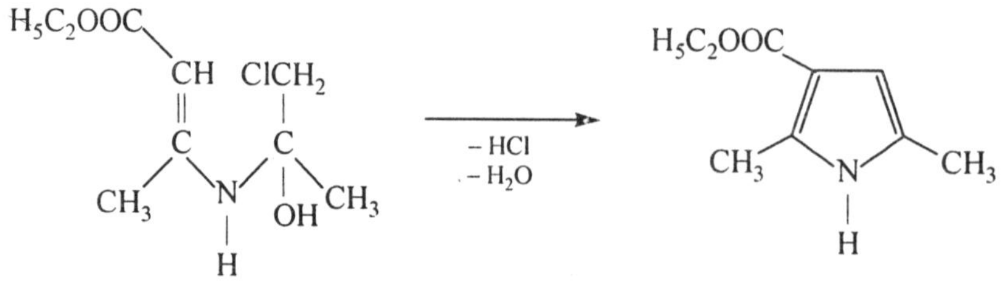

The scope of this method has been recently extended by McDonald[11] by the use of α-halo derivatives of other aldehydes.

(*iv*)The Knorr synthesis affords a yield of about 50%, therefore, to augument the yield Khimura and coworkers[12] have suggested the following

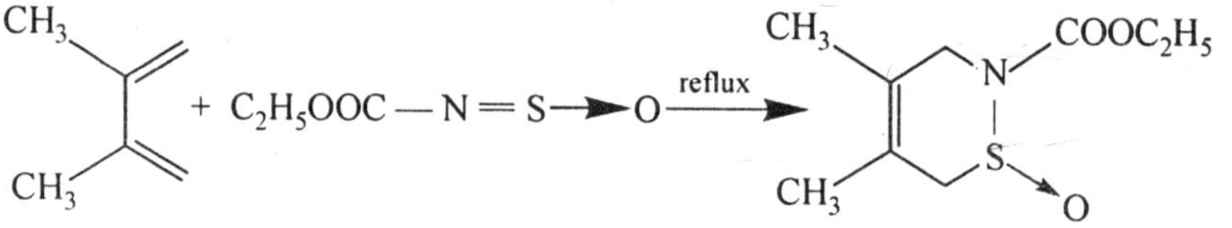

2,3-Dimethylbutadiene

2-Ethoxycarbonyl-3, 6-dihydro-4,5-dimethyl-1, 2-thiazine-1-oxide

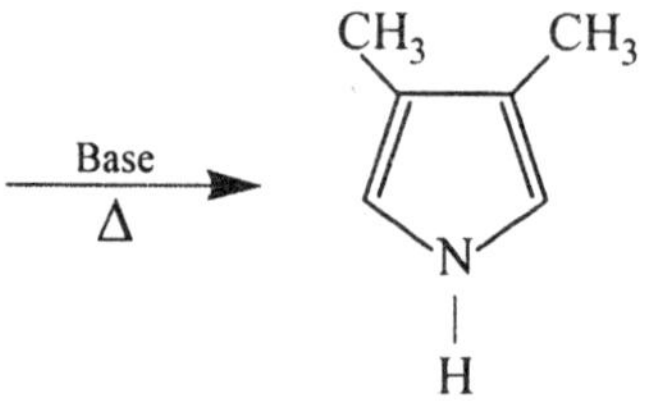

sequence of reactions to obtain pyrrole derivatives from 2,3-dimethylbutadiene and ethyl N-sulfinyl carbamate. Initially a Diels-Alder adduct is formed which decomposes in the presence of base to form 3,4-dimethylpyrrole.

4. *From Substituted Azetidines:* Substituted azetidines rearrange photochemically to pyrroles[17] as depicted below:

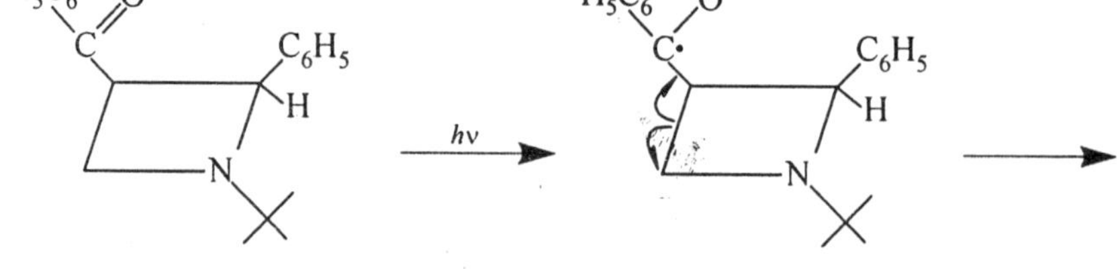

1-*t*-Butyl-3-benzoyl-2-phenylazetidine

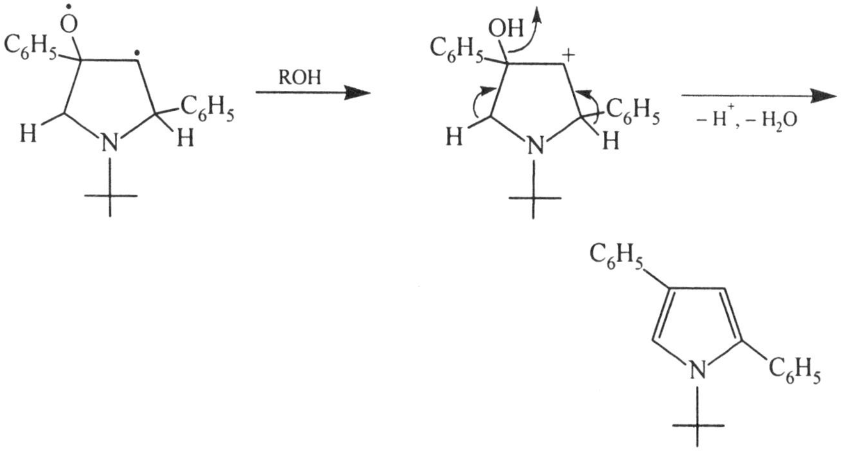

1-*t*-Butyl-2, 4-diphenylpyrrole

5.   *The Piloty-Robinson Synthesis:* This is the monocyclic equivalent of the Fischer-indole synthesis. This consists of treating ketazines with strong acids to give pyrroles through a [3,3] sigmatropic rearrangement of the tautomeric divinyl hydrazine.[14,15]

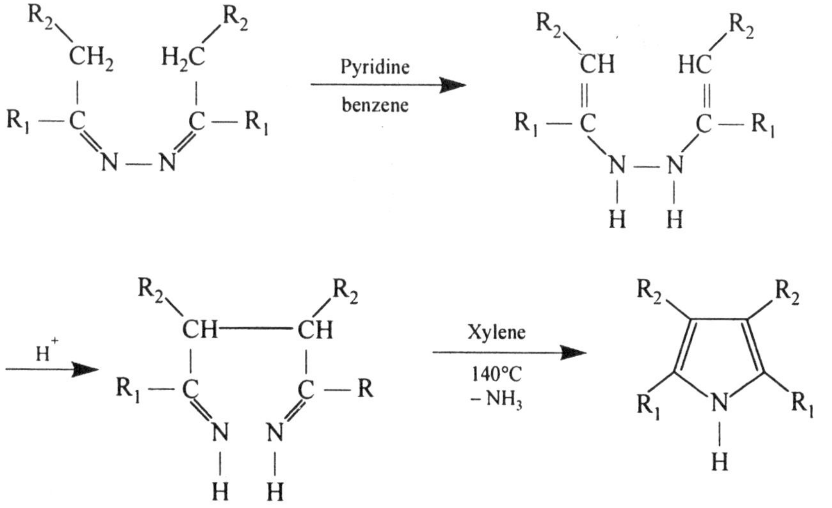

6.   *From Mesionic Ring Systems:* Mesionic ring systems (28) have been known to cycloadd to a number of electron-deficient dipolarophiles. The resulting adduct on heating yields a pyrrole derivative. For instance, the adduct obtained from anhydro - 2, 4-diphenyl-3-methyl-1, 3-oxazolium-5-oate and DMAD, loses carbon dioxide to afford the pyrrole derivative in an excellent yield.

7. *Synthesis of 3-Pyrroline:* 2, 5-Dihydropyrrole or 3-pyrroline (29) is an important starting material for pheromones. It has recently been prepared[18] in 65% yield by the reaction of *cis*-1, 4-dichloro-2-butene with hexamethylene tetramine followed by acid hydrolysis and finally reaction with diazabicyclo [5.4.0] undec-7-ene (DBU).

5.1.3 Chemical Reactions

Pyrrole undergoes reactions of electrophilic substitution, condensation and ring opening. It is very reactive and behaves both as an acid and a base. Its properties are somewhat similar to phenol.

1. *Protonation:* The proton attached to the nitrogen atom undergoes a rapid exchange in acid and alkali. Similar exchange of proton attached to the carbon atom occurs only in more acidic conditions, the α-protons exchange at twice the rate of the β-protons. Pyrrole on treatment with acids has been known to provide a mixture of polymers (pyrrole-red) but under controlled conditions a trimer can be isolated though in low yield.[19] Its formation has been explained in a sequence of several steps.

Two cations are possible by the protonation of pyrrole, the one formed by protonation at position-3 is involved in trimer formation.

2.   *Reaction with Bases:* The pKa for the loss of the N—H hydrogen of pyrrole is 17.5, which is larger than imidazole, therefore, the former is relatively a weak acid than imidazole. It is a weaker acid than phenol but corresponds in acid strength to ethanol. It reacts with potassium but not sodium to liberate hydrogen and to form the corresponding salt. The acidity of pyrrole can be enhanced by putting electron-withdrawing groups at the 3-position because in that case the anion can be stabilized by resonance.
In this respect it has some resemblance to phenol, in which case the phenoxide ion is stabilized by resonance.

3.   *Alkylation and Arylation:* The position of substitution in pyrrole alkylation depends on two factors, namely the nature of the cation (or catalyst) and the solvating capacity of the solvent. Potassium- and sodium-salts of pyrrole react with alkyl halides to give good yields of the corresponding N-alkyl pyrroles.[20] Pyrrole thallium salts[21] have instead been found to give better yields of N-substituted products, compared to pyrrole potassium salt. The metal salts unlike pyrrole also react with acyl chloride, ethyl chloroformate, etc. In general electron-withdrawing groups that favor the dissociation of the salt also favor N-alkylation.

N-arylation of electron-deficient pyrroles can be affected[22] by cross-coupling of the pyrrole with aryl boronic acid in the presence of copper(II) acetate.

92%

This is a case of C—N cross-coupling and has been the subject of studies in recent years.[23]

Alkyl- and benzyl-substituents in N-substituted pyrroles migrate to the 2- and 3-positions in the pyrrole ring at high temperatures (500-600°C). N-(1-phenylethyl) pyrrole on heating to 600°C, affords a mixture of 2- and 3-(1-phenylethyl) pyrrole equation (5.2).[24] This process constitutes a useful synthetic approach to such isomers. The mechanism is considered to be a homogeneous unimolecular process proceeding *via* a cyclic transition state. This has been confirmed by the large negative entropies of activation obtained for these reactions. Furthermore, the substitution at the 2-position is obtained from the N-substituent while that at the 3-position from the corresponding 2-position.[24a]

$$ (5.2) $$

4.   *Electrophilic Substitution:* Pyrrole is an π-excessive heterocycle, i.e., π-electron density is greater on its carbon atoms than in the benzene molecule. Various studies[25,26] have indicated that electrophilic substitution on π-excessive heterocyclic compounds occur in a similar manner to benzenoid compounds. The mechanism is also the same as suggested for benzene involving a "*Wheland intermediate*". Pyrrole, itself undergoes electrophilic substitution predominantly at the 2-position, however, if this position is blocked than substitution occurs at other positions. The preference for substitution at the 2-position may be explained by considering the following resonance structures:

It is seen that there are three resonance forms for substitution at the 2-position whereas there are only two forms for substitution at the 3-position. This can be further demonstrated from the energy profile diagram (Fig. 5.1) for substitution at the 2- and 3-positions in pyrrole.

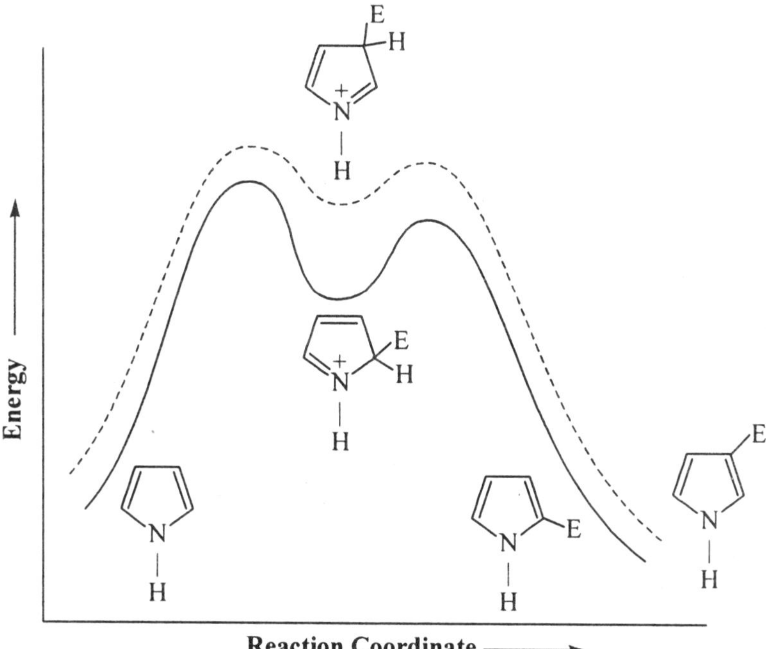

Reaction Coordinate ⟶

Fig. 5.1 Potential energy diagram for electrophilic attack on pyrrole.

Halogenation: Pyrrole is extremely reactive towards halogens,[27] chlorination (SO_2Cl_2), bromination (Br_2/CH_3COOH) and iodination (I_2/KI_3) yield the corresponding tetrahalo derivatives, equation (5.3). It is very difficult to prepare

$$\text{(5.3)}$$

the mono-halopyrroles except under very special conditions. The halopyrroles are very unstable compounds and decompose readily in air and light. In substituted pyrroles the vacant position is generally attacked during halogenation, equation (5.4).

$$(5.4)$$

Mono-halogenation also occurs in the presence of electron-withdrawing groups on the pyrrole ring.

Nitration: Electrophilic aromatic nitration[28] is usually carried out with nitric acid generally in the presence of sulfuric acid. This reagent causes extensive decomposition of the pyrrole ring resulting in the formation of tar. With acetyl nitrate ($Ac_2O + HNO_3$) at lower temperature ($-10°C$), 2-nitropyrrole is obtained in 55% yield, with some contamination of the 3-isomer.[29]

Sulfonation: Pyrroles form resinous material with sulfuric acid at ordinary temperatures but in the presence of a mild sulfonating agent such as pyridine sulfur trioxide complex at 100°C pyrrole forms, pyrrole-2-sulfonic acid in about 90% yield.

The Friedel Crafts Reaction: N-acylation of pyrroles is accomplished by reaction of the alkali-metal salts with acid chlorides. But pyrrole reacts rapidly in the Friedel-Crafts acylation with acid chloride even in the absence of a catalyst, to yield 2-acetylpyrrole (C-acylation). Acetic anhydride in the presence of an acid scavanger such as an amine also gives 2-acetylpyrrole, equation (5.5).

$$(5.5)$$

Acetylation in the presence of acetic anhydride and acetic acid results in 2-acetylpyrrole, whilst the addition of Lewis acids such as BF_3 or $ZnCl_2$ ensures C-acylation.

Gatterman-Koch Reaction: Five-membered heterocyclic rings are formylated under a variety of conditions, a behavior analogous to anilines and phenols. The original reaction using a mixture of anhydrous hydrogen cyanide and dry hydrogen chloride gas in ether or chloroform solution fails with pyrrole itself. Formylation of pyrrole may be achieved by heating pyrrole with phosphorus oxychloride and dimethyl formamide. The intermediate is hydrolyzed in the presence of a mild base to 2-pyrrolecarbaldehyde. This is known as the *Vilsmeier-Haack reaction.*[30,31]

The electrophilic reagent in the Vilsmeier reaction appears to be a chloroimmonium ion (31). The mechanism proceeds in the following steps:

5.   *Reaction with Dichlorocarbenes and Nitrenes:* Reaction of pyrrole with dichlorocarbene has generally been studied by generating the reagent in protic and basic media, and sometimes with dichlorocarbene generated by thermolysis of sodium trichloroacetate. In this case pyrrole gives a mixture of a ring expansion product[32] and 2-pyrrolecarbaldehyde.[33] On boiling pyrrole in chloroform with a base, a halocyclopropyl intermediate is initially formed by the addition of chlorocarbene[34, 35] to the 2, 3 C—C double bond. This intermediate then rearranges as shown to yield the final product.

3-Chloropyridine

2-Pyrrolecarbaldehyde

Though furan and thiophene undergo normal addition reaction with carbenes derived from copper- or light-catalyzed decomposition[36] of diazoacetic ester, but this ester reacts with pyrrole to give α-substituted pyrrolacetic esters, equation (5.6).

Ethyl pyrrole acetate

Ethoxycarbonylnitrene (obtained by the pyrolysis of azidoformic acid) gives rise to 2-amino-N-ethoxycarbonylpyrrole[37] according to the steps indicated below: The addition of nitrene yields homoazopyrrole which rearranges to a 2, 1-adduct and subsequent ring opening yields the product.

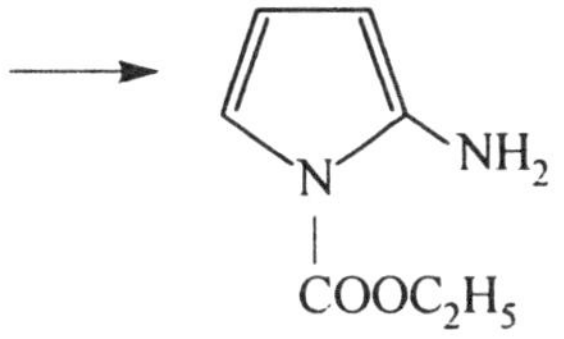

6. *Formation of 1-Azafulvene:* Flash photolysis of dialkylaminopyrroles and thermolysis of 2-pyrrylmethylphenyl sulfoxide at 65°C in solution leads to the formation of azafulvene.[38, 39] Thus compound (33) on heating yields 1-azafulvene which has been trapped by thiophenol.

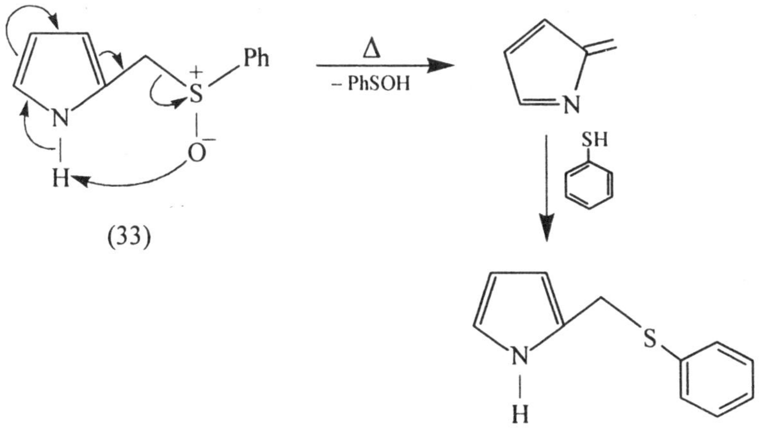

7. *Preparation of Indole:* 1-Methyl-2-vinylpyrrole serves as a typical starting material for the preparation of indoles.[40] This synthesis requires the following sequence. Reaction of 1-methyl-2-vinylpyrrole with methyl propiolate leads to the direct formation of 1-methyl-6, 7-dihyroindole-9-carboxylic ester.

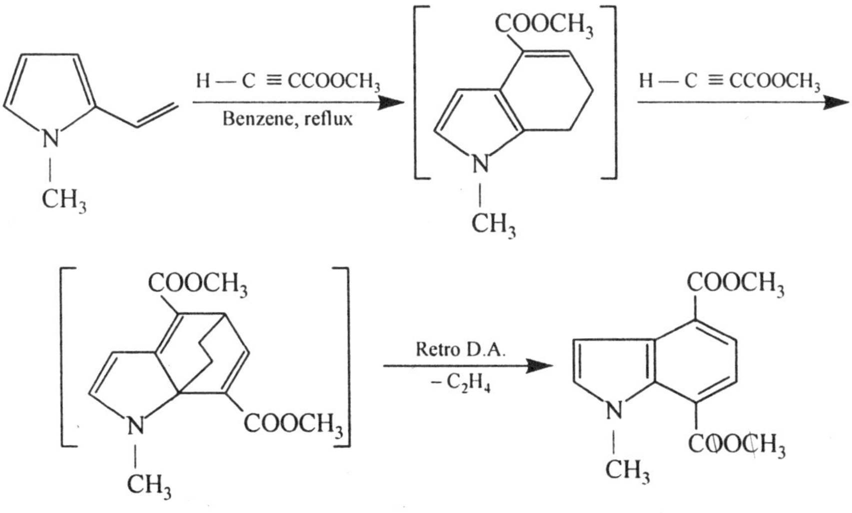

This compound reacts further with a second molecule of methyl propiolate *via* $(4\pi + 2\pi)$ cycloaddition. The adduct loses ethene by retro Diels-Alder extrusion to form an indole derivative.

8.   *The Diels-Alder Reaction:* Because of its aromatic character, pyrrole is less likely to function in reactions typical of a diene. Furan, however, behaves to be more diene-like. Depending on the nature of pyrrole and experimental conditions employed, it appears to follow two different pathways, i.e. a (4 + 2) cycloaddition or a Michael type addition at the α-position.[41]

When pyrrole reacts with benzyne, the only product isolated is 2-phenylpyrrole. Benzynes react with N-alkyl-or N-alkoxycarbonylpyrroles to afford azanorbornadiene derivatives (34) by the Diels-Alder type cycloaddition.

(34)

Pyrroles bearing electron-withdrawing groups at the 1-position tend to reduce the availability of the lone-pair to the ring and thus N-carbomethoxypyrroles will react with highly reactive dienophiles such as dimethylacetylenedicarboxylate (DMAD) to facilitate the formation of a 1:1 Diels-Alder adduct (35)[42]. The yields of the adduct are modest. The adduct (35) reacts further with DMAD to give another adduct (36)[43].

(35)

(36)

Typical dienophiles such as DMAD and maleic anhydride undergo Michael-type addition with pyrrole, equation (5.7) with pyrrole and its N-alkyl derivatives.[43-45]

$$(5.7)$$

The addition of Lewis acids considerably improves the yield of (4 + 2) cycloaddition products between 1-methoxycarbonylpyrrole and diethylacetylenedicarboxylate (DEAD). For instance, using aluminum chloride and DEAD in a ratio of 5:1 results in almost quantitative yield of the cycloadduct. This observation also suggests that the cycloaddition reaction is not concerted rather takes place in two discrete steps.

Recently[48,49] properly substituted pyrroles such as methyl 1-pyrrolylfumarate have been shown to undergo an intramolecular Diels-Alder adduct formation, equation (5.8).

$$(5.8)$$

9. *Reaction with Reducing Agents:* Pyrroles are not reduced by nucleophilic reducing agents such as Zn/CH_3COOH, lithium aluminum hydride, sodium ethoxide or sodium/liq ammonia but can be reduced in acidic media in which the species under attack is most probably the protonated cation. The products are Δ^3-pyrrolines.[50] Catalytic hydrogenation (Pt, Pd, Raney Ni) leads to the formation of pyrrolidine derivative.

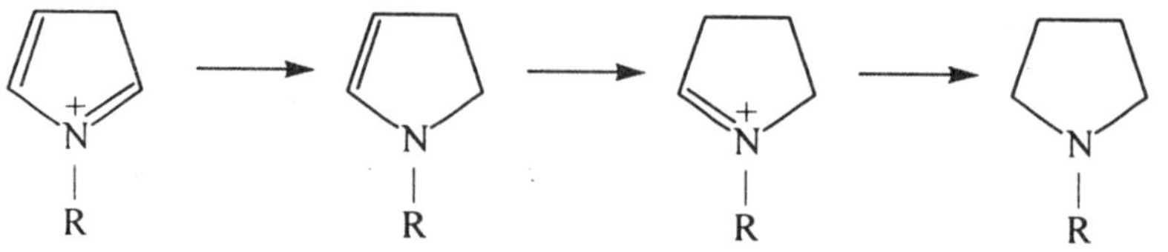

10. *Reaction with Oxidizing Agents:* Simple pyrroles are generally easily attacked by oxidizing agents. On standing in air and light, pyrrole autoxidizes to a reddish brown color. Ozonolysis of pyrroles[51] at low temperatures affords aldehydes in poor yields which have been attributed to 2, 3 - and 2, 5-addition reactions of oxygen followed by the complete breakdown of the ring. When the ring does survive under these conditions maleinimide derivatives are the common products (equation 5.9).

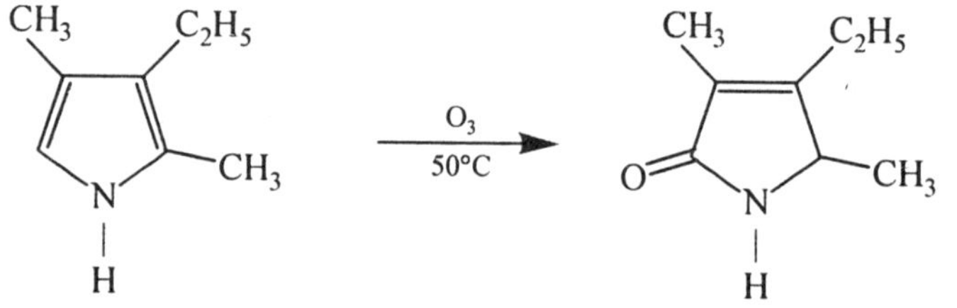

$$(5.9)$$

With strong hydrogen peroxide, pyrroles yield 2-oxo- Δ^3 -pyrrolines,[52] equation (5.10).

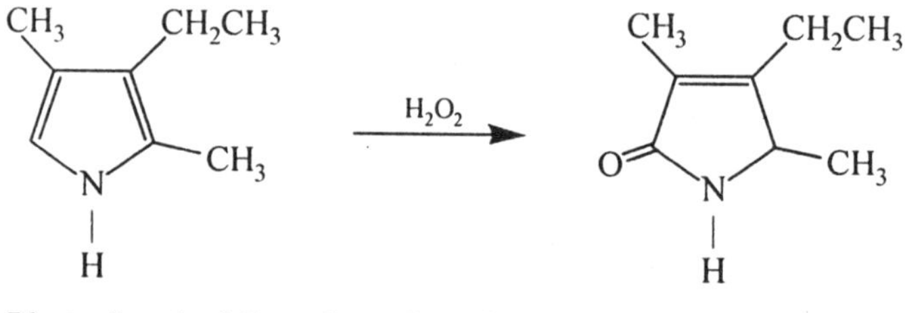

$$(5.10)$$

11. *Photochemical Reactions:* Pyrrole and its simple derivatives, 2, 4- and 2, 5-dimethylpyrroles have been irradiated in the gas phase but only degradation products were obtained.[53] In the condensed phase pyrroles isomerize in several ways and the most frequently observed reaction leads to 1, 2- and 1, 3-isomerization. Thus N-benzylpyrrole on irradiation in methanol solution or in the vapor phase gives a mixture of 2-benzyl- and 3-benzylpyrroles. It was also demonstrated that both these products were photostable.[54] N-acetylpyrrole, rearranges to 2-acetylpyrrole *via* a dipolar intermediate.[55]

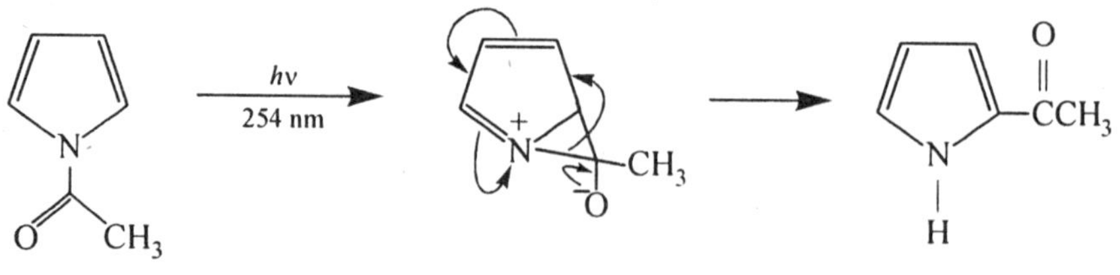

12. Under appropriate conditions N-methylpyrrole reacts with *n*-BuLi to yield a dianion rejecting the earlier beliefs that such metals almost exclusively attack the α-position. The first example of a nucleophilic attack in the pyrrole system has been reported. Thus 1-methyl-2, 5-dinitropyrrole undergoes substitution by methoxide ion or piperidine under milder conditions. There have been similar studies for thiophene system as well.[56]

5.2.4 Derivatives of Pyrrole

A number of side-chain derivatives of pyrrole are known and the chemistry of some of the important ones will be discussed here.

Alkyl- and Aryl-Pyrroles:[57] Alkyl pyrroles can be prepared by heating pyrrole with sod. alkoxide in an autoclave at 200-270°C. Pyrrole, for instance, yields 2-alkylpyrrole, but if the α-position is blocked then substitution takes place at the β-position. Several types of alkyl pyrroles have also been obtained by reducing the corresponding acyl pyrroles under Wolff-Kishner conditions, 1-alkyl-and 1-aryl-pyrroles also rearrange to give rise to 2-substituted pyrroles.

3, 4-*di*- *t*-Butylpyrrole has been prepared from 3, 4 -*di-t*-butylthiophene, on treatment with tosylnitrene with the resultant formation of N-tosyl-3, 4-*di*-t-butylpyrrole. Its hydrolysis yields the substituted pyrrole.[58] 3, 4-*di-t* Butylthiophene on the other hand, on oxidation with *m*-chloroperbenzoic acid yields 3, 4-*di t*-butylthiophene-1, 1-dioxide which on flash vaccum photolysis give rise to 3, 4-*di t*-butylfuran.

Alkyl pyrroles are generally more reactive and more basic than pyrrole itself. Chlorination of alkyl pyrroles initially gives rise to a product containing chlorine at all the unsubstituted positions in pyrrole and eventually the attack occurs at the 2-methyl group. It contrast bromination can be controlled in that the substitution ceases at the monobromo methyl stage. The nitration of 1-methylpyrrole (HNO_3/Ac_2O) gives a mixture of 2- and 3-nitropyrroles in a 2:1 proportion. 2-Phenyl-1, 3-oxazepine rearranges to N-formyl-2-phenylpyrrole among other products on heating[59] (equation 5.11).

$$\text{(5.11)}$$

Both 1-alkyl- and 1-aryl-pyrroles give the Michael type adduct with DMAD. A mild and effective method for obtaining N-acyl and N-alkylpyrroles is to carry out the reaction under phase transfer conditions.

Aminopyrroles: Not very many aminopyrroles are known. Substituted aminopyrroles have been prepared by reducing the corresponding nitropyrrole derivatives. There is little information available about the diazotization and coupling reactions of simple aminopyrroles.

Hydroxypyrroles: 2-Hydroxypyrrole exists in the tautomeric form, pyrrolone. It is colorless liquid, b.p. 75°C/0.05 mm. 3-Hydroxypyrrole has not yet been prepared.

Pyrrole Aldehydes and Ketones: These derivatives are stable compounds and thus do not autoxidize or polymerize. The carbonyl group in pyrroles is less reactive than that in aryl ketones. 2-Pyrrolecarbaldehyde is obtained from pyrrole *via* the Vilsmeier reaction.

Pyrrole aldehydes possess some general properties corresponding to aromatic aldehydes. The formyl group can be reduced to the corresponding alcohol by sodium borohydride or lithium aluminum hydride. Unlike aromatic aldehydes, however, they do not respond to Tollens' reagent or Fehling's solution. Furthermore, they do not take part also in benzoin, Perkin or Cannizzaro reactions. This failure has been ascribed to the deactivation of the carbonyl group through resonance:

Pyrrole ketones may be obtained by acylating pyrroles directly with acetic anhydride on heating. They may also be obtained by a modification of the Gatterman-Koch reaction. The pyrrole ketones undergo several reactions typical of ketones, for instance, they are reduced by lithium aluminum hydride, oxidized with NaOCl, and undergo condensation with hydrazine.

Pyrrole Carboxylic Acids and Esters: 2-Pyrrolecarboxylic acid is prepared by the oxidation of pyrrole-2-carbaldehyde using silver oxide. The pyrrole acids are seldom prepared by the oxidation of alkylpyrroles. The acid, however, may be prepared by the oxidation of the corresponding pyrrole ketones with sodium hypoiodate. The main feature of the acids is the ease with which the carboxyl group can be decarboxylated on heating (equation 5.12).

$$(5.12)$$

The pyrrole-2-carboxylic esters are readily hydrolyzed and decarboxylated in comparison to the 3-isomer. Accordingly, ethyl 3-pyrrolecarboxylate has been obtained in the following steps.

5.2.5 Biologically Active Compounds

Pyrrole occurs is tobacco smoke and the reduced compound pyrrolidine, is found in carrot green, in the amino acid proline $\left(\begin{array}{c}\text{pyrrolidine-COOH}\end{array}\right)$ and in various biologically active compounds, for instance, the alkaloid funebrine (37) also contains the pyrrole ring.[61]

(37)

Porphyrins:[62] A porphin (38) is a compound formed by linking four pyrrole groups at the α-position *via* the methene bridges. This cyclic tetrapyrrole structure was first suggested by Küster in 1912.

(38)

The porphin molecule is a highly conjugated system. There are normally 22π electrons but only 18 of these are included in any one delocalization pathway. This compiles with the Hückel's rule of $(4n + 2)$ π electrons for aromaticity. It is a stable structure but does not occur naturally. The peripheral positions are numbered from 1 to 8 while the methene bridges are designated by the Greek letters α, β, γ, δ. The pyrrole rings are assigned letters A, B, C and D as shown in structure (38). Substituted porphins are called porphyrins. These are widely distributed in nature both in the free form and as the more complex structures. Porphyrins are highly colored substances and include chlorophylls, vit. B_{12}, bile pigments and prodigiosins.

Since many substitution patterns of (38) are possible, many porphyrins are named indicative of the nature of the substituents.[63]

Etioporphyrins: These are derivatives of porphin in which the vacant positions in each pyrrole nucleus are substituted by four $-CH_3$ and four $-C_2H_5$ groups. Four of such isomers referred to as I, II, III and IV are known. Etioporphyrin (39) is prepared by treating bromine solution in acetic acid with *t*-butyl-4-ethyl-3, 5-dimethylpyrrole-2-carboxylate.

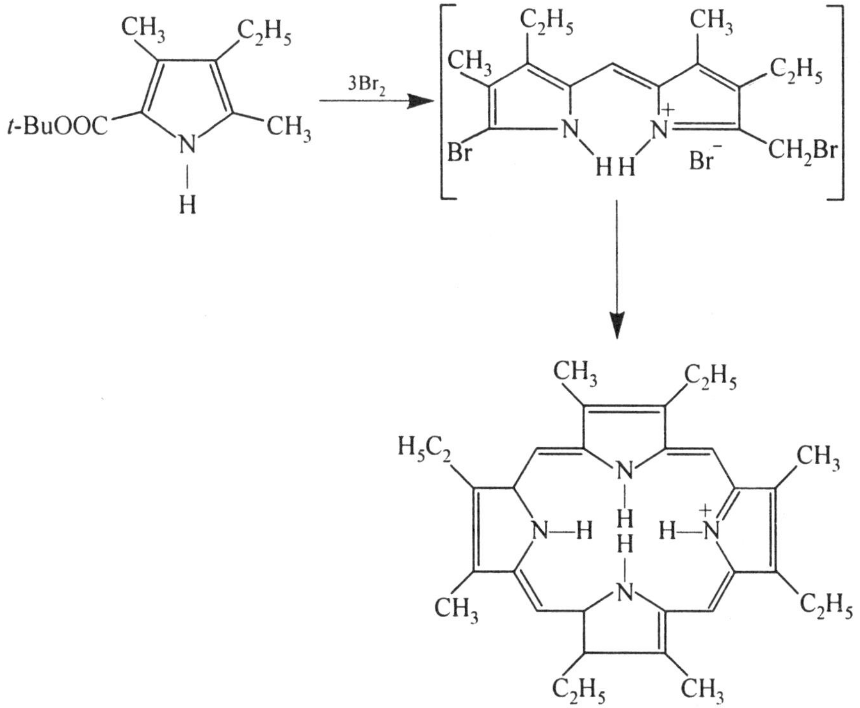

(39)

Protoporphyrin: The protoporphyrin (40) is related to etioporphyrin in which two ethyl groups have been replaced by vinyl groups and the remaining two by propionic acid residues. In other words, they contain three different side chains. Other porphyrins of this type are uroporphyrin, coproporphyrin, etc.

Hemin: The red pigment haemoglobin constituent of the red corpuscles of blood is an important derivative of porphin. It consists of heme, the porphyrin moiety containing a ferrous atom and the protein globin. Oxidation of the ferrous atom in heme to the ferric state occurs resulting in the formation of hemin (41).

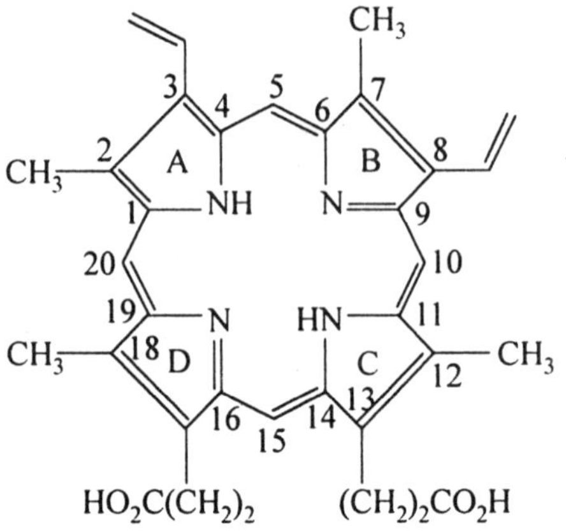

(40)

Hemin was first synthesized by Fischer and Ziele[55] in 1929. An average human adult synthesizes 8g of haemoglobin daily to replace the breakdown in the body.

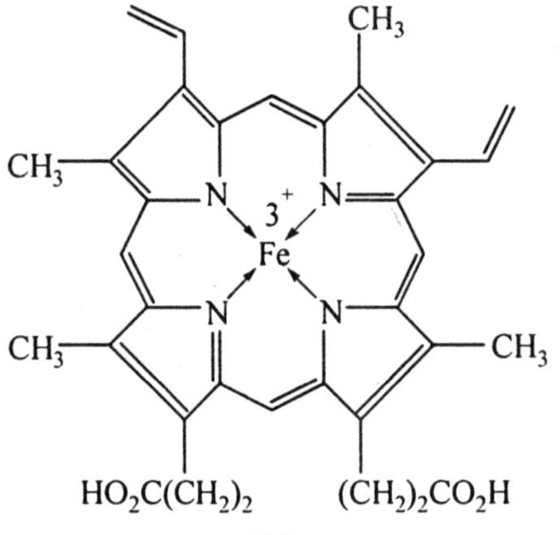

(41)

Chlorophylls: The green coloring matter which occurs in leaves was given the name chlorophyll. It has been shown to be a prophyrin in which four pyrrole nitrogens are complexed to magnesium. Chlorophyll (42) is essential for the process of photosynthesis. Stokes in 1888 showed that chlorophyll was a mixture and about eight different chlorophylls are known nowadays. Chlorophyll (*a*) and (*b*) can be extracted in large quantities from the leaf tissue. These were first separated by a classical application of chromatography. Chlorophyll-*b* differs from chlorophyll-*a* in having a formyl rather than a methyl substructure at C-3.

Their total synthesis was achieved in 1960.

Vitamin B$_{12}$: It is also known that cyanocobalamine is effective in the treatment of pernicious anemia. It crystallizes from water in beautiful deep red hydrate crystals. The structure (43) was elucidated by Professor Hodgkin.[65] The

CH$_2$OH

Phytol

(42)

Chlorophyll-*a*   Chlorophyll-*b*
R = CH$_3$ R = CHO

nitrogen atoms of pyrroles are complexed with cobalt which in turn is covalently bonded to a cyano group. The molecule contains a large planar group containing four reduced pyrrole rings which resemble porphyrin. It is, however, different from porphyrin system as it lacks one methine bridge. Its synthesis was worked out by Woodward and coworkers.[66]

(43)

***Bile Pigments*:** Heme compounds are particularly susceptible to oxidation at the carbon bridges linking the pyrrole rings, such an oxidation reaction leads to rupture of the carbon bridge resulting in the formation of open-chain tetrapyrrole structure. These are collectively known as bile pigments. These were first characterized from the pigments of animal biles - the green biliverdin (44) and its yellow reduced product billirubin (45).

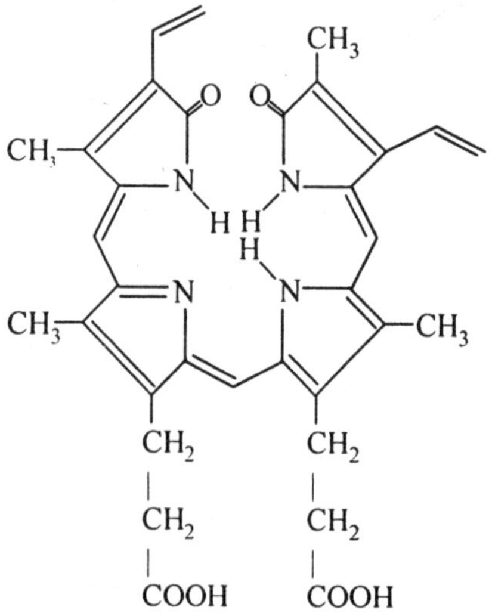

(44) 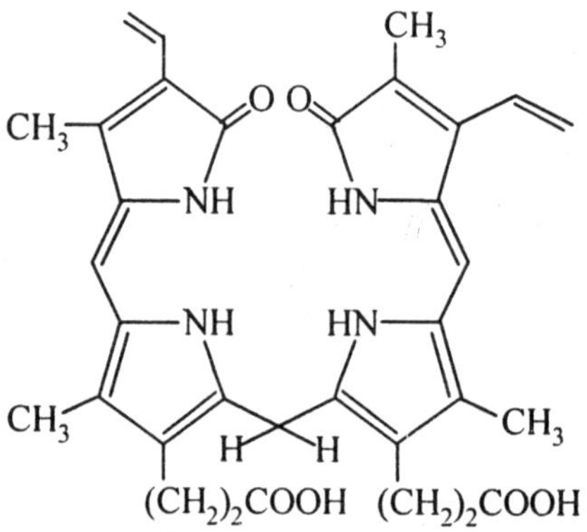   (45)

5.3 FURANS

The chemistry of furan itself started with the discovery of furan containing compounds. The aromatic furan ring system though not found in animal metabolism but occurs in secondary plant metabolites especially in terpenoids. Furan 2-carboxylic acid was first obtained by Scheele in 1780 from the dry distillation of mucic acid.[67] Furfural an important chemical was obtained in 1832. Furan was obtained much later in 1870 by heating barium fureate with soda lime. Many furans give a characteristic green color when a pine splinter soaked in HCl is exposed to furan vapors. Some furans also respond to the *Ehrlich test* but the color tests should be interpreted with caution.

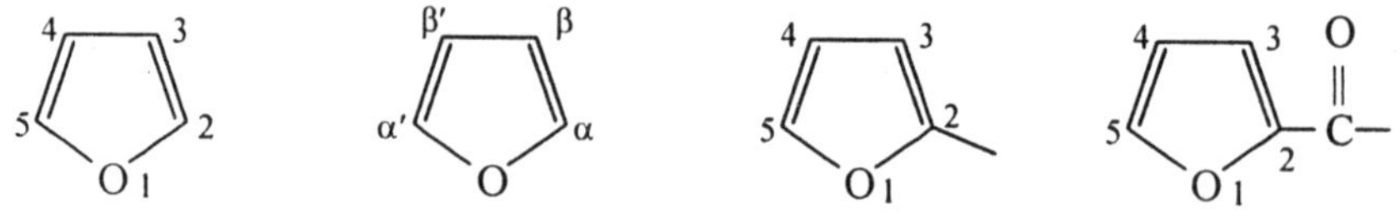

(46)   (47)   (48)   (48a)

Furoyl

The numbering of furan (46) commences from the oxygen atom which is assigned position-1. The substituents are numbered in Arabic letters or less commonly by the Greek letters (47). The group derived from furan is designated as furyl (48). Two dihydrofurans, namely 2, 3-dihydrofuran (49) and 3, 4-dihydrofuran (50) are possible. The fully saturated compound is known as tetrahydrofuran (51).

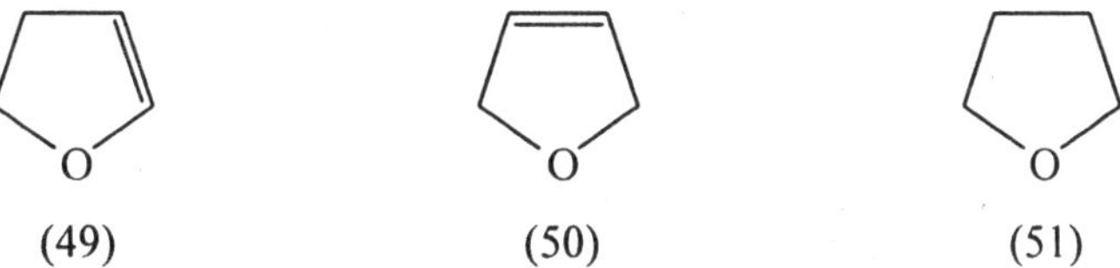

(49)   (50)   (51)

5.3.1 Physical and Spectroscopic Properties

Furan is a colorless liquid, b.p. 31.5°C. It possesses a chloroform-like odor and is soluble in most organic solvents but is only slightly miscible with water. Microwave spectroscopy has shown that furan is planar and has the following molecular parameters.

The 2, 3- and 3, 4-bond distances are 1.36 Å and 1.43Å respectively which implies that the bond lengths are intermediate between C—C single bond and C—C double bond. The C—O—C angle is 106.8°, C—C—C angle is 106° while C—C—O is 110.7°. This has often been taken as a criterion for the aromaticity of furan and thus can be represented by the following contributing resonance structures, the lone-pair on the oxygen atom is donated to the ring carbon atoms.

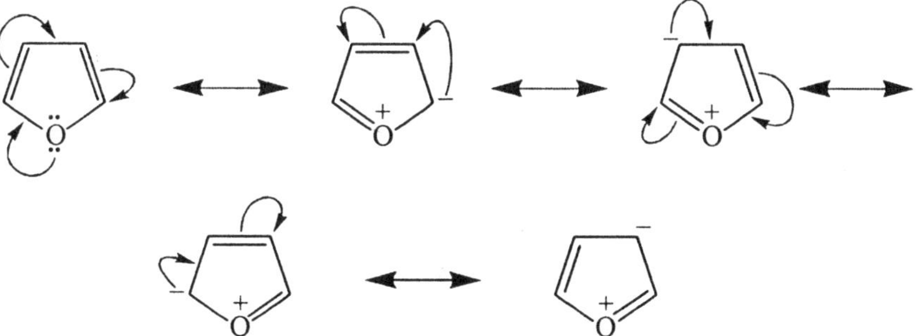

The aromatic character of a molecule is dependent on the availability of the lone-pair of electrons for resonance and this in turn will depend on the eletronegativity of the hetero atom. More electronegative atoms will have a greater hold on the lone-pair which will therefore, be localized. Furan, as a result is less aromatic than pyrrole or thiophene. Its resonance energy calculated from the thermochemical data is 15.8 Kcal/mole, a value intermediate between pyrrole and thiophene. The aromatic sextet in furan is made up of one p_z orbital contributed by each of the four carbon atoms and the lone-pair provided by the hetero atom. Thus similar to pyrrole, there are six electrons in five orbitals.

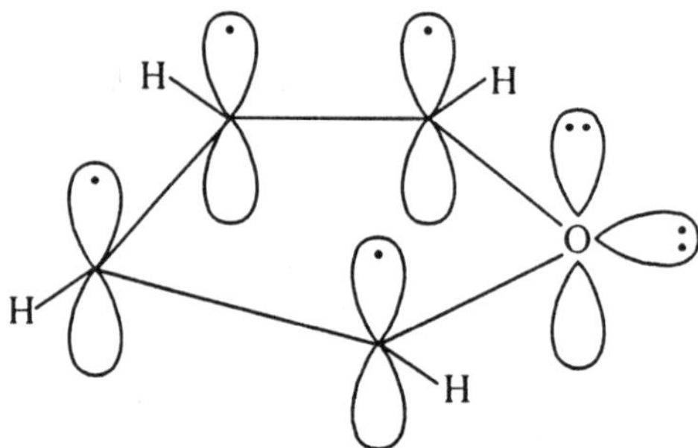

The π-electron densities have been calculated for furan and are found to be 1.800, 1.017 and 1.083 for position 1, 2 and 3 respectively. This would mean that the electrophilic attack on furan should take place at position-3. This is, however, in sharp contrast to the resonance structures shown in the preceding paragraph. According to these the electrophilic attack is favored at position-2. This controversy has been resolved by the localization energy calculations which favor attack at positions 2 and 5. It may be noted that there is a good empirical relationship between the localization energies and the free valence numbers for the same position. The smaller the localization energy the greater is the number of free valences. In general, an electrophilic reagent will preferably attack the position having the largest free valence numbers.[68]

The dipole moment of furan is only 0.70 D whereas that of tetrahydofuran (THF) is 1.73 D. This difference clearly indicates that the oxygen atom acquires a negative charge in furan and withdraws electrons inductively from the ring. Although pKa of furan has not been measured through it is much lower than that of aliphatic ethers as is evident from the fact that the latter is soluble in conc. hydrochloric acid while the former is not.[69] In the *i.r.,* the C – H stretching vibrations appear at 3159 and 3128 cm^{-1} respectively.

5.3.2 Synthetic Methods

A number of methods are available for the synthesis of furans.

1.   *From Carbohydrates:* The most important source of furan is furfural. The latter is available by acid hydrolysis[71] of polysaccharides present in oat, husks and corn cobs. These are degraded to pentoses which are subsequently converted to furfural on further treatment with acid.

Furan is obtained by passing the vapor of the aldehyde over nickel catalyst. Alternatively furan can be obtained by the decarboxylation of 2-furoic acid which is obtained by the air oxidation of furfural in the presence of cuprous and silver salts.[72]

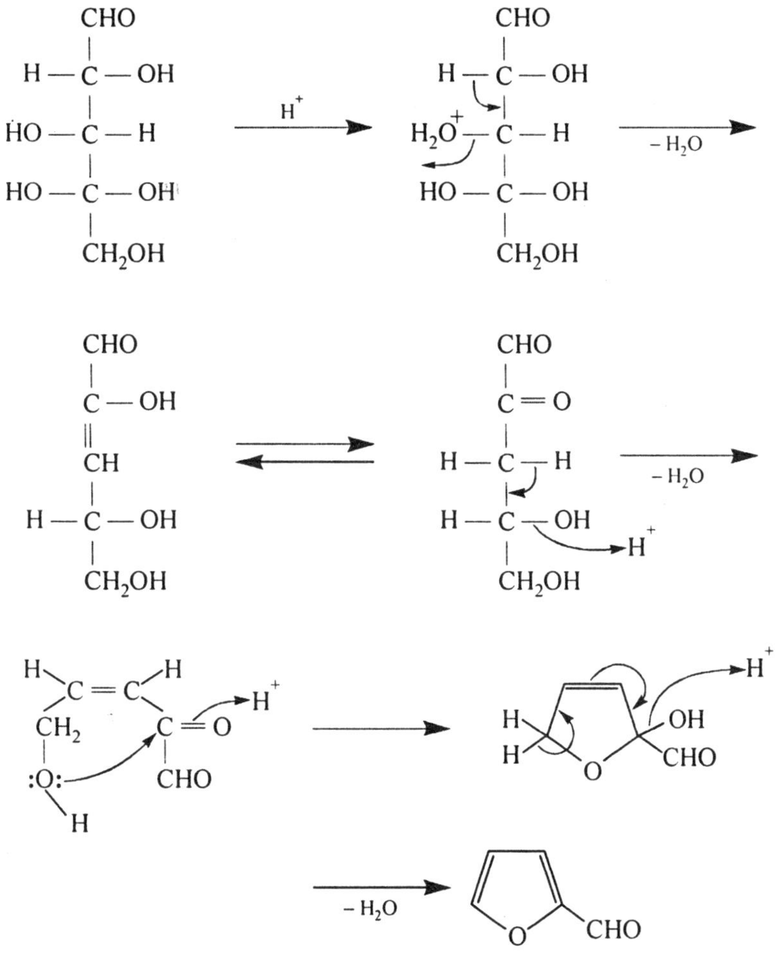

$\sim 100\%$

Furfural can also be converted directly to furan by its gas phase decarbonylation in the presence of palladium and charcoal, equation (5.13).

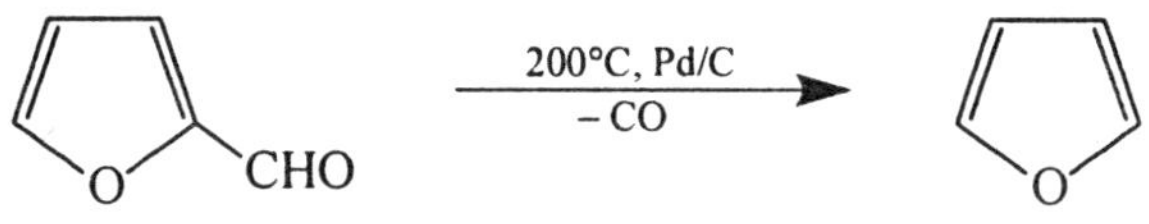

$$\text{Furfural} \xrightarrow[-\text{CO}]{200°C,\ Pd/C} \text{furan} \qquad (5.13)$$

Furfural

2.   *Ring Closure Methods:* Appropriately substituted open-chain compounds cyclic to give furan derivatives.

(*i*)   *The Paal-Knorr Synthesis*: The cyclization of 1, 4-diketones under acidic reagents results in furan formation. The requisite dicarbonyl compounds may be obtained by standard methods.[73]

The mechanism probably proceeds *via* the formation of a mono-enol.[74] Sterically hindered diketones do not cyclize to furans according to this reaction.[75] β-Chloroallyl ketones which are readily obtained by alkylation of enamines or enolate anions, on treatment with acid furnish furans.[76]

(*ii*) *The Fiest-Benary Synthesis:* In its simple form, this synthesis involves an aldol condensation of an α-haloketone or α-haloaldehyde with a, β-keto ester (or a β-diketone) in the presence of sodium hydroxide. The resulting furan contains an ester substituent at the β-position. The ester anion attacks the carbonyl group of α-chloroketone followed by the formation of an intermediate and cyclization takes place by intramolecular displacement of the chloride ion and finally loss of water.

(iii) *Cyclization of Allenals and Alkenes*: Treatment with silver nitrate or silver borontetrafluoride ($AgBF_4$) in CH_3CN of allenals and alkenanes affords furans in high yields.

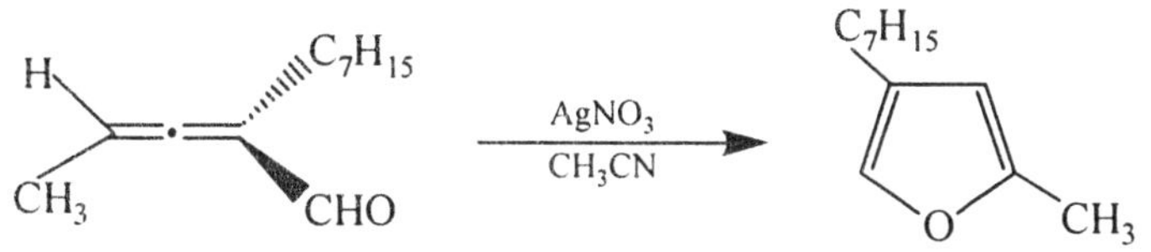

This method is also applicable to 2, 5-bridged furano macrocyclic compounds. This is illustrated for the conversion of (52) to (53).

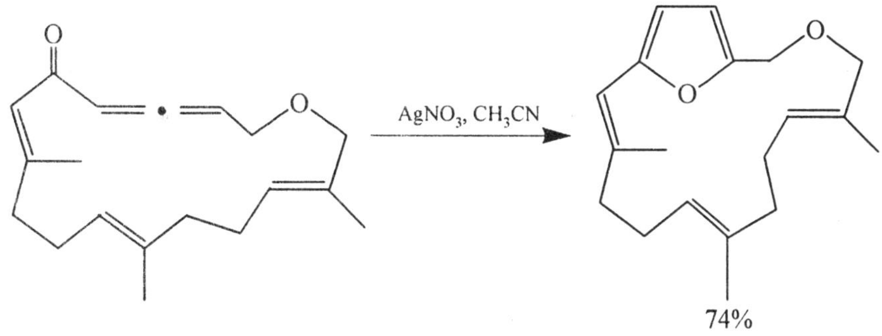

(52)

74%
(53)

3. *From other Heterocyclic Systems*: Oxazoles give Diels-Alder adducts with acetylenic dienophiles which suffer *cheletropic* loss of a nitrile to yield furans.[79]

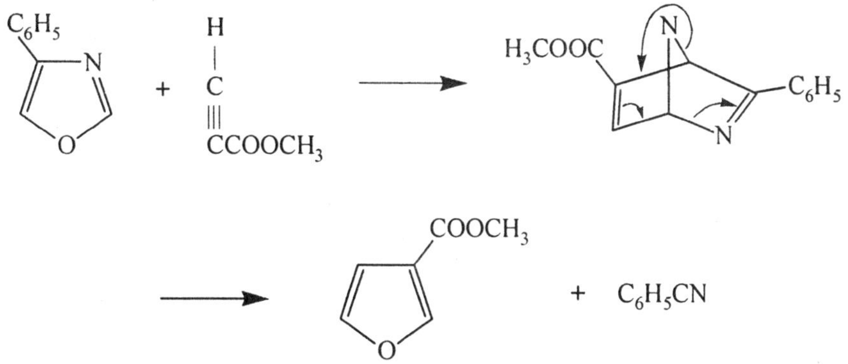

Methyl 3-furoate

Pyrylium salts on treatment with aqueous acidic hydrogen peroxide suffer oxidation and ring contraction, yielding furans. Nenitzescu and Balaban[80] described the following example:

4. 3-Furan derivatives can also be prepared from the Diels-Alder adducts.[81] Thus 3-halofuran can be obtained from furan and maleic anhydride in the following steps:

Another interesting example includes the following for the preparation of 3, 4-trifluoromethylfuran.

5. *Dewar Furan:* Dewar benzene was synthesized in the early sixties. Dewar
furan (54) has now been prepared rather by a long procedure.[82]
Dewar pyrrole[83] and thiophene have also been successfully obtained.

(54)

5.3.3 Chemical Reactions

The furan ring behaves chemically as a typical diene ether which at the same
time is resonance stabilized.

1. *Reaction with Acids:* Furan is readily hydrolyzed by acids, but the
acidification of furan to yield 1, 4-diketo derivatives is not satisfactory from
the preparation point of view.[70, 81] Mild conditions should be employed to
achieve the isolation of even succinaldehyde otherwise the protonated intermediate undergoes polymerization. Furans containing electron-withdrawing
groups are more stable to acids.

2. *Alkylation:* Alkylation of furan may be accomplished with alkenes and
mild Lewis acid catalysts such as boron trifluoride or phosphoric acid. The
reaction of furan with isobutylene in the presence of phosphoric acid on
Kiesulguhr gives a mixture of 2- and 3-*t*-butylfuran.[84, 85]

3.   *Electrophilic Substitution:* The π-excessive heterocyclic compounds have higher electron density on the ring carbon atoms than in the case of benzene. Furan analogous to pyrrole undergoes electrophilic substitution with exceptional ease and even more so than benzene. The conditions under which electrophilic substitutions are carried out need to be carefully controlled. Furthermore, the mechanism of reaction is often not so simple as is characteristic of the benzenoid system. The 2-position of the furan nucleus is much more reactive towards electrophilic attack than the 3-position.

***Halogenation*:** Direct halogenation of furan proceeds readily and leads to a mixture of either mono-and poly-substituted products or resins. Gradual treatment of furan with 1.6 mol equivalent of chlorine in dichloromethane at $-40°C$ yields a mixture of 2-chloro (64%), 2, 5-dichloro- (28%) and 2, 3, 5-trichlorofuran (8%), while increasing the proportion of chlorine yields tetrachlorofuran and its dichloride.

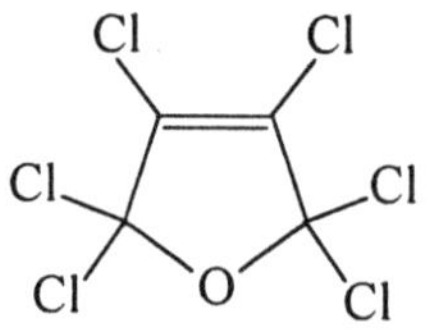

The yield of 2-chlorofuran may be increased by using less quantity of chlorine.

Bromination (Br$_2$/dioxane) at $-5°C$ gives 2-bromofuran in good yield.[86] Direct bromination of furan gives ill-characterized products but its reaction with dioxane dibromide at $-5°C$ give 2-bromofuran in reasonable yields.

***Nitration*:** Nitration of furan is achieved with mild nitrating agents[87] such as acetyl nitrate (a mixture of acetic anhydride and nitric acid) at -5 to $-30°C$ to yield 2-nitrofuran. The reaction proceeds largely by an addition-elimination mechanism. It is an advantage in special cases to use a mixture of acetyl nitrate and pyridine. Further reaction affords 2, 5-dinitrofuran. A low yield (14%) of 2-nitrofuran has been reported by treatment of furan with nitronium tetrafluoroborate.[88]

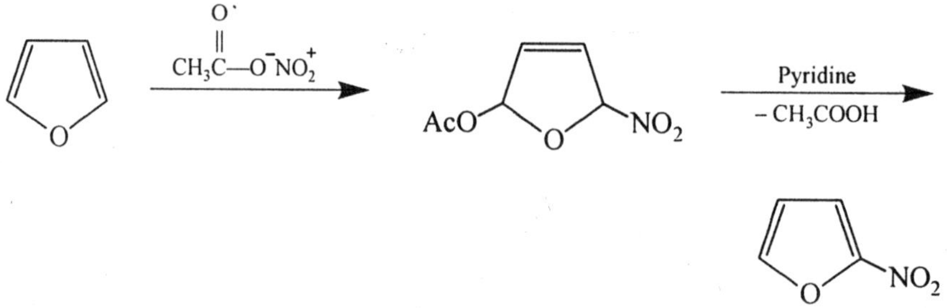

Sulfonation: Furan like pyrrole is resinfied in the presence of sulfuric acid, but it can be sulfonated by the pyridine – SO_3 complex[89] to give 2-furan sulfonic acid in 90% yield. If furan is treated with a three-fold excess of pyridine–SO_3 complex for 5 hr at 35-40°C, then furan 2, 5-disulfonic acid is obtained in 80% yield. Furan 2-carboxylic acid, however, may be sulfonated with oleum.

The Friedel-Crafts Reaction: Carboxylic acid anhydrides or acyl halides normally react with furans in the presence of Lewis acid catalysts, but more reactive anhydrides such as $(CF_3CO)_2O$ can effect uncatalyzed substitution. Furan reacts with acetic anhydride in the absence of a catalyst or a mild Lewis acid like $ZnCl_2$ or $SnCl_4$ to yield 2 - acetylfuran.[90] The Friedel-Crafts alkylation has not been so successful partly because of polymerization caused by the catalyst and partly because of polyalkylation. Furans containing electron-withdrawing groups have been successfully alkylated.[91]

Gatterman-Koch Reaction: Furan reacts under the Gatterman-Koch conditions[92] to give furfural. It can also be formylated by the Vilsmeier method at low temperature.

Furan undergoes phenylation and diazo coupling with benzenediazonium salts unlike pyrrole and thiophene.

5.   *Reaction with Carbenes and Nitrenes:* Carbenes add across the 2, 3-carbon double bond[93] in furan. Ultraviolet irradiation of a solution of methyl diazoacetate in furan affords a carbene adduct which on heating to 160°C undergoes valence tautomerization[94] to *cis-trans-* muconaldehyde (56). Not many reactions of furans with nitrenes, have been reported.[95]

(55)

(56)

6. *The Diels-Alder Reaction:* The diene character of furan is clearly demonstrated in the case of the formation of Diels-Alder adducts with a suitable dienophile.[96, 97] But due to the aromatic character of furan ring and the inherent strain in the adduct, the cycloadduct is sensitive thermally towards reverting to the starting materials. As a result with the use of powerful dienophiles like maleic anhydride, DMAD or maleimide respectable yields of the products are obtained. In the reaction of furan and maleic anhydride at 25°C, a 90% yield of the *exo*-adduct (58) and a small quantity (5-6%) of the *endo*-adduct (57) are formed,[98] equation (5.14). The *exo*-adducts are thermodynamically more stable than the *endo*-adducts.

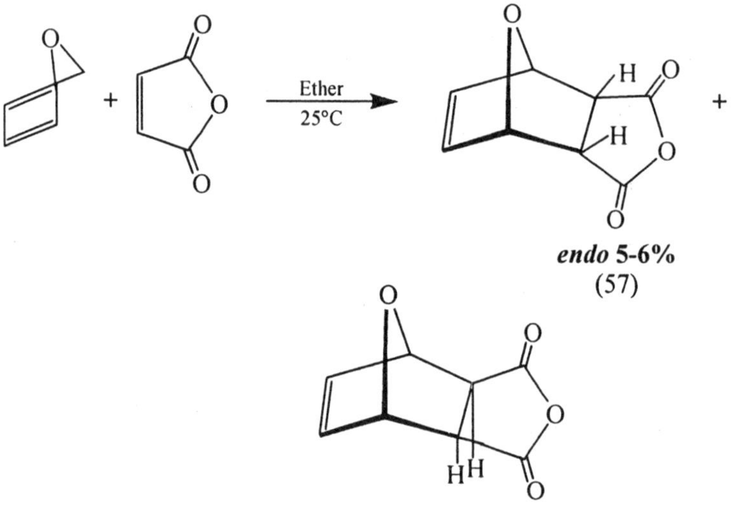

endo **5-6%**
(57)

exo **90%**
(58)

$$(5.14)$$

With maleimide the formation of the adduct is controlled by temperature. At 25°C the kinetically favored *endo*-adduct predominates while the *exo*-adduct at 90°C. The *endo*-adduct also converts to the *exo*-adduct on heating.[99] Furan also reacts with cyclopropene at room temperature to yield a 1:1 mixture of *exo*-(60) and *endo*-(59) adducts.[100]

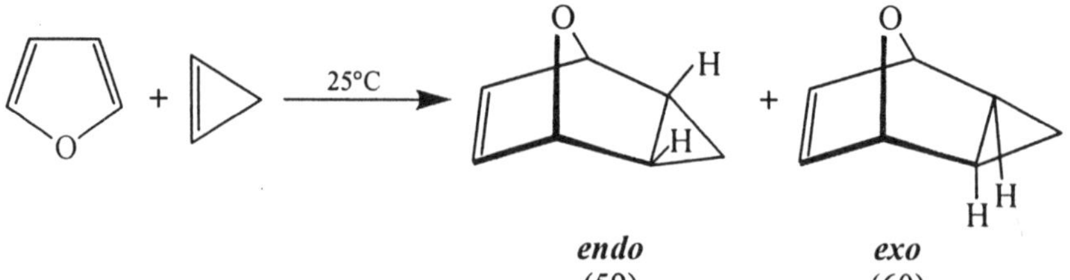

endo
(59)

exo
(60)

The addition of maleic acid in water to furan occurs slowly and the adducts revert thermally to the starting materials. Initially the *endo*-adduct (61) is formed four times faster than the *exo*-isomer (62), but after several days the ratio is 1:1. Ethylenic dienophiles with only one electron-withdrawing group add to furans slowly at room temperature. Both *endo*- and *exo*- adducts are obtained in low-to-moderate yields. At higher temperatures the reverse reaction, as well as polymerization of the dienophile generally takes place. However, by using very high pressures it is possible to overcome these difficulties to obtain excellent yields of the D-A adducts.[97]

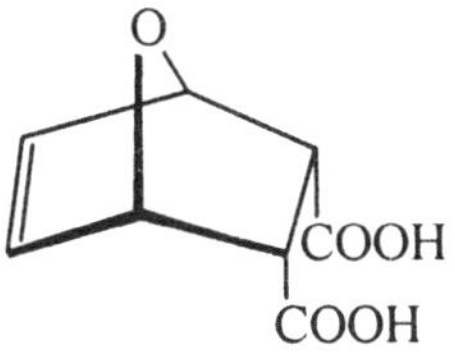

endo
(61)

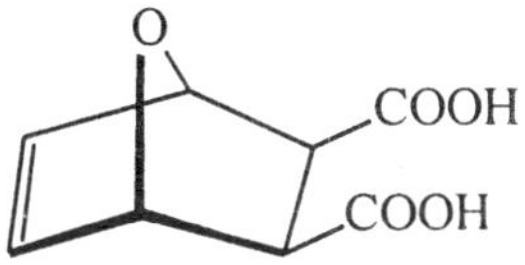

exo
(62)

Furan reacts with acetylenic dienophiles rather under milder conditions to yield oxanorbornadienes (equation 5.15).

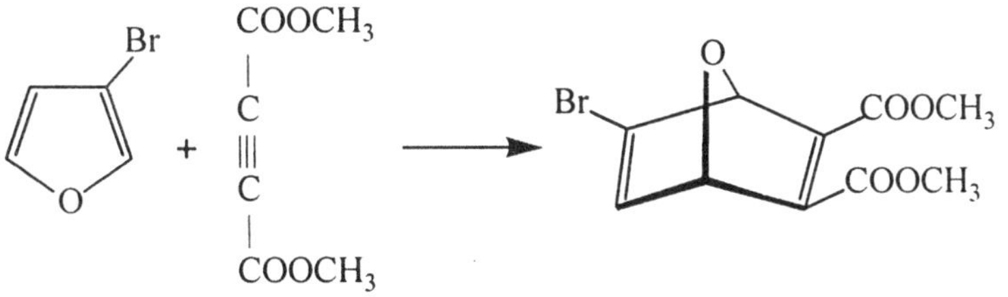

$$(15.5)$$

The norbornadienes so formed have found applications in organic synthesis.[98-101]

The adduct (63) on reduction yields 3, 4-Trifluoromethylfuran.

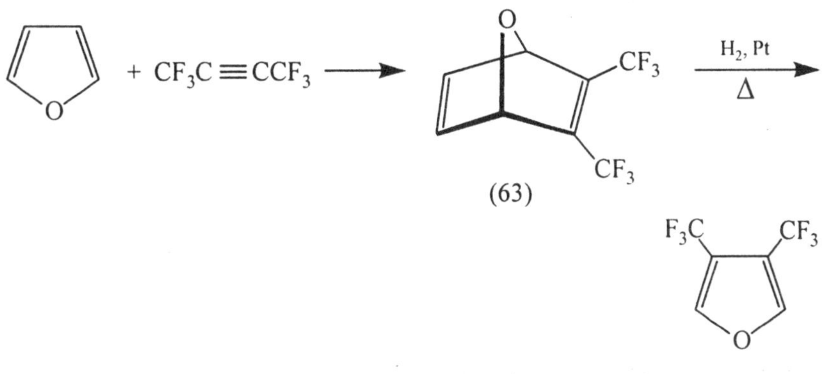

(63)

2, 5-Dimethylfuran and DMAD form the oxanorbornadiene (63a) on heating, but 2, 5-diphenylfuran forms a similar adduct with DMAD in the presence of Lewis acid only.[102]

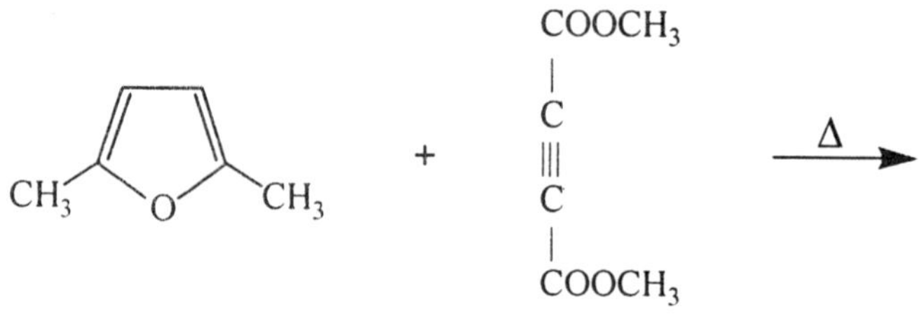

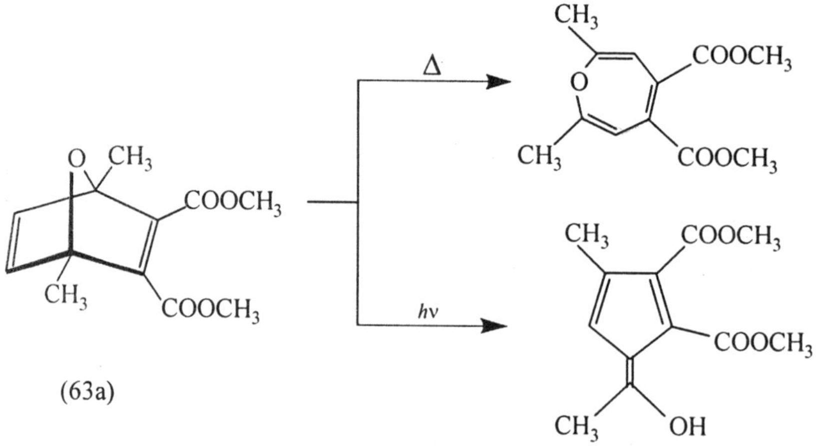

(63a)

The oxanorbornadiene (63a) on irradiation yield oxaquadricyclane which on thermolysis[103-106] afford oxepins or fulvenes[102] on further irradiation.

Furan fails to react with ethyl propiolate at room temperature while at 130°C the adduct (64) is obtained in a poor yield (9%). 2, 5-Dimethylfuran, on the other hand, reacts with ethyl propiolate at 95–120°C to give initially the adduct (65) which undergoes retro Diels-Alder reaction with the loss of acetylene.[103]

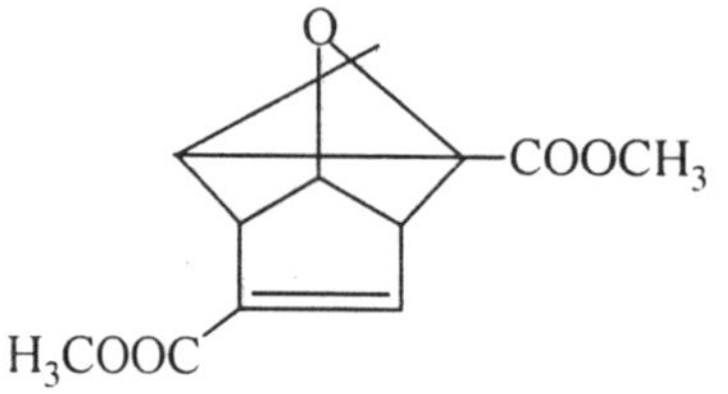

(64) 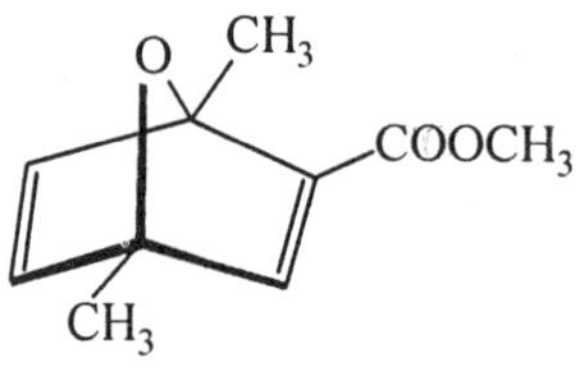   (65)

Arynes, generated, in *situ* may be trapped by furan or alkyl furan to yield the Diels-Alder adduct[104-107] from dehalogenation of *o*-dihalobenzene in the presence of Lithium amalgam, lanthanum metal[106] or *n*-butyllithium. The resulting adduct naphthalene oxide can be converted into useful products.

La, I₂ (Catalyst)
THF 25°C, 4 hr

100%

n-BuLi

**2,3-Dimethoxy –1,4-dihydro-
1,4-epoxynaphthalene**

Li(Hg)

H₂, Ni

HCl

1, 4-Dihydroepoxinaphthalene

(66)

Furan also adds to 1, 5-naphthodiyne to yield the *bis*-adduct which on hydrogenation and dehydration with HCl leads to the hydrocarbon chrysene in 93% yield.

(67)

H₂　　　HCl

Properly substituted furans $(67)^{108}$ also undergo intramolecular D.A. reaction to (68).

Benzene, 50°C

(67)   (68)

7. *Reaction with Reducing Agents:* The outcome of the reaction of furan reduction depends on the catalyst, solvent, temperature, etc. Reduction of furan takes place readily with Raney nickel at 125°C and 100 atom to afford THF. In the presence of Pt/Al or Cu/Al catalysts at 150°C, the side-chain is reduced selectively,[109] equations (5.16). With platinum in acetic acid, furan yields *n*-butanol.[110]

H_2.Pt, Al   (5.16)

8. *Reaction with Oxidizing Agents:* Furan is particularly sensitive to oxidation and is not stable in the presence of air or oxygen, such reactions often involve 1, 4-addition of the oxidant to the diene system. Air oxidation (or reaction with singlet oxygen) gives the transannular peroxide, i.e. a bridged peroxy adduct which has been isolated at − 100°C equation (4.17). Further reaction leads to succinaldehyde. When this reaction is conducted in methanol, the lactone (69) is obtained. 2, 5-Dimethylfuran reacts with *t*-butyl hydroperoxide to give homogeneous *bis* peroxide and in the presence of dil. sulfuric acid to give the hydroperoxide.[111, 112]

Air   (5.17)

(69)

4–Hydroxy-2–butenolide

Electrochemical oxidation of 2, 5-dimethylfuran carried out in methanol yields a mixture of *cis-* and *trans-*2, 5-dihydro-2, 5-dimethoxyfurans.[113] With *m*-chloroperbenzoic acid (2 moles) alkyl substituted furans undergo ring expansion.[114]

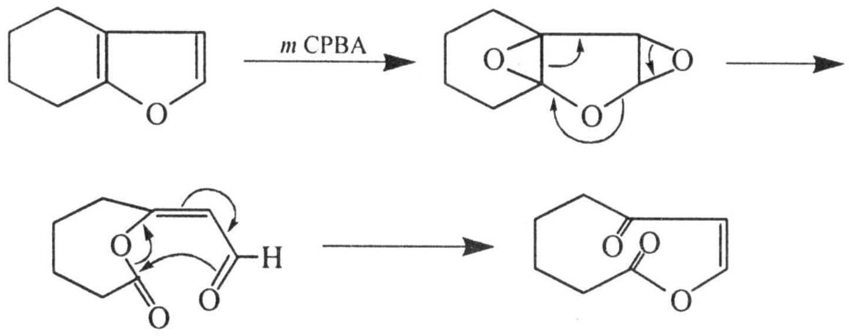

9.   *Metallation of Furans:* Grignard reagents in the furan series are difficult to prepare and their use has been superseded by the more versatile lithio-compounds. Furan readily undergoes hydrogen-lithium interconversion at the 2-position on treatment with *n*-butyllithium to give 2 - lithiofuran.[115] This enters into reaction with other reagents to give useful products.

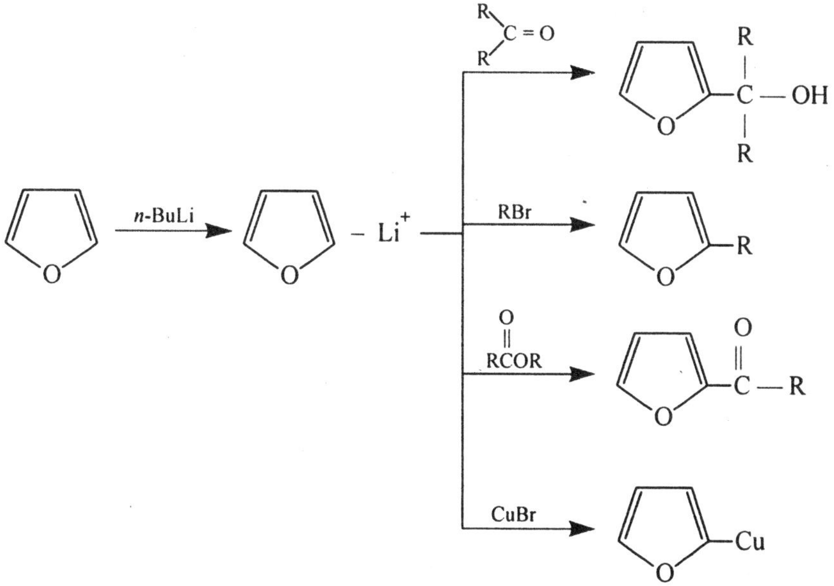

10. *Reaction Under Sonication:* In order to prepare heterocyclic acyl cyanides, an appropriate carboxylic acid chloride is treated with solid potassium cyanide in acetonitrile and the mixture is subjected to sonication (44 KHz, 2000 W). Thus 2 - furoyl chloride can be rapidly converted to 2 - furoyl cyanide in 76% yield.[116]

The *Barbier reaction*, i.e. one step coupling of an organic halide with a carbonyl compound is usually accomplished with magnesium but lithium has been found to improve yields significantly. This reaction, however, gives good yields and in shorter times under sonication (50 KHz, 60 W).[117]

11. *Photochemical Reactions:* The Photochemistry of furan appears to have attracted very little attention. Photosensitized decomposition of furan in the gas phase leads to decarbonylation as the principal product,[118] i.e. cyclopropene and carbon monoxide.

Ketones add to furan photochemically to yield oxetane derivatives.[119, 120] Irradiation of a mixture of benzene and furan gives largely the D.A. adduct (70) along with some minor products. This adduct subsequently undergoes a cope rearrangement to give (71) and ($2\pi + 2\pi$) photoaddition to afford (72).[121]

12. *The Tishchenko Reaction:* This is a dimerization reaction of aldehydes to yield the corresponding esters in the presence of aluminum alkoxides.[122] Recently this reaction has been achieved in the presence of heterogenous catalyst as well.

**2-Furylmethyl
2-furancarboxylate**

5.3.4 Derivatives of Furan

Alkyl-and Aryl-furans: Various alkyl- and aryl- substituted furans are known to have been synthesized by rather indirect methods. Direct alkylation under the Friedel-Crafts reaction is not successful. 2-Methylfuran is obtained from furfural by catalytic reduction using copper chromite. It has also been obtained by the following sequence of reactions:

The best method for the preparation of 2, 5-dimethylfuran (94%) is the distillation of hexan-2, 5-dione from acidic ion-exchange resins.[124]

The physical and chemical properties of alkylfurans parallel those of furan itself. The electron-donating alkyl group increases the basicity of the oxygen atom and thus enhances the reactivity of the ring towards electrophilic attack. Bromination with N-bromosuccinimide yields 2-(bromomethyl) furan; 2, 5-dimethylfuran gives the Diels-Alder adduct with DMAD while 2, 5-diphenylfuran does so on heating.[125]

Halofurans: Halofurans are unstable compounds which decompose on standing. Direct halogenation to give monohalo furan derivatives is difficult. Halofurans are obtained by decarboxylation of the appropriate halofuran carboxylic acid and reaction of chloromercury furans with iodine, 2, 5-Dibromofuran (78%) may be prepared[126] from sodium 5-bromofuran

2-carboxylate by heating in water with Br_2/KBr. Tetrafluorofuran has been prepared from tetrahydrofuran and cobalt (II) fluoride at 100-120°C.[127]

The monohalogens are colorless liquids with a sweet odor and are sparingly soluble in water. The carbon-halogen bond in halofuran is quite stable and is not replaced by the usual nucleophiles. This reaction, however, takes place when the halogen atom is attached to the side-chain as in 2-(chloromethyl) furan. With aqueous cyanide ion the reaction seems to take place through an intimate ion-pair (73) and the attack of cyanide ion[128] gives the normal product (74) and the rearranged product (75). Polar solvents and nucleophiles like CN^-, OR^-, $OC_6H_5^-$ favor rearrangement product while other nucleophiles offer normal product.[129] 2 - Iodo- and 3 - iodofurans form Grignard reagents which have the usual properties.

Aminofurans: All attempts to prepare 2-aminofuran including the hydrolysis of 2-acetamidofuran or reduction of 2-nitrofuran have not resulted in the formation of isolable products. The unequivoacal way for the synthesis of substituted aminofurans is by the Curtius degradation of the appropriate furoic acid azide.

The aminofurans are colorless liquids and possess an odor like that of pyridine. 2-Aminofuran is unstable but 2-aminofuran substituted with electron-withdrawing groups are stable compounds and exist as the amino tautomers but their reactions are not characteristics of aromatic amines.[130] 2-Amino-3-cyano-4-methylfuran undergoes a spontaneous Diels-Alder reaction[131] to (76), equation (5.18).

$$(5.18)$$

(76)

Hydroxyfurans: The 2- and 3-hydrofurans are characterized by extreme stability. 2-Hydroxyfuran is formally the enolic form of 2(3H)-furanone ($\Delta^{4,5}$ -butenolide) but this and 2 (5H) - furanone ($\Delta^{2,3}$ - butenolide) show no evidence of enolic behavior. Their chemistry has been comprehensively reviewed.[132]

(77)   (78)   (79)

2(3H)-Furanones (78) may be obtained from enolizable-γ-keto acids i.e. levulinic acid by slow distillation or by treatment with acetic anhydride. Treatment of 2-lithiofuran with ethyl borate and hydrolysis of the resultant boronic ester yields 2(3H)-furanone.[133] The 2(3H)-furanones on treatment with bases are converted to 2(5H)-furanones as the latter are thermodynamically more stable and are also conjugated. The 2(5H)-furanones are widespread in nature and some examples are found in the sesquiterpenes.[134] 3-Hydroxyfuran (80) is formally the enolic form of 3(2H)-furanone (81).[135]

(80)   (81)

Furan Aldehydes and Ketones: 2-Furan carbaldehyde, *furfural*, is commercially available and acts as a key starting material for many furan synthesis. Besides it is a precursor of the very widely used solvent tetrahydrofuran. It is colorless liquid b.p., 161.7°C and it possesses an aromatic pungent odor. It is also miscible with organic solvents. Because of its easy availability and industrial significance, its chemistry has been extensively studied. Catalytic reduction (H$_2$, Cu chromite), yields 2-methylfuran or furfural alcohol (H$_2$/Ni) depending on the catalyst employed. It can be oxidized by air or alkaline potassium permanganate solution to furoic acid. Electrophilic substitution occurs more readily on furfural than benzaldehyde and the electrophile occupies position-5. It also behaves as an aromatic aldehyde in most condensation reactions. Furfural resins with phenol find industrial applications. Furfural also adds to 2, 3-dimethyl-2-butene to yield an oxetane derivative.[136]

Furylketone (2-furylmethyl ketone) can be obtained by the acetylation of furan with acetic anhydride in the presence of stannic chloride. The reactions of simple furyl ketones, in general, parallel those of the aldehydes. 2-Acetylfuran forms enamines which on heating gives *o*-dialkylaminophenols.[137]

Furan Carboxylic Acids and Esters: Furan carboxylic acids are often obtained by ring closure methods such as the Paal-Knorr and Feist-Bernary synthesis. 2-Furoic acid may be obtained by the catalytic oxidation of furfural. 2-and 3-furoic acids can also be obtained from the corresponding lithiofurans and carbon dioxide.

2-Furoic acid is a colorless crystalline compound, m.p. 133°C and is sparingly soluble in water. It is stronger acid (pKa 3.6) than the 3-isomer (pKa 4.5) or benzoic acid (pKa 4.20) because of the inductive effect of the oxygen atom. Similarly 2 - furoic acid is much more easily decarboxylated than the 3-isomer. The electron-withdrawing effect of the carboxyl group decreases electron-density on the ring but nitration (HNO_3, H_2SO_4) and sulfonation (H_2SO, SO_3) can be normally carried out. The electrophilic attack occurs at the 5-position. The 3 - furoic acid can be obtained by selectively decarboxylating 2, 3- or 2, 4-furan dicarboxylic acids.

5.3.5 Naturally and Biologically Active Compounds

Simple monocyclic furan derivatives are found in natural products although as a group such substances are not of wide spread occurrence. 2-Methylfuran, for instance, finds its origin in oil of turpentine. Furfural, furfural alcohol and 5-methylfurfural occur in *Iodium fragile* and roasted coffee. 4, 5-Dimethylfuran-3-carbaldehyde occurs in oil of cloves. The more complex ones such as (+)-1pomeamarone (82), linderene (83), evadone (84), etc. occur naturally. Furacin (85) or 5-nitrofurfural semicarbazide is a useful antibacterial.

(82)

(83)

(84)

(85)

Khellin (86) a medicinal agent has been used for a variety of pharmacological indications including hypertension, renal and biliary colic and stomach disorder.[138]

(86)

5.4 THIOPHENES*

Thiophene was discovered by Victor Meyer in 1882 during his ingenious experiments on the investigation of aromatic compounds. Following Meyer's

---

*The spelling without the terminal *e* is common in Great Britain.

pioneering work, Steinkopf and his collaborators established the classical concepts of thiophene chemistry as a close heterocycle similar to the parent of aromatic system, benzene. Thiophene occurs in coal tar and a large number of thiophene derivatives occur in plants and in animal metabolism. Thiophene responds to the *indophenine test*, i.e. on treatment of 1 ml of dil. solution of thiophene in benzene with a few milligrams of isatin and 1 ml of conc. Sulfuric acid gives an intense blue color. Analogous to other five-membered heterocyclic compounds, the hetero atom in thiophene (87), is also numbered as 1. The nomenclature of thiophene in Greek letters (88) is not in vogue any more.

(87)   (88)

The reduced thiophenes are given the trivial names, for instance, 2, 3-dihydrothiophene (89) is designated as 2-*thiolene*, the 3, 5-dihydrothiophene (90) as 3-*thiolene* and the completely reduced tetrahydrothiophene (91) as *thiolene*. It may be pointed out that though systematic names have been recommended for use, the trivial names of the different radicals derived from thiophene itself are more common.

(89)   (90)   (91)   2-Thienyl   3-Thienyl

2-Thenyl   3-Thenal or   Thenoyl
                     Thenylidene

5.4.1 Physical and Spectroscopic Properties

Thiophene is a colorless liquid, b.p. 80°C (benzene b.p. 84°C), the freezing point is − 38.3°C (benzene 5.5). The dielectric constant of thiophene is 16, in comparison to 2.27 for benzene. It is miscible with water but soluble in most organic solvents. The microwave spectroscopy has revealed that thiophene molecule is planar and its molecular dimensions are known with a great accuracy.[139] The bond lengths, except C—S bond, in thiophene are intermediate between that of C—C single and C $=$ C double bonds. However, the results of microwave studies indicate a surprisingly high double bond character of the C—S bond also, not easily evident from the usual Kekule structure. The molecule can thus be expressed as a resonance hybrid of the following contributing structures:

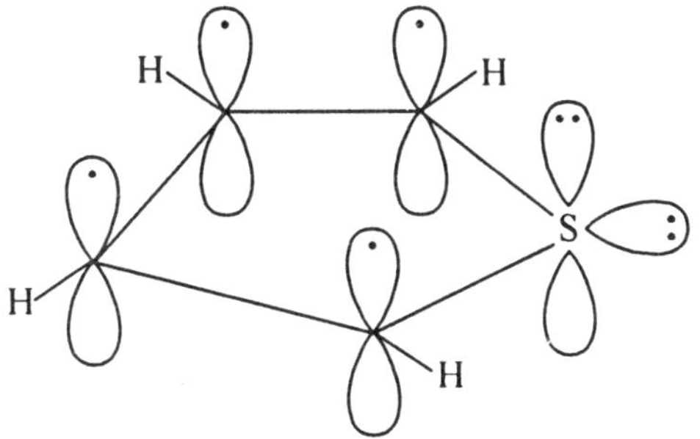

The C—S—C internal angle is 92.5° and the C—S—C angle is 111.5°. The resonance energy and aromaticity of thiophene has been a subject of considerable debate and has been found to be in the range of 29-34 Kcal/mole. This large value also indicates that there is considerable conjugation in thiophene. The magnitude of resonance energy of thiophene is higher than that for pyrrole or furan. This is in agreement with the fact that sulfur atom is less electronegative than oxygen and nitrogen (O > N > S) and as a consequence can release electrons into the ring to form a π- sextet of electrons required for aromaticity. Thiophene is therefore aromatic and also π-electron excessive. From the molecular orbital viewpoint thiophene is a planar molecule with sp^2 hybridized carbon atoms. Each of the four carbon atoms of the ring contribute one p_z electron while the sulfur atom contributes two p- electrons to form the π-electron cloud above and below the ring. Further evidence that the aromatic character for thiophene is greatest as compared to pyrrole and furan has been obtained from *n.m.r.* studies. The chemical shifts of the α- and β-hydrogens of thiophene have been determined through a study of deuterated thiophenes[140] and the order is found to be furan > pyrrole > thiophene and being 1.05 ppm, 0.125 ppm and 0.057 ppm respectively. The minimum value of thiophene corroborates the fact that thiophene has more aromatic character than pyrrole and furan. The fact that the resonance of β-hydrogens occur at lower fields than that of the α-hydrogens in these heterocyclic rings has been attributed to lower electron density at the α-hydrogens.[141]

The electrophilic attack on thiophene takes place preferably at the 2- and 5-positions. The π-electron density and localization energy have been calculated for thiophene and as in the case of furan, the electrophilic attack is governed by the latter parameter.

The dipole moment of thiophene has been calculated to be 0.51 D and that of tetrahydrofuran is 1.90 D. The difference observed is of the same order as in the case of furan. It is thus concluded that the electrons on the sulfur atom are involved in resonance with the ring.

Pyrrole is a weak base while furan is extremely weak and thiophene, in contrast, seems to be devoid of basic properties.

The C—H stretching frequencies in thiophene are 3110 and 3063 cm^{-1} while in *n.m.r.* the ring protons absorb at 7.18 and 6.99 ppm for α- and β-protons respectively.

5.4.2 Synthetic Methods

Thiophene is commercially available nowadays however, a large number of synthetic pathways are available for its preparation.

1.   *From Sodium Succinate:* The classical method for preparing thiophene in the laboratory consists of heating a mixture of sodium succinate and phosphorus trisulfide (the product of the reaction of red phosphorus and sulfur), equation (5.19). Thiophene is manufactured commercially from paraffin hydrocarbons and sulfur. This process involves the cyclization of butane, butene or 1, 3-butadiene with elemental sulfur.

$$^+Na\ ^-OOC—CH_2CH_2—COO^-\ Na^+ \xrightarrow[180°C]{P_2S_3} \text{thiophene} \qquad (5.19)$$

The reactants are preheated to about 600°C and then the vapors are passed through a reaction tube and the exit gases are rapidly cooled. Some cracking and polymerization also occur at this high temperature even though the contact time is extremely short. The unreacted materials are recycled and then redistilled thiophene of 99% purity is obtained. This process may be employed to prepare 3-methylthiophene.

The substituents R and R$_1$ may be H, alkyl or aryl.

2.   *Ring Closure Methods:*

(i)   From Unsaturated Compounds: This approach involves the use of methyl thioglycolate and its condensation with dimethyl fumarate[142] in the presence of a base. An acetylenic compound may also be used to obtain a thiophene derivative.[143, 144]

$$(5.20)$$

(ii) The Paal-Knorr Synthesis: It is the most general method for the preparation of substituted thiophenes. As in the case of pyrrole and furan, an enolizable, 1, 4-diketone is heated with phosphorus pentasulfide to give rise to 2, 5-disubstituted thiophenes.[145] The yields in this reaction are only moderate to good. This procedure has been utilized to prepare cyclobutathiophenes (92)[146], the cyclophanes (93)[147] and some non-classical thiophenes such as thieno (c) pyrrole (94)[148] in which sulfur is tetracovalent and present in a ring.

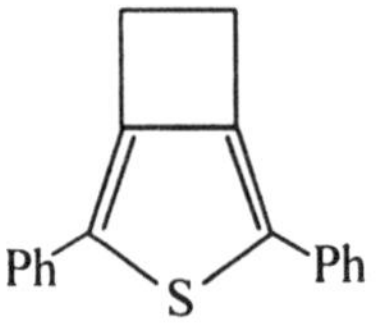

Ph   S   Ph

(92)

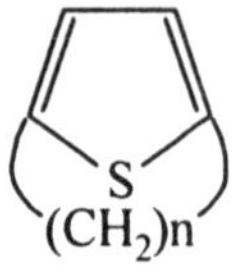

S

(CH₂)n

(93)

Ph   Ph

CH₃—N   S

Ph   Ph

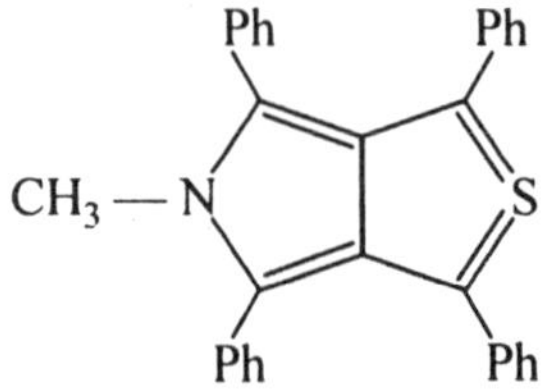

(94)

(iii) The Hinsberg Method: This method involves the reaction between 1, 2-dicarbonyl compound and diethylthiodiacetate in the presence of a strong base. The mechanism of the reaction involves two consecutive aldol condensation between the reactants resulting in the formation of the half ester.

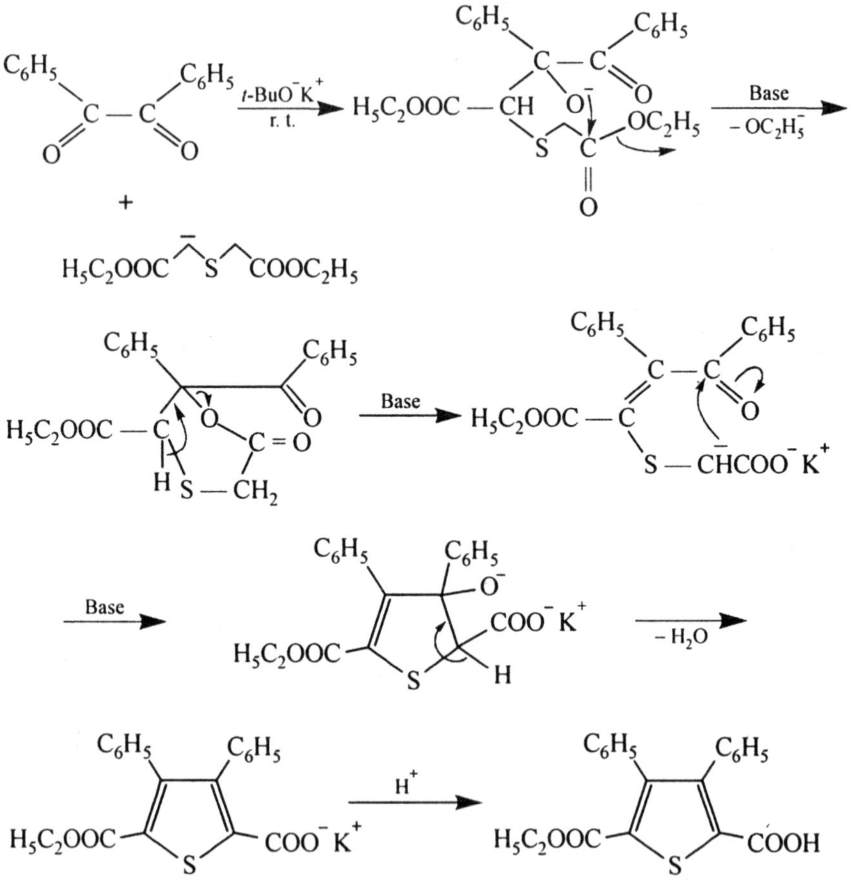

A related method[149] makes use of a *bis*-ylide in place of the thiodiglycolate ester and thus enabling isolation of highly strained cyclobutathiophenes (95) and (96).

(95)   (96)

3.   *Thienylation of alkenes:* Thiophene derivatives have been prepared by reacting 2-thenoyl chloride with an alkene in the presence of Pd(II) acetate in refluxing xylene.[150] Under these conditions 2-thenoyl chloride and methyl methacrylate give the thiophene derivative E-methyl 3-(2-thienyl) methyl methacrylate (97), equation (5.21).

2-Thenoyl Chloride

(97)

The alkene insertion is largely regiospecific with the acid halide adding to the least hindered portion of the alkene.

4. *By Simmons - Smith Reaction:* According to this procedure divalent sulfur adds to a carbenoid intermediate. Thus treatment of α-oxoketene dithioacetals with $Zn/Cu/CH_2I_2$ on refluxing affords 2-ethylthio-4-phenylthiophene equation (5.22)[151].

5.4.3 Chemical Reactions

Thiophene contains a sulfur atom which belongs to the second row elements of the periodic table. It can expand its valence shell by using the empty *d*-orbitals in hybridization. Consequently instead of having six electrons in five orbitals (as in the case of pyrrole and furan) thiophene by using *pd²* hybrid orbitals might have six electrons in six orbitals. This fact probably accounts for the extreme stability of thiophene and its resemblance to benzene in physical and chemical properties.

1. *Reaction with Acids:* Thiophene is very stable to the action of acids, however, very strong acids may bring about oligomerization. Orthophosphoric acid under milder conditions has been found to give a trimer (98) of the following structure. This trimer is different from the obtained in the case of pyrrole.

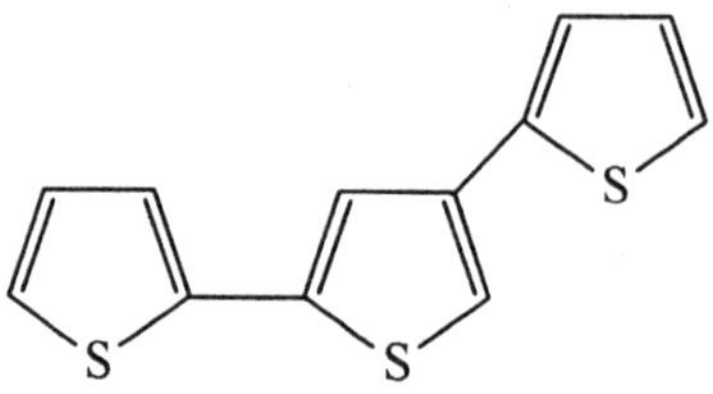

(98)

2. *Electrophilic Substitution:*[152] Because of unsymmetrical charge distribution all the three 5-membered heterocyclic compounds are very reactive towards electrophilic reactions. Thiophene is more reactive than benzene, the reactivity order: pyrrole > furan > thiophene > benzene generally applies and yields of products are generally high. In other words, thiophene is the least reactive of the three but more so than benzene. The majority of the reactions follow the same mechanistic pathway as those of benzene and the preferential attack takes place at position-2.

Halogenation: Thiophene reacts with chlorine to yield a mixture of substituted as well as addition products. Bromination occurs rapidly even in the dark at $-30°C$. But clorination with sulfuryl chloride yields 2-chlorothiophene. In contrast bromination (Br_2/CH_3COOH) always gives substitution product exclusively.[153a] Iodination (I_2/HgO) results in mono- and diiodothiophenes. 2-Bromothiophene may also be obtained by refluxing thiophene with N-bromosuccinimide.

The interaction of heteroaromatic (as well as aromatic) bromides with bases such as Li, Na, KNH_2, LDA etc. results in rearranged products *via* a mechanism called the base-catalyzed **halogen dance mechanism.**

Nitration: The nitration of thiophene under milder conditions HNO_3/Ac_2O) gives a mixture of 2- and 2, 5-dinitrothiophenes, the former being the major product.

Thiophene derivatives are nitrated more readily than the analogus benzene derivatives but less so than furan. In an interesting example for the nitration of 2-furyl-2'-thienyl ketone, the only product obtained was 5-nitro2-furyl-2'-thienyl ketone due to nitration of the furan ring[154], equation (5.23).

$$(5.23)$$

Sulfonation: Thiophene on sulfonation (95% H_2SO_4) at room temperature gives thiophene 2-sulfonic acid readily. This reaction, indeed, provides the basis for its isolation from coal tar. It is noteworthy that such severe sulfonating conditions cannot be employed to pyrrole or furan. Chlorosulfonation also proceeds satisfactorily though in a lower yield of the product, thiophene 2-chlorosulfonic acid.

The Friedel-Crafts Reaction: Acetylation of thiophene can be carried out using a wide variety of conditions. With acetyl chloride in the presence of stannic chloride, thiophene yields 2-acetylthiophene. Thiophene also reacts with phthalic anhydride in the presence of aluminum chloride to yield 2, 2′ - thienoylbenzoic acid (equation 5.24).

$$\text{(5.24)}$$

Direct alkylation of thiophene occurs in the presence of an alkene and 80% sulfuric acid. Among alkenes, ethylene fails to react, propylene reacts slowly while isobutylene reacts readily. The α-selectively decreases in this reaction because of the increasing high vigor of the intermediate carbocation, as a result a mixture of the isomers is obtained.

Several alkyl thiophenes and 2-phenylthiophene give normal coupling products rather than phenylation with 2, 4-dinitrobenzenediazonium ion.[185]

Vilsemeier Formylation: Formylation proceeds rapidly on thiophene with DMF and $POCl_3$.[156]

3.   *Reaction with Carbenes and Nitrenes:* The thermal and photochemical addition of diazoacetate to π-excessive heterocycles to give cyclopropanated product is a well known process. Similarly carboethoxy carbene adds to C_2—C_3 bond of thiophene to give a cyclopropane compound (100) which can be opened with an acid to yield thiophene β-acetic acid ester.

$$\text{(100)}$$

Ethyl thiophene β-acetate

Ethoxycarbonylnitrene surprisingly reacts to give a 1, 4-adduct (101) which loses sulfur to form N-carboethoxypyrrole.

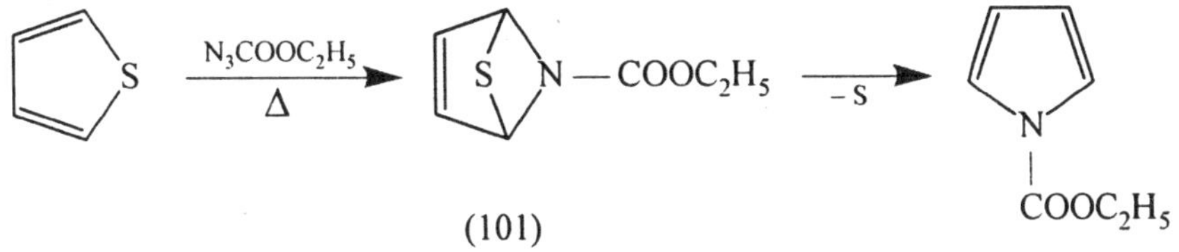

(101)

4. *The Diels-Alder Reactions:* Diels-Alder reactions of thiophene are well documented although not thoroughly investigated.[157] Acetylenic dienophiles are particularly effective and the reaction often involves the chelotropic expulsion of sulfur from the initially formed unstable intermediate (102) as illustrated here to give a benzene derivative.

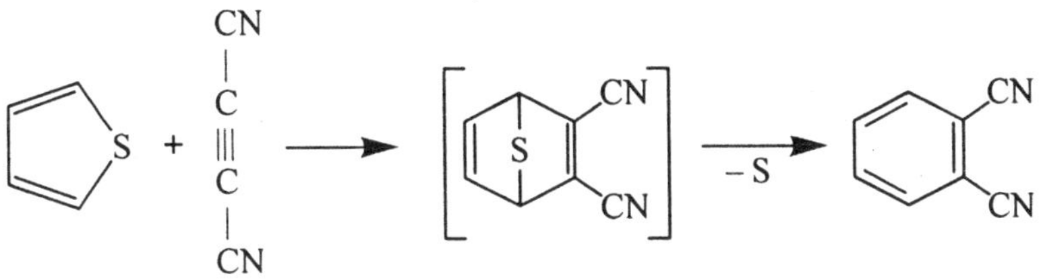

(102)

It does not even form the cycloadduct with the powerful dienophile like benzyne. However, benzyne, obtained from diphenyliodonium 2-carboxylate does undergo reaction with thiophenes first by addition to the sulfur and β-carbon followed by loss of an acetylene molecule to give benzo [b] thiophene.[158]

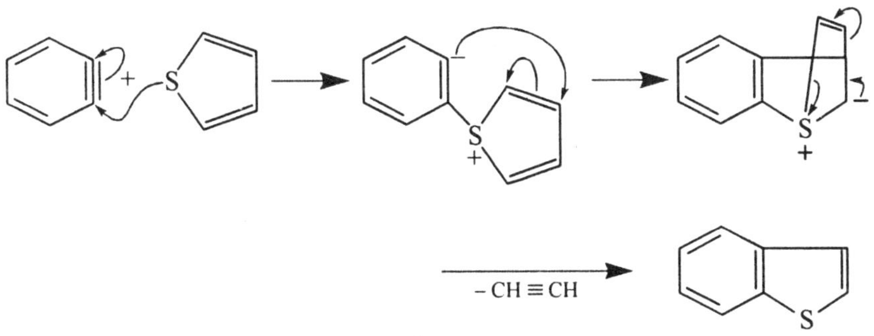

Certain highly substituted thiophenes have been known to undergo $AlCl_3$ catalyzed (2 + 2) cycloaddition with dicyanoacetylene.[159]

Contrary to earlier belief, thiophene has been shown to undergo Diels-Alder reaction with acetylenes and other dienophiles. With maleic anhydride, thiophene produces the *exo*-adduct in a variety of solvents at 100°C and high pressure.

5.   *Reaction with Reducing Agents:* Reduction of thiophene with sodium in a mixture of methanol and ammonia at $-40°C$ gives rise to a mixture of reduced products,[160] equation (5.25). Palladium-charcoal catalyzes the hydrogenation of thiophene to tetrahydrothiophene in 70% yield so long as two parts of the

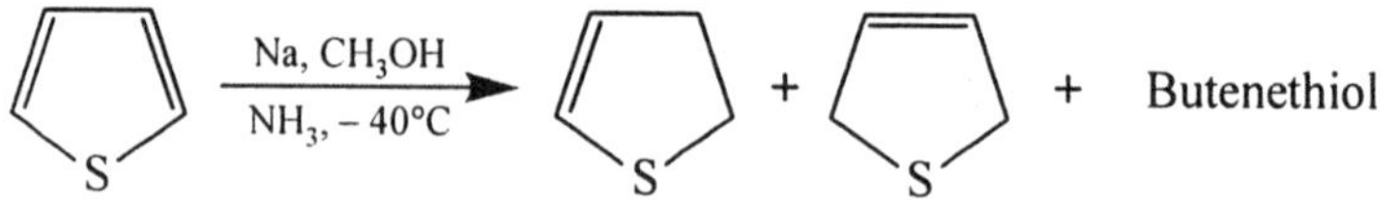

catalyst are employed for one part of thiophene.[161] Raney nickel in acetic anhydride results in hydrogenative desulfurization of thiophene to paraffinic hydrocarbon *n*-butane. This is the most efficient and valuable reaction both as a means of structure determination of thiophenes and for the synthesis of saturated compounds.[162]

6.   *Reaction with Oxidizing Agents:* Thiophene ring system generally resists the action of moderate oxidizing agents. The ring, however, breaks down to maleic acid and oxalic acid when oxidized by nitric acid. Oxidation by peracids such as perbenzoic acid is known to attack the sulfur atom. The initial product, thiophene sulfoxide cannot be isolated because of its dimerization and further oxidation to (103). The thiophene-1, 1-dioxide or thiolane (104) is also reactive but is isolable. Very vigorous oxidation of thiophene ring results in its breakdown to maleic and oxalic acids and sulfur.

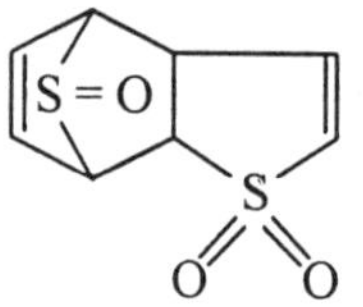

(103) 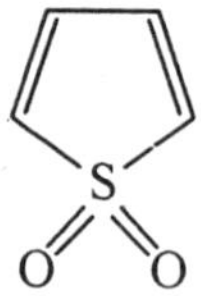   (104)

7.   *Reaction with Nucleophiles:* At a first glance one might anticipate that while electrophilic substitution of thiophene proceeds much more rapidly than benzene, the nucleophilic substitution should be less effective. This, however, is contrary to observation. Every positional combination of the nitro- and halo-thiophenes activates the halide towards nucleophilic substitution.

Chlorothiophenes have been obtained from the corresponding bromothiophenes in a normal displacement reaction, equation (5.26).

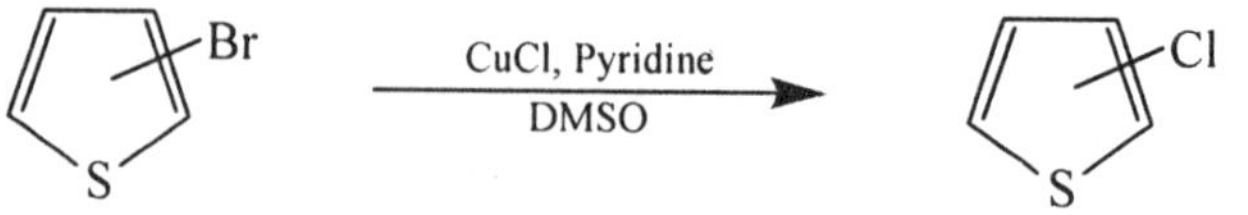

Thiophene reacts with *n*-butyllithium in a manner similar to pyrrole and furan at the α-position to give thienyllithium reagent,[163] which can be used in organic synthesis. Of equal applicability and of synthetic value is the halogen-metal exchange reaction wherein nucleophilic attack occurs at the halogen. While fluoro- and chloro-thiophenes are generally inert, bromo- and iodo-derivatives react almost instantaneously with *N*-butyllithium at –70°C in ether.[164-166]

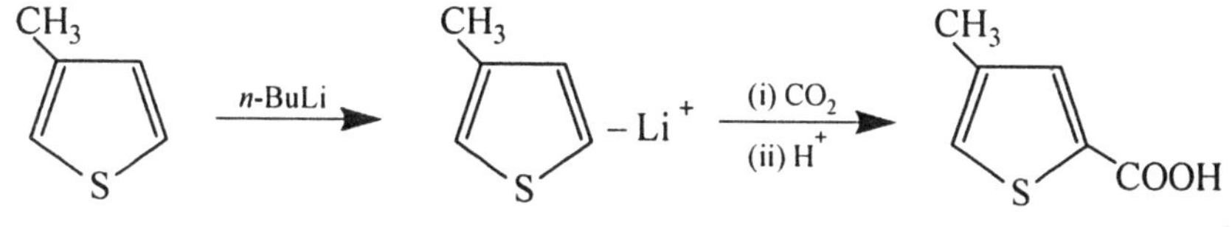

Dibromothiophene undergoes halogen migration *via* the *base-catalyzed halogen dance* (BCHD) mechanism as illustrated below:

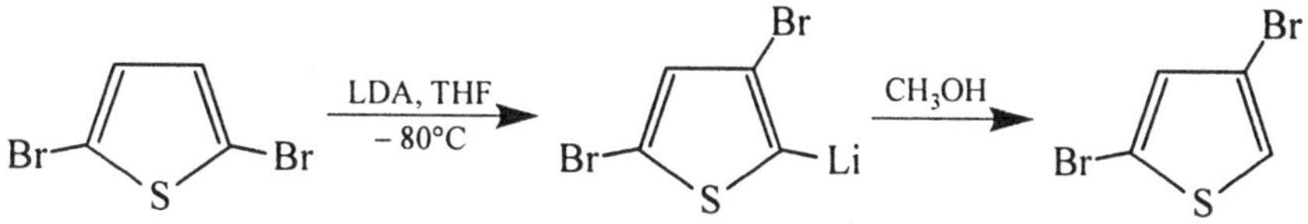

The starting compound 2, 5-dibromothiophene undergoes an initial metallation in the presence of LDA and a rearranged lithium intermediate is obtained. This on quenching with methanol forms 2, 4-dibromothiophene.

8.   *Reaction with Free Radicals:* The Gomberg-Bachmann reaction offers one of the best method for obtaining arylthiophenes,[169, 170] because the method is simple. Most other arylating agents show a 90-95% preference for α-over β-substitution except for phenylazotriphenylmethane which yields a mixture of α- to β-phenylthiophene in a ratio of 4:1. Thiophenes show a slight increase in reactivity towards radicals in comparison to benzene.[171]

   Attempted phenylation, on the other hand, using benzoyl peroxide preferentially yields 2, 2' - bithienyl (103) by an addition-elimination mechanism.[172]

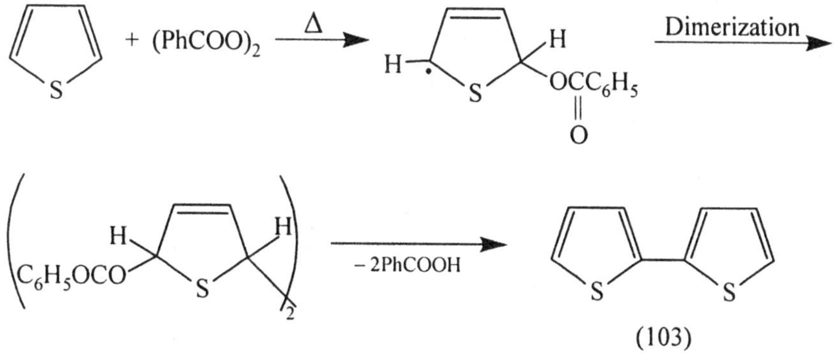

(103)

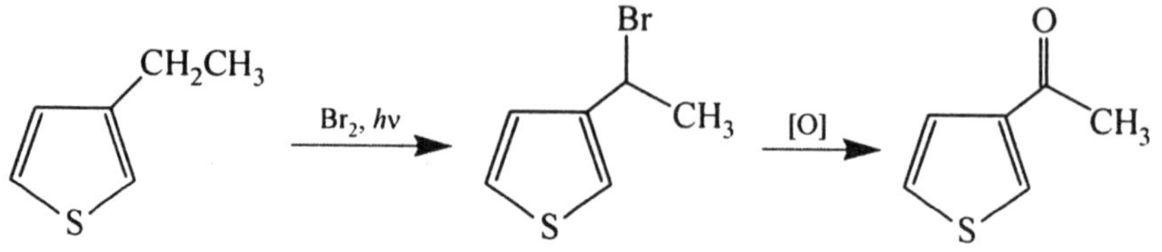

9. Dehydrothiophenes: The formation of a five-membered hetaryne i.e. *thiophyne* was predicted[173] in 1902. There existence still has not been shown with certainty. In contrast to benzyne and related six-membered arynes, the *ortho* halothienyl metal derivatives (104) are remarkably stable and are useful intermediates in synthesis. The ring strain, in such systems is not the only cause for the inability to form hetarynes because cyclopentyne can be made. It is possible that if a hetaryne is formed at all, it reacts by less energetic pathway to give other products.[173] Reinecke, *et.al.*[174] have provided the only evidence for thiophyne formation by flash vacuum photolysis of the anhydride of thiophene 2, 3-dicarboxylic acid and trapping it with suitable reagents.

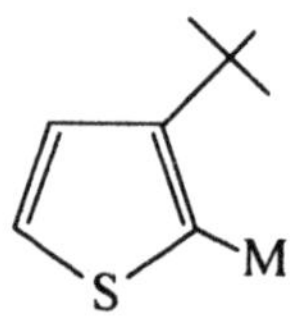

(104)

10. *Reactions of 3, 4- di-t- Butylthiophene:* Sterically crowded 3, 4- *di-t* butylthiophene is produced by an intramolecular reductive coupling of α, α' -diketosulfides in the presence of $TiCl_4$ and zinc followed by acid catalyzed dehydration of the resulting 3, 4-dihydroxythiolanes. Thus *bis* (2- *t*-butyl - 2 - oxoethyl) sulfide (105) leads to *cis* - 3, 4- di *t*-butyl -3, 4-dihydroxythiolane (106) which on subsequent dehydration gives 3, 4-*di t* butylthiophene (107) in 89% yield.[175] This compound behaves similar to thiophene in several of its reactions but there are no side products obtained. For instance, nitration (HNO_3, AcOH, Ac_2O, r.t.) gives 2-nitro derivative in 74% yield. Bromination with N-bromosuccinimide the 2-bromo derivative is the sole product. Oxidation with *m*-chloroperbenzoic acid at room temperature yields the sulfone and in the presence of aluminum chloride (107) rearranges to 2, 4-di *t*-butylthiophene.

(105) → [TiCl$_4$, Zn / THF, 0°C] → (106) → [TsOH / benzene reflux, 1hr] → (107)

11. *Photochemical Reactions:* The photosensitization of thiophene has been investigated by Wynberg and coworkers.[176] The most studied reactions are the isomerisation of 2-substituted thiophenes to their 3-isomers. Although a great deal of information is at hand but the mechanism of this reaction is not fully understood. The photorearrangement of 2-phenylthiophene to 3-phenylthiophene has been shown to proceed according to the following mechanism *via* a number of valence bond isomers of thiophene.

Alternative mechanisms[177] have also been suggested for this reaction as well as for the photoisomerization of *bis* (trifluoromethyl) thiophenes.[178] Similarly 2-(2-thienyl) thiophene rearranges to give 2-(3-thienyl) thiophene.[179] Photocycloaddition reactions are also known in the thiophene series. For instance, photogenerated singlet oxygen does not attack thiophene unlike pyrrole and furan but reacts with alkylthiophenes.[180] certain unsaturated compounds like alkenes, arynes and ketones also add involving (4 + 2) and (2 + 1) reactions, equation (5.26)[181, 182]

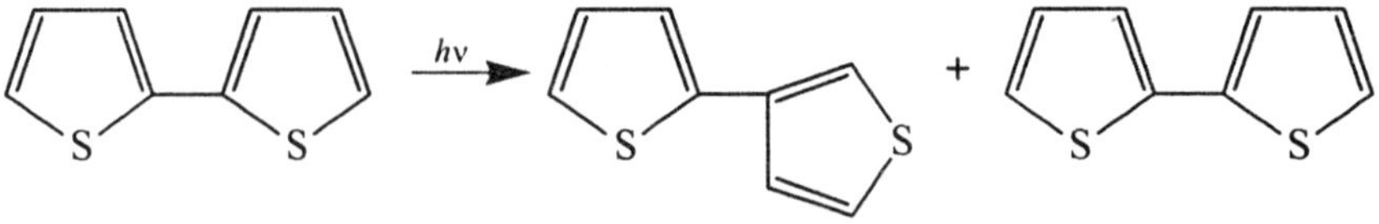

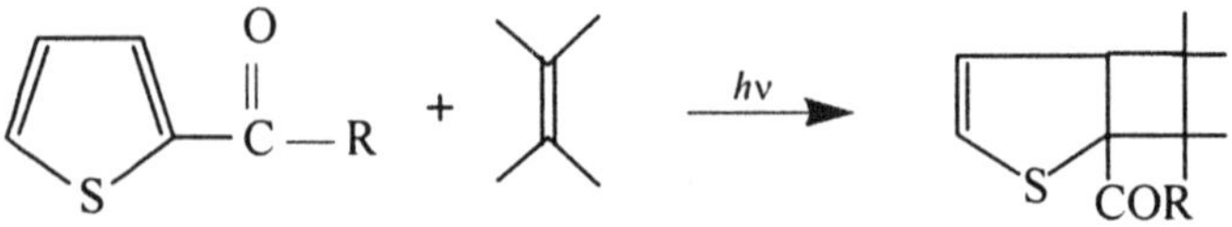

5.4.4 Derivatives of Thiophenes

Allyl- and Aryl-Thiophenes: Alkylthiophenes have been prepared by several methods. Alkylation using an alkene in the presence of an acid occurs readily but it is not a useful synthetic method. 2-Alkylthiophenes can be obtained conveniently instead by the Wurtz-Fittig reaction, between 1-iodothiophene and an alkyl iodide.

The Clemmensen and Wolff-Kishner reductions of acylthiophenes have been widely employed to obtain alkylthiophenes. Several ring closure methods using appropriate ketones or esters in the presence of phosphorus trisulfide have been used, as demonstrated below (equations 5.27 and 5.28).

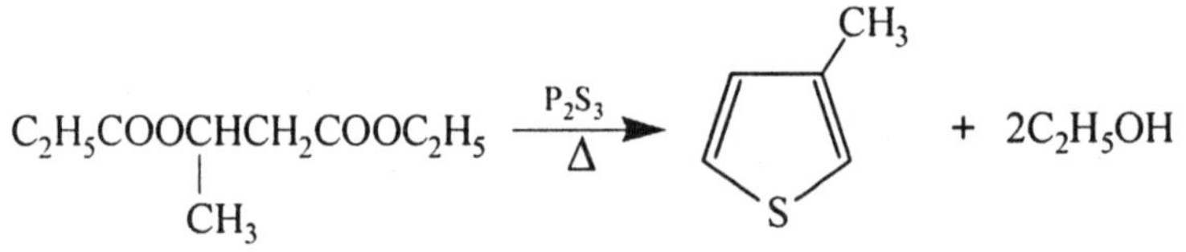

$$C_2H_5COOCHCH_2COOC_2H_5 \xrightarrow[\Delta]{P_2S_3} \quad + \quad 2C_2H_5OH \qquad (5.27)$$

$$CH_3CCH_2COOC_2H_5 \xrightarrow[\Delta]{P_2S_3} \quad + \quad 2C_2H_5OH + H_2O \quad (5.28)$$

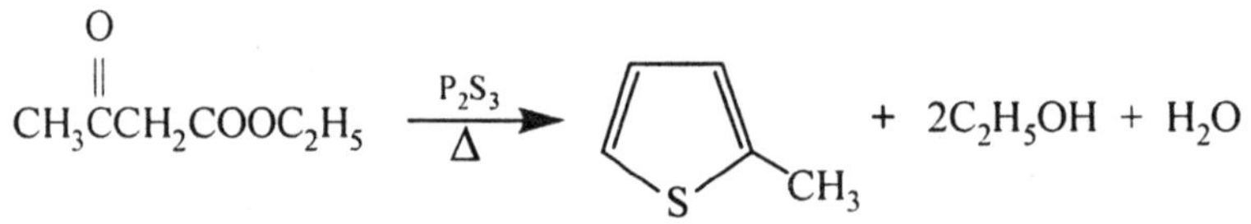

2-Vinylthiophene is obtained in the following steps:

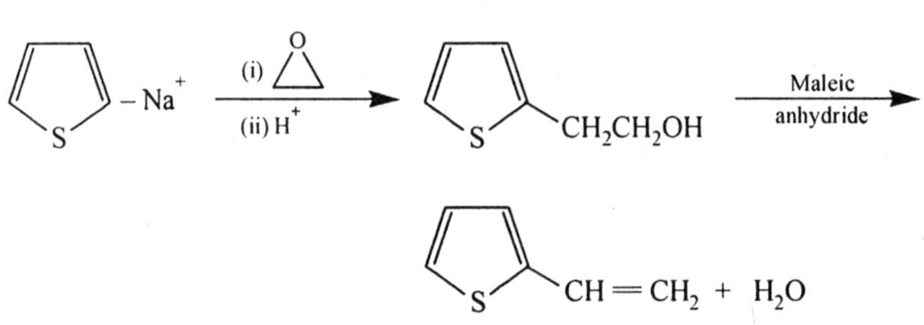

2-Methylthiophene is easily attacked by electrophiles, thus its nitration (HNO_3, Ac_2O) yields 2-methyl-5-nitrothiophene (70%) and 2-methyl-3-nitrothiophene (30%). For the 3-methylthiophene, the incoming electrophile takes position 2- or 5.

The arylthiophenes have been prepared by cyclizing the appropriate keto acids with phosphorus pentasulfide. Recently[183] Pd catalyzed nucleophilic arylation of thiophene has been achieved. Bromobenzene and 2-thiophenecarboxamide on refluxing or *o*-xylene using palladium acetate yields 2, 3, 5, triphenylthiophene in high yields.

83%

Such derivatives are used as biologically active compounds. Tetra-phenylthiophene is obtained by heating two molecules of stilbene with sulfur. It is very stable molecule and does not decompose upto 460°C.

Halothiophenes: 2-Chloro- or bromo-thiophenes are the only ones prepared by direct electrophilic substitution, 2-chlorothiophene is a colorless liquid, b.p. 128°C. It undergoes the usual substitution reactions. It also reacts readily with magnesium to give good yield of theinyl Grignard reagent. The best yield of 2-bromothiophene is obtained by refluxing thiophene with N-bromosuccinimide. 3-Bromothiophene is obtained from the easily prepared 2, 3, 5-tribromothiophene by treating with Zn/CH_3COOH. 3-Bromothiophene does not form the Grignard reagent.

Aminothiophenes: Most aminothiophenes have been prepared by the reduction of the corresponding nitrothiophenes and Hofmann rearrangement of the acid amides. The latter reaction, however, fails for 2-thienamide in contrast to the 3-isomer.

2-Aminothiophene is a liquid b.p. 77-70°C/11 mm, and decomposes readily in air but is stable under inert atmosphere. 2 - Aminothiophene can be diazotized and subsequently couples with β - naphthol to form (108).

(108)

Hydroxythiophenes or Thienols: One of the most marked differences between benzene and thiophene chemistry is the extreme instability of the thiophene analogs of compounds like aniline and phenol. 2 – and 3 – Hydroxythiophenes have been prepared by the reaction of thienyl Grignard reagents with oxygen in the presence of iso-propyl magnesium bromide. Similarly thienyllithium derivatives with hydroperoxides also yield thienols.

Both the isomers are soluble in alkali and give deep green color with $FeCl_3$. The tautomerism of 2-thienols with the corresponding 3-thienols seems to have been well studied. In the 2-hydroxythiophene series, all the three tautomers are noted, and in general are non-aromatic.

On the other hand, for 3-hydroxythiophenes only two tautomers are possible, and in general the hydroxy form is more favored.

Thiophene Aldehydes and Ketones: 2-Thiophene carbaldehyde is obtained by the Vilsmeier reaction on thiophene. It is a liquid b.p., 198°C and resembles benzaldehyde in smell and many of its chemical properties. Among other reactions it reacts smoothly in benzoin condensation.

2,2′-Thenoin

2,2′-Thenil   **2,2′-Thenilic acid**

2-Acylthiophene has been obtained in 90% yield by the acylation of thiophene with acetyl chloride in the presence of $AlCl_3$ while the 3-isomer is obtained as follows:

The acylthiophenes are reduced by Wolff-Kishner and Clemmensen reductions, oxidized by hypochlorite and undergo reactions with hydroxylamine, phenyl hydrazine and Claisen ester condensation. They also undergo the Reformatsky reaction.

Thiophene Carboxylic Acids and Esters: The hypochloride oxidation of theonyl compounds results in the corresponding theonic acids. The 2- or 3-methylthiophenes are also converted to the acids in high yield with aqueous dichromate.[184] 2-thiophenecarboxylic acid m.p. 129-130°C resembles benzoic acid. It can be nitrated (HNO_3, Ac_2O) to yield 3-nitro-thiophene. The acids decarboxylate in the presence of copper-quinoline catalyst.

Carbonylation of pyrrole, furan, thiophene or benzofuran yields the corresponding 2-carboxylic acid esters with CO/C_2H_5OH[185], equation (5.29).

$$(5.29)$$

X = N, O, S

5.4.5 Naturally and Biologically Active Compounds

The vitamin biofin (109) is an essential growth factor for a number of micro-organisms for animals. It was first isolated from egg yolk but it is widely distributed and is easily obtained from milk concentrates.

(109)

REFERENCES

1. A. Grossauer, *Die Chemie der Pyrrole*, Springer-Verlag, Berlin (1974), p.21.
2. R. A. Jones, *Adv. Heterocyclic Chem.*, **11**, 383 (1970).
3. W. S. Wilcox and J. H. Goldstein, *J. Chem. Phys.*, **20**, 1656 (1952).
4. C. A. Coulson and H. C. Longet-Higgins, *Rev. Sci. Paris*, **85**, 929 (1947); R. D. Brown, *Aust. J. Chem.*, **8**, 100 (1955).
5. For a review see, H. J. Anderson and C. E. Loader, *Synthesis*, 353 (1985).
6. H. El. Khodem *et al.*, *J. Heterocyclic Chem.*, **9**, 1413 (1973).
7. H. Fischer, *Org. Synth. Coll.* Vol., **2**, 202 (1943), also see, G. G. Kleinspehn, *J. Am. Chem. Soc.*, **77**, 1546 (1955); G. G. Kleinspehn, and A. H. Corwin, *J. Org. Chem.*, **25**, 1048 (1960).
8. C. Paal, *Ber.*, 18, 367 (1885); L. Knorr, *ibid.*, **18**, 299, 1568(1885).
9. L. F. Miller and R. E. Banbury, *J. Med. Chem.*, **13**, 1022 (1970).
10. H. S. Broadbent, *J. Heterocyclic Chem.*, **5**, 757 (1968).
11. M. W. Room and S. F. McDonald, *Canad. J. Chem.*, **48**, 1689 (1970).
12. K. Ichimura, S. Ichikawa and K. Imamura, *Bull. Chem. Soc. Japan*, **49**, 1157 (1976).
13. A. Padwa R. Gruber and L. Hamilton, *J. Am. Chem. Soc.*, **89**, 3077 (1967); T.Y. Chen., *Bull. Chem. Soc. Japan*, **41**, 2540 (1968).
14. H. Posvic, R. Dombro, H. Ito and Ti. Telinski, *J. Org. Chem.*, **39**, 2575 (1974).
15. J. E. Baldwin and W. Kaiser, *Tet. letters*, 2185 (1964).
16. Y. K. Shin, J. I. Youn, J. S. chin, T. H. Park, M. H. Kin and W. J. Kin, *Synthesis*, 753 (1990).
17. D. Van Leusen, E. Flentge and A. M. Van Leusen, *Tetrahedron*, **47**, 4639 (1991); also see D. H. R. Barton, J. Kervagovet and S. Z. Zard, *ibid*, **46**, 7587 (1990).

18. J. S. Warmus, G. J. Dilley and A. I. Meyers, *J. Org. Chem.* **58,** 270 (1993).

19. K. Hafner and W. Kaiser, *Tet. letters*, 2185 (1964).

20. R. A. Jones and C. F. Candy, *J. Org. Chem.*, **36,** 3993 (1971).

21. H. Heaney and S. V. Ley, *J. Chem. Soc.*, *Perkin Trans.*, **1,** 499 (1973).

22. S. Yu, J. Saenz and J. K. Sriramgam, *J. Org. Chem.*, **67,** 1699 (2002).

23. J. P. Wolfe and S. L. Buchwald, *J. Org. Chem.*, **65,** 1144 (2000); J. F. Hartwig, Angew. Chem. Int. Edin. (Engl), **37,** 2047 (1998); D. M. T. Chan et al.,

24. J. M. Patterson, L. T. Burka and M. R. Boyd., *J. Org. Chem.*, **33,** 4033 (1968).

24a. I. A. Jacob and M. B. Jensen, *J. Phys. Chem.*, **68,** 3066 (1964) and references cited therein.

25. G. Marino in, *Advances Heterocyclic Chemistry*, A. R. Katritzky and A. J. Boulton (Eds.) Vol. 13, Academic Press, New York (1971), pp. 253-314.

26. M. D. Rosa, *J. Org. Chem.*, **47,** 1008 (1982), *ibid.*, **43,** 2639 (1978).

27. H. M. Gilow and D. E. Bourton, *J. Org. Chem.*, **46,** 222 (1981).

28. For a comprehensive review, see, G. A. Olah and S. J. Kuhn, *Friedel Crafts and Related Reactions*, Vol. 3, Part 2, Interscience, New York (1964).

29. A. R. Cooksey, K. J. Morgan and D. P. Morrey, *Tetrahedron*, **26,** 6101, 5101 (1970).

30. NG. PH. Bou-Hoi and J. Searle, *Jr. J. Org. Chem.*, 20, 608 (1955).

31. S. S. Dizey, *Synthetic Reagents*, Vol. J. Billis Haywood Ltd., New York (1974), p. 54.

32. R. L. Jones and C. W. Rees, *J. Chem. Soc.* (C), 2249 (1969).

33. E. R. Alexander, A. B. Herrick and T. M. Roder, *J. Am. Chem. Soc.*, **72,** 2760 (1950);

34. A. Gambacorta *et al.*, *Tetrahedron*, **27,** 985 (1971).

35. M. Nagazaki, *Bull. Chem. Soc. Japan*, **76,** 1169 (1955).

36. G. M. Badger, J. A. Ellix and G. E. Lewis, *Aust. J. Chem.*, **20,** 1777 (1967);

37. K. Hafner and W. Kaiser, *Tet. letters*, 2185 (1964).

38. R. A. Barcock, N. A. Moorcraft, R. C. Shorr, J. H. Young and L. S. Fuller, *Tet. Letters*, 1187 (1993).

39. A. D. Abell and J. C. Litten, *ibid.*, 30005 (1992).

40. R. A. Jones, T. A. Saliente and J. S. Arques *J. Chem. Soc. Perkin Trans* I, 2541 (1984); R. A. Jones, M. P. Marriot, W. P. Rosenthal and J. S. Arques, *J. Org. Chem.*, **45,** 4515 (1980).

41. For a comprehensive review see, R. A. Jones and G. P. Bean, *The Chemistry of Pyrroles*, Academic Press, New York (1977), pp. 146 and 256.

42. R. C. Bansal, A. W. McCulloch and A. G. McInnes, *Canad. J. Chem.*, **53,** 138 (1975), A. G. Schultz. M. Shen and R. Ravichandran, *Tet. letters*, 1767, (1981), E. Vogel *et al.*, *Angew. Chem. Ind. Edin* (Engl.), **22,** 778 (1983).

43. C. K. Lee, C. S. Hahn and W. E. Noland, *J. Org. Chem.*, **43,** 3727 (1978).

44. K. A. Shafiq, H. Plieninger and D. Wild, *Chem. Ber.*, **106,** 812 (1973).

45. R. A. Jones and J. S. Arques, *Tetrahedron*, **37,** 1597 (1981).

46. R. C. Bansal, A. W. McCulloch and A. G. McInnes, *Canad. J. Chem.*, **47,** 2391 (1969); *ibid.*, **48,** 1472 (1970).

47. J. Sauer, *Angew Chem. Int. Edin.* (Engl.), **6,** 16 (1947).

48. M. E. Jung and J. C. Rohloff, *Chem. Comm.*, 630 (1984).

49. A. Eddaif, *Tet. letters*, 2779 (1984).

50. C. B. Hudson and A. V. Robertson, *Tet. letters*, 4015 (1967).

51. J. B. Wibaut, *Ned. Akad. Wetenschap Proc. Sen.* (B.), **68,** 117 (1965), C. A. **63,** 11306 (1965).

52. G. P. Gardini and V. Bocchi, *Gazzetta*, **102,** 91 (1972).

53. E. Chung wu and J. Heicklen, *Canad. J. Chem.*, **50,** 1678 (1972); *J. Am. Chem. Soc.*, **93,** 3432 (1971).

54. J. M. Patterson and L. T. Burka, *Tet. letters*, 2215 (1969).

55. J. H. Babler and K. P. Spina; *Tet. letters*, 1659 (1984); *J. Org. Chem.*, **47,** 3668 (1982); *ibid.*, **45,** 5336 (1980); K. Utimoto H. Miwa and H. Nozaki, *Tet. letters*, 4277 (1981).

56. D. J. Chadwick, *Chem. Comm.*, 790 (1974).

57. H. Shizuka *et al.*, *Mol. Photochemistry*, **1,** 135 (1969).

58. J. Nakayama, Y. Singihara, K. Terada and E. L. Clennan, *Tet. letters*, 4473 (1990).

59. T. Mukai, *Tet. letters*, 1067 (1975); W. C. Guida and D. J. Mathre, *J. Org. Chem.*, **45,** 3172 (1980).

60. G. H. Ellames, C. T. Hewkin, R. F. W. Jackson, D. I. Smith and S. P. Standen, *Tet. letters*, 3471 (1989).

61. R. F. Raffauf, T. M. Zennie, K. D. Onam, and P. W. Le Quesne, *J. Org. Chem.*, **49,** 2714 (1984).

62. *Porphyrins and Metalloporphyrins*, (*Ed.*), K. M. Smith, Elsevier, Amsterdam (1975).

63. J. M. Orten, *Rec. Chem. Progr.* (Kresge-Hocker Sci. Lib.), **17,** 259 (1956).

64. R. B. Woodward, *Angew. Chem.*, **72,** 651 (1960); R. B. Woodward *et al.*, *J. Am. Chem. Soc.*, **82,** 3800 (1960).

65. D. C. Hodgkin *et al.*, *Nature*, 176, 325 (1955).

66. R. B. Woodward *et al.*, *J. Am. Chem. Soc.*, **82,** 3850 (1960); R. B. Woodward *et al.*, *Pure Appl. Chem.*, **33,** 145 (1973).

67. A. P. Dunlop and F. N. Peters, *The Furans*, Reinhold, New York (1953).

68. C. Sandorfy and P. Yvon, *Bull. Chim. Soc. France*, 131 (1950).

69. R. D. Brown and B. A. W. Coller, *Aust. J. Chem.*, **12,** 152 (1959).

70. For a review see, P. Bosshard and C. H. Eugster, *Adv. Heterocyclic Chem.*, **7,** 377 (1966).

71. R. Adams and V. Voorhees, *Org. Syn. Coll. Vol.*, **1,** 280 (1941).

72. D. M. Burness, *ibid.*, **4,** 628 (1963).

73. E. C. Kornfield and R. G. Jones, *J. Org. Chem.*, **19,** 1671 (1954).

74. L. D. Krasnoslobodskaya and Ya. L. Gold'barf, *Russ. Chem. Rev.*, **38,** 389 (1969).

75. L. E. Edelman *J. Am. Chem. Soc.*, **72,** 5765 (1950).

76. E. J. Nienhouse, R. M. Irwin and G. R. Finni, *J. Am. Chem. Soc.*, **89,** 4557 (1967).

77. Y. Fukuda, H. Shiragami, K. Utimoto and H. Nozaki, *J. Org. Chem.*, **56,** 5816 (1991).

78. J. A. Marshall and E. D. Robinson, *ibid.*, **55,** 3450 (1990).

79. R. Lakhan and B. Ternai, *Adv. Heterocyclic Chem.*, **17,** 99 (1974).

80. A. T. Balaban and C. D. Nenitzescu, *Ber.*, **93,** 599 (1960).

81. A. Meyeyers, *Heterocycles in Organic Synthesis*, Wiley, New York (1974), p.222.

82. D. Wirth and D. M. Lemal, *J. Am. Chem. Soc.*, 104, 847 (1982); Also see, I.G. Pitt, *Chem, Comm.*, 131 (1978).

83. J. A. Barltrop and A. C. Day *Chem. Comm.*, 131 (1978).

84. K. Atsumi and I. Kuwajima, *J. Am. Chem. Soc.*, **101,** 2208 (1979).

85. L. I. Belen'kii, Izvest. *Akad, Nauk. SSSR.*, *Ser. Khim.*, 344 (1975).

86. A. P. Terent'ev *et al.*, *Zhur. Obshchel Khim.*, **24.** 1265 (1955).

87. J. G. Michels *et al.*, *Med. Pharm. Chem.*, **5,** 1042 (1962); C.A., **58,** 5602 (1962).

88. G. Olah, S. Kuhn and A. Mlinko, *J. Chem. Soc.*, 4257 (1956).

89. L. A. Kazitsyna, *Uch. Zap. Moskov. Gosu. Univ.*, **131,** 5 (1950); C.A., **47,** 10518 (1953).

90. C. K. Dien and R. E. Lutz, *J. Org. Chem.*, **21,** 1096 (1956).

91. N. Elming, *Acta. Chem. Scand.*, **6,** 605 (1952).

92. J. G. Traynelis J. J. Miskel, Jr. and S. R. Sowa, *J. Org. Chem.*, **22,** 1269 (1957).

93. E. Muller, H. Kessler, M. Fricke and H. Suhro, *Tet. letters*, 1047 (1963).

94. G. O. Schenck and R. Steimmetz, *Ann.*, **668,** 19 (1963).

95. D. W. Jones, *J. Chem. Soc.*, *Perkin*, **1,** 2728 (1972).

96. A. P. Dunlop and F. Peters, *The Furans*, Reinhold, New York, (1963), pp. 54-64.

97. W. G. Dauben and H. O. Krabbenhoft, *J. Am. Chem. Soc.*, **98.** 1993 (1976); D. D. Sternbach *et al.*, *Tet letters*, 591 (1985); Pilaszlo and J. Lucchetti, *Tet. letters*, 4387 (1984); P. F. Schuda and J. M. Bennett, *ibid.*, 5525 (1982).

98. H. Stockman, *J. Org. Chem.*, **26**, 2025 (1961).

99. H. Kwart and I. Burchuk, *J. Am. Chem. Soc.*, **74**, 3094 (1952).

100. R. W. La Roschelle and B. M. Trost, *Chem. Comm.*, 1353 (1970).

101. F. A. L. Anet, *Tet. letters*, 1219 (1982).

102. Y. Nishiyama *et al.*, *Tetrahedron*, **59**, 6609 (2003).

103. H. Prinzbach *et al.*, *Chem. Ber.*, **109**, 2823 (1976).

104. A. W. McCulloch and A. G. McInnes, *Canad. J. Chem.*, **52**, 143 (1974).

105. M. B. Stinger and D. Wege, *Tet. letters*. 591 (1985).

106. D. D. Sternbach, D. M. Rossana and K. D. Onan, *Tet. letters*, 591 (1985).

107. M. V. Sargent *et al.*, *Aust. J. Chem.*, **53**, 267 (2000).

108. A. B. Smith, N. J. Liverton, N. J. Hrib, H. Sivaramakrishna and K. J. Winzenvberg, *J. Am. Chem. Soc.*, **108**, 3040 (1986).

109. N. I. Shuikin and I. F. Bel'skii, *Dokl. Akad. Nausk, USSR*, **125**, 345 (1959), C.A., 53, 20015 (1959).

110. A. A. Gorman, I. R. Gould and I. Hamblett, *Tel. letters*, 1087 (1980); I. L. Doerr and R. E. Wilette, *J. Org. Chem.*, **38**, 3878 (1973).

111. N. A. Milas, R. L. Peeler and O. L. Mageli, *J. Am. Chem. Soc.*, **76**, 2322 (1954).

112. D. Seebach, *Chem. Ber.*, **96**, 2712 (1963).

113. A. J. Baggaley and R. Brettle, *J. Chem. Soc.*, (C), 969 (1968).

114. K. Manfredi, S. B. Gingerich and P. W. Jennings, *J. Org. Chem.*, **50**, 535 (1985).

115. V. Ramanathan and R. Levine, *J. org. Chem.*, **27**, 1216 (1962).

116. T. Ando, T. Kawate, J. Yamawaki and T. Hanafusa, *Synthesis*, 637 (1983).

117. J. L. Luche and J. C. Domiano, *J. Am. Chem. Soc.*, **102**, 7926 (1980).

118. R. Srinivasan, *J. Am. Chem. Soc.*, **89**, 1758 (1967); *Pure, Appl. Chem.*, **16**, 65 (1968).

119. A. Zamojski and T. Kozluk, *J. Org. Chem.*, **42**, 1089 (1977).

120. G. S. Hammond and N. J. Turro, *Science*, **142**, 1541 (1963).

121. D. Bryce-Smith, J. C. Berridge and A. Gilbert., *Chem. Comm.*, 611 (1975).

122. H. Hattori *et al.*, *Chem. Comm.*, 1000 (2001).

123. T. Saeguza and T. Ueshima, *J. Org. Chem.*, **33**, 3310 (1968).

124. L. T. Scott and J. O.Naples, *Synthesis*, 209 (1973).

125. P. E. Verkade *et al.*, *Rec. Trav. Chem.*, **74**, 763 (1965).

126. Z. N. Nazavova, *Zhur. Org. Khim.*, **70**, 1341 (1974).

127. J. Burdon, J. C. Tetlow and D. F. Thomas, *Chem. Comm.*, 48 (1966).

128. M. M. Joullie, S. Divald, M. C. Chun and M. M. Joullte, *J. Org. Chem.*, **41**, 2835 (1976).

129. L. A. Spurlock and R. G. Fayter., *J. Org. Chem.*, **34**, 4035 (1969).

130. F. Ebinito, *Heterocycles*, **2**, 391 (1974).

131. J. L. Isidor, M. S. Brookhart and R. L. Mckee, *J. Org. Chem.*, **38**, 612 (1973).

132. Y. S. Rao, *Chem. Rev.*, **76**, 625 (1976).

133. A. B. Hornfeldt, *Arkiv, Kemi.*, **29,** 229 (1969).

134. F. M. Dean, *Naturally Occurring Oxygen Ring Compounds*, Butterworths, London (1963).

135. For a review on the tautomerism of hydroxyfurans, see, A. R. Katritzky *et al.*, *Adv. Heterocyclic Chem.*, Suppl. **1,** 214 (1976).

136. T. S. Cantrell, *J. Org. Chem.*, **42,** 3774 (1977).

137. M. R. Boyd. T. M. Harris and E. J. Wilson, *Synthesis*, 545 (1971).

138. S. A. Nash and R. B. Gammill, *Tet. letters*, 4003 (1987).

139. D. Christenson *et al.*, *J. Mol. Spectroscopy*, 7, 58 (1961).

140. R. A. Hoffman and G. Gronowitz, *Arkiv. Kemi.* **15,** 45 (1959).

141. R. J. Abraham and H. J. Bernstein, *Canad. J. Chem.*, **37,** 1056 (1959).

142. D. Binder and P. Stanetty, *Synthesis*, 200 (1977).

143. S. Gronowitz, *Adv. Heterocyclic Chem.*, **1,** 1 (1963).

144. H. Fiesselmann and F. Thoma, *Chem. Ber.*, **89,** 1907 (1956) and references cited therein.

145. D. E. Wolf and K. Polkers, *Org. Reactions*, **6,** 410 (1951).

146. P. J. Garratt and S. B. Neoh, *J. Org. Chem.*, **40,** 970 (1975).

147. H. Nozaki, T. Koyama and T. Mori, *Tetrahedron*, **25,** 5357 (1969).

148. M. P. Cava and M. V. Lakshmikantham, *Acc. Chem. Res.*, **8,** 139 (1975).

149. P. J. Carvatt and D. N. Nicolaideo, J. Org. Chem., **39** 2222 (1974).

150. A. Kasahara, T. Izumi and T. Ogihara, *J. Heterocyclic Chem.*, **26,** 597 (1989).

151. L. Bhatt, A. Thomas, H. Ila and H. Junjappa, *Tetrahedron*, **47,** 10377 (1992).

152. G. Marino; in *Advances in Heterocyclic Chemistry*, A. R. Katritzky and A. J. Boulton, (*Eds.*), Vol. 13, Academic Press, New York (1971); pp. 235–314.

153a. S. O. Leuosson, *Ark. Kemi*, **11,** 373 (1957).

153b. P. Rocca *et. al.*, *J. Org. Chem.*, **58,** 7832 (1993); J. F. Bunnett, *Acc. Chem. Res.*, **5,** 139 (1972).

154. H. Gilman and R. V. Young, *J. Am. Chem. Soc.*, **56,** 464 (1934).

155. P. Linda, G. Marino and S. Santini, *Tet. letters*, 4223 (1970).

156. M. Bartle, S. T. Gore, R. K. Mackie and J. M. Tedder, *J. Chem. Soc.*, Perkin I, 1636 (1976).

157. D. N. Reinhoudt *et al.*, *Tel. letters*, 4777 (1976), D. N. Rheinhoudt and C. G. Kouwenhoven, *Tetrahedron*, **30,** 2093 (?1974).

158. D. D. Muzza and M. G. Reinecke, *Chem. Comm.*, 124 (1981).

159. R. H. Hall H. J. Den Hertog and D. N. Reinhoudt, *J. Org. Chem.*, **45,** 967 (1982).

160. A. J. Birch and D. T. McAlan, *J. Chem. Soc.*, 2556 (1951).

161. R. pettit, *Tet. letters*, 11 (1960).

162. A. I. Meyers, *Heterocycles in organic Synthesis*, Wiley Interscience, New York, (1974), pp. 15,228, 243 and 266.

163. S. Melander, *Arkiv. Kemi*, **11**, 397 (1957).

164. P. Moses and S.Gronowitz, *Arkiv. Kemi.* **18,** 119 (1962).

165. S. Gronowitz and B. Holm, *Acta. Chem. Scand.*, **30**, 505 (1976).

166. S. Gronowitz *et al.*, *Chem. Scripta*, **5**, 217 (1974).

167. H. Frohlich and W. Kalt, *J. Org. Chem.*, **55**, 2993 (1990).

168. E. C. Taylor and D. E. Vogel, *ibid*, 50, 1002 (1985).

169. S. Gomberg and W. E. Bachmann, *J. Am. Chem. Soc.*, **46**, 2339 (1924).

170. G. Vernin and J. Metzger, *J. Org. Chem.*, **40**, 3183 (1975).

171. J. Gegani, M. Polloti and A. Tundo, *Ann, Chim.* (Rome), **51**, 434 (1961).

172. F. Minisci and O. Porta, *Adv. Heterocyclic Chem.*, **16**, 123 (1974).

173. M. G. Reinecke, H. W. Addickes and C. Pyun, *J. Org. Chem.*, **36**, 2690 (1971); *ibid.*, **36**, 3820 (1971).

174. M. G. Reinecke and J. G. Newkom, *J. Am. Chem. Soc.*, **87,** 3998 (1965).

175. J. Nakayama, S. Yamaoka and M. Hoshino, *Tet. letters*, 1161 (1988).

176. H. Wymberg *Acc. Chem. Res.*, **4,** 65 (1971).

177. H. Wymberg and H. Van Driel, *J. Am. Chem. Soc.*, **87**, 3998 (1965).

178. Y. Kobayaski, K. Kwada and H. Nozaki, *Tet. letters*, 1917 (1984).

179. H. Wymberg, R. M. Kellog, J. K. Dik and H van Driel, *J. Org. Chem.*, **35,** 2737 (1970).

180. C. N. Skold and R. H. Schlessinger, *Tet. letters*, 791 (1970); H. H. Wasserman and W. Strehlow, *ibid.*, 795 (1970).

181. H. J. Kuhn and K. Gollnick, *Tet. letters*, 1909 (1972).

182. T. S. Cantrell, *J. Org. Chem.*, **39**, 2242 (1974).

183. M. Miura *et. al.*, *J. Am. Chem. Soc.*, **124,** 5286 (2002). Also see N. Miyaura (Ed.), Cross Coupling Reactions, springer, Berlin (2002).

184. L. Friedman, D. L. Fishel and H. Shechter, *J. Org. Chem.*, **30,** 1453 (1965).

185. R. Jaouhari and P. H. Dixneuf, *Tet. letters*, 6315 (1986).

REVIEW PROBLEMS

5.1 Predict the major product of the following reactions:

(*a*)   isoprene (CH₃-substituted butadiene) + S $\xrightarrow{600°C}$

(*b*)   thiophene-2-OC(CH₃)₃ $\xrightarrow{H^+}$

(*c*)   NCCH(Br)–CH₂–CH(Br)CN $\xrightarrow{RNH_2}$

(d)

(e)

(f)

(g)

(h)

(i)

(j)

(k)

(*l*) [furan-CO-CH₂Br structure] → (i) NaBH₄ / (ii) Alkali

(*m*) [epoxycyclohexane with CH₂COCH₃ structure] → H⁺ / Δ

(*n*) [2,5-dimethylpyrrole (N-H) structure] → CHCl₃, NaOH / Δ

(*o*) [N-methyl-3-vinylpyrrole] + [dimethyl acetylenedicarboxylate, COOCH₃—C≡C—COOCH₃] → Δ

(*p*) [2-furyl 2-thienyl ketone structure] → HNO₃, Ac₂O / − 5CO

(*q*) [3-methoxyfuran] + [acrolein, CH₂=CH—CHO] → Δ

(*r*) [2-methylene-2H-pyrrole] + [pyrrole-2-CH₂NMe₂ (N-H)] →

(*s*) [2-(oct-1-ynyl)cyclohexanone, C≡C C₆H₁₃] → (i) PdCl₂, r. t. / (ii) H⁺

(*t*) [structure] + [structure] SO$_3$Ph $\xrightarrow{\text{r. t.}}$

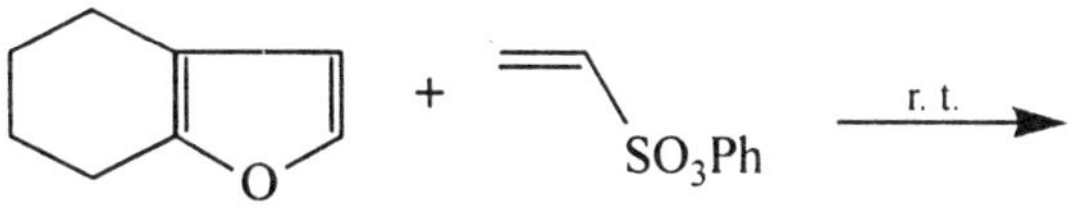

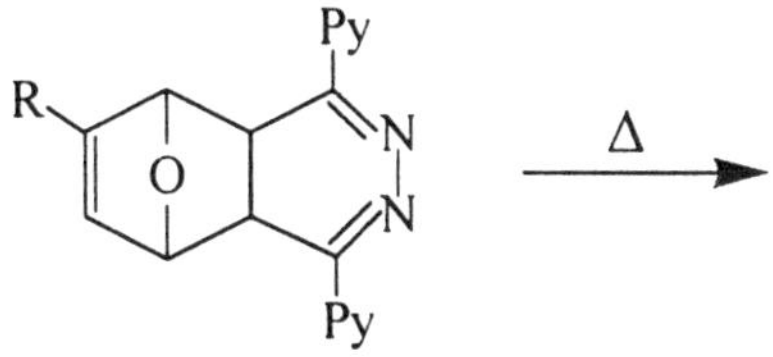

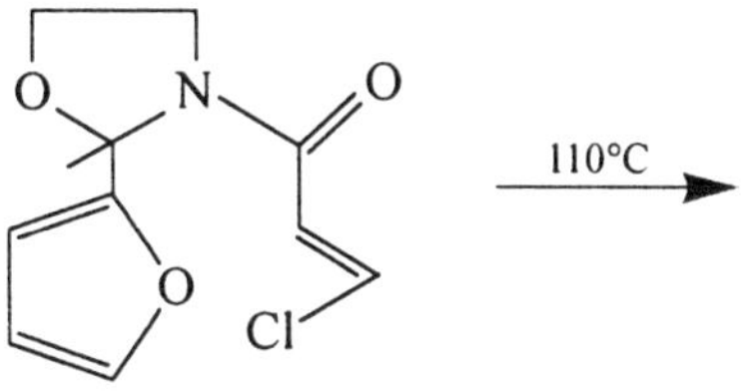

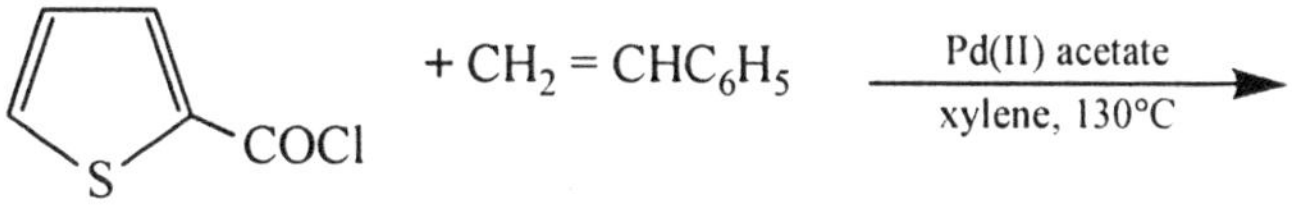

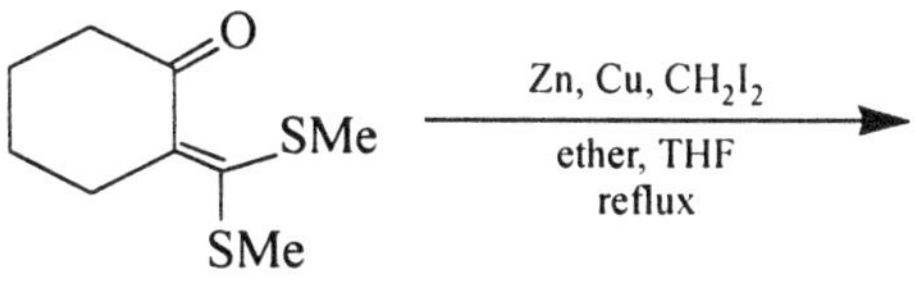

(*u*) [structure, R, Py, N, N, Py] $\xrightarrow{\Delta}$

(*v*) [structure] $\xrightarrow{110°C}$

(*w*) [structure] COCl + CH$_2$ = CHC$_6$H$_5$ $\xrightarrow[\text{xylene, 130°C}]{\text{Pd(II) acetate}}$

(*x*) [structure] SMe / SMe $\xrightarrow[\substack{\text{ether, THF} \\ \text{reflux}}]{\text{Zn, Cu, CH}_2\text{I}_2}$

(*y*) [structure] CH$_3$ + [structure] N — Br $\xrightarrow[\text{CCl}_4]{h\nu}$

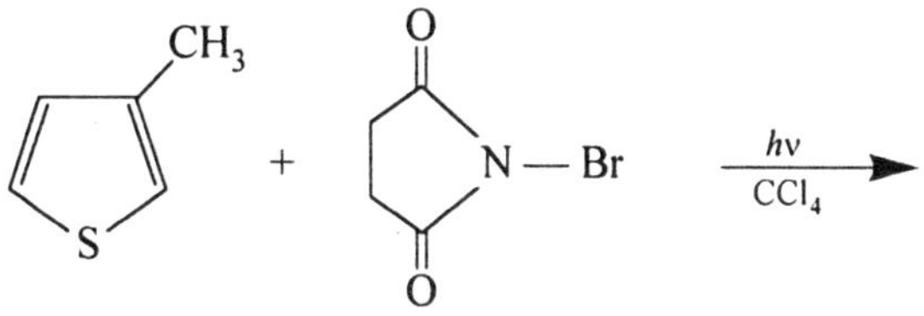

5.2 Suggest a resonable mechanism for each of the following reactions:

(*a*)
$\begin{array}{c} \text{COOC}_2\text{H}_5 \\ | \\ \text{COOC}_2\text{H}_5 \end{array}$ + C$_2$H$_5$OOCCH$_2$ S CH$_2$COOC$_2$H$_5$

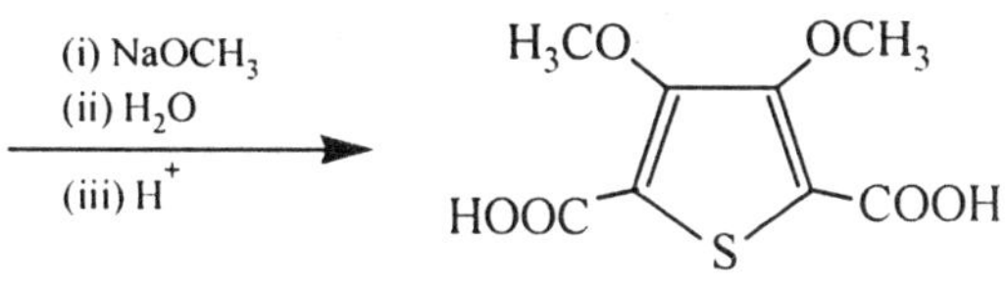

$\xrightarrow[\substack{\text{(ii) H}_2\text{O} \\ \text{(iii) H}^+}]{\text{(i) NaOCH}_3}$ [structure: H$_3$CO, OCH$_3$, HOOC, S, COOH]

(b) [chemical structure: 1-methyl-3-acetyl-... azetidine with H_5C_6 and H_3C] $\xrightarrow{h\nu}$ [pyrrole with C_6H_5]

(c) $CH_3\overset{O}{\overset{\|}{C}}CHCH_3$ (with Cl substituent) $+\ CH_3\overset{O}{\overset{\|}{C}}CH_2\overset{O}{\overset{\|}{C}}CH_3 \xrightarrow{\text{Base}}$ [furan with CH_3, CH_3 and acetyl CH_3]

(d) [structure with H_5C_6, C_6H_5, epoxy lactone] $\xrightarrow[-CO_2]{\Delta}$ [furan with C_6H_5 and C_6H_5]

(e) [N-COOCH$_3$ pyrrole] $+$ $CH_3OOC\overset{|}{\underset{|}{C}}\equiv C\overset{|}{\underset{|}{COOCH_3}}$ (dimethyl acetylenedicarboxylate) $\xrightarrow{200^\circ C}$ [product: H_3COOC and $COOCH_3$ substituted pyrrole, N-COOCH$_3$]

(f) [bicyclic epoxide with CH_3, OH] $\xrightarrow{h\nu}$ [furan product with CH_3, CH_3]

(g) [N-Ph pyrrole with SCH_2COOH] $\xrightarrow{\text{PPA}}$ [thieno-pyrrole product with S, N-Ph] (*h*) [structure: epoxide-fused cyclohexene diester] $\xrightarrow{\Delta}$ [furan with COOCH$_3$ groups]

5.3 Offer explanation for the following observations:

(*a*) Furan is least aromatic in comparison to pyrrole and thiophene.

(*b*) π-Excessive heterocycles have higher reactivity than benzene towards electrophilic substitution.

(*c*) Pyrrole-2-carbaldehyde does not take part in benzoin or Perkin condensation while furfural does.

(*d*) Furan undergoes electrophilic substitution much faster than benzene.

(*e*) Pyrrole 2-carbaldehyde does not respond to Tollens'reagent.

(*f*) 2-Nitrofuran is less reactive towards acids than furan itself.

(*g*) Pyrrole is weakly basic.

(*h*) Furan cannot be directly alkylated in the Friedel-Crafts reaction.

(*i*) Pyrrole has a higher boiling point than furan.

(*j*) Pyrrole undergoes halogenation at a rate comparable to phenol.

5.4 Discuss the diene character of pyrrole, furan and thiophene in the Diels-Alder reaction.

5.5 By drawing resonance structures, show that the electrophilic attack in thiophene takes place at the 2- and not 3 - position.

5.6 What do you understand when we say that five-membered heterocyclic compounds and cyclopentadiene are aromatic?

5.7 Discuss the similarities between furfural and benzaldehyde.

5.8 Discuss two methods each for the preparation of 2 - alkyl derivatives of pyrrole, furan and thiophene.

5.9 Write all the steps of the Vilsmeier reaction for the formylation of pyrrole.

5.10 Write a note on the intermolecular Diels-Alder reaction in five-membered heterocyclic compounds.

5.11 List the reactions with suitable examples to demonstrate the similarities between thiophene and benzene.

5.12 How would you prepare 2, 4 - dimethyl - 3 - cyanofuran by the Fiest-Bernary method?

Chapter 6

Six Membered Heterocyclic Compounds with One Hetero Atom

The methine (– CH=) group in a benzene ring can be replaced by a hetero atom such as nitrogen, oxygen or sulfur to give an important class of heterocyclic compounds. The trivalent nitrogen, for instance, replaces one methine group to give pyridine (1). The introduction of the nitrogen atom does not alter the aromatic character of the resulting ring. A similar change for oxygen, a divalent atom can be considered only if the lone-pair is involved in bonding and the hetero atom has a positive charge. This happens in the case of

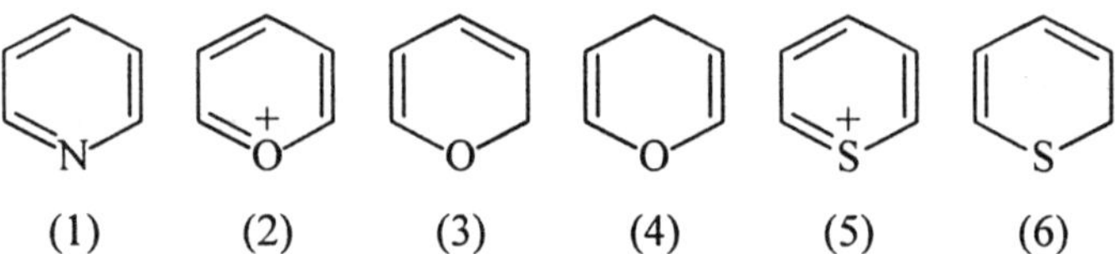

(1)   (2)   (3)   (4)   (5)   (6)

pyrylium salts (2) which are not very stable through they have some aromatic character. The oxygen containing six-membered heterocycles which do not possess aromatic sextet are the α- and γ-pyrrones (3) and (4) respectively. These compounds though simple have not been prepared but their derivatives are known. The sulfur analogues of (2) and (3) have been obtained and are designated as thiopyrylium salts (5) and thiopyran (6). In this chapter a discussion of the chemistry of these compounds will be undertaken.

6.1 PYRIDINES

Pyridine (1) is the simplest and perhaps the best known heterocyclic compound. The credit for the discovery of pyridine goes to Anderson who first obtained it from bone oil. It occurs in coal tar as well. The structural relationship between pyridine and benzene was first recognized independently by körner in 1869 and Dewar in 1971. Pyridine ring system is highly distributed in nature as pyridine derivatives and in many important alkaloids.

Like other heterocyclic ring systems the nitrogen atom in pyridine is assigned position-1. The presence of this atom introduces an element of asymmetry into the aromatic ring and as a consequence there are three monosubstituted pyridines. The positions in a mono-substituted pyridine can be designated either by the numbering system or by the Greek alphabet.

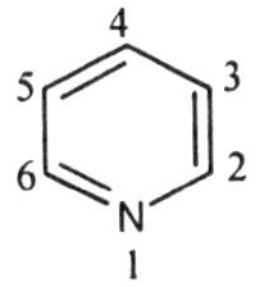

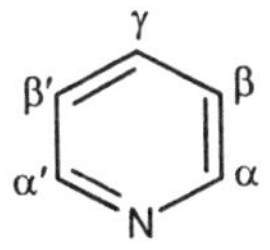

The alkyl derivatives have been given both systematic as well as trivial names.

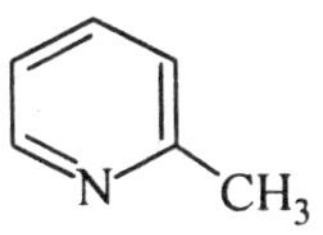

2-Methylpyridine
(α-Picoline)

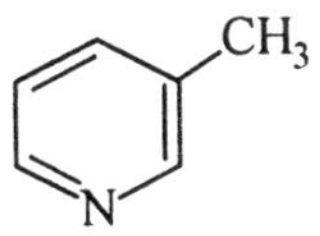

3-Methylpyridine
(β-Picoline)

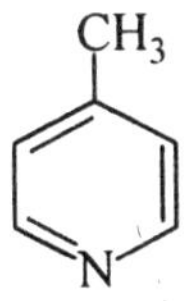

4-Methylpyridine
(γ-Picoline)

There are, in contrast, six dimethyl pyridines which are called by the general name *lutidines*, and an equal number of trimethyl substituted derivatives known as *collidines*.

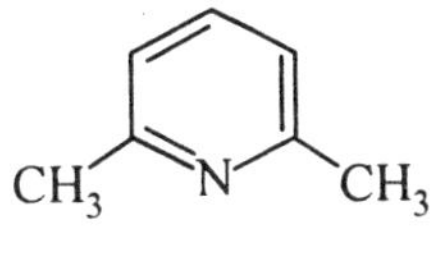

2,6-Dimethylpyridine
(2,6-Lutidine)

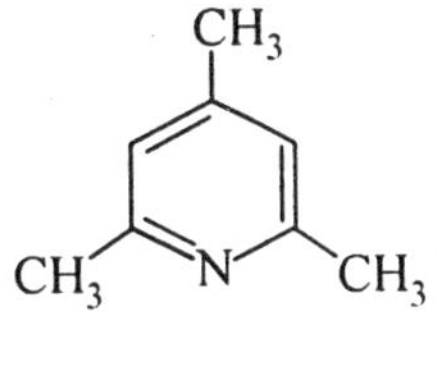

2,4,6-Trimethylpyridine
(2,4,6-Collidine)

In all, pyridine has a total of nineteen possible methyl substitution products while benzene has only thirteen. Three reduced pyridines are also known, *i.e.* dihydropyridine (7), tetrahydropyridine (8) and hexahydropyridine (9) which is also known as piperidine.

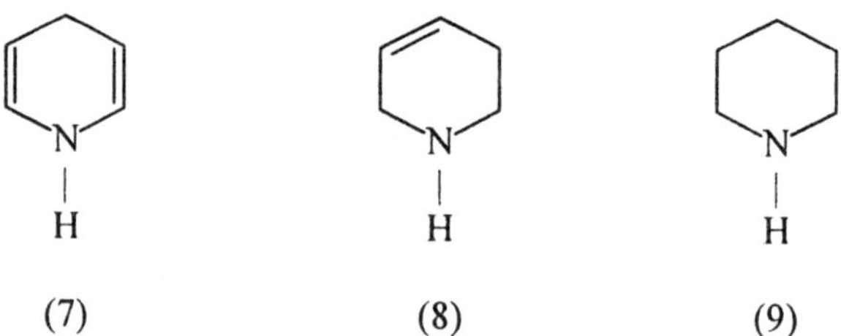

(7)   (8)   (9)

6.1.1 Physical and Spectroscopic Properties

Pyridine is a colorless liquid b.p. 115°C and freezing point − 42°C whereas benzene boils at 80°C and freezes at − 5.5°C. A large intermolecular association in pyridine because of its greater polarity accounts for its higher boiling point. The presence of alkyl groups on the ring increases the boiling point, for instance, 2, 6-lutidine (b.p. 144°C), 2, 4-lutidine (b.p. 157°C) and 2, 4, 6-collidine (b.p.170°C). But the isomer with an alkyl group adjacent to the nitrogen atom boils lower than the one having at the 2- or 3- position. For example, 2-methylpyridine boils at 129°C while the 3-isomer at about 15°C higher. This lowering of boiling point is explained in terms of the hindrance to association offered by the groups in the vicinity of the nitrogen atom.

In marked contrast to benzene, pyridine is completely soluble in water and most organic solvents. Its solubility in water is due to the presence of excellent H - bonding between pyridine and water. Pyridine has a characteristic unpleasant odor. It is dried by keeping it over potassium hydroxide or barium oxide. The molecular dimensions of pyridine have been obtained from microwave spectroscopy. The C − C and C − H bond lengths are extremely close to those of benzene (C − C, 1.39 Å, C − H, 1.09 Å). The C–N–C angle is 116.7° and the C–C–N angle is 124.0°. In benzene each carbon atom is present in sp^2 hybridized state while in pyridine this is slightly disturbed as the bond angles are not quite 120°. Introduction of a nitrogen atom in the benzene ring permits far more resonance structures and the negative charge is localized on the nitrogen atoms in contrast to the five-membered heterocyclic ring.

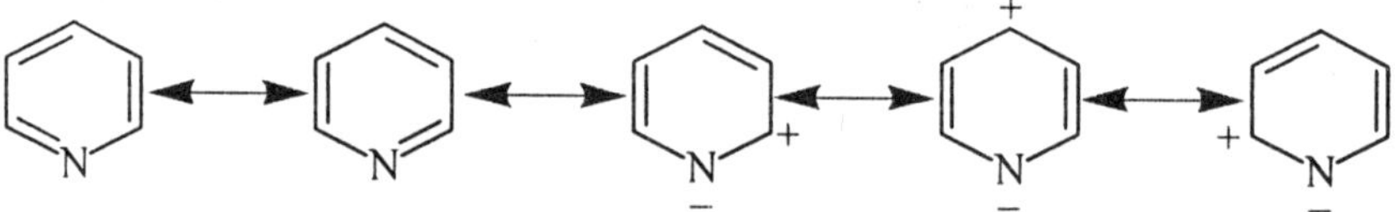

The resonance energy of pyridine from combustion data has been obtained to be 31.9 ± 2.0 Kcal/mole which is less than that of benzene. Pyridine possesses a sextet (6π - electrons) five of them provided by the five carbon atoms and the sixth by the nitrogen atom and the molecular orbital picture may be represented as follows:

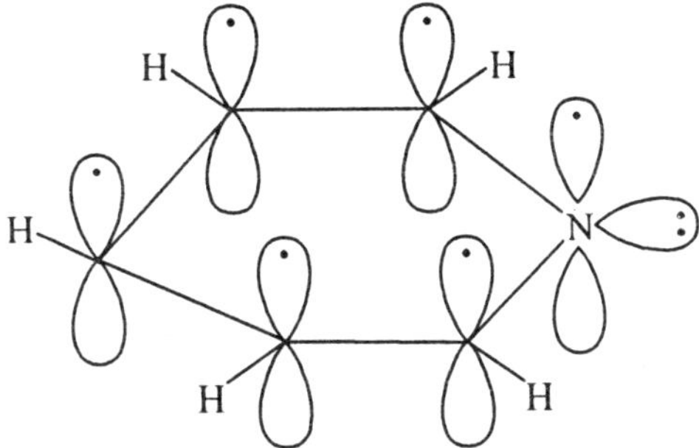

The nitrogen lone-pair is located in an sp^2 hybridized orbital which is perpendicular to the π-system of the ring. A consequence of this structural feature is that the lone-pair on the nitrogen atom is not associated with the ring and is available for protonation. Pyridine thus behaves as a tertiary base and has pKa of 5.17. The basicity becomes more pronounced if electron-donating groups are present on the ring at 2-and 6-positions, because they alter the electron availability on the nitrogen atom by resonance. Substituents at positions 3- and 5-can only act by inductive effects and their influence is rather less pronounced. The alkyl pyridines have the following pKa values:

	2-Methylpyridine	3-Methylpyridine	4-Methylpyridine
pKa	5.97	5.68	6.02

These values shows that the basicity does not differ much. A similar situation obtains for phenylpyridines.

	2-Phenylpyridine	3-Phenylpyridine	4-Phenylpyridine
pKa	5.3	4.8	5.5

In both cases, however, the trend is towards electron-donation by the groups. The pKa value of 2, 6-*di-t*-butylpyridine is 0.8 less than pyridine. This is explained by steric crowding around the nitrogen atom and obstructing donation of electrons.

Pyridine is considerably more basic than pyrrole (pKa 0.4), but less so than an aliphatic tertiary amine. This is ascribed to the fact that as the nitrogen atom becomes progressively more multiply bonded, the lone-pair is accommodated in an orbital which has more s-character (sp^2 as compared to sp^3 in aliphatic amines). These electrons are thus drawn closer to the nitrogen nucleus and held more tightly by it ; thereby becoming less available for forming a bond with a proton with the resultant drop in the basicity of pyridine as compared to aliphatic amines.

The charge separation in pyridine is supported by its large dipole moment value, 2.23 D which is larger than piperidine, 1.17D or pyrrole, 1.81 D. In pyridine electrons are attracted towards the nitrogen atom, it being more electronegative than carbon. As discussed in the previous chapter, the situation in pyrrole is exactly reverse because the electron pair is involved in the aromatic sextet.

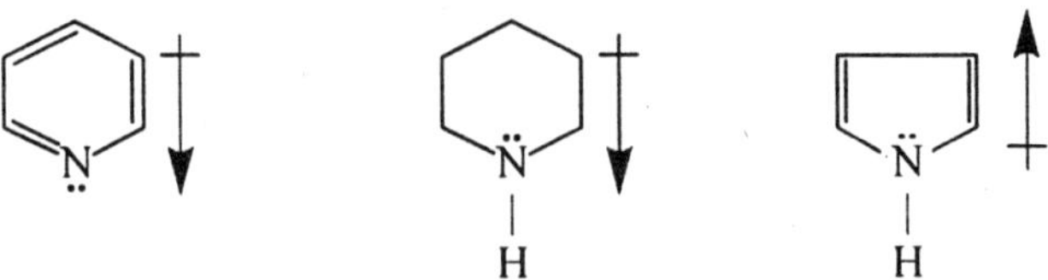

The presence of an electron-donating group at C-4 will enhance the charge separation and as a consequence 4-aminopyridine has a dipole moment of 4.40 D. On the contrary electron-withdrawing groups will have the opposite effect and 4-cyanopyridine has much decreased value of 1.60 D.

There is a considerable positive charge present at 2, 4-and 6-positions of pyridine which should make electrophilic attack on pyridine rather difficult. This has been further corroborated by the calculation of π-electron density at different positions in pyridine. These values being lower at 2-position (0.881) and 4-position (0.941) demonstrate that there is a relay of electrons from the ring towards the nitrogen atom compared to 3 - position (1.082).

The proton *n.m.r.* of the various C – H bonds give chemical shifts at δ 8.52, δ 7.16 and δ 7.55 for the α, β-and γ-protons respectively. The main features of mass spectrum of pyridine consists of intense ions corresponding to M^+ and $(M - MCN)^{+\cdot}$.

6.1.2 Tautomerism in Pyridines

In pyridine, similar to majority of heteroaromatic compounds, tautomeric structures involve group such as – XH linked directly to the pyridine ring, X may be oxygen, nitrogen or sulfur. Such tautomeric structures involve the transfer of a proton to the ring nitrogen atom in pyridine.

A tautomeric (10) structure between 2-hydroxypyridine and pyridon-2-one (11) is one between a benzenoid and a non-benzenoid structure. The compound (10) possessing an aromatic sextet is aromatic but (11) also contains a six π-electron structure in a cyclic *p*-orbital system provided the carbonyl group contributes no electrons. This structure, therefore, is aromatic to certain

(10)   (11)

extent though not as aromatic as (10). Therefore, there is a tendency for 2-hydroxypyridine to exist considerably in the benzenoid form. Furthermore, oxygen is a highly electronegative atom, consequently the polarized forms (11a) and (11b) will be favored. Thus both 2-and 4-hydroxypyridines exist largely in the pyridinone forms.

(11)   (11a)

The electronic distribution in the ring may be changed significantly from one tautomer to another leading to large differences in non-specific polar solvents. Since polar solvents are expected to stabilize polar forms, a predominance of the hydroxy tautomer (10) is predicted in solvents less polar than water as well as in the vapor phase. This has been supported by experimental observations and spectroscopic investigations, that at equilibrium both 2-and 4-hydroxypyridines exist as such rather than as pyridinones. 3-Hydroxypyridine is in equilibrium with an approximately equal amount of the zwitterion.

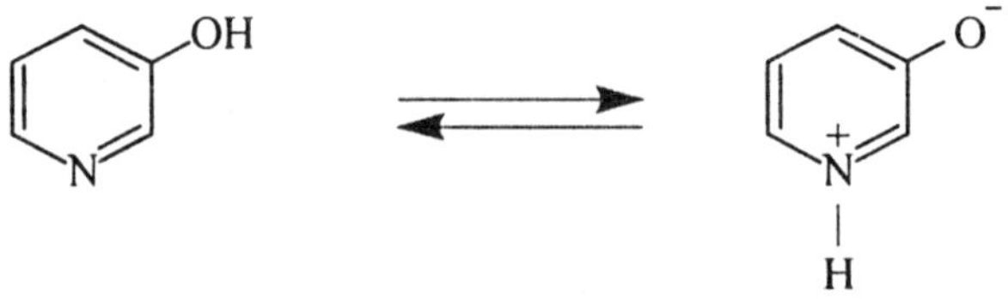

IR investigations of 3-hydroxypyridine-1-oxide (12) show that these compounds exist in the C-hydroxy form although strong association occurs specially in the solid state.

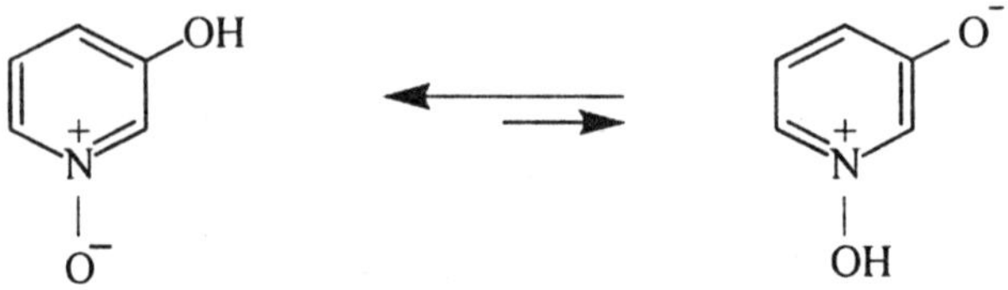

(12)   (12a)

In the amine-imine system (13) and (14) the mobile proton can in principle be

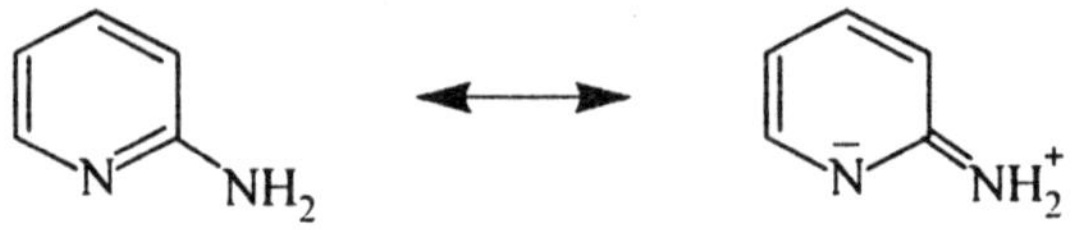

(13)   (13a)

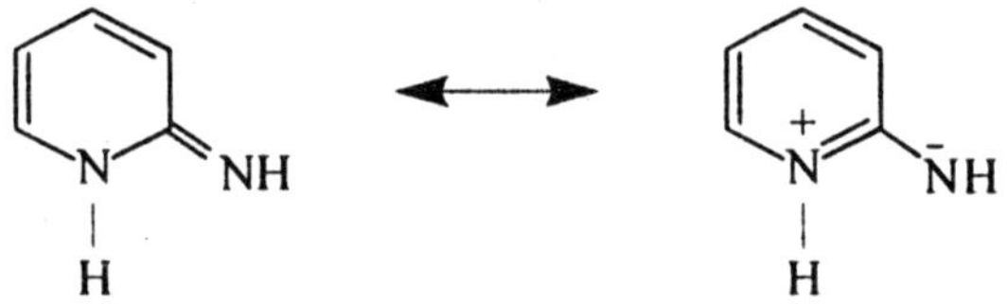

(14)   (14a)

located at either of the two basic nitrogen atoms as in (15). The resonance

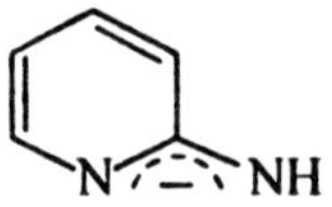

(15)

form with aromatic structure is polar in the imine (14a) but non-polar in amine (13). Consequently the amine structure should be favored in non-polar solvents.

This seems generally the case in water and other polar solvents. The importance

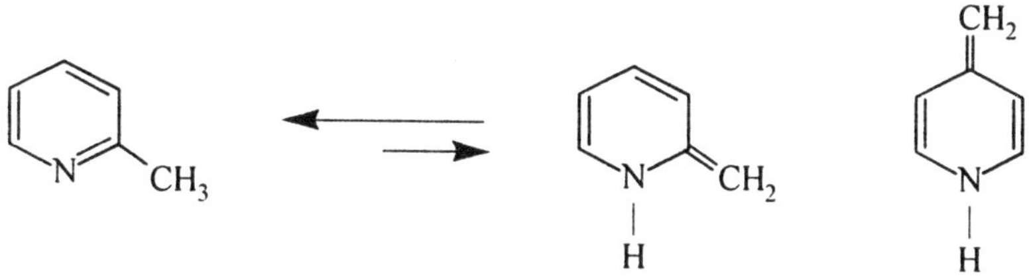

(16)   (16a)   (17)

of tautomeric structures and chemical reactivity is obvious from their tautomeric equilibrium. Between (13) and (13a) the electrophile should attack on the ring nitrogen atom (13a) while the electrophile attacks the exocyclic nitrogen atom (14a). Substitution of a methyl group at the 2-and 4-positions in pyridine, the equilibrium is greatly favored to the methyl form (16). There is a little aromaticity associated with the methylene tautomers (16a) and (17). These tautomers are thus present in very small amounts.

The effect of other substituents on tautomeric equilibria can often be predicted from general chemical principles. For instance, an electron-withdrawing group adjacent to a ring nitrogen atom tends to decrease its basicity, so a tautomer with a proton at that nitrogen atom would be correspondingly destabilized, and the equilibrium displaced towards alternative forms. Substituents may also favor one tautomer over another by intramolecular hydrogen-bonding. Thus 2-Pyridylacetone can exist in the keto (18), enol (18a) or methide-enamine (18b) forms.

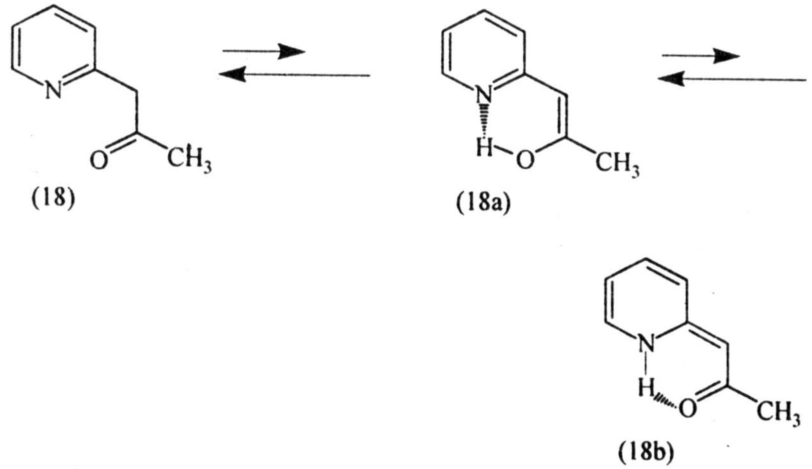

(18)   (18a)

(18b)

The enolic form (18a) is favored because of the presence of intramolecular hydrogen-bonding. Tautomer (18b) lacks an aromatic benzenoid structure.

Thiol-thione tautomerism parallels that of the corresponding hydroxy – oxo systems although a thiol group is more active than a hydroxy group, the thioamide resonance (19) to (19a) is strongly polarized towards the latter form; because of the weak C = S π-bond in (19). This causes the basicity of the nitrogen

(19)   (19a)   (19b)

atom to increase, therefore the tautomerism in 2-pyridinithiol-2-pyridinethione favors the thione (19) over the thiol (20) form to an extent slightly greater than that found in hydroxy-oxo equilibria. Similarly 4-pyridinethiol (20) exists mainly in the theine form (20a).

(20)   (20a)

In contrast, in the gas phase 2-and 4-pyridinethiols predominate over thiones.

6.1.3 Synthetic Methods

Pyridine and several of its simple alkyl derivatives were obtained from coal tar in which they occur in enough quantity. Nowadays, pyridine is being obtained synthetically. It can be produced commercially by the gas phase interaction of a mixture of crotonaldehyde, formaldehyde, steam, air and ammonia on silica-alumina catalyst at 400°C in a yield of about 60-70%. A large number of synthetic procedures, however, exist for the preparation of pyridines in the laboratory.

1.   *The Hantzsch Synthesis* : This is probably the most important synthesis of pyridine and involves the condensation of an aldehyde with two moles of a β-dicarbonyl compound and ammonia.[3]

(21)

(22)

(21)   (22)

Michael type
addition

H^+ transfer

$- H_2O$

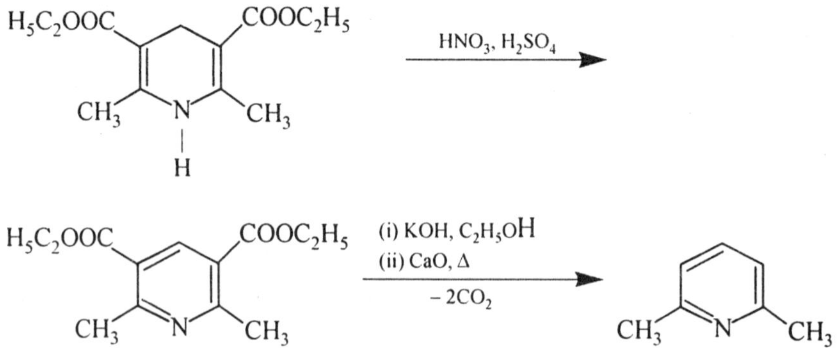

Ammonia reacts with acetoacetic ester to yield β-aminocrotonic ester (21) while the aldehyde molecule undergoes a base catalyzed condensation with a second molecule of the ester to produce alkylidene acetoacetic ester (22). Addition of (21) across the double bond of (22) takes place in a Michael type reaction and subsequent cyclization *via* dehydration yields a 1, 4 – dihydropyridine. This is then oxidized *in situ* by a mixture of HNO_3 and H_2SO_4 to give 2, 6-dimethylpyridine. The aldehyde carbon in this reaction becomes the pyridine γ-carbon.

Several variations of the Hantzsch synthesis are known. According to one, two molecules of aldehydes (in place of two ester molecules) can be condensed with one that of the keto ester under the same experimental conditions.

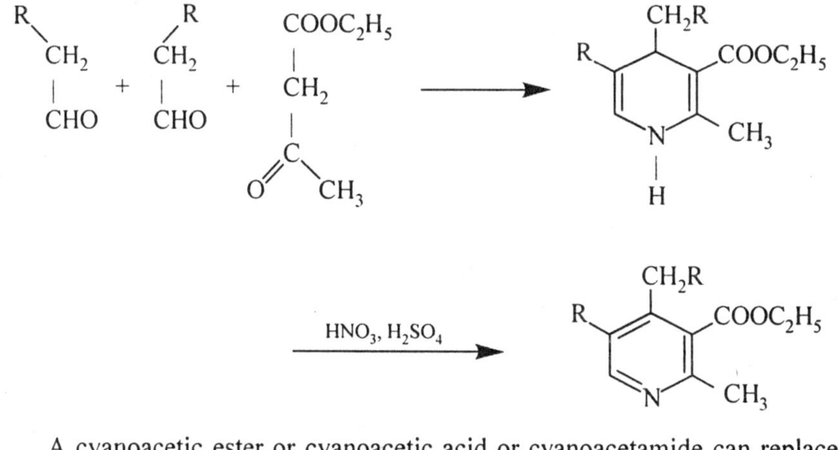

A cyanoacetic ester or cyanoacetic acid or cyanoacetamide can replace acetoacetic ester in the Hantzsch synthesis. This is also known as the ***Guareschi-Thorpe synthesis.*** In another modification ammonia can be replaced by hydroxylamine and two molecules of esters can be condensed with one of the aldehyde.

2. ***From Other Ring Systems :*** Various heterocyclic ring systems such as furans, pyrylium salts, pyrroles, pyrones and condensed ring systems rearrange to pyridines under appropriate conditions.

Furans and some of their dihydro-and tetrahydro-derivatives have been converted to pyridines under a variety of conditions. In these synthesis the furan derivative has an appropriate group at the 2-position to provide the fifth carbon atom while the ramaining are derived from furan. Thus 2 –acetylfurans react with ammonia ammonium chloride at high temperatures to give 3-hydroxypyridines.[4]

$R = CH_3$

2, 4, 6-Trisubstituted pyridines are obtained from pyrylium salts and ammonia, amines or hydrazines. the pyrylium salts can be prepared easily. Generally an excess of the amine is used with mild heating in the absence of a solvent. The procedure is limited to the availability of the starting pyrylium salts.

Pyrones on reaction with excess ammonia or amines are converted to pyridones, by the replacement of ring oxygen by $= NH$ group.[5] In this reaction aliphatic amines give good yields while aromatic amines are seldom employed.

The conversion of pyrrole to 3-chloropyridine[6] by treatment with chloroform and a base is well known. The yield of the product in this reaction is though poor (15%), equation (6.1).

$$\text{(pyrrole)} \xrightarrow[\Delta]{\text{CHCl}_3,\ \text{NaOH}} \text{(3-chloropyridine)} \qquad (6.1)$$

Generally oxidation of quinoline in air below 190°C forms quinolinic acid while above this temperature nicotinic acid is obtained.

3.   *Electrochemical Methods* : A large number of synthetic methods for pyridine utilizing electrochemical techniques have been recently reviewed.[7] The reported studies involve electrochemistry with cathode transformations. These methods give reduced products.

4.   *From Cyclic Ketones* : Annelated, *i.e.* one ring built over another pyridine are most conveniently prepared by the condensation of cyclic ketones with an appropriate nitrogen-containing compound to build up the pyridine ring. Thus condensation of cyclopentanone or cyclohexanone with β-aminoacrolein produces 2, 3-fused pyridines.[8]

5.   *From 2-Methyleneallyl Anions* : These anions prepared[9] by dimetallating isobutylene with *n*-butyllithium in tetramethylenediamine (TMEDA) under argon at 25°C on heating with benzonitrile yields a pyridine derivative[10], 2, 6-diphenyl-4-methylpyridine. 2, 4, 6-Trisubstituted pyridines such as 2, 6-diphenyl-4-methylpyridine have also been obtained using cycloimmonium ylides on reaction with substituted benzylidieneacetophenones.[11]

Acetophenone, N, N, N-trimethylhydrazonium fluoroborate on pyrolysis rearranges to 2, 6-disubstituted pyridines.

6. *From 1, 5-Diketones* : Reaction of 1, 5-diketones with hydroxylamine leads to pyridines.[12]

7. *Sonication Method* : 4-Aryl-1, 4-dihydropyridine on oxidation with manganese dioxide on bentonite clay without solvent under sonication as the energy source is converted to 4-arylpyridine[13] in a high yield (92%).

8. *From Aza Wittig Reaction* : The Aza Wittig reaction of iminophosphorane prepared by the reaction of triphenylphosphorus on ethyl-2-azido-5 -phenyl-2, 4-butadiene carboxylate followed by its reaction with isocyanate yields a carbodiimide. This on thermal electrocyclic ring closure yields (23) which aromatizes to 2-aminopyridine derivative.

(23)   (24)

9. 6-Benzyl-3, 5-dichloro-2*H*, 1, 4-oxazin-2-one on refluxing with phenylacetylene in toluene yields a highly substituted pyridine.[15]

The yield of 2, 6-dichloro-3-benzyl-5-phenyl pyridine in this reaction is 95%.

6.1.4 Chemical Reactions

Simple pyridines are basic in nature, extremely stable and possess penetrating odors. The ring remains intact in most of its chemical reactions. Therefore, they are much used specially pyridine itself as solvents.

1. *Reaction with Acids* : The strength of an organic base depends on the electron density at the nitrogen atom and its availability for donation. The electron pair on the nitrogen atom in pyridine is available for extra bonding and it displays most of the typical properties of a tertiary amine. Pyridine forms crystalline salts with most protic acids. With HCl it forms pyridinium chloride, equation (6.2).

Metallic ions such as aluminum, berrylium, boron, etc. which are electron acceptor form tetrahedral salts with pyridine. Aluminum chloride, for instance, forms a complex in which the unshared sp^2 electrons on the nitrogen atom fills the *sp²* orbital of aluminum, structure (25). The formation of such a complex is hindered if alkyl substituents are present at the 2- and 6-positions.[17] Pyridine

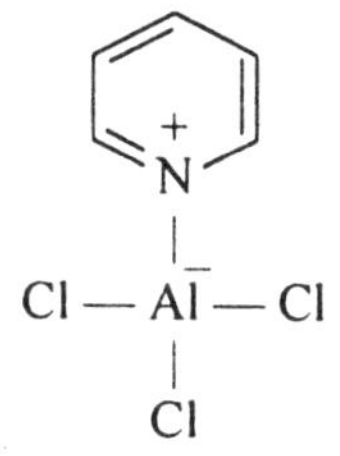

(25)

is frequently employed as a catalyst for acylating of phenols, alcohols and amines using an acyl chloride or anhydride. The actual acylating species is the reactive acylpyridinium salt and the mechanism of acylation involves the following steps:

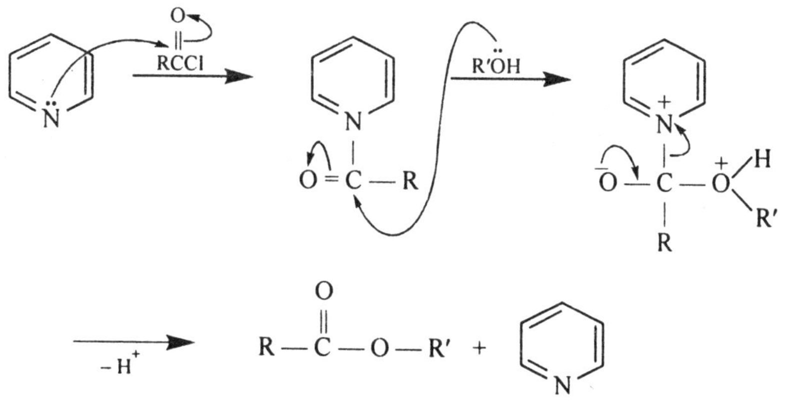

Acyl, sulfonyl as well as anhydrides readily react with pyridine to form quaternary salts. These salts are not usually isolated but they function as acylating and sulfonating agents.

2.   *Electrophilic Substitution* : Substitution by electrophilic reagents is a special reaction which typifies the chemistry of benzene and its derivatives. Pyridines, on the other hand, with an electronegative nitrogen atom causes π-deficiency at carbon atoms (it is a π deficient heterocycle) and is thus considerably deactivated towards electrophiles. The electrophilic substitution may thus be accomplished only with extreme difficulty and only under severe experimental conditions. A further deactivation is caused when the reaction is carried out in acidic medium because of the protonation of the nitrogen atom resulting in the formation of pyridinium ion. The partial rate factors for electrophilic substitution of pyridine[18] are of the order of 10^{-6} which are similar to those for nitrobenzene, pyridine also undergoes substitution at the 3-position.

This can also be seen from the following resonance structures, as electronegative nitrogen atom does not require to bear a positive charge. Therefore, the attack takes place at the 3-position. This trend is also consistent with electron density calculations.

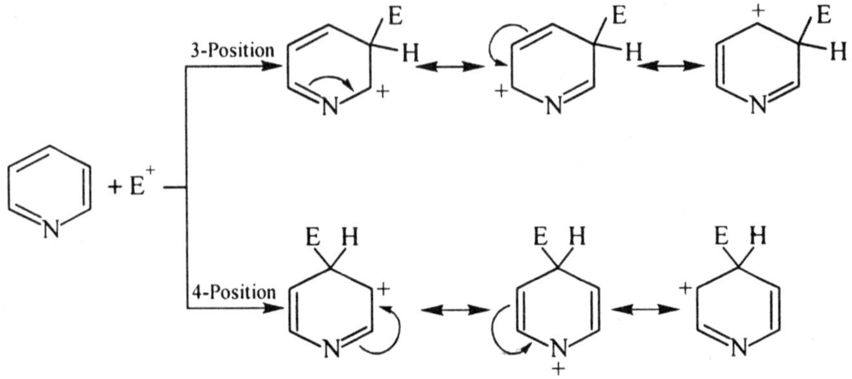

Halogenation : In view of the electron-withdrawing property of the pyridine ring, it is not surprising that electrophilic halogenation occurs only under vigorous conditions. The reaction of chlorine with pyridine in the presence of a large excess of aluminum chloride gives 3-chloropyridine (30-35%) alongwith 3, 5-dichloropyridine.[19] Bromination of pyridine on the other hand, may be accomplished easily by using bromine in oleum.[20]

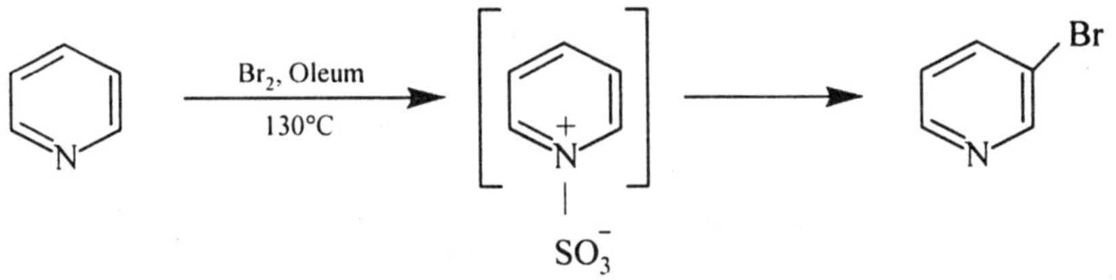

Vapor phase (300°C) chlorination or bromination of pyridine gives a complex mixture of products. At such a high temperature, probably halo free radicals are formed and attack at the 2-position becomes of major importance. Iodine and oleum at 320°C with pyridine produce 3-iodopyridine in only 18% yield.

Nitration : Nitration of pyridine probably involves its conjugate acid and thus requires vigorous conditions.[21, 22] 3-Nitropyridine is obtained in a poor yield on nitration (HNO$_3$/H$_2$SO$_4$/300°) of pyridine. Low yields are also obtained on nitrating 2-methyl- and 2, 4-dimethylpyridines but 2, 6-dimethylpyridine is nitrated (KNO$_3$/Oleum/110°C) at 3-position in a yield greater than 65%. Pyridine cannot be nitrated on ring carbon atom using nitronium fluoroborate because such reactions lead to N-nitropyridinium fluoroborate.[23]

Sulfonation : Pyridine is resistant to sulfonation by conc. sulfuric acid or oleum but presence of catalytic amount of mercuric sulfate facilitates sulfonation after prolonged heating at 230°C to form pyridine 3-sulfonic acid (75-85%).[24] The three methylpyridines are similarly monosulfonated at high temperature, but 2, 6-di-*t*-butylpyridine is sulfonated rather easily[25] with sulfur trioxide at −10°C. Presumably in this case the steric hindrance of the bulky *t*-butyl groups prevents the formation of the normal pyridine-sulfur trioxide adduct. The adduct (25) is employed as a sulfonating agent in organic synthesis.

(25)

At higher temperature, 2, 6-*t*-butylpyridine sulfonic acid is cyclized to the sulfone (26).

(26)

The Friedel-Crafts Reaction : The Friedel-Crafts reaction is a very useful reaction in the benzenoid series but it is unknown in pyridine chemistry. Pyridine forms a complex with aluminum chloride which is highly unreactive towards the attack by carbocations.

Mercuration : The mercuration of pyridine may be carried out easily by heating pyridine with mercuric acetate at 170-180°C. A salt is initially formed by the interaction of pyridine with mercuric acetate at room temperature which rearranges to give 3-pyridylmercuriacetate. This may be converted to pyridine mercurichloride and finally into 3-bromopyridine by treating with bromine.[26]

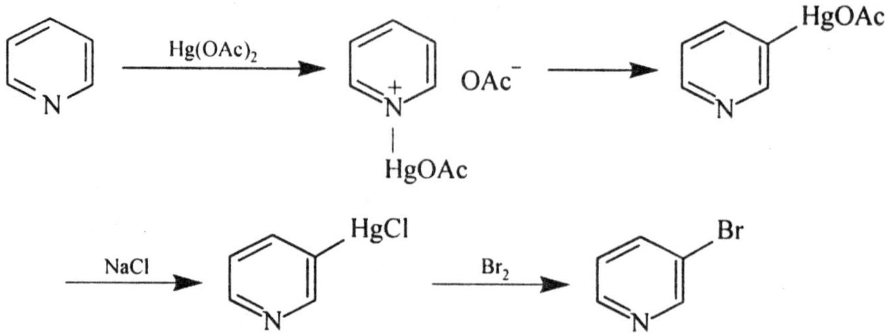

3.   *The Diels-Alder Reaction* : The Diels-Alder reaction is uncommon with aromatic compounds though some react under certain conditions. Pyrrole itself does not undergo the Diels-Alder reaction but pyrroles with electron-withdrawing groups present at position-1 react with dienophiles to form a 1:1 adduct. Pyridine does not afford the normal Diels-Alder adduct[27,28] but may react with powerful dienophiles like dimethylacetylenedicarboxylate (DMAD) to form a 1:2 adduct (28). Initially a "zwitterion" (27) is formed between pyridine and DMAD to give (27a) which after cyclization and rearrangement gives the final adduct (28).[29, 30]

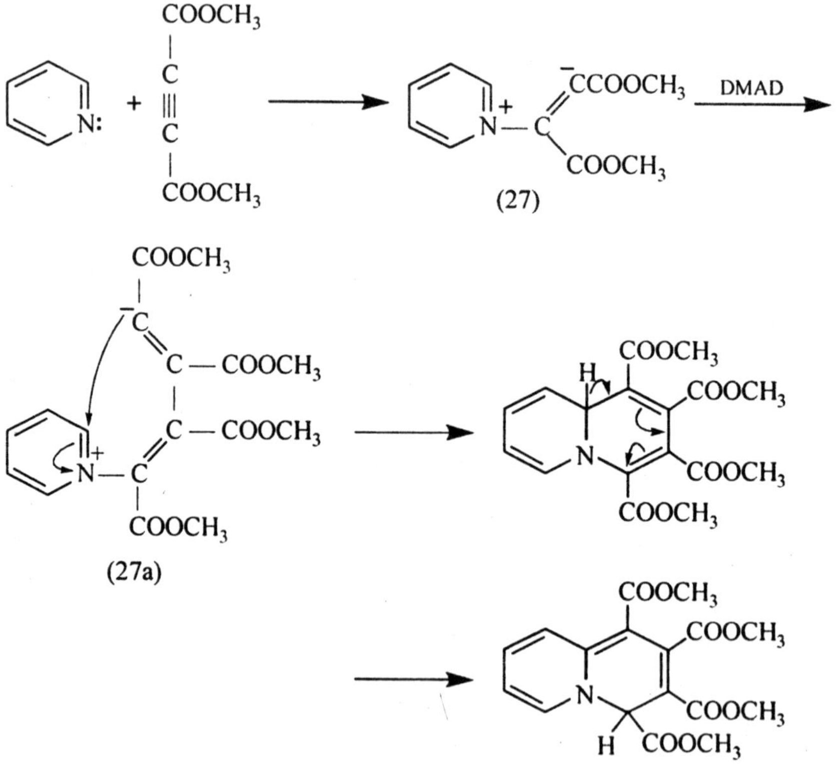

But pyridine derivatives undergo cycloaddition under milder conditions. Thus 2-nitrosopyridine[31] reacts with 2-methyl-1, 3-butadiene to afford the adduct (29) and N-vinyl pyridinium tetrafluoroborate reacts with cyclopentadiene[32] to produce (30).

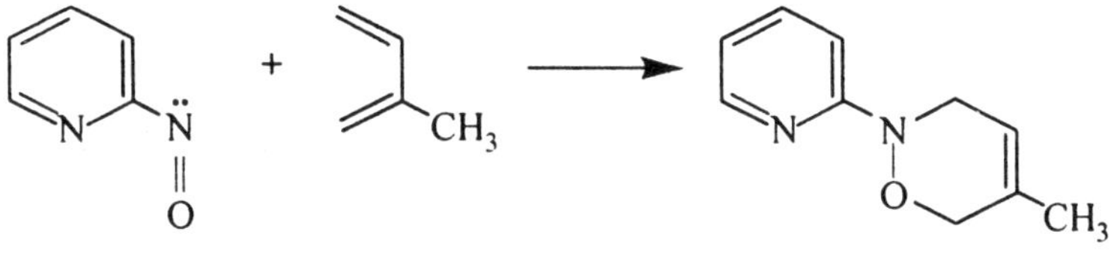

(29)

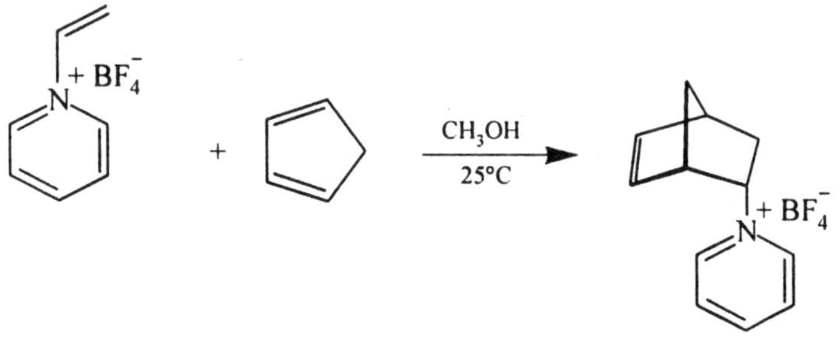

(30)

4.   *Quaterinization* : Certain alkyl and aryl halides and sulfates react with pyridine to give quaternary N-substituted salts by a displacement reaction. Methylation is effected by methyl iodide and N-acyl pyridinium salts are formed with acyl halides and iodine. Steric effects of the substituents influence quaternization.[33]

5.   *Reaction with Reducing Agents* : Pyridines are more susceptible to the action of reducing agents than benzene derivatives. Thus pyridine substituted with phenyl on or away from the ring (equation 6.3) results in the reduction of pyridine.

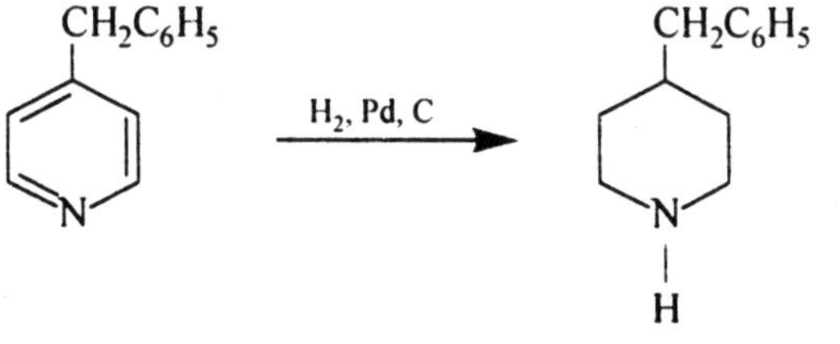

$$\text{(6.3)}$$

Since pyridine reacts with nucleophiles rather easily, it may be reduced by nucleophilic reducing agents such as complex hydrides.[34] The more reactive pyridine derivatives such as pyridinium salts may even by reduced by more weakly nucleophilic reducing agents such as sodium dithionite.[35]

These reductions result in 1, 2 -and / or 1, 4-dihydropyridines and finally even may give piperidine. Reduction with lithium aluminum hydride gives a mixture of dihydropyridines and finally piperidine. Piperidine is also obtained by reducing pyridine with Raney nickel (H_2, 120°C). Sodium borohydride does not apparently react with pyridine and its simple derivatives but readily reduces those pyridines containing electron-withdrawing substituents.

Reduction of pyridine to piperidine can readily be achieved in acidic conditions because it is the protonated form and not the neutral molecule that is being reduced.[36]

Reduction of pyridine may also be accomplished by "one-electron process", for example by a metal like sodium. Sodium in ethanol reduces pyridine completely to piperidine although in some cases 1, 2, 3, 6-tetrahydropyridine is also produced.[37]

6.   *Reaction with Oxidizing Agents* : Oxidizing agents are considered as electron acceptors, therefore, it is anticipated that pyridine ring being π – deficient will be reluctant to oxidation. In some cases the side chain may be oxidized to the corresponding carboxyl group with the survival of the ring itself. The nitrogen atom, on the other hand, is a center of high electron density and thus can be easily oxidized even by alkaline hydrogen peroxide or various peracids to pyridine N-oxide, an important derivative of pyridine, equation (6.4).

$$ (6.4) $$

Pyridinium chlorochromate (PCC) obtained in the following manner functions as a mild oxidizing agent[37] equation (6.5).

$$HCl + Cr\,O_3 \longrightarrow HCr\,O_3^+Cl^- \longrightarrow \underset{}{NH^+[Cr\,O_3Cl]^-}$$

$$(6.5)$$

7. *Nucleophilic Substitution* : In its reaction pyridine resembles a benzene ring carrying a strong electron-withdrawing group. Pyridine is unreactive towards electrophilic substitution but undergoes nucleophilic substitution rather readily. A useful reaction of this type is the preparation of amino-pyridine from pyridine and sodium amide and is known as the *Chichibabin* reaction. Amino pyridines are versatile intermediates for the synthesis of biologically active compounds.[38] As regards the mechanism of this reaction, all the available evidence is consistent with the steps shown below: The reaction is initiated by attack of the nucleophile at C-2 or C-4, in the

second step a hydride ion is eliminated which reacts with aminopyridine to evolve hydrogen. Chichibabin – like reactions also occur with organolithium compounds. This amination reaction shows diminished reactivity in the case of heteroaromatics containing a methyl group bonded to an annular carbon. This is interpreted to mean that the methyl group undergoes ionization under the experimental conditions and thereby hinders the addition of amide ion to the annular carbon atom, a necessary step in the Chichibabin reaction.

Compounds having pKa in the range of 5-8 have usually been aminated. They include pyridines, quinolines, isoquinolines and benz- and -naphth-imidazols. Outside this pKa range the reaction proceeds with difficulty. Studies have also revealed that stronger the base easier it is to aminate. With regards to the electronic effects, electron-withdrawing groups present on pyridine ring enhance amination because the π electron density is decreased making the heterocyclic ring more susceptible to nucleophilic attack. In contrast, electron-donating groups hinder the reaction, as they increase electron density on the ring. This is supported by the fact that pyridine is aminated at 110°C but a temperature of 170°C is required to aminate 2-aminopyridine. While 2, 4, 6-triaminopyridine requires a temperature of 200°C. Preferential attack at 2-position of 3-substituted pyridine is a general phenomenon in the Chichibabin reaction.[42] However, when bulky groups are present at the 3-position then amination at the 6-position becomes competent. For instance, *n*-butylpyridine produces 2-and 6-amination pyridine in a ratio of 4:1. The rate of nucleophilic substitution is increased by benzo annelation. A striking example is the easy amination of 1-methylbenzimidazole.

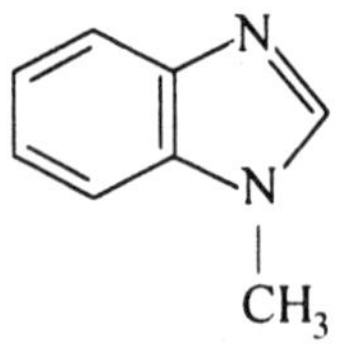

Recently it has been reported that bromopyridines also undergo an efficient nucleophilic substitution in a palladium phosphine catalyzed reaction.

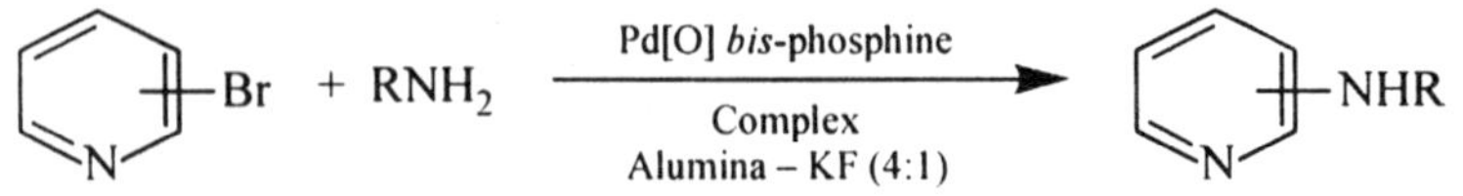

Pyridinium salts being more reactive towards nucleophiles than pyridines themselves, are able to react with less powerful nucleophiles.[43] Hydroxylation, for instance, takes place rapidly by the reaction of a pyridinium salt with alkaline potassium ferricyanide and this reaction constitutes a simple route to N-substituted 2-pyridones.

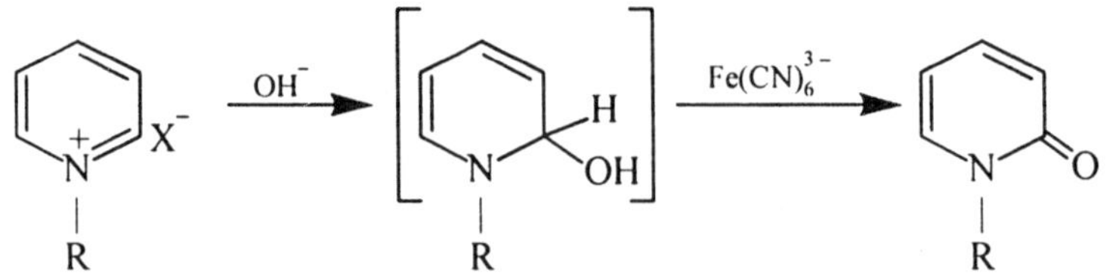

N-alkoxypyridinium salts may react with nucleophiles with the formation of an aldehyde.[44]

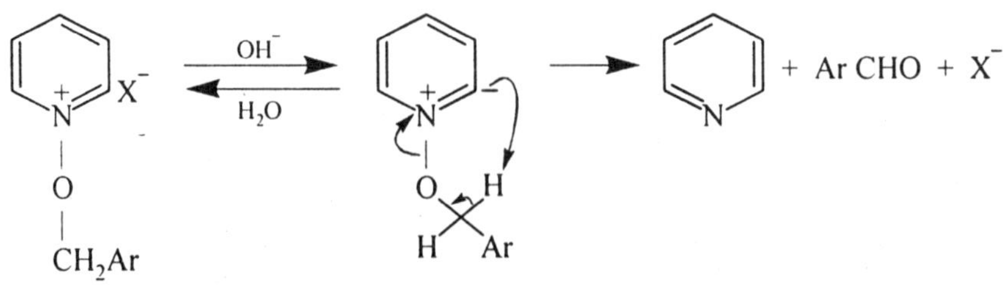

8. *Hetaryne Formation* : Benzyne (31) is the prototype of a well known kind of bidentate reactive intermediate also known as aryne or dehydrobenzene which can formally be generated by removing the vicinal atoms from a parent aromatic nucleus. The resulting species can be represented by several structures. A *hetaryne* (31 or 32) is also a bidentate intermediate with a triple bond in the

$$(6.7)$$

nucleus containing the hetero atom. Only one type of dehydrobenzene is known though there are, two possible hetarynes namely, 2, 3-hetarynes and 3, 4-hetarynes. These are generated by a variety of elimination reactions. The 3, 4-hetarynes (32) are produced, equation (6.7) by the reaction of $NaNH_2$ in liq. ammonia on 3-or-4-halopyridine[41-45] or from 3-bromo-4-chloropyridine by treatment with lithium amalgam[46] (equation 6.8) or from the thermal decomposition of pyridine diazonium carboxylates,[47] (equation 6.9).

$$(6.8)$$

$$(6.9)$$

The corresponding 2, 3-hetaryne cannot be obtained from 3- or 4-halopyridines by heating with $NaNH_2$[31] and liq NH_3, since 2-aminopyridine is obtained as the sole product in 85% yield.[48] The hetaryne may be obtained from 3-bromo-2-chloropyridine with lithium amalgam or by lead tetraacetate oxidation of aminotriazolopyridines.[49] Several reagents such as furan,[49] tetracyclopentadienone, etc. which are employed for trapping benzyne can be used to trap pyridynes as well. 3, 4-Pyridyne (32) cycloadds to furan to

---

* The term was coined by T. Kauffmann and F. P. Boetteher, *Ber.*, 95, 949 (1962).

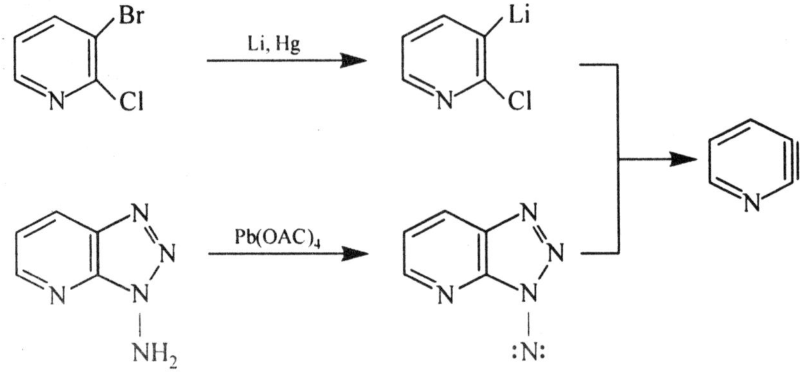

give the adduct (32a) and to tetraphenylcyclopentadienone to yield (33) which decarboxylates to give tetra-phenylisoquinoline.[50]

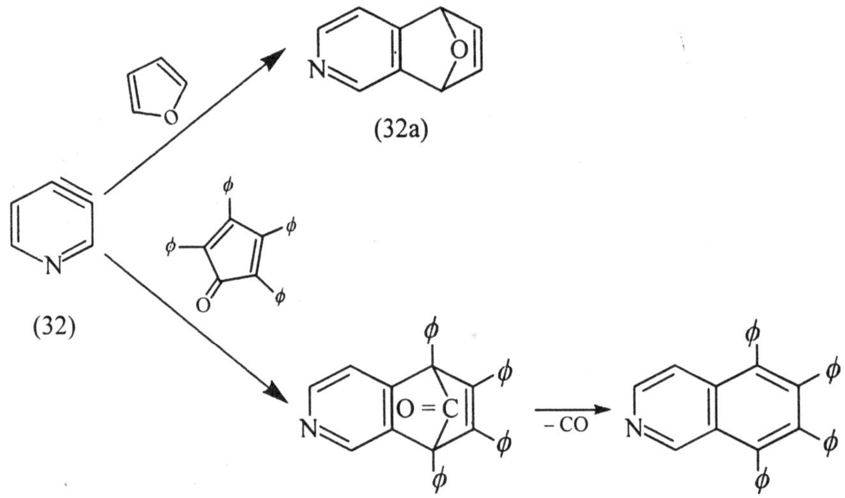

(32a)

(32)

(33)

In contrast to benzyne, 3, 4-Pyridyne forms no adducts with anthracene, dimethylfuran or norbornadiene. Though with cyclopentadiene however, a (2 + 2) and not a (4 + 2) adduct (34) is formed which is unusual.[51]

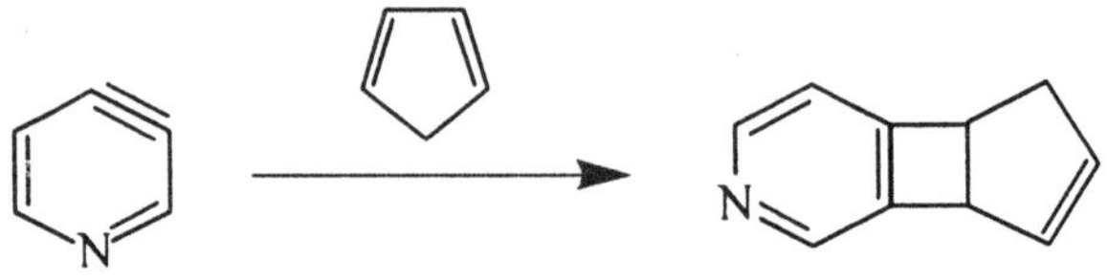

(32)

(34)

A 1, 3-dipolar addition of 3, 4-pyridyne takes place with (35) to give (36).

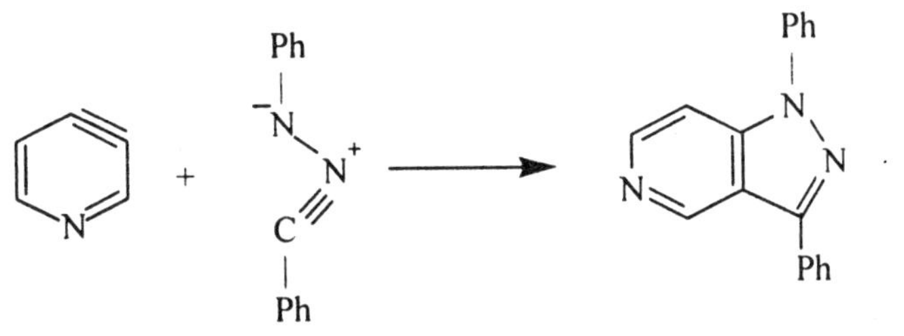

(32)   (35)   (36)

Coupling[52] of (32) leads to diazobiphenylene (37).

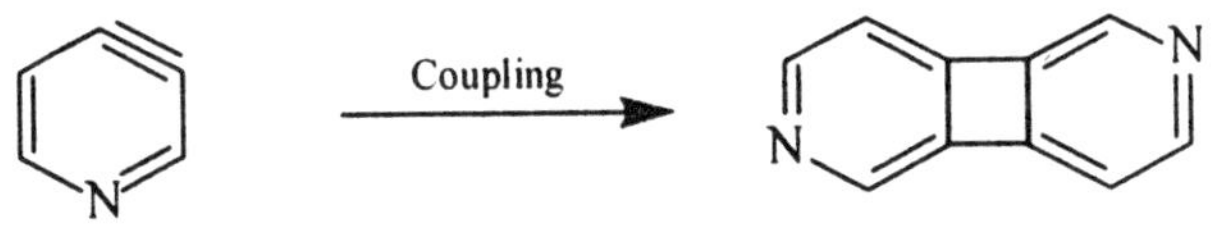

(32)   (37)

2, 3-Pyridyne similar to 3, 4-pyridyne cycloadds to furans[53] and tetraphenyl cyclopentadienone[54] to give the corresponding adducts.

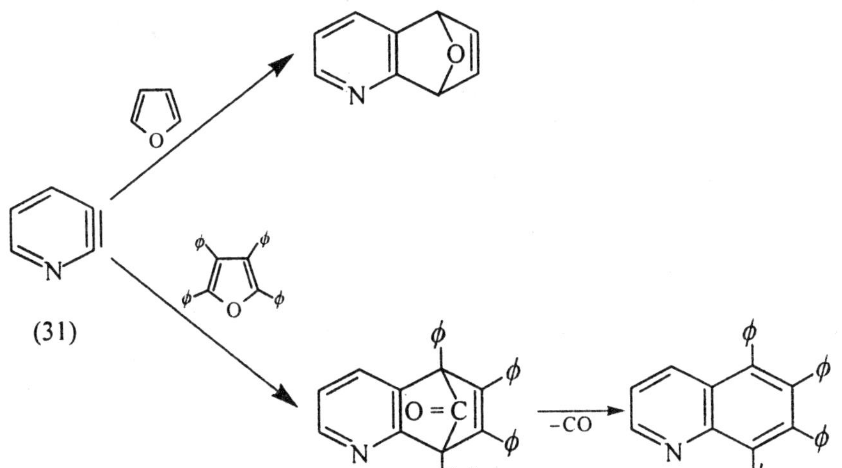

(31)

With thiophene, 2, 3-pyridyne undergoes a (3 + 2) cycloaddition involving the sulfur atom and the β – carbon of thiophene. The intermediate adduct (38) loses acetylene to produce pyridothiophene (39).

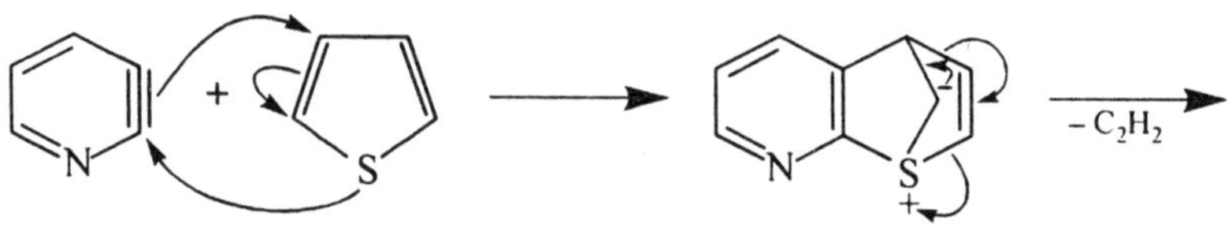

(31)   (38)

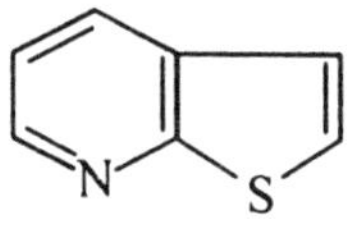

(39)

9. Ring-Opening Reactions :[55] A reaction which is without its analogy
in benzene chemistry involves the ring opening reactions of pyridine.
Pyridine is not cleaved by acids and is stable in aqueous acids upto 300°C.
Interestingly pyridine is cleaved during its reduction with sodium in 95%
ethanol resulting in a small amount of glutaconic dialdehyde $OHC(CH_2)_4CHO$.
By far the most important cleavage reactions are those which take place with
pyridinium salts and a base. Pyridinium salts react readily with nucleophiles
and the reaction in many cases is followed by ring opening resulting in the
formation of gluconaldehyde derivative. The reaction in its general form is
shown below :

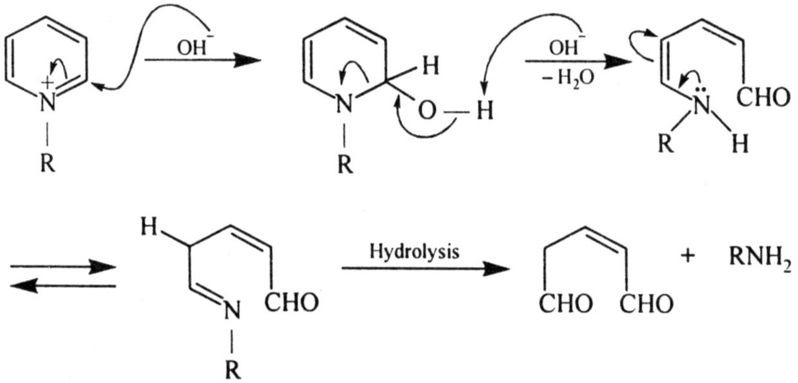

2, 4-Dinitrophenylpyridinium chloride opens on treatment with base.[56]
Halogenopyridines in some cases are also opened similarly in the presence of
amide ion, while in others ring-opened products recyclize to some other type
of ring system. This is demonstrated by the following two examples.[57, 58]

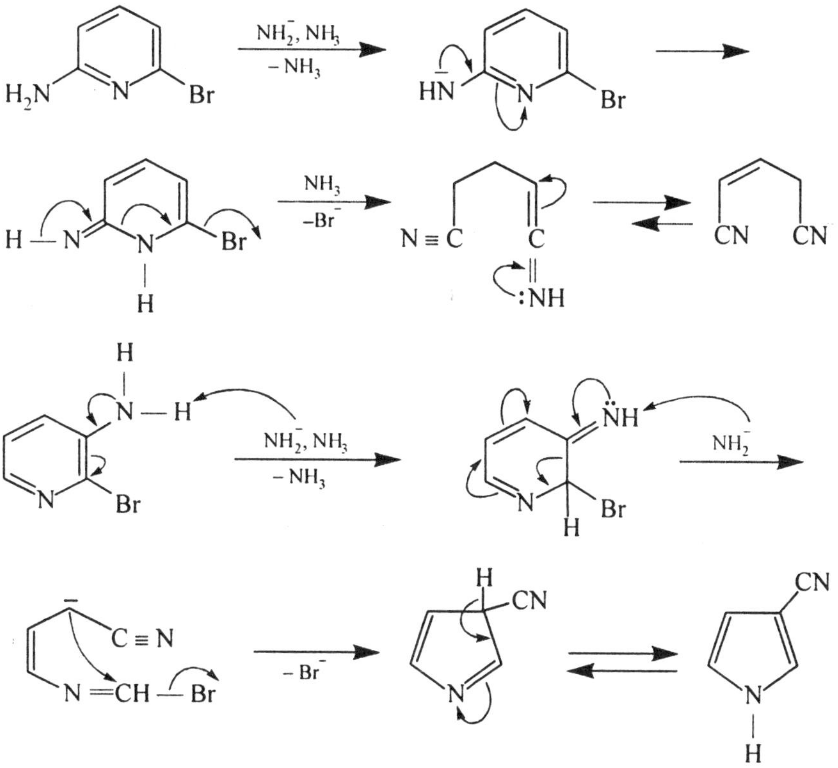

10. *Reactions with Free Radicals* : Pyridine undergoes substitution with a variety of free radicals analogous to benzene. The principal feature of these reactions is the predominant formation of 2-substituted pyridines. The attack at the 3-position takes place only if the 2-and 4-positions are already substituted. Alkylation of pyridine may be achieved by thermolysis of dialkyl peroxides[59], diacyl peroxides[60] or oxidative decarboxylation of carboxylic acids[62], lead tetrabenzoate[62], or aza-or diazo-precursors.[62, 63]

11. *Photochemical Reactions* : Pyridine itself has not been much subjected to photochemical irradiation. It has, however, been found that on irradiation (2537 Å) in butane it is converted[64] to *"Dewar pyridine"* (40) which is very unstable and non-planar. Compound (40) decomposes back to pyridine with a half life of about 2 min. There is another type of *"Dewar pyridine"* (41) but this is still unknown.[65]

Yet another type of valence isomer an azabenzvalene (42) has been proposed as an intermediate during the conversion of N-methylpyridinium chloride by irradiation of its aqueous solution to give a fused aziridine (43)[66].

Stilbenes on irradiation yield phenanthrenes and in the same manner aza-and diaza-stilbenes give rise to aza-and diaza-phenanthrenes.[67, 68]

12. *The Claisen Rearrangement* : The mechanism of the Claisen rearrangement of benzyl vinyl ethers has not been clearly established. The Claisen rearrangement has been applied to 2 –, 3-and 4-(hydroxymethyl) pyridines. Triethyl orthoacetate being employed to generate the intermediate ketene acetals. In all cases the major product resulted from normal rearrangement, equation (6.10). The rearrangement is more facile in 3- than 2-and 4-pyridyl systems. The authors[69, 70] suggest that the electron deficient pyridine ring promotes rearrangement of the electron rich ketene acetal units compared to the benzylic compounds.

$$(6.10)$$

6.1.5 Derivatives of Pyridine

Alkyl- and Aryl-pyridines : Many alkylpyridines are obtained from coal tar and from the pyrolysis of natural products, for instance, 2-methylpyridine from anabasine, 3-methylpyridine from nicotine, 4-methylpyridine from spartein, etc. The reaction of RX with pyridine in strong base (*n*-BuLi) yields predominantly the 2-alkyl isomer. On the other hand, in the presence of free metal (Li or Mg) alkylation of pyridine by RX occurs exclusively at the 4-position.[71] 2, 6-di-*t*-butyl pyridine can be obtained by the following sequence of reactions.[71a]

The decarboxylation of alkyl substituted carboxylic acids is another method for the preparation of alkylpyridines but the yields are unsatisfactory.

2-Picoline is a liquid, b.p. 129.4°C, 3- and 4-picolines have higher boiling points. All the three picolines are miscible with water while ethylpyridines and lutidines are only partially miscible. The hydrogen atoms in toluene are only weakly basic while in 2-and 4-methylpyridines are much more acidic and are potential carbanion sources. Deprotonation of 2-methylpyridine by an organolithium gives 2-pyridylmethyllithium[72] which is a useful nucleophile in the manner of a Grignard reagent.[73] It reacts with aldehydes and ketones to give alcohols which can be dehydrated to 2-vinylpyridines. This intermediate also reacts with CO_2, alkyl halides, etc.[74]

2-Alkylpyridines are useful starting materials for the synthesis of bicyclic systems, for instance with α-halogenocarbonyl compounds, they give indolizines,[75] equation (6.11).

The electron deficiency caused by the electronegative nitrogen atom at the 2-and 4-positions of pyridine also permits nucleophilic reagents to add in a Michael type fashion to the vinyl substituent attached at 2-and not at 3 – position.[76] 2-Vinylpyridine on ozonolysis gives formaldehyde and pyridine-2-carboxyaldehyde. 2- and 4-Alkylpyridines are oxidized by a variety of oxidizing agents to the corresponding carboxylic acids. 3-Methylpyridine, in contrast, is not oxidized and also is unreactive towards phenyllithium and it rather furnishes 3-methyl-2-phenylpyridine.[77] The reaction of 3 –alkylpyridines are expected to be similar to a benzene derivative having a methyl group *meta* to an electron- attracting group.

2-Arylpyridines are best prepared from appropriate aryllithium and pyridine and 4-arylpyridines are obtained *via* the Hantzsch synthesis using an aryl substituted starting material. 3-Phenylpyridine is best obtained from 3-aminopyridine and reacting it with pentyl nitrite in benzene.[78] Monophenylpyridines are high boiling liquids. On oxidation with potassium permanganate/H_2SO_4, they yield pyridine carboxylic acids while with alkaline potassium permanganate, benzoic acid is obtained. This result differs from that of quinoline where the benzene ring is always destroyed to yield pyridine carboxylic acid. Electrophilic substitution always takes place in the benzene ring.

Halopyridines : All of the monohalopyridines are known and can be prepared by the direct halogenation of pyridine or replacement of the diazonium group by the Sandmeyer reaction.[79] The 2-and 4-derivatives may also be obtained by the displacement of the hydroxy group with phosphorus pentachloride.

3-Halopyridine cannot be obtained in a similar manner. 3-Bromopyridine may be obtained by mercuration of pyridine followed by bromination.

2-Chloropyridine is a liquid b.p., 168°C, while the 3-isomer boils at 148°C. They are pleasant smelling liquids in comparison to the parent compound pyridine and are also more basic. The most important reactions of halopyridines are nucleophilic substitution and elimination to give pyridynes. The halo group at the 2- and 4-positions are more reactive than that at the 3-position and readily undergo hydrolysis in the presence of a base. 4-Chloro-3-nitropyridine is rather very labile and it is hydrolyzed even in warm water. Halopyridines do not readily form Grignard reagents by direct reaction with magnesium. Pyridyl Grignard reagents and pyridyllithium compounds are both prepared by reaction of halopyridines with alkylmagnesium halides or alkyllithium compounds, *i.e.* by *"entranchment method"*. The cyano group displaces the halogen at the 3-position much more rapidly than at the 2- and 4-positions to form nicotinonitrile.[80] 2-and 3-Bromopyridines undergo photochemical *SRNi* reaction in liq. ammonia with potassium enolate of acetone,[81] equation (6.12) and many other ketones to yield pyridine derivatives *via* a free radical mechanism.

3-Pyridylacetone

$$(6.12)$$

Aminopyridines : 2-Aminopyridine is most conveniently prepared by the Chichibabin reaction. In this reaction a small amount of the 4-isomer is also obtained. All isomeric halopyridines can be converted to the corresponding aminopyridines by the action of ammonia and a suitable catalyst at elevated temperatures (equations 6.13 to 6.15).

$$(6.13)$$

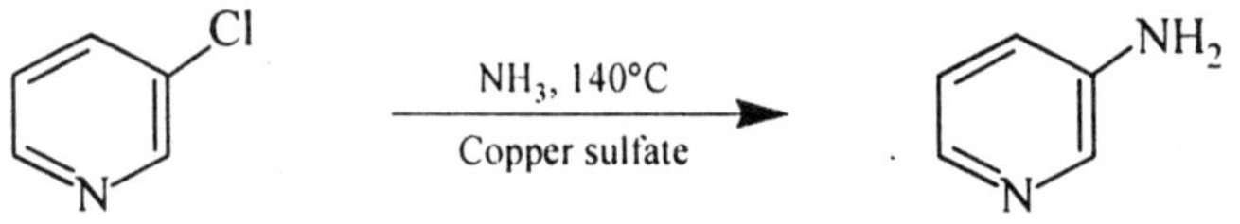

$$\text{(6.14)}$$

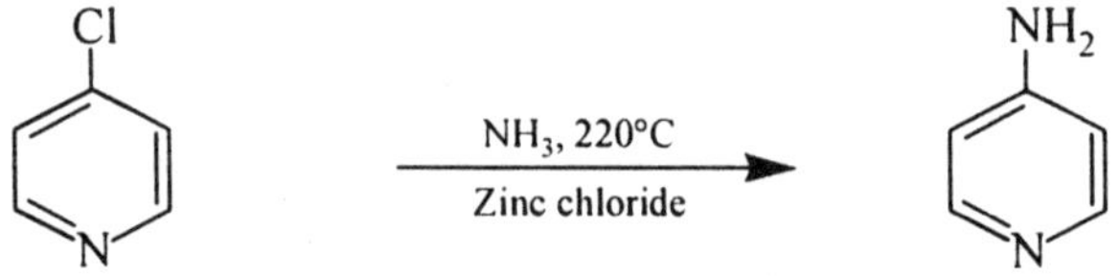

$$\text{(6.15)}$$

All the three mono aminopyridines can also be obtained by the Hofmann reaction on the corresponding pyridine carboxamides.

The mono aminopyridines are solids. 2-aminopyridine (m.p. 50°C), 3-aminopyridine (m.p. 65°C) and 4-aminopyridine (m.p. 157°C). They are all quite basic and form stable crystalline salts by protonation of the ring nitrogen atom. There exists also a marked difference in the properties of 3-amino and 2-and 4-aminopyridines because the latter display tautomerism analogous to hydroxypyridines. Electrophilic substitution on 2-aminopyridine is

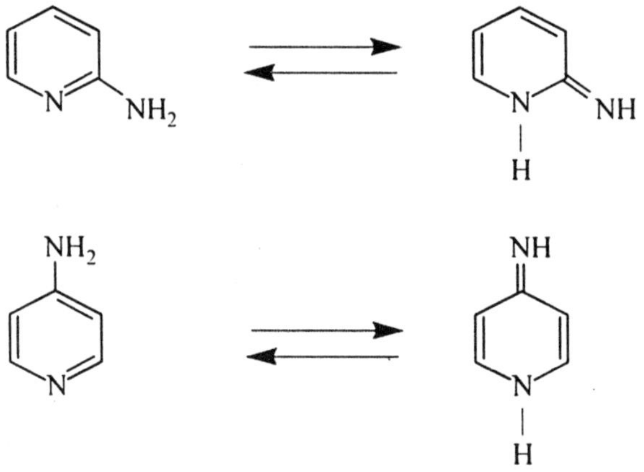

controlled by the $-NH_2$ group and this directs the incoming group to the 5-position.[82] 3-Aminopyridine behaves as a normal aromatic amine and like aniline can be diazotized and gives all the usual reactions of a diazonium salt. 3-Aminopyridine condenses[83] with ethyl-2-benzoyl 2-bromoacetate to yield an excellent anti- inflammatory agent. The 2-and 4-aminopyridines can be diazotized only under strongly basic conditions.

Pyridine N-Oxide : Pyridine itself is stable towards the action of oxidizing agents but towards peracids it behaves, in a manner characteristic of tertiary amines. The unshared pair of electrons on the nitrogen atom links an oxygen atom in a coordinate linkage forming an amine oxide. With perbenzoic acid, pyridine yields pyridine N-oxide[83] in a yield of 85%, equation (6.16).

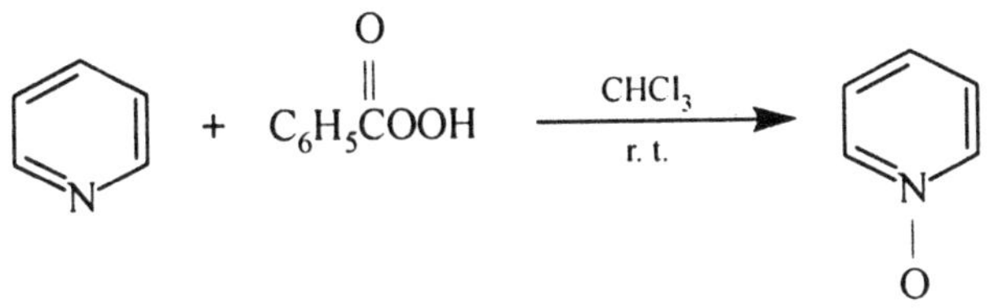

$$\text{(6.16)}$$

2-, 3-and 4-Picoline N-oxides, 2, 3-, 2, 4-and 4, 6-lutidine N-oxides and isomeric halo N-oxides can be prepared in a similar manner by direct oxidation.

Pyridine N-oxide is a colorless solid, m.p. 66°C and is soluble in water and most organic solvents, 2-Picoline N-oxide boils at 127°C, 3-picoline N-oxide at 133-36°C, and 4-picoline N-oxide at 186-88°C. Pyridine is weakly basic (pKa 5.17) whereas pyridine N-oxide has a pKa 1.90, thus the drop in basicity is considerable. That the donation of electrons from the oxygen into the ring is significant is apparent from the dipole moment value. The dipole moment of pyridine N-oxide is 4.30 while that of pyridine is 2.23 D.

4.30D   2.23D

This value is considerably less than anticipated, *i.e.* the sum (6.6 D) of the dipole moment of pyridine and $N^+ - O^-$ bond moment. Therefore, there is a significant contribution from the following resonance structures. These

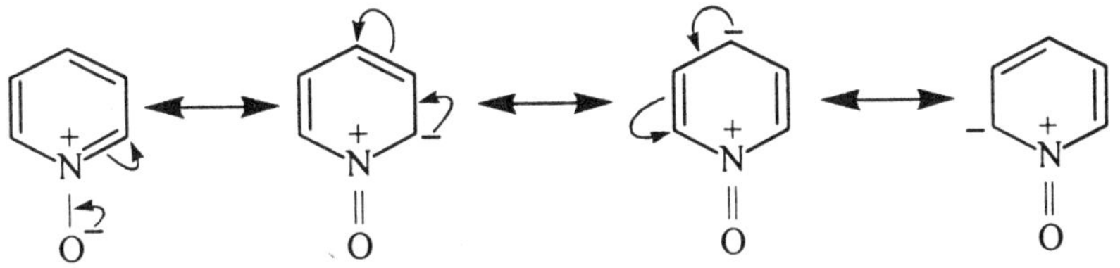

structures also imply an increased susceptibility of the ring in pyridine N-oxide to electrophilic substitution This is also corroborated by π-electron density calculations. The pyridine N-oxide molecule is quite versatile because of its polarizability. The N-oxide group can function both as an electron-acceptor and as an electron-donor depending on the demand of the reagent. This dichotomous behavior leads to some interesting reactivity patterns in N-oxide chemistry. As a consequence both electrophilic and nucleophilic substitutions occur with ease at the 2-and 4-positions. Chlorination of pyridine N-oxide with sulfuryl chloride gives a mixture of 2-and 4-isomeric chloropyridine. 2-Picoline N-oxide with phosphorus oxychloride yields 72%

4-chloro-2-picoline. Direct chlorination and bromination to form halopyridine N-oxide is difficult.

Pyridine N-oxides are nitrated more easily than the parent pyridines. Nitration (HNO_3, H_2SO_4) of pyridine N-oxide gives a high yield of 4-nitropyridine N-oxide[84]. The corresponding 2-isomer is not obtained by direct nitration[85] but by the oxidation of 2-aminopyridine N-oxide, equation (6.17). The N-oxide can be removed by catalytic hydrogenation.

This procedure also reduces the $-NO_2$ group. In contrast, the sulfonation of pyridine N-oxide is not easy and takes place only under severe conditions (oleum, $HgSO_4$/220-240 °C) and 3-pyridine N-oxide sulfonic acid is obtained.

$$(6.17)$$

This drastic change in orientation in sulfonation is probably due to the formation of the species (44) in which the electron-donating property of the oxygen atom is suppressed.[86, 87]

(44)

Pyridine N-oxide reacts with acetic anhydride on heating to yield 2-acetoxypyridine. The rate is slightly decreased by the addition of sodium acetate and no gaseous products such as methane or carbon dioxide are evolved.[88] These observations suggest the following mechanism for this reaction.[89]

2-Methylpyridine N-oxide similarly interacts with acetic anhydride. 2- and 4-Methylpyridine N-oxides similar to other reactive methylene compounds can be nitrosated[90] and participate in the Claisen condensation.[91] The anions derived from pyridine N-oxides serve as useful intermediates.

Pyridine N-oxides can be reduced by a wide variety of reagents to regenerate the parent pyridines, such reagents include catalytic hydrogenation over Raney Nickel[9], iron and acetic acid[94] or sodium borohydride.[94] The reagents of choice, however, are the trivalent phosphorus compounds such as triphenylphosphine,[91] triethylphosphite, dialkylsulfoxylates,[96] phosphorus trichloride in chloroform[97] and aluminum iodide[98] in acetonitrile and ammonium formate/Pd-C in methanol,[99] pyridine N-oxide is also deoxygenated to pyridine on

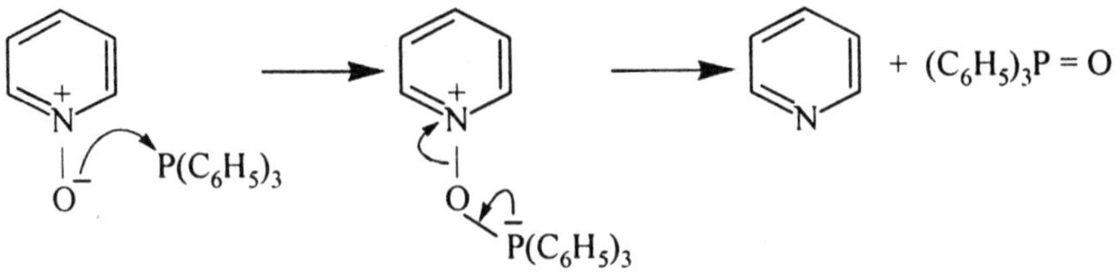

irradiation,[100] or on treatment with hexamethyldisilane ($Me_3Si - SiMe_3$) in hexamethyl-phosphoric triamide [$(Me_2N)_3 PO$] (HMPT) as solvent at 0°C.[100] Analogous to certain halopyridines, appropriately substituted halopyridine N-oxides react with strong bases to form 2, 3-and 3, 4-pyridyne N-oxides. Accordingly 2-chloro or 3-bromopyridine N-oxides result in 2,3-pyridine N-oxides which on treatment with potassium amide in liq. ammonia give a mixture of 2-and 3-aminopyridine N-oxides (5-10% yield).[101] On theother hand,

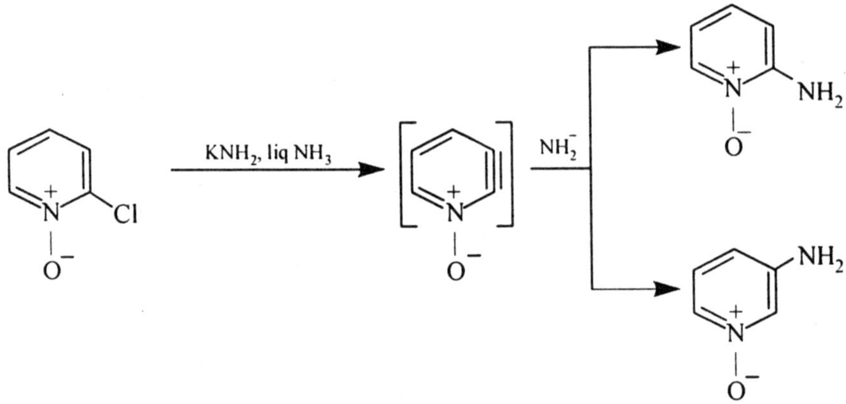

3-bromo-2-ethoxypyridine N-oxide under these conditions yields the 3, 4-pyridyne. In the presence of a secondary amine in water the pyridine N-oxide ring is opened on ultraviolet irradiation to lead to an unsaturated nitrile,[102]

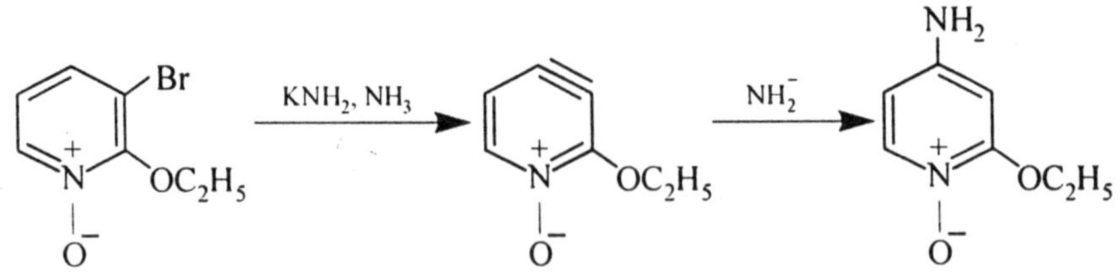

equation (6.18). The reactivity of a 2-and 4-alkylpyridine can be increased by conversion to the corresponding pyridinium salt or pyridine N-oxide.

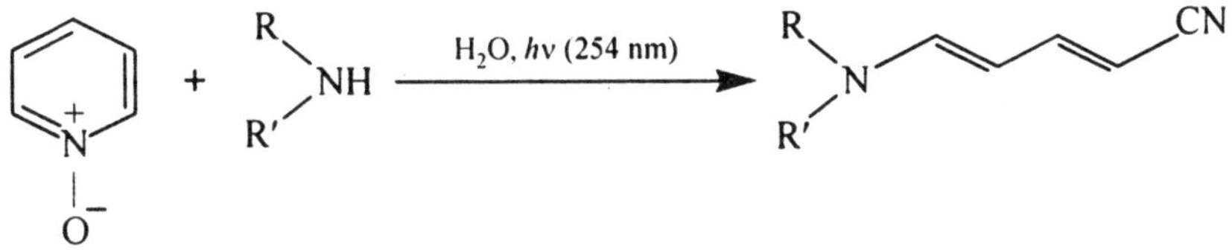

$$(6.18)$$

The result of this side-chain activation is generally reflected in the need of a much milder base to affect condensation,[103] equation (6.19).

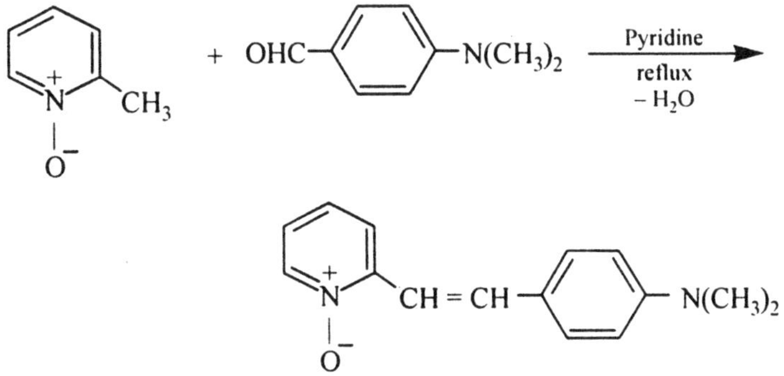

$$(6.19)$$

An alkyl nitrite and sodamide in liq. NH_3 with α – benzylpyridine N – oxide provides an oxime.

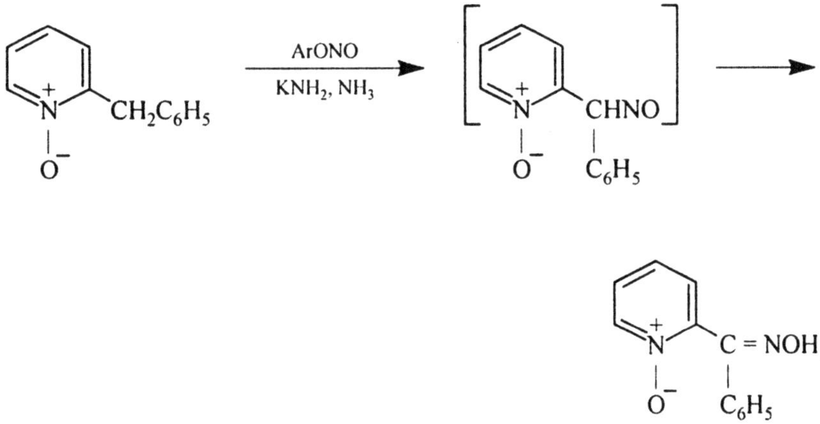

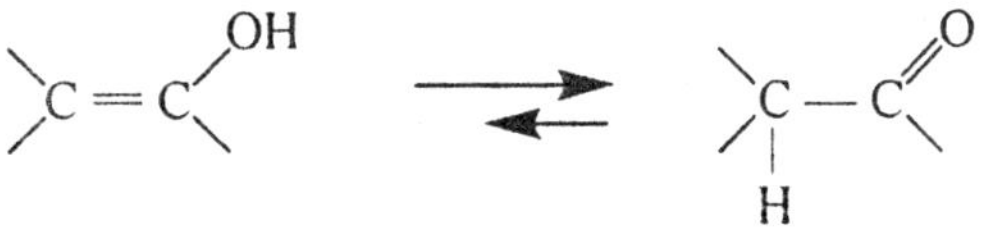

Hydroxypyridines (Pyridinols) : The keto-enol tautomerism is well known in organic chemistry. The keto form is believed to be the more stable than the enol form. 2-and 4-Hydroxypyridines also exist in tautomeric forms with 2-and 4-pyridones.

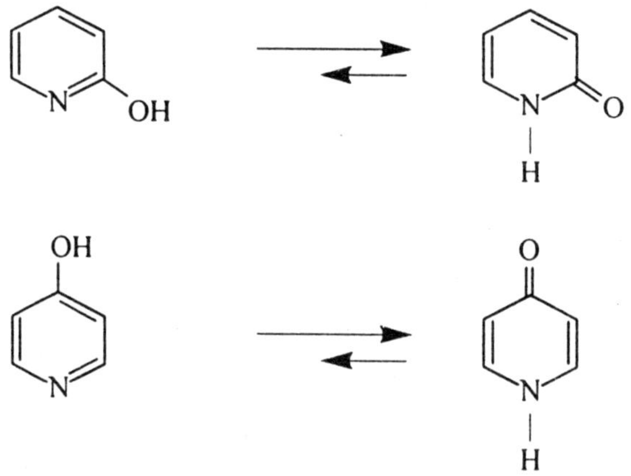

Comparison of the ultraviolet spectra of the reference compounds 2-methoxypyridine and 1-methylpyridine clearly establishes that in polar solvents the 1-oxo form is preferred.[104]

Furthermore, the keto forms behave as amides. The case with 3-hydroxypyridine is different as it exists to a considerable extent in solution as the pyridinium oxide zwitterion.

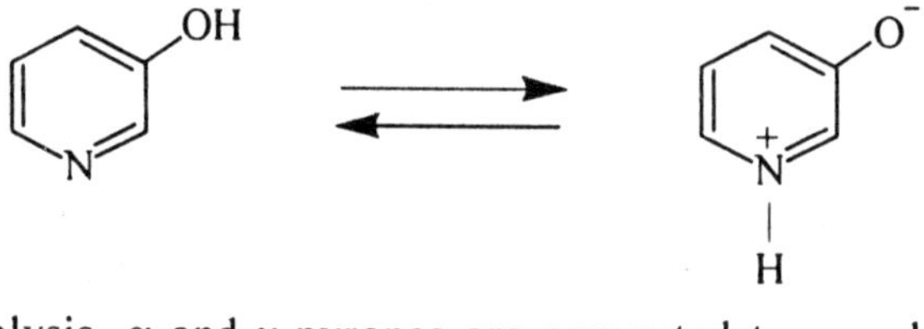

On ammonolysis, α-and γ-pyrones are converted to α-and γ-pyridones respectively.

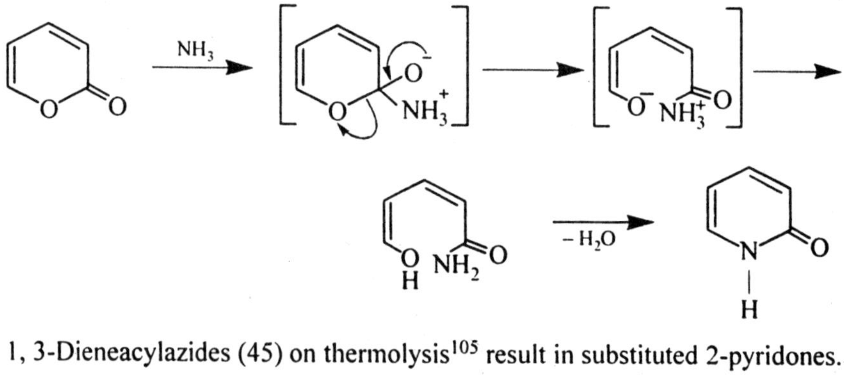

1, 3-Dieneacylazides (45) on thermolysis[105] result in substituted 2-pyridones.

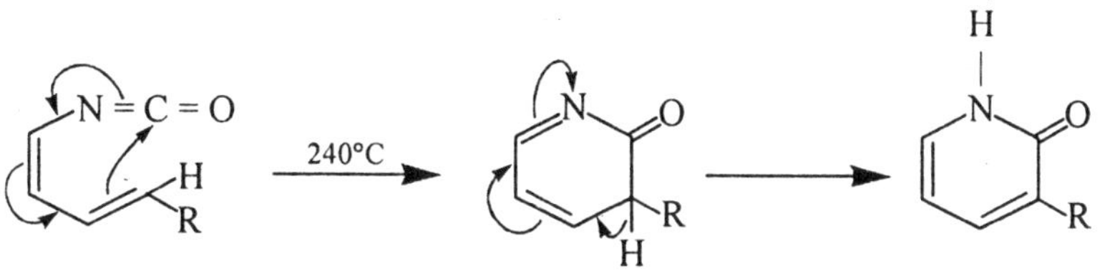

(45)

Substituted 2-pyridones may also be obtained by the thermal rearrangement[105] of propargylic pyrrolidine *pseudo* urea. The latter can be obtained[106] by potassium alkoxide catalyzed reaction of secondary propargyl alcohol and cyanopyrrolidine.

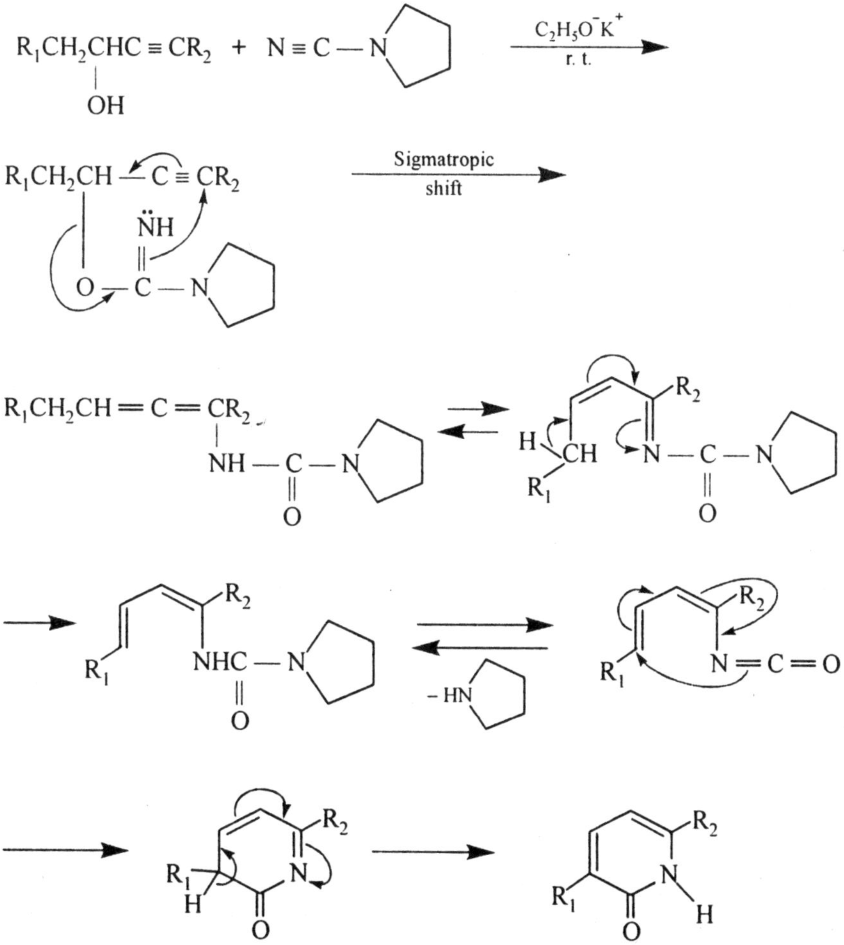

2-and 4-Pyridinols behave similarly in their chemical properties. Unlike 3-pyridinol, however, they are less phenolic in character. On treatment with diazomethane, the 2-isomer yields mainly the O-methyl while the 4-isomer yields the N-methyl derivative. Phosphorus pentachloride and thionyl chloride can replace the hydroxyl group by chlorine.

3-Pyridinol behaves as a typical phenol and gives a purple color with ferric chloride. It undergoes electrophilic substitution such as halogenation, nitration and sulfonation with predominant attack at the 2-position. 3-Pyridinols also form ethers similar to phenols.

2-Pyridone leads to bicyclo [2.2.0] pyran-2-one on direct irradiation,[107] equation (6.20)

$$(6.20)$$

Pyridine Aldehydes and Ketones : Pyridine aldehydes have been prepared by a variety of methods. 2-Pyridinecarbaldehyde may be prepared by the ozonolysis of 2-stilbazole and reduction of the ozonide, equation (6.21).

$$(6.21)$$

Substituted pyridinecarbaldehydes have also been prepared by pyrolyzing 2-*bis* (methylthio) methylpyridine S-oxide.[108]

The pyridyl aldehydes closely resemble arylaldehydes largely because a carbonyl group cannot interact mesomerically with the pyridine nitrogen atom. It thus adds sodium bisulfite, undergoes benzoin condensation and the Cannizzaro reaction, equation (6.22).

2-Pyridinecarbaldehyde   **Pyridoin**

$$(6.22)$$

Pyridyl ketones are obtained by treating cyanopyridines with Grignard reagent and subsequent hydrolysis.[109] A number of ketones have been prepared

by the Japp-Klingemann reaction. This involves the reaction of a diazonium salt with a pyridinecarboxylic acid to form an aryl hydrazone and the ketone is obtained on hydrolysis.[109] Pyridyl ketones give reactions expected of aromatic

ketones. They condense with the usual carbonyl reagent to form well defined derivatives. They also behave normally towards the action of oxidizing agents. Thus 2-acetylpyridine undergoes the normal iodoform reaction. 2-Acetyl-pyridine also undergoes the Darzens condensation with α-halogeno esters.[110]

Pyridinecarboxylic Acids : The pyridinecarboxylic acids and their derivatives are present in many natural products (alkaloids, vitamins, coenzymes). They are obtained by the oxidation of the corresponding mono alkylpyridines, quinoline and isoquinoline and hydrolysis of cyanopyridines. The *p*Ka values of the acids indicate that they exist mainly in zwitterionic form[111] in aqueous solution and this may be in part responsible for their facile decarboxylation. These acids are slightly more stronger than benzoic acid.

2-Pyridinecarboxylic acid   **Nicotinic acid**   **Isonicotinic acid**

6.1.6 Naturally and Biologically Active Compounds

Pyridine and piperidine derivatives occur in nature abundantly. Nicotinic acid was isolated from yeast and also from rice bran. Its biological significance as an additional member of the vitamin B complex was recognized in 1937. It is formed *in vivo* from the essential amino acid tryptophan in animals. Nicotinamide is essential for curing the disease pellagra in human beings. It occurs in liver and is excreted in the human urine as glycine conjugate nicotinuric acid (45). Protein enzymes function as catalysts in chemical

(45)

reactions of living systems, in many cases they react with a substrate only when a second component referred to as coenzyme is also present. One of the

(46)

more important coenzymes is NAD (nicotinamide adenine dinucleotide) (46). With the aid of other enzymes, NAD^+ takes part with great ease in electron transfer reactions. The nicotinamide portion abstracts a H^- ion and is reduced to the 4, 4-dihydropyridine counterpart ($NADH^+$) and the substrate undergoing oxidation. NAD functions as a reducing agent in the following manner, equation (6.23). The pyridylmethanol derivatives are the B_6 or pyridoxime group of vitamins,[112] and constitute three members namely pyridoxine (47), pyridoxamine (48) and pyridoxal (49) all of which occur in combined form.

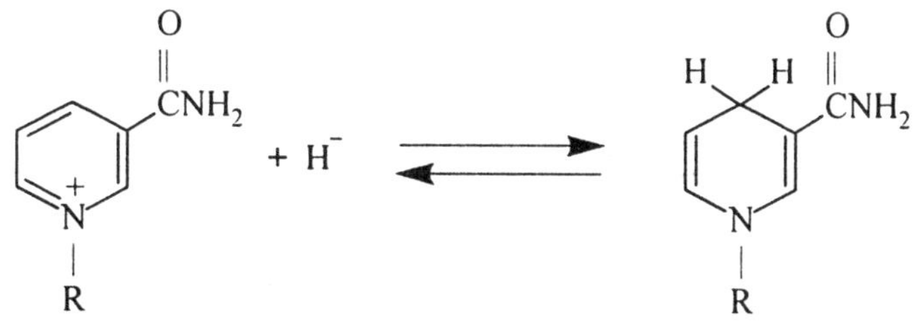

$$(6.23)$$

The most important role of these substrates in biological systems centers around transmission and decarboxylation reactions.

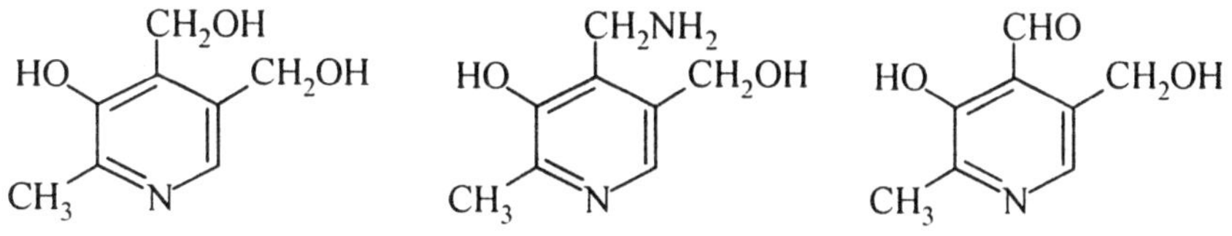

(47)   (48)   (49)

The pyridine ring occurs in nature in many alkaloids and the best known of these are the tobacco alkaloids such as nicotine (50), anabasine (51) and vicirine (52). Another important pyridine alkaloid, buchananine (53) is obtained

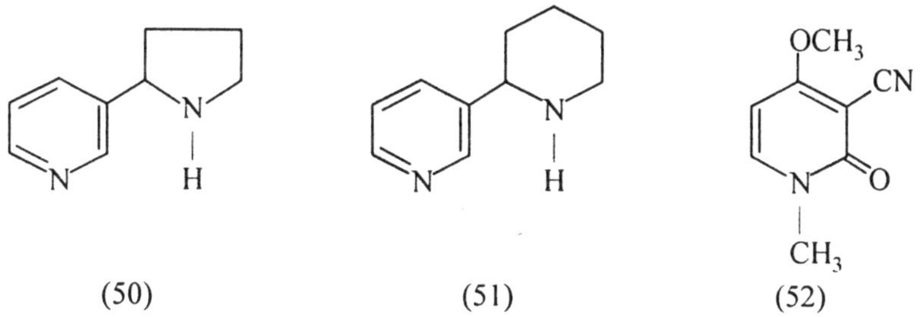

(50)   (51)   (52)

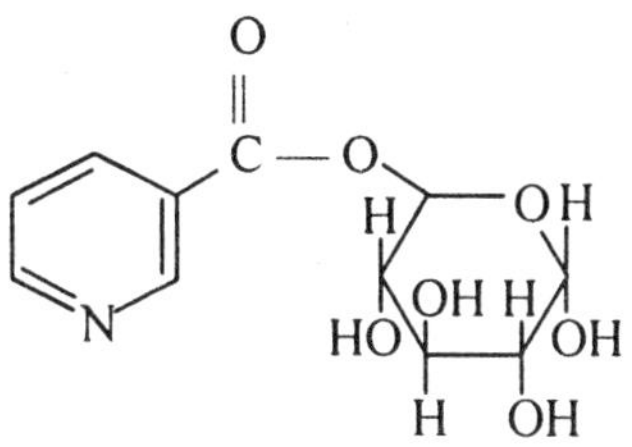

(53)

from a plant which is used for the treatment of rickets in children[113] and ascididemin[114] (54) a pentacyclic alkaloid has anti-leukemic activity.

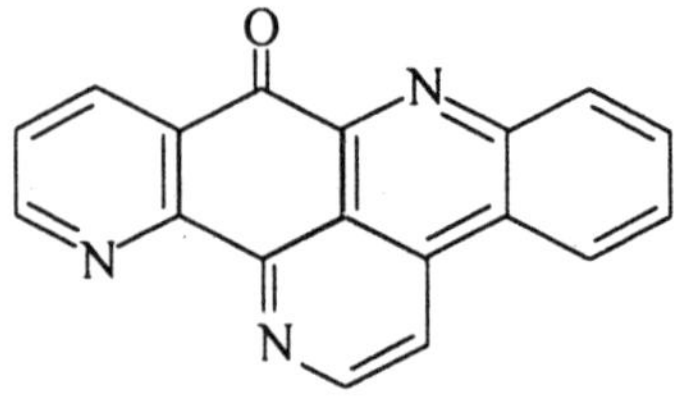

(54)

Alkaloid cribrochalinamine oxide (55) has been isolated from a marine sponge
cribrochalina.

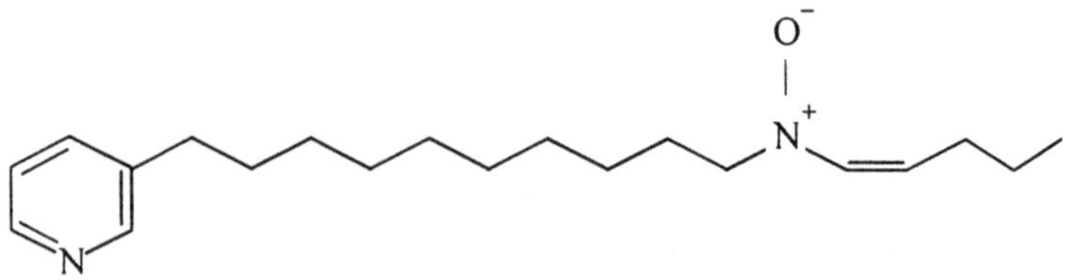

(55)

6.2 PYRYLIUM SALTS

The aromatic oxygen and sulfur containing nuclei comprise the pyrylium (2)
and the thiopyrylium (5) salts. These salts are quite stable, (5) being more
stable than (2). Valency requirements do not permit the existence of an
uncharged oxygen or sulfur as in the case of pyridine. Structures (56) and (57)
are 1-benzopyrylium (chromylium) and 2-benzopyrilium (isochromylium) ions.
Structure (58) represents the xanthylium ion. In naming these ions the
numbering commences from the hetero atom. These salts are very reactive
and many of their derivatives are the basis of important natural products. The
first pyrylium salt was prepared by Kostanecki and Rossbach in 1896 from
the reaction of 1, 3, 5-triphenylpentane-1, 5-dione and sulfuric acid. They
were unable, however, to isolate the compound but was later recognized by
Dilthy as 2, 4, 6-triphenylpyrylium salt. Many of the pyrylium salts have been
obtained in a crystalline state. The stability of these salts may be ascribed to
the π-system of the pyrylium salt which has an aromatic sextet (6π-electrons)
and thus possesses an aromatic character similar to that of pyridine and benzene
despite the fact that they have a positive charge. The aromatic character of the
pyrylium ring is also reflected in the *n.m.r.* spectra of these compounds where
the ring protons show a marked downfield shifts, indicating a strong ring
current.[116]

(2)   (5)   (56)

(57)   (58)

The properties of pyrylium compounds are different than those expected of benzene. For instance, benzene is easily attacked by electrophiles but is resistant to the attack of nucleophiles. The pyrylium salts on the other hand, display a behavior of the opposite order, *i.e.* they are easily attacked by nucleophiles but not by electrophiles.

6.2.1 Synthetic Methods[117]

The synthesis of the pyrylium compounds isolated as salts involves the formation and cyclization of an appropriately unsaturated 1, 5-dicarbonyl precursor.[118]

1. *From 1, 5-Dicarbonyl Derivatives* : The simplest pyrylium perchlorate may be obtained from the sodium salt of glutanaldehyde in the presence of

perchloric acid[119] at –20°C. 1, 3, 5-triphenyl-2pent-1, 5-dione prepared in the following manner cyclizes to 2, 4, 6-triphenylpyrylium salt on treatment with acid.

Chlorovinyl ketones[120] similarly react with acetophenone to give an intermediate which cyclizes to pyrylium salt.

2. *The Hantzsch Type synthesis* : A general pyrylium sysnthesis is closely related to Hantzsch pyridine synthesis. This involves the reaction of an aromatic aldehyde with two moles of aryl methy ketone in acetic anhydride with the

$$-H_2O \rightleftharpoons \qquad \xrightarrow[\text{Oxidation}]{FeCl_3}$$

(59)

resultant formation of a pyran[121, 122] (59) which on oxidation yields the pyrylium system. 1-Benzopyrylium compounds can be prepared[123] as depicted in equation (6.24).

$$\xrightarrow{FeCl_3,\ HCl}$$

(6.24)

6.2.2 Chemical Reactions

1. *Attack by Nucleophilic Reagents* : The pyrylium salts are very susceptible to nucleophilic attack but the degree of susceptibility varies widely. A nucleophilic attack takes place at C-2, C-4 or C-6 positions as is evident from the various resonance structures of the pyrylium ion. Of these sites, C-2

and C-6 positions are favored than C-6 because the intermediate (60) is more conjugated than (61). The unsubstituted pyrylium cation is attacked by water even at 0°C but the 2, 4, 6-trimethylpyrylium cation is stable in water at 100°C.

$$\xrightarrow{NUC^-}$$

(60)

$$\xrightarrow{NUC^-}$$

(61)

Hydroxide ion, however, adds rapidly resulting in the formation of 2-hydroxy-2*H*-pyran. The adduct (62) initially formed is interconvertible with the ring opened derivatives (63).

(62)

(63)

Certain other nucleophiles also open the pyrilium ring.[124, 125]
Substituted pyrylium salts are deprotonated at the 2-and 4-positions, though much faster at the latter position, (equation 6.25).

$$(6.25)$$

Ammonia and primary amines react with pyrylium salts to afford pyridine derivatives and quaternary pyridinium salts respectively.[126]

With secondary amines the adducts cannot cyclize to pyridines,[126] equation (6.26), as there is no hydrogen available.

$$(6.26)$$

Pyrylium salts react with certain reactive methyl compounds leading to ring fission and subsequent cyclization to yield benzenoid products. This is illustrated[127] for the reaction of 2, 4, 6-triphenylpyrylium cation with nitromethane.

A number of interesting photochemical reactions of pyrylium compounds have also been known.[128, 129]

The 1-benzopyrylium system is very resistant to electrophilic attack even in the benzene ring but undergoes attack by a nucleophile at the 2-position. Addition of hydroxide ion gives as 2*H*-1-benzopyran-2-ol which is in equilibrium with a ring opened isomer. The xanthylium salts are attacked by nucleophiles at the 9-position.

With ammonia or primary amine 1-benzopyrylium salts yield only a ring opening product which does not cyclize. On the other hand, 2-benzylpyrylium

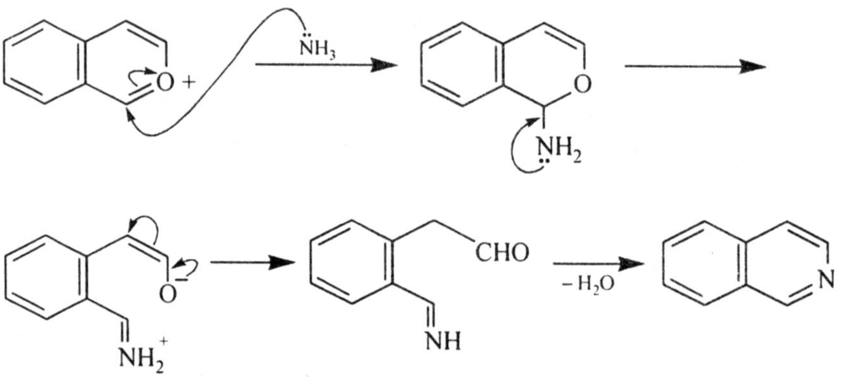

salts recyclize after ring-opening to give isoquinoline. The properties of xanthylium salts are different from those of pyrylium salts. They exist in the form of a cationic structure.

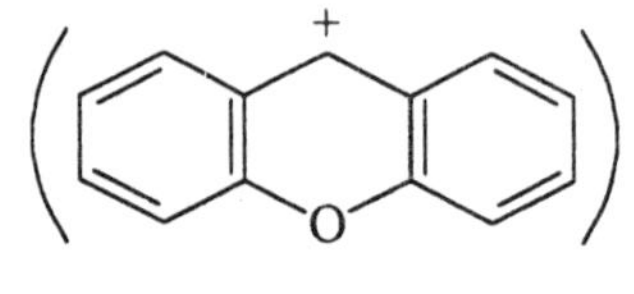

6.3 α - AND γ - PYRONES

Pyrones are six-membered heterocyclic compounds containing one oxygen atom in the ring and five sp^2 hybridized carbons. Two isomeric pyrones namely α-pyrone (64) and γ-pyrone (65) are possible. They are also known by other names such as 2*H*-pyrone and 4*H*-pyrone respectively.

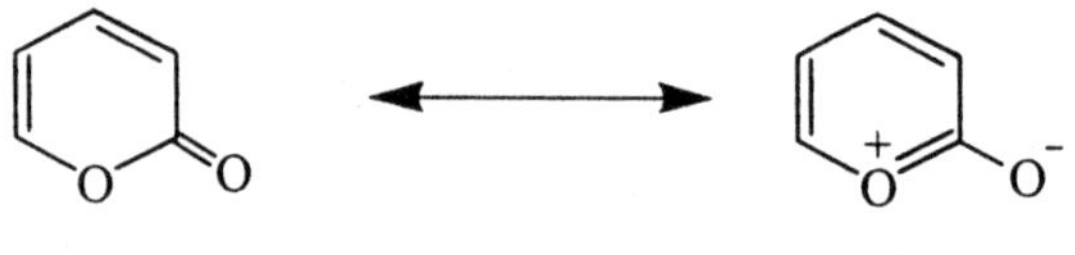

(64)   (64a)

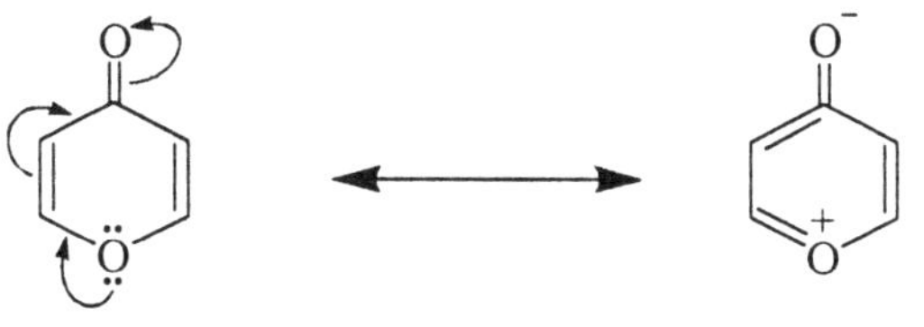

(65)   (65a)

Benzannulated pyrone derivatives of α-pyrone are coumarin (66) isocoumarin (67) while that of γ-pyrone are chromone (68) and xanthone (69). Both types of pyrones display characteristics which partake some reaction of alkenes and some those of arenes. The α-pyrones behave more in the former manner and

(66)   (67)

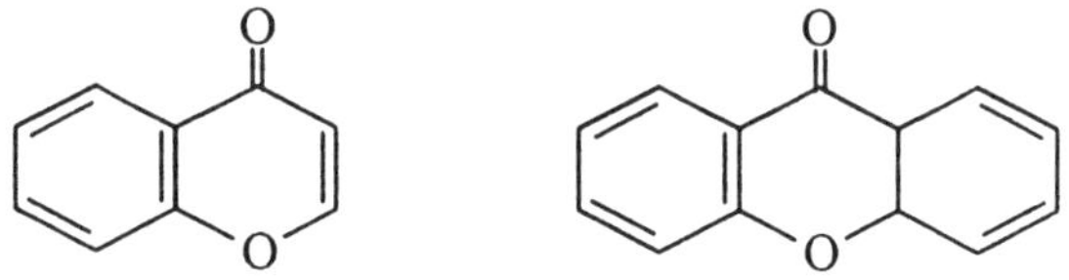

(68)   (69)

they can be viewed as conjugated enol-lactones, the extent to which they are additionally stabilized by resonance (64a) is not clear. On the contrary the reactions of γ-pyrones are observed to be of the pyrylium betaine structure (65a). Therefore, γ-pyrones have a greater degree of aromaticity than α-pyrones. This has been supported by spectroscopic data.[130] The pyrones are insignificantly basic and their basicities are comparable to those of urea and nitroanilines.

6.3.1 Synthetic Methods

A simple synthesis of α-pyrone involves the decarbonylation of malic acid in the presence of conc. sulfuric acid to form fumaric acid. Two molecules of the acid self-condense *in situ* to coumalic acid. This on decarboxylation

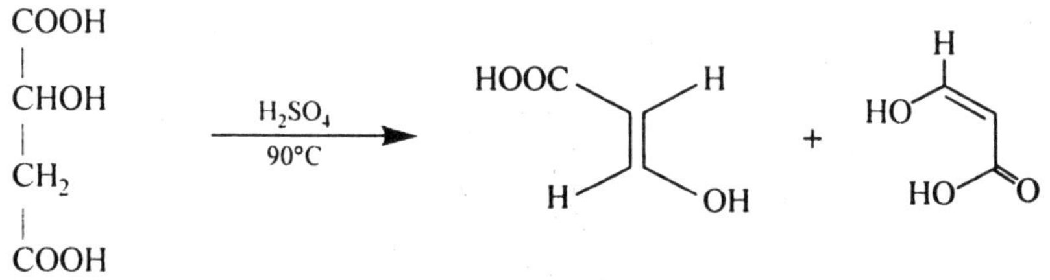

Coumalic acid

yields α-pyrone.[131] Substituted pyrones can be obtained by reacting ethyldiazoacetate with an appropriate keto ester to form an intermediate of type (70). This subsequently forms the 4, 6-disubstituted α-pyrone. The ester group can be removed by hydrolysis and decarboxylation.

(70)

6.3.2 Chemical Properties

Both α-and γ-pyrones are potentially aromatic but the latter are more stable than the former. α-Pyrone itself polymerizes slowly on standing while the γ-pyrones are quite stable crystalline substances.

1.   *Electrophilic Substitution* : The aromatic character of pyrones is reflected in their reactions towards electrophilic substitution. Both α-and γ-pyrones undergo substitution at the 3-or 5-positions, *i.e. o*-or *p*-to the carbonyl group. The ease of electrophilic substitution is enhanced by the presence of alkyl substituents on the ring.

Halogenation : α-Pyrone is brominated at the 3-position *via* an addition elimination reaction and is not the result of direct electrophilic substitution.

The ring can be nitrated and sulfonated.[132] Reaction of α-pyrone with nitronium tetrafluoroborate, reversibly forms a *o*-nitro salt which is slowly converted to 5-nitro-α-pyrone.

Bromination of γ-pyrone yields a mixture of mono-and di-bromo γ-pyrones *via* a perbromide formation.

Acetylation (RCOCl/CF$_3$COOH) of γ-pyrones gives rise to 3-and 5-substituted products.[133]

The methyl groups at the 2-position in both α-pyrones are activated,[134] and the resulting anion reacts with carbonyl compounds to yield alcohols as illustrated by the following examples (equations 6.27 and 6.28)

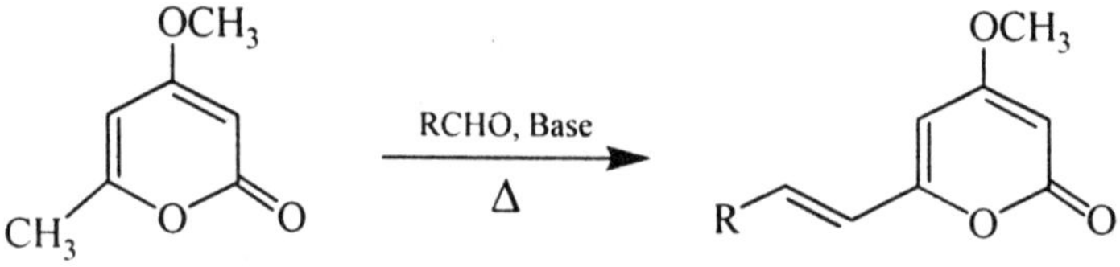

$$(6.27)$$

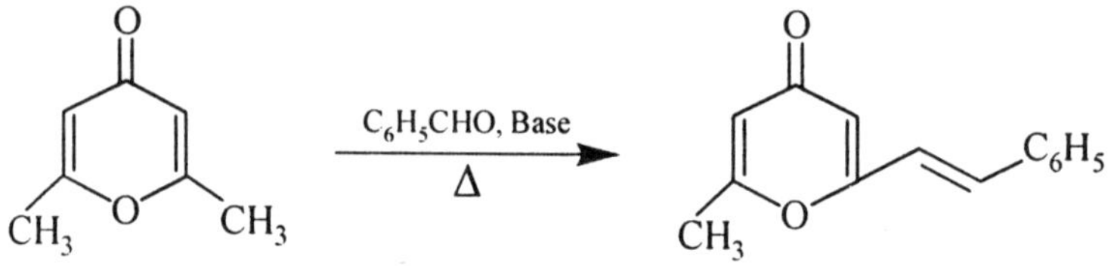

$$(6.28)$$

2.   The Diels-Alder Reaction: The nature of the α-pyrone ring is reflected in the Diels-Alder reaction in which it functions as a diene. For instance, with maleic anhydride and α-pyrone forms an adduct (71) which suffers a pericyclic decarboxylation to form a new diene which reacts with a second

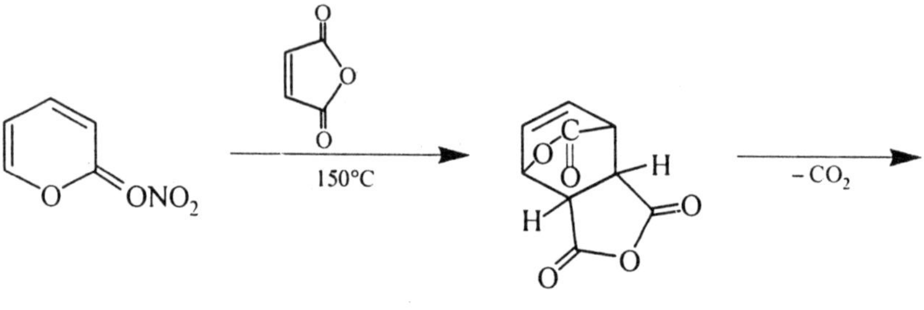

(71)

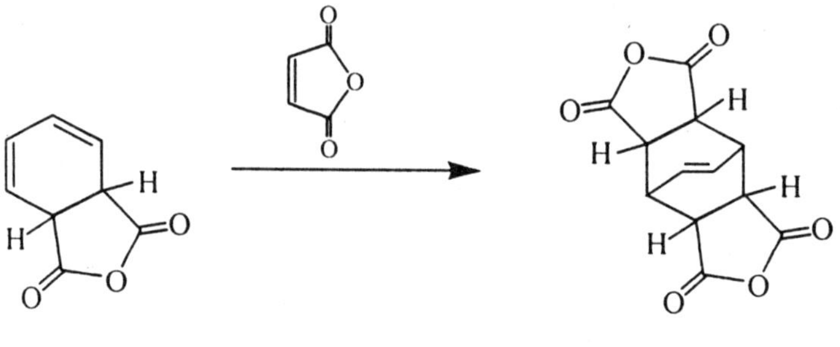

(72)

molecule of maleic anhydride to form a new adduct (72).[135] There seems to be no case of a γ-pyrone serving as a diene in the Diels-Alder reaction.

3. *Reaction with Reducing agents:* Catalytic hydrogenation (H_2, Ni) equation (6.29) and (6.30) occurs at the C – C double bonds to give saturated lactones.[135] Hydrogenation further supports the olefinic nature of both α-and γ-pyrones.

$$(6.29)$$

$$(6.30)$$

Sodium borohydride does not react with pyrones but lithium aluminum hydride breaks the α-pyrone ring to form an acid but an alcohol is obtained in the case of γ-pyrone, equation (6.31) due to reduction of the carbonyl moiety.

$$(6.31)$$

4. *Nucleophilic Reactions:* In marked contrast to electrophilic substitution, pyrones are easily attacked by nucleophilic reagents. Weak nucleophiles add at the 2-position while strong ones at the 6-position.[136]

Grignard reagents react preferentially at the carbonyl group to yield initially an alcohol which is converted to a substituted pyrylium salt on acidification.

5. *Photochemical Reaction* : Pyrones have been widely investigated photochemically and they have been shown to give interesting products. On photolysis α – pyrone loses carbon dioxide to yield cyclobutadiene.[137]

Irradiation of appropriately substituted α-pyrones in benzene results in the formation of a mixture of dimers. In a more recent study it has been reported that a single dimer is obtained if the irradiation is carried out in the presence of

(72)

a sensitizer.[138] On irradiation 2, 6-dimethyl-4-pyrone forms a cage structure (head-to-tail) equation (6.32). Interestingly this dimer (72) reverts to the monomer on treatment with acid.[139]

REFERENCES

1. *"Comprehensive Heterocyclic Chemistry,"* A. J. Boulton and A. Mckillop, (Eds.), Vol. 2, Pergamon Press, Oxford (1984).
2. For a review on pyridine synthesis see, F. Krohnke, *Synthesis,* **22,** 1(1976).
3. N. H. Cantwell and E. V. brown, *J. Am. Chem. Soc.,* **74,** 5967 (1952).
4. W. Gruber, *Canad. J. Chem.,* **31,** 564 (1953).
5. J. Fried, in *Heterocyclic Compounds,* R. C. Elderfield, (*Ed.*), Vol. I, Wiley, New York (1950), p. 343.

6. E. R. Alexander, A. B. Herrick and T. M Roder, *J. Am. Chem. Soc.,* **72,** 2760 (1950).

7. J. E. Tooney, Jr., *Adv. Heterocyclic Chem.,* **37,** 167 (1984).

8. E. Breitmaier and E. Bayer, *Tet. letters,* 3291 (1970); E. Breitmaier S. Gassenmann and E. Bayer, *Tetrahedron,* **26,** 5907 (1970).

9. R. B. Bates W. A. Beavers and B. Gordon *J. Org. Chem.,* **44,** 3800 (1979); J. Klein and A. Medlik, *Chem. Comm.,* 275 (1973).

10. R. B. Bates, B. Gordon, P. C. Keller and J. V. Rund, *J. Org. Chem.,* **45,** 168 (1980).

11. R. S. Tewari *et al., J. Heterocyclic Chem.,* **17,** 953 (1980); R. S. Tewari and A. K. Dubey, *Ind. J. Chem.,* **19B,** 153 (1979).

12. G. R. Newkome and D. L. Fishel, *Chem. Comm.,* 916 (1970).

13. C. Alvarez, E. Delgado, O. Garcia, S. Medina and C. Marquez, *Synth. Comm.,* **21,** 619 (1991).

14. P. Molina, P. M. Presnede and P. Alarcon, *Tet. Letters,* 379 (1988).

15. L. Meerpoel, G. Deroover, K. Van Aken, G. Lux and G. Hoornaert, *Synthesis* 765 (1991).

16. Yu. I. Chumakov and V. P. Sherstyuk, *Tet. Letters,* 129 (1988).

17. H. C. Brown and R. H. Horowitz, *J. Am. Chem. Soc.,* **77,** 1733 (1955).

18. A. R. Katritzky and J. B. Ridgewell, *J. Chem. Soc.,* 3753 (1963).

19. D. E. Pearson, W. W. Harogrove, J. K. T. Chow and B. R. suthers, *J. Org. Chem.,* **26,** 789 (1961).

20. H. J. Den Hertog *et al., Rec. Trav. Chim.,* **81,** 804 (1962).

21. See ref. 1.

22. H. J. Den Hertog J. Overhoff, *Rec. Trav. Chim.,* **49,** 552 (1930).

23. G. A. Olah. J. A. Olah and N. A. Overrchuk, *J. Org. Chem.,* **30,** 3373 (1965).

24. H. J. den Hertog *et al., Rec. Trav. Chim.,* **77,** 963 (1968); S. M. McElvain and M. A. Goese. *J. Am. Chem. Soc.,* **65,** 2223 (1943).

25. H. C. van der Plas and H. J. den. Hetrog, *Rec. Trav. Chim.,* **81,** 841 (1962).

26. C. D. Hurd and C. J. Morrissey, *J. Am. Chem. Soc.,* **77,** 4658 (1955).

27. For a review see, R. M. Acheson, *Adv. Heterocyclic Chem.,* **1,** 125, (1963).

28. R. Huisgen, *Topics in Heterocyclic Chemistry,* R. N. Castle, (*Ed.*), Wiley Interscience, New York (1969), Chapter VIII.

29. R. M. Acheson and G. A. Taylor, *J. Chem. Soc.,* 1691 (1960).

30. E. Winterfeldt, *Angew. Chem. Int. Edin.,* (Engl.), **6,** 413 (1967).

31. H. Laboziewicz, K. R. Lindfors and T. H. Kejonen, *Heterocycles,* **29,** 2327 (1989).

32. A. R. Katritizky and O. Rubio, *J. Org. Chem.,* **48,** 4017 (1983).

33. L. W. Deady and D. C. stillman, *Aust. j. Chem.,* **29,** 1745 (1960).

34. R. E. Lyle and P. S. Anderson, *Adv. Heterocyclic Chem.,* **6,** 45 (1966).

35. U. Eisner and J. Kuthan, *Chem. Rev.,* **72,** 1 (1972).
36. M. Ferles and J. Pliml, *Adv. Heterocyclic Chem.,* **12,** 43 (1970).
37. For a comprehensive review see, G. Piancatellai A. Seettri and M. D. Auria, *Synthesis* 245 (1982).
38. R. Benigni, *et al., Chem. Rev.,* **100,** 3697 (2000).
39. For a recent review see, C. K. McGil and A. Rapple, in *"Advances in Heterocyclic Chemistry",* A. R. Katritzky, vol. **44,** Academic Press, London (1988); T. Vajda and K. Kovacs, *Rec. trav. Chem.,* **80,** 47 (1961).
40. R. A. Abramovitch *et al., Canad. J. Chem.,* **43,** 725 (1965); R. A. Abramovitch and J. G. Saha, *Adv. Heterocyclic Chem.,* **6,** 229 (1966).
41. B. Basu *et al., Tet. Letters,* **43,** 7967 (2002).
42. R. A. Abramovitch C. S. Grain and G. A. Boulton, *J. Chem. Soc. (B),* **901,** (1969).
43. R. A. Abramovitch, and A. R. Vinntha, *J. Chem. Soc. (B),* **131,** (1971).
44. R. E. Manning and F. M. schasefer, *Tet. Letters,* 213 (1975).
45. For a review on hetarynes, see M. G. Reinecke, *Tetrahedron,* **38,** 427 (1982).
46. H. J. deb Hertog and H. C. van der Plase, *Adv. Heterocyclic Chem.,* **4,** 122 (1965); G. W. Gribble and M. G. Saulniener, *Tet. letters,* 643 (1962).
47. R. J. Mortens and H. J. den Hertog, *Tet. letters,* 643 (1962).
48. R. J. Mortens and H. J. den Hertog, *Rec. Trav Chim.,* **83,** 621 (1964).
49. G. W. J. Fleet and I. Fleming, *J. Chem. Soc. (C),* 1758 (1969).
50. T. Saski and K. Kanematsu and M. Uchide, *Bull. Chem. Soc., Japan,* **44,** 858 (1971).
51. T. Kauflmann, J. Hansen, K. Udluff and R. Wirthwein, *Angew. Chem. Int. Edin.* (Engl.), **3,** 650 (1964).
52. J. M. Kramer R. S. Berry, *J. Am. Chem. Soc.,* **99,** 8336 (1972).
53. B. J. Wakefield and J. D. Cook *J. Chem. Soc. (C).,* 1973 (1969).
54. M. Mallet *et al., Comp. rend (C),* **274,** 219 (1972).
55. For a review see, J. Becher, *Synthesis,* 589 (1980).
56. J. Becher, L. Finsen and I. Winckelmann, *Tetrahedron,* **37,** 2375 (1981).
57. H. N. M. Vander lans, H. J. den Hertog and A. van Velomuizen, H. J. den Hertog *et al., Tet. letters,* 1875 (1971).
58. H. J. den Hertog R. J. Martens. H. C. Vanderplas and J. Bon, *Tet. letters.,* 4325 (1966).
59. R. A. Abramovitch and K. Kenashuke, *Canad. J. Chem.,* **45,** 509 (1967).
60. S. Goldschmidt and M. Minisinger, *Chem. Ber.,* **87,** 956 (1954).
61. R. Bernardi, T. Caronna, R. Galli, F. Minisci and M. Perchinunno., *Tetrahedron,* **27,** 3573 (1971); *Tet. letters,* 645 (1973).
62. D. H. Hey *et al., J. Am. Chem. Soc.,* 3963, (1965).
63. R. A. Abramovitch and J. G. Saha, *J. Chem. Soc.,* 2175 (1964).

64. O. L. Chapman, C. L. McIntosh and J. Pacansky., *J. Am. Chem. Soc.,* **95**, 614 (1973).

65. K. E. Wilzbach and D. J. Rausch, *J. Am. Chem. Soc.,* **92**, 2178 (1972). Also see, Y. Ogata and K. Takagi, *J. Org. Chem.,* **43**, 944 (1978).

66. L. Kaplan J. W. Pavlik and K. E. Wilzbach, *J. Am. Chem. Soc.,* **94**, 3283 (1972).

67. P. L. Zumler and R. A. Dybas, *J. Org. Chem.,* **35**, 125, 3825 (1970); C.E. Loader and C. J. Timmons, *J. Chem. Soc., (C),* 1078 (1966).

68. G. Galiazzo, P. Bortolus and G. Gauzzo, *Tet. Letters,* 3717 (1966).

69. W. J. Le Noble and P. J. Timmons. *J. Chem. Soc., (C),* **86**, 1649 (1964).

70. C. R. Costin, C. J. Morron and H. Rapport, *J. Org. Chem.,* **41**, 535 (1976).

71. G. R. Newkome and D. C. Hager, *J. Org. Chem.,* **47**, 599 (1982).

71a. H. C. Brown and B. Kanner, *J. Am. Chem. Soc.,* **88**, 986 (1966); H. C. Brown and W. A. Murphy, *ibid.,* **73**, 3308 (1951).

72. O. F. Beumel *et al.,* synthesis, 43, (1974).

73. L. E. Tenebaum, in *Pyridine and its Derivatives,* E. Klingsberg, (Ed.), part II, Interscience, New York (1961).

74. C. Osuch and R. Levine, *J. Am. Chem. Soc.,* **78**, 1723 (1956).

75. E. T. Borrows and D. O. Holland, *Chem. Rev.,* **42**, 615 (1948).

76. W. E. Doering and R. A. N. Weil, *J. Am. Chem. Soc.,* **69**, 2461 (1947).

77. A. D. Miller, C. Osuch, N. N. Goldberg and R. Levine, *J. Am. Chem. Soc.,* **69**, 2461 (1947).

78. J. I. G. Cadogan, *J. Chem. Soc.,* 4257 (1962).

79. T. Talik *et al., Synthesis.,* 293 (1974); J. I. G. Cadogan D. A. Roy and D. M. Smith, *J. Chem. Soc. (C),* 1249 (1966).

80. R. Jujo, *J.Pharm. Soc. Japan.* **66**, 21 (1946); C. A. **45**, 6200 (1951).

81. A. P. Komin and J. F. Wolfe, *J. Org. Chem.,* **42**, 2481 (1977).

82. B. A. Fox and T. L. *Threlfall, Org. Synth.,* **44**, 34 (1964).

83. E. Abignente, P. De Capraris, E. Fattorusso and L. Mayol, *J. Heterocyclic Chem.,* **26**, 1875 (1989).

84. E. Ochaiai, *J. Org. Chem.,* **18**, 534 (1953); H. J. den Hertog and J. Overhoff, *Rec. Trav. Chim.,* **69**, 468 (1950).

85. H. C. Brown, *J. Am. Chem. Soc.,* **79**, 3565 (1975).

86. A. R. Katritzky, *Q. Rev.,* **10**, 395 (1956).

87. M. van Ammes and H. J. Den Hertog, *Rec. Trav. Chim.,* **78**, 586 (1959).

88. J. M. Markgraf *et al., J. Am. Chem. Soc.,* **85**, 958 (1963).

89. (*a*) S. Oae and E. Kozuka, *Tetrahedron.,* **21**, 1971 (1965); (*b*) Z.C. Chen and P. J. Stang, *Tet. letters,* 3923 (1984).

90. T. Kato and Y. Goto, *Chem. Pharm. Bull. (Japan)* **11**, 461 (1963).

91. R. Adams and . Miyano, *J. Am. Chem. Soc.,* 1263 (1958).

92. A. R. Katritzky and A. M. Monro, *J. Chem. Soc.,* **76**, 3168 (1974).

93. J. M. Essery and schofield, *J. Chem. Soc.,* 1263 (1958).

94. H. C. Brown and B. C. Subba Rao, *J. Am. Chem. Soc.*, **78**, 2582 (1956).
95. E. Howard and W. F. Olszewski, *J. Am. Chem. Saoc.*, **81**, 1483 (1959).
96. H. Kagan and S. Motoki, *J. Org. Chem.*, **43**, 1267 (1978).
97. A. R. Katritzky and J. M. Lagowski, *Chemistry of the Heterocyclic N-oxides*, Academic Press, London (1971), pp. 200-202.
98. D. Konwar, R. C. Boruah and J. S., Sandhu, *Synthesis*, 337 (1990).
99. R. Balicki, *ibid*, 645 (1989).
100. G. P. Spence, E. C. Taylor and O. Buchardt, *Chem. Rev.*, **70**, 231 (1970); also see J. R. Hwu and J. M. Wetzel, *J. Org. Chem.*, **50**, 400 (1985).
101. R. J. Martens and H. J. de hertog, *Rec. Trev. Chim.*, **83**, 621 (1964).
102. J. Becher, *Synthesis*, 589 (1980).
103. L. Pentimalli, *Tetrahedron*, **14**, 151 (1961).
104. A. R. Katritzky and J. M. Lagowski, *Adv. Heterocyclic Chem.*, **1**, 339 (1963).
105. F. Eloy and A. Deryckere, *J. Heterocyclic Chem.*, **7**, 1191 (1970); also see, J. H. MacMillan and S. S. Washburn, *J. Org. Chem.*, **38**, 2982 (1973).
106. L. E. Overman, *J. Am. Chem. Soc.*, **102**, 747 (1980); L. E. Overman and S. Tsuboi, *ibid.*, **99**, 2813 (1977).
107. E. J. Corey and J. Streith, *J. Am. Chem. Soc.*, **86**, 950 (1964).
108. G. R. Newkome J. M. Rohinson and J. B. Sawer *Chem. Comm.*, 410 (1974).
109. R. L. Frank and J. Philips, *J. Am. Chem. Soc.*, **71**, 2804 (1949).
110. A. T. Balaban *et al.*, *Adv. Heterocyclic Chem.* **10**, 241 (1969).
111. For instance, see H. P. Stephenson and H. Sponer, *J. Am. Chem. Soc.*, **79**, 2050 (1957).
112. For a review see, S. A. Harris, E. F. Harris, and R. W. Burg, E. Kirk-Othmer, *Encyclopaedia of Chemical Technology*, 2nd ed., Vol. 16, Interscience (New York), 1968, p. 806.
113. K. M. Smith *et al.*, *J. Org. Chem.*, **45**, 4999 (1980).
114. F. Bracher, *Heterocycle*, **29**, 2093 (1989).
115. S. Matsunaga, K. Shinoda and N. Fusetani, *Tet. Letters*, 5933 (1993).
116. A. R. Katritzky *et al.*, *J. Chem. Soc.*, **1**, 46 (1964).
117. For a review see, J. E. Toomey, Jr, in *Advance in Heterocyclic Chemistry,"* Vol., **37**, A. R. Katritzky, Academic Press, London (1984).
118. A. T. Balaban *et al.*, *Adv. Heterocyclic Chem.*, **10**, 249 (1969).
119. F. Klages and H. trager, *Ber.*, **86**, 1327 (1953).
120. G. Gischer and W. Schroth, *Z. Chem.*, **3**, 260 (1963).
121. G. N. Dorogeenko *et al.*, *Zh. Obsch. Khim.*, **35**, 632 (1965).
122. G. N. Doroogeenko and S. V. Krivum, *ibid*, **34**, 105 (1964).
123. S. Wawzonek in, *Heterocyclic Compounds*, R. C. Elderfield, (*Ed.*), Vol. 2, Wiley, New York), Chapter 9.
124. G. Kobrich and D. Wunder, *Ann.*, 654, 131 (1962).

125. G. Rio and Y. Fellion, *Tet. letters.,* 1213 (1962).

126. A. T. Balaban, *Tetrahedron* (supplement), **7**, 1 (1966); C. Toma and A. T. Balaban, *Tetrahedron* (supplement), **7**, 9 (1966).

127. D. Dimroth *et al., Chem. Ber.,* **90**, 1634, 1668 (1957).

128. J. A. Barltrop, K. Dawes, A. C. Oay, S. J. Nuttall and A. J. H. Summers, *Chem. Comm.,* 410 (1973); *ibid;,* 1240 (1972).

129. J. W. Pavlik and J. Kwong, *J. Am. Chem. Soc.,* **95**, 7914 (1973), J. W. Pavlik and E. L. Clennen, *ibid.,* **95**, 1697 (1973).

130. W. H. Pirkle and M. Dines, J. *Heterocyclic Chem.,* **6**, 313 (1969).

131. R. H. Wiley and N. R. Smith, *Org, Synth.,* **31**, 23 (1951).

132. W. H. Pirkle and M. Dines, *J. Heterocyclic Chem.,* **6**, 313 (1969).

133. N. P. Shusherina *et al., Zhur. Obschch. Khim.,* **31**, 2794 (1961).

134. L. L. Woods, *J. Org. Chem.,* **27**, 696 (1962).

135. J. Fried and R. C. Elderfield., *J. Org. Chem.,* **6**, 566 (1941).

136. G. Vogel, *Chem. and Ind.,* 989, 1829 (1962).

137. W. M. Horspool, in *Photochemistry, Bryce-Smith, (Ed.)* Vol. 5, Chemical Society (London), 1974, p. 385.

138. M. van Meerbeck S. Toppet and P. C. Deschryver, *Tet. letters.,* 2247 (1972).

139. P. Yates and M. J. Jorgenson, *J. Am. Chem. Soc.,* **85**, 2956 (1963).

REVIEW PROBLEMS

6.1 Predict the major product of the following reactions:

(a) CH_3I reaction omitted — see structures

(*a*) [3-aminoacrolein/enaminone structure] + [cyclohexanone structure] $\xrightarrow[\Delta]{NH_4OAC}$

(*b*) [3-methylpyridine structure] $\xrightarrow[\text{(ii) } CH_3I]{\text{(i) } NaNH_2, \text{ liq.}NH_3}$

(*c*) [4-phenylpyridine structure] $\xrightarrow[\Delta]{KMnO_4, H^+}$

$\xrightarrow[\Delta]{KMnO_4, OH^-}$

(d) 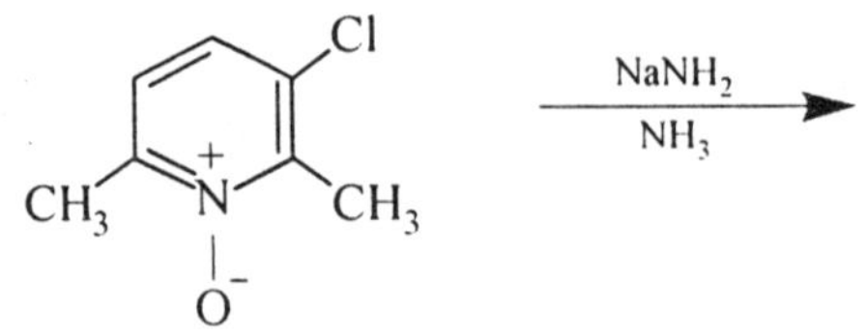
Cl
CH3
N
CH3
O
NaNH2
NH3

(e) 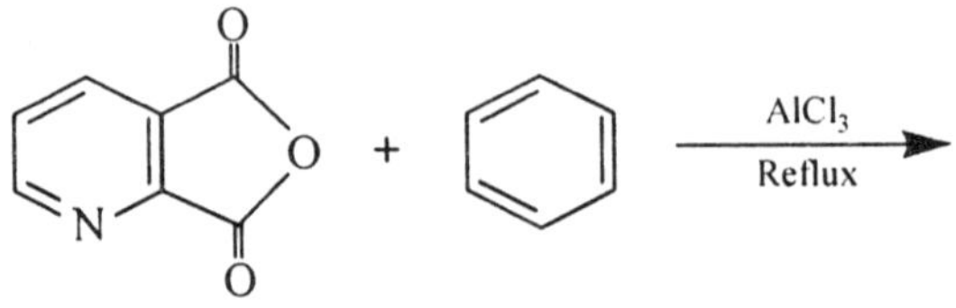
O
O
N
O
+
AlCl3
Reflux

(f) 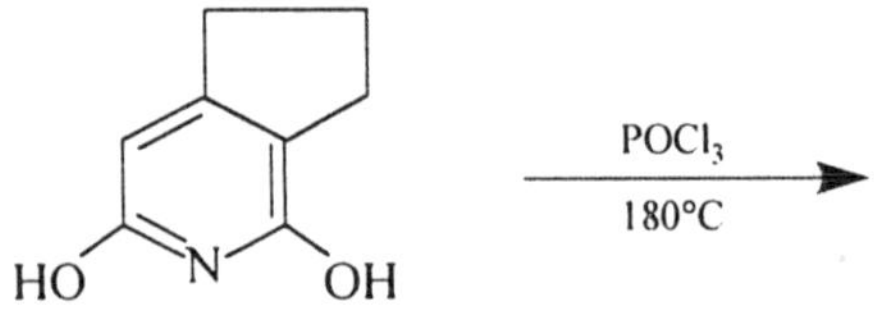
HO
N
OH
POCl3
180°C

(g) 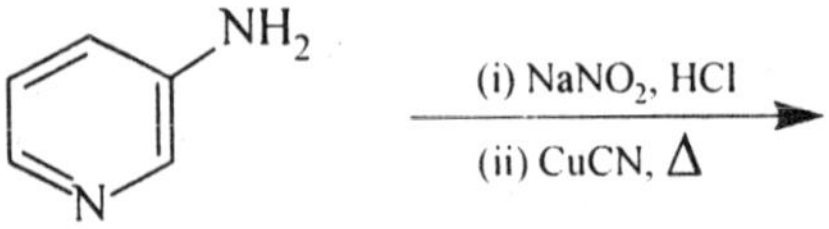
NH2
N
(i) NaNO2, HCl
(ii) CuCN, Δ

(h) 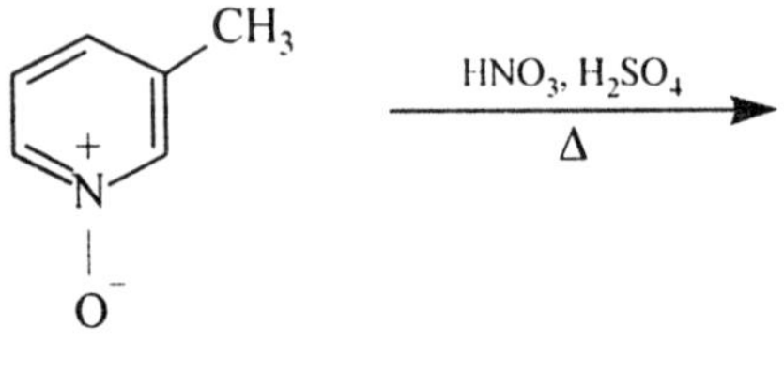
CH3
N
O
HNO3, H2SO4
Δ

(i) 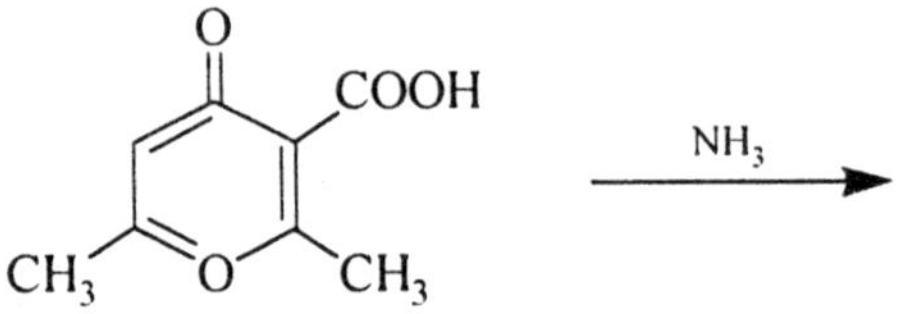
O
COOH
CH3
O
CH3
NH3

(j) 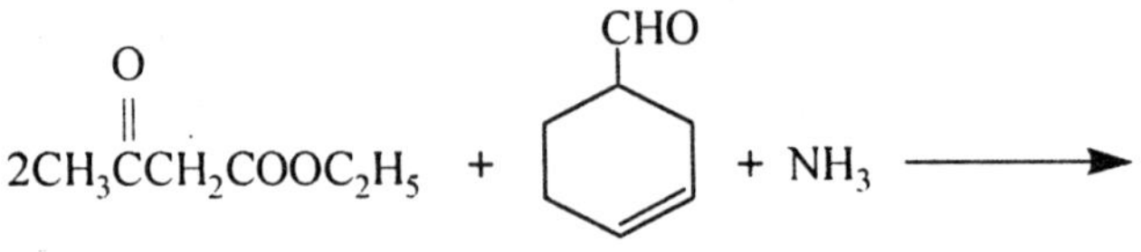
O
2CH3CCH2COOC2H5
+
CHO
+ NH3

(k) [pyridine with 4-CH$_2$Cl] + CH$_2$=CHCN $\xrightarrow[\text{HCON(CH}_3)_2]{\text{NaH}}$

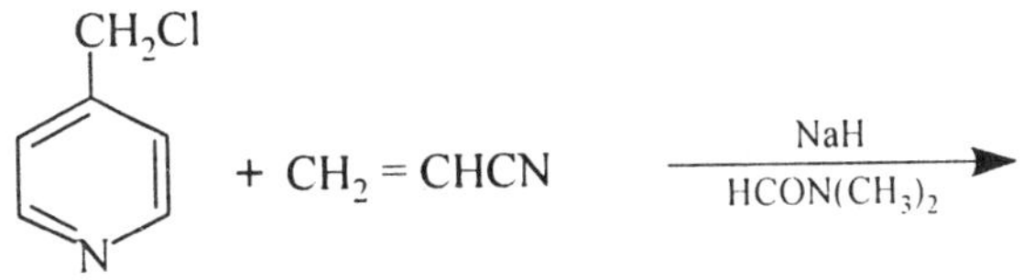

(l) [2-(CH$_2$Li)pyridine] + ClCOOC$_2$H$_5$ $\xrightarrow{\text{Ether}}$

(m) [2-(MgBr)pyridine] $\xrightarrow[\text{(ii) H}_2\text{O, H}^+]{\text{(i) HC(OC}_2\text{H}_5)_3}$

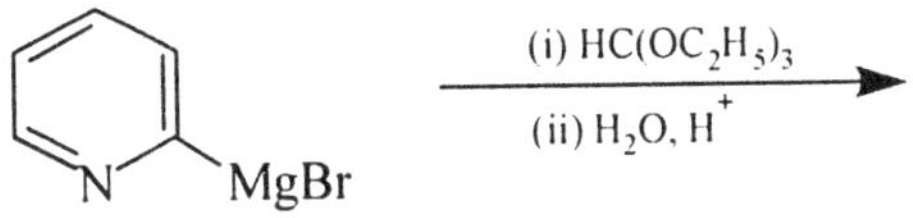

(n) [3-Br-2-Cl-pyridine] $\xrightarrow[\text{(ii) furan}]{\text{(i) } n\text{-BuLi, (C}_2\text{H}_5)_2\text{O}}$

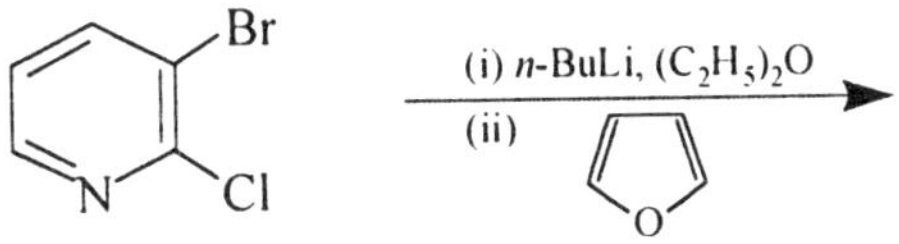

(o) [2-(CH$_2$Li)pyridine] $\xrightarrow[\text{(ii) H}_2\text{O, H}^+]{\text{(i) 2-chlorobenzonitrile}}$

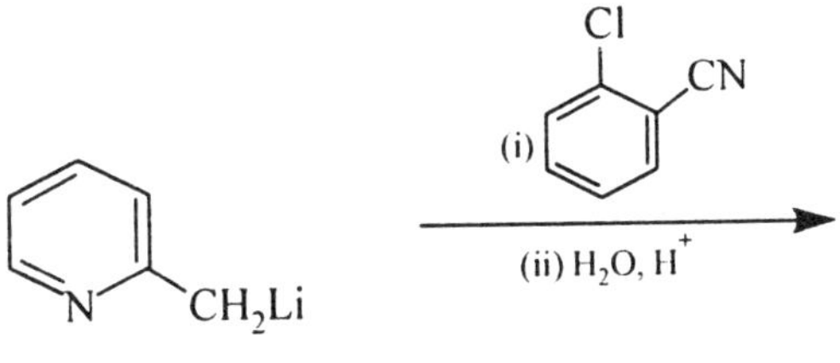

(p) [2-phenyl-4H-pyran-4-one] $\xrightarrow{\text{C}_6\text{H}_5\text{NH}_2}$

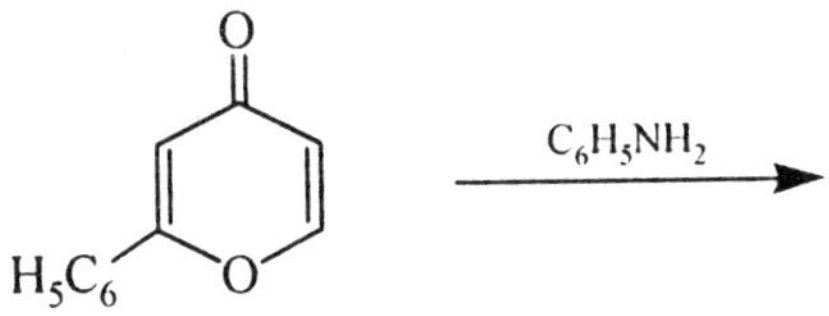

(q) [4H-pyran-4-one] $\xrightarrow[\text{(ii) 70\% HClO}_4, -40°\text{C}]{\text{(i) CH}_3\text{MgBr}}$

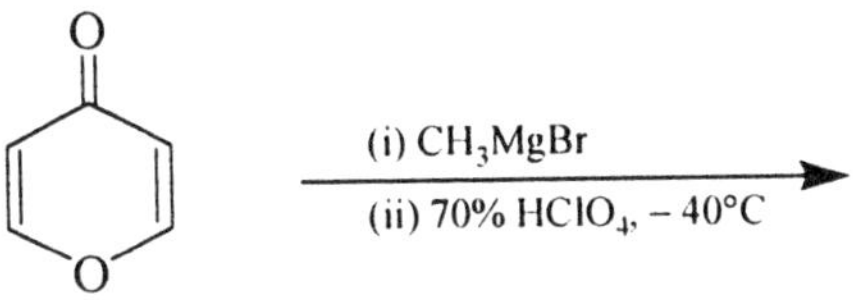

(r) [pyridine] $\xrightarrow{\text{Ph (CO}_2)_2}$

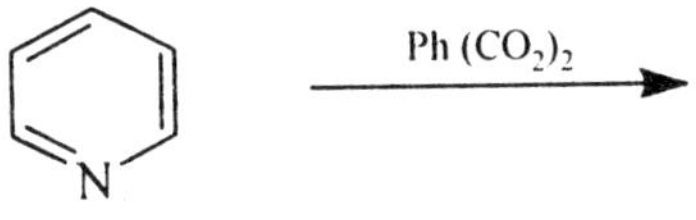

(s) [structure: pyridin-3-ylmethyl vinyl ether with OC_2H_5] $\xrightarrow[\text{195°C}]{\text{Propionic acid}}$

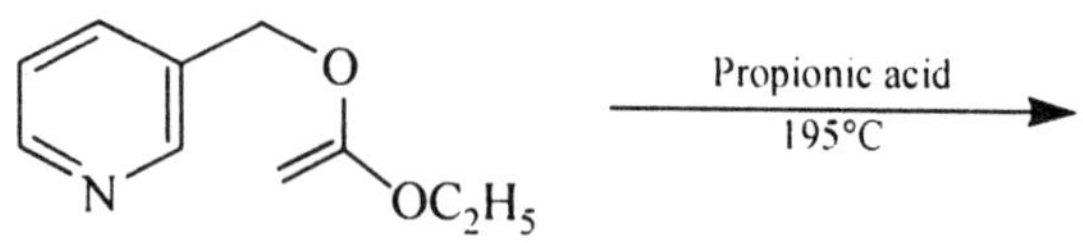

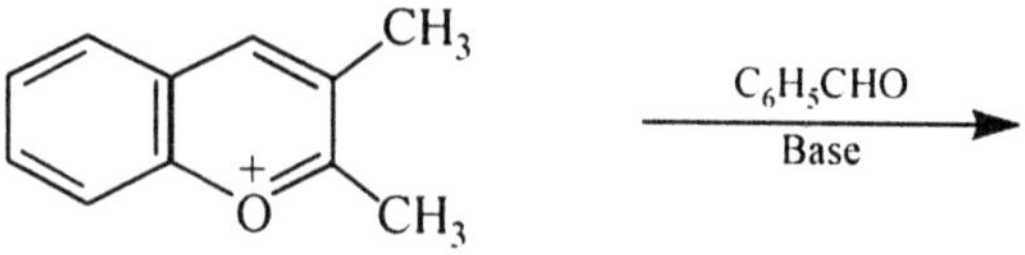

(t) [structure: 2,3-dimethylbenzopyrylium cation] $\xrightarrow[\text{Base}]{C_6H_5CHO}$

(u) [structure: 3-acetylpyridine] $\xrightarrow[\Delta]{NH_2NH_2,\ OH^-}$

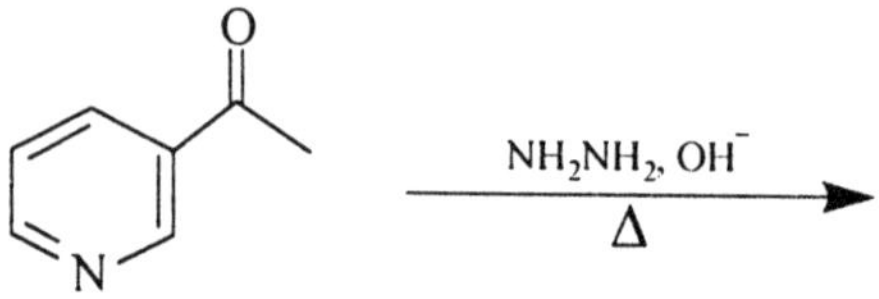

6.2 Suggest an appropriate mechanism for each of the following reactions:

(a) [structure: pyridin-4-yl phenyl ketoxime, H_5C_6, $C=N$, OH] $\xrightarrow{PCl_5}$ [structure: 3-(benzoylamino)pyridine, $NHCC_6H_5$, O]

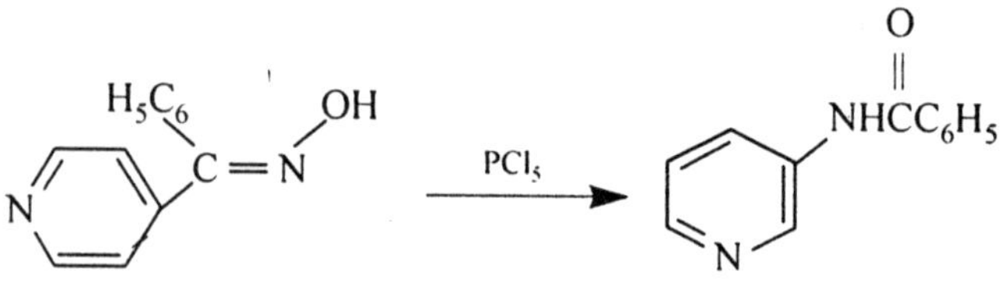

(b) [structure: 2H-pyran-2-one] $\xrightarrow{NH_3}$ [structure: 2-hydroxypyridine, N, OH]

(c) [structure: 2-methylpyridine N-oxide] $\xrightarrow{h\nu}$ [structure: 2,5-dimethylpyrrole, CH_3—...—CH_3, N—H]

(d) [structure: 2-bromo-3-aminopyridine, NH_2, Br] $\xrightarrow{KNH_2,\ NH_3}$ [structure: 3-cyanopyrrole, CN, N—H]

(e) [structure: 4-methoxypyrylium cation, OCH_3] $\xrightarrow{OH^-}$ [structure: 4H-pyran-4-one, O]

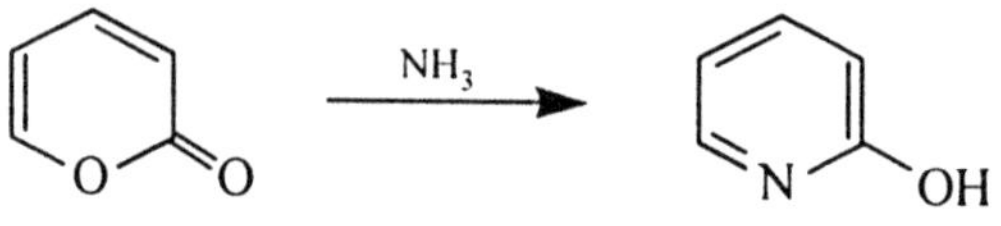

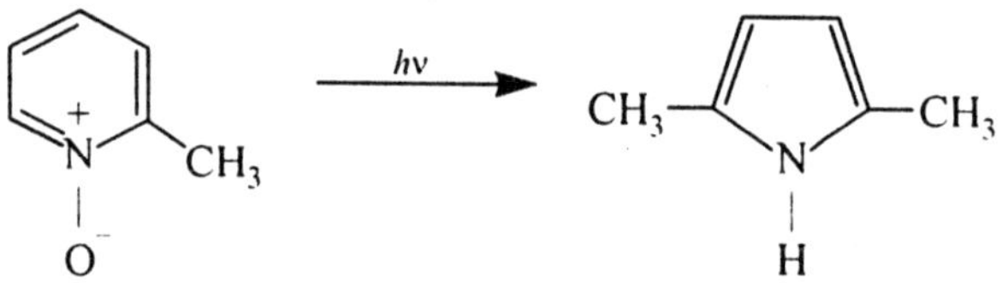

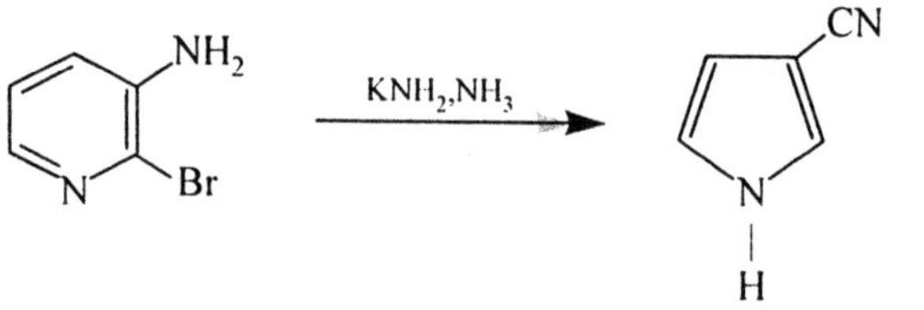

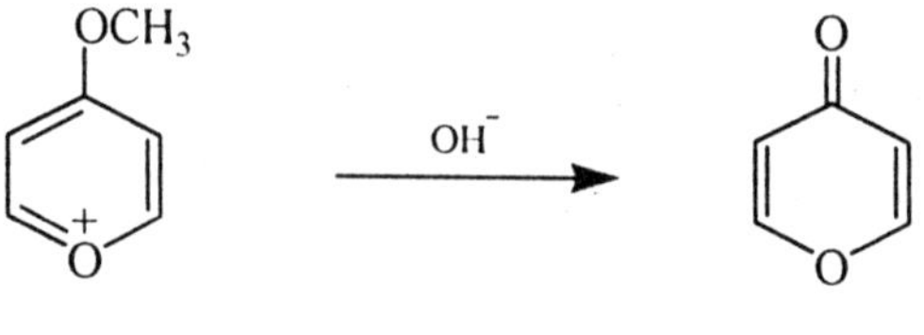

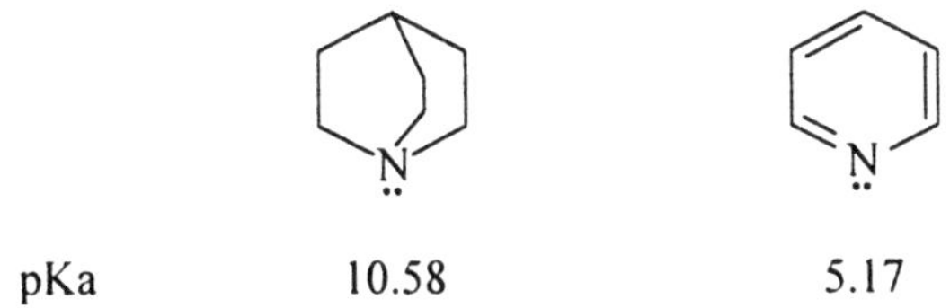

6.3 Offer explanation for the following observations:

(*a*) Bromine atom in 3-bromopyridine but not in bromobenzene can be replaced by sod. methoxide in methanol.

(*b*) Quinuclidine is more basic than pyridine.

pKa   10.58   5.17

(*c*) A nucleophile attacks exclusively at C-9 position in a xanthylium salt to give xanthydrol.

(*d*) Pyridine but not benzene dissolves in water.

(*e*) 4-Aminopyridine has a larger dipole moment (4.4D) than 4-cyanopyridine (1.6 D).

(*f*) 2, 6-Lutidine boils (140°C) at a lower temperature than 3, 5-lutidine (b.p. 175°C).

(*g*) Electrophilic substitution in 4-phenylpyridine occurs in the benzene ring.

(*h*) 4-Chloro-3-nitropyridine hydrolyzes readily even in warm water.

(*i*) Alkyl substtituents at 2-and 6-positions in pyridine reduce salt formation with acids.

6.4 In contrast to pyridine itself, pyridine N-oxide readily undergoes electrophilic substitution principally at the 4-position. How do you account for this reactivity and orientation?

6.5 Nitration for pyridine N-oxide occurs at 4-position while sulfonation at 3-position though both are carried out in the presence of sulfuric acid.

6.6 Discuss the action of different reducing agents on pyridine.

6.7 Explain the ease of electrophilic substitution on pyridine N-oxide as compared to pyridine.

6.8 Discuss the preparation and properties of picoline N-oxide.

6.9 Describe Hantzsch synthesis of pyridine and variations of this procedure.

6.10 Discuss the formation and reactions of pyridines.

6.11 Discuss the chemistry of α-pyrones.

6.12 With appropriate chemical reactions bring out the similarities between pyridine 2-carbaldehyde and benzaldehyde.

6.13 Illustrate with appropriate examples how pyridine compares with nitrobenzene in electrophilic substitution.

6.14 1-Benzopyrylium system is inert to the attack of electrophilic reagent. Explain. Would you expect the following ion to undergo nitration? Comment.

Chapter 7

Bicyclic Ring Systems Derived from Pyrrole, Furan and Thiophene

The three parent five-membered π-excessive heterocyclic rings namely pyrrole, furan and thiophene can condense with a benzene ring in two distinctive ways represented by structures (1) and (2). The resultant bicyclic heterocyclic

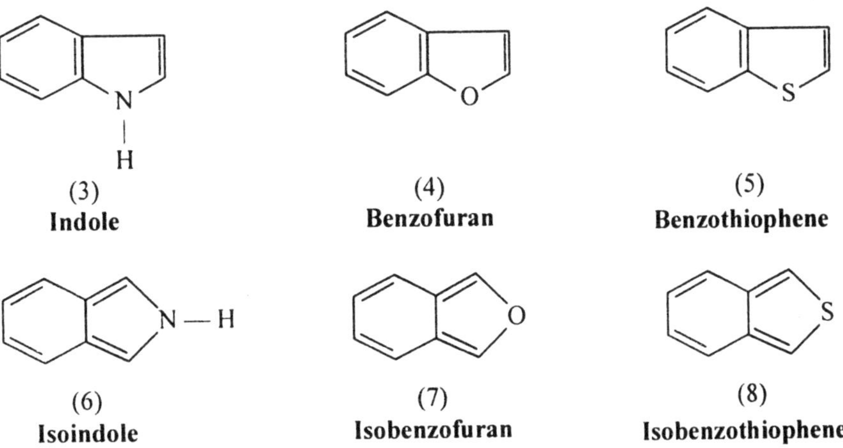

(1)   (2)

X = NH, O or S

rings possesses properties different from the individual ring systems. Following six possible structures (3-8) are obtained in doing this. The stability of these heterocycles is dependent on the 2-electrons which the hetero atom contributes

(3)
Indole

(4)
Benzofuran

(5)
Benzothiophene

(6)
Isoindole

(7)
Isobenzofuran

(8)
Isobenzothiophene

to the π-system. The only resonance forms which can be written for the compounds in the series (3-5) which do not involve destruction of the benzenoid 6π-system are the following:

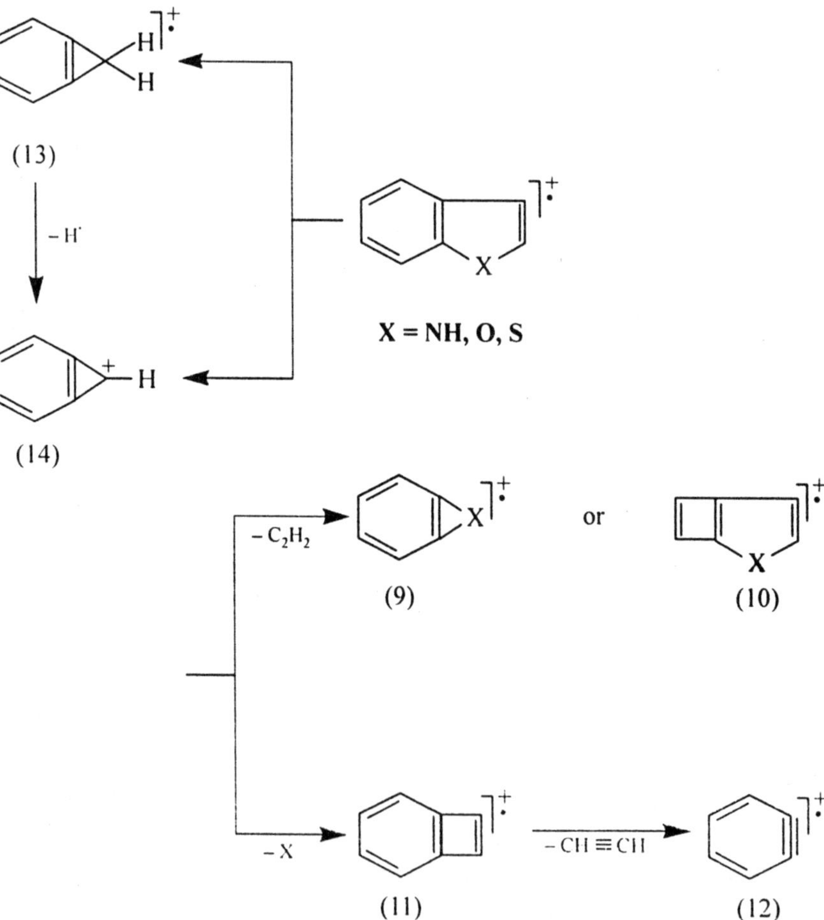

Ring systems (3) to (5) are aromatic and undergo typical electrophilic substitution reactions. Their derivatives occur in a number of natural products and are constituent of many alkaloids. Compounds (6) to (8) are unstable and very reactive. Their existence has been confirmed by trapping them with various reagents. Though they are stable at low temperature but all of them have been synthesized.

The fragmentation patterns exhibited by the benzo [b] fused derivatives of the five-membered heterocyclic compounds parallel the monocyclic compounds. The molecular ion is also the base peak. The fragment (9 to 12) has only been noted as a minor ion in the *m.s.* spectrum of benzo [b] thiophene.

The loss of the hetero atom X = S, is also only slightly observed in this case. In contrast, the ions (13 and 14) are rather prominent in the spectra of all the benzo [b] heterocycles.

7.1 INDOLES

Indole was first prepared by Baeyer in 1866 by zinc dust distillation of oxindole. It finds an important place in chemistry because of its relationship to the naturally occurring dye, indigo. The chemical degradation of this dye yields indoxyl, oxindole and finally indole. Indole is found in coal tar and in essential oils (jesamine oil, orange oil) of many plants. It also occurs in amino acids (tryptophan) as a plant growth hormone (indole 3-acetic acid), in alkaloids (brucine, psilecene) and dye stuff (indigo). The indole derivatives are important in their own right. The indole (3) ring comprises of a benzene ring fused to the 2-and 3-positions of a pyrrole nucleus.

The IUPAC name of indole is 1*H*-benzo [b] pyrrole, it being the b-face benzo-fused isomer. The atoms are numbered as shown commencing from the nitrogen atom and going counter-clockwise around the two condensed rings.

(3)

The bridgehead carbons are assigned positions 3a and 7a. Positions-2 and-3 are also sometimes designated as α and β respectively. Many substituted indoles have acquired trivial names such as skatole, gramine, tryptophan, etc.

The tautomeric forms of indole are known as indolenines; 3*H*-indolenine (15) and 2*H*-indolenine (16). Tautomerism in indole is similar to that in pyrrole where the two tautomers are called pyrrolenines. Structure (15) is anticipated to be stable but structure (16) corresponds to an *o*-quinonoid which would obviously be much less stable. However, the loss of aromatic resonance energy of indole to indolenine is much less than that for pyrrole to pyrrolenine.

(15)   (16)

7.1.1 Physical and Spectroscopic Properties

Indole and the simple alkyl indoles are colorless crystalline solids. Indole melts at 52°C and boils at 254°C. It is soluble in most organic solvents. While pure indole has a very pleasant smell and is used as a perfume base, impure indole and skatole (2-methylindole) have unpleasant odour. Many indoles are quite stable in air with the exception of those which carry a simple alkyl group at the 2-position, for instance, 2-methylindole autoxidizes readily even when stored in a dark brown bottle. Indole responds to the Ehrlich test and gives blue color. The molecular dimensions of indole have not yet been determined.

Indole is a planar molecule with a conjugated system of 10 π-electrons, two from the nitrogen and eight from the carbon atoms and is thus a π-excessive molecule. It is considered a resonance hybrid of the following cannonical forms:

(1a)   (1b)

(1c)   (1d)

The resonance energy is of the order of 47-49 Kcal/mole. From the resonance structures it may be noted that structures (1a, 1b) and (1d) are the only ones in which the benzenoid 6π-system is preserved. The electrophilic attack as a result takes place at 3-position. Presence of a high electron density at the 3-position has also been supported by the calculation of π-electron densities and by the molecular orbital methods. Indole like pyrrole is a very weak base, pKa-3.63, alkyl indoles are stronger bases, for instance, 2-methylindole (pKa-0.28) and 1-methylindole (pKa-2.32). 3-Methylindole (pKa-4.6), in contrast is a weaker base than indole. Furthermore, the cation of 3-methylindole cannot hyperconjugate with the ring.

7.1.2 Synthetic Methods

Through the years different methods have been devised[1] for the synthesis of indole and its derivatives. These methods differ in their range and applicability. A number of general methods are also known in which the pyrrole ring is formed through ring closure reactions.

1.   *The Fischer-Indole Synthesis:* This is probably the most widely investigated synthesis of indole and has been reviewed.[2] It involves an acid-

catalyzed rearrangement of a phenylhydrazone of an aldehyde or ketone, with the elimination of a molecule of ammonia.[3] The conventional catalysts employed are zinc chloride, polyphosphoric acid or a Lewis acid (BF_3)[4] equations (7.1 and 7.2) and it is likely that the mechanistic details of the

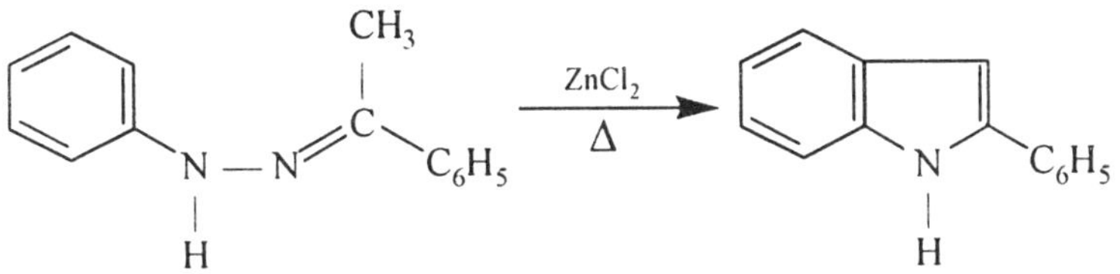

reaction depend on reaction conditions and the nature of the hydrazone substrate. It is, therefore, possible that no definite mechanism applies to all the conditions. The following mechanism of the reaction has however, been widely investigated and involves the following steps:

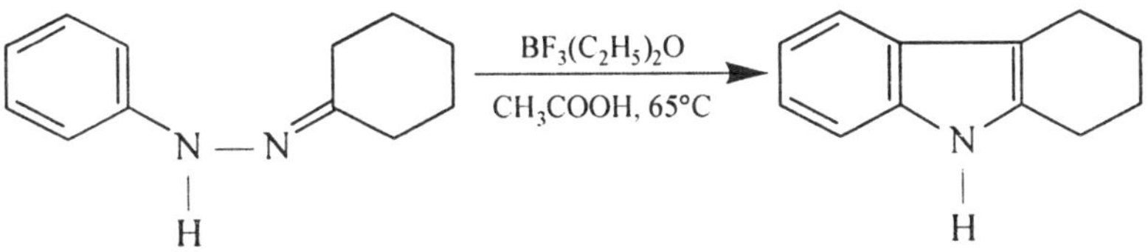

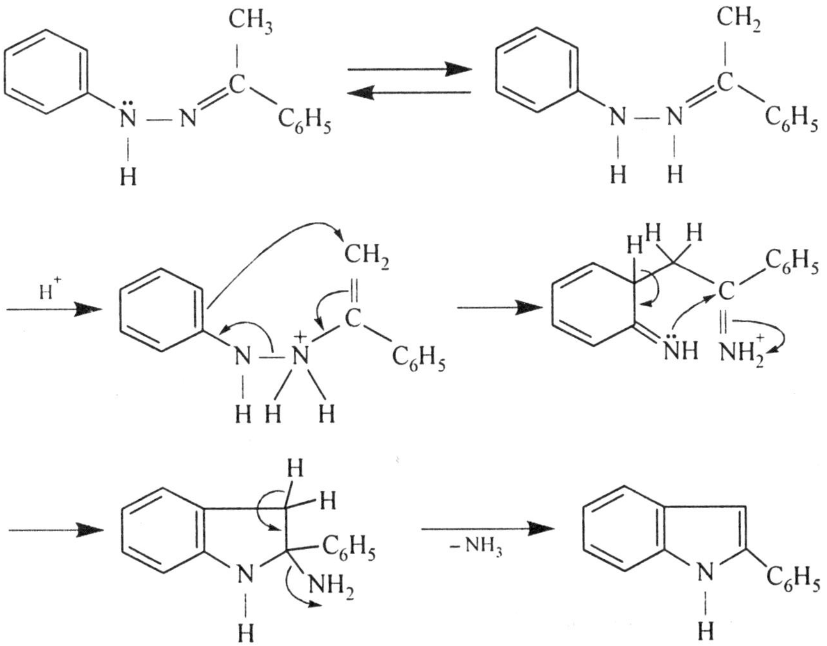

The scheme outlined above explains the various steps most satisfactorily and involves an *o*-Claisen type rearrangement of a tautomer of the hydrazone. The various postulated intermediates in the sequence of reactions have been isolated by different workers.[7, 8]

Recently Hughes and Zhao[9] have suggested on the basis of kinetic and solvent isotope effects that a ring protonation occurs in conc. acids while in dilute acids no much protonation takes place. This was evidenced by the absence of solvent isotopic effect and lack of deuteration of the ring.

An interesting feature of the Fischer-indole synthesis which is of considerable practical importance is the ratio of the two possible indoles which can be formed from an unsymmetrical phenyl hydrazone. In many cases a mixture of the two products is obtained. The treatment of methyl ethyl ketone phenylhydrazone (17) yields, for instance, 2, 3-dimethylindole (18) and 2-ethylindole (19).

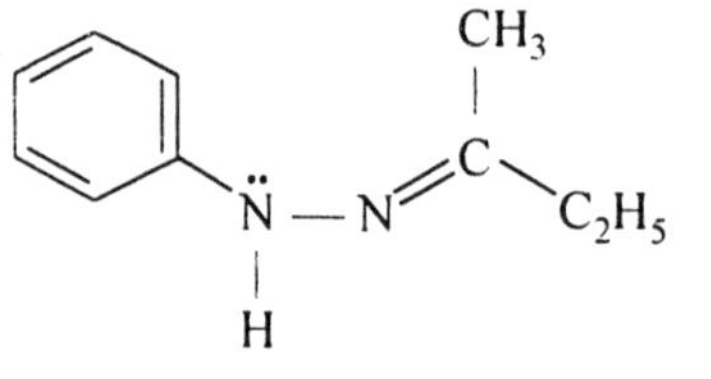

(17)

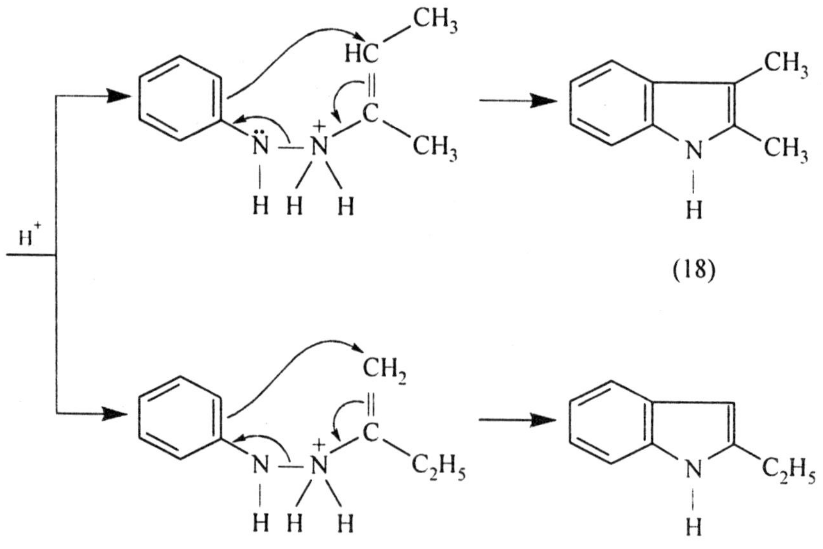

(18)

(19)

The cyclization depends on the steric interactions and are affected by the size of the acid employed. With hydrochloric acid, (18) is the major product while with phosphoric acid, (19) is obtained as the major product.[10]

A second interesting feature is the migration of the group that takes place during the reaction as is exemplified by the treatment of acetophenone 2, 6-dimethylphenylhydrazone[11] with an acid.

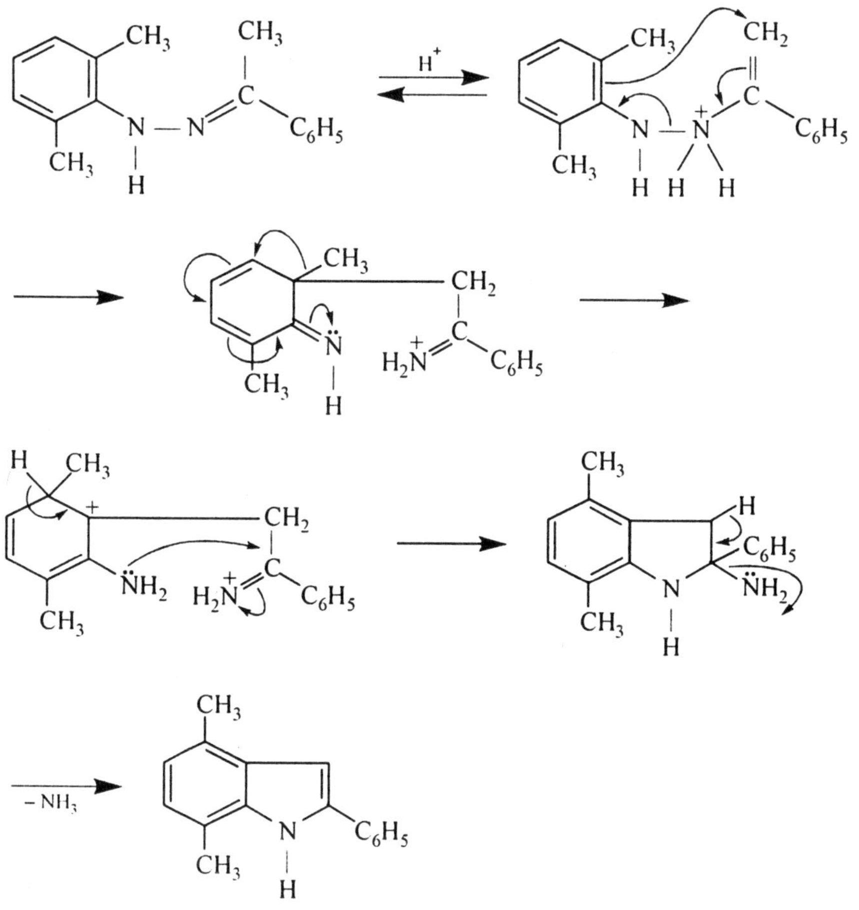

A 1, 2-shift of the methyl group has taken place in the above case. In some cases a 1, 4-shift could also have been observed but the steric effects seem to play an important role[12] to decide between the two modes of shifts.

An important extension of the Fischer-indole synthesis is the Japp-Klingemann reaction[13] which provides an access to a number of phenyl hydrazones which are not obtained by direct methods. This reaction involves a coupling of an arenediazonium salt with an activated carbanion derived from an α, β-carbonyl precursor. This is illustrated for the preparation of tryptamin (20) from piperidine 2-or-3-carboxylic acid.[14]

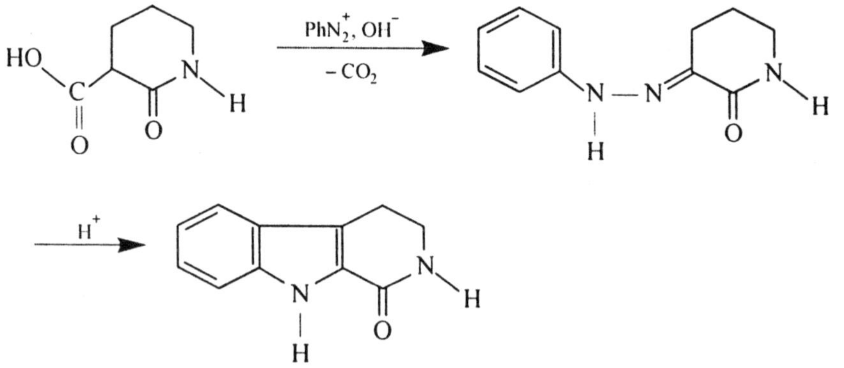

(20)

2.   *The Madelung Synthesis:* This involves the cyclic dehydration of an acyl
o-toluidine in the presence of a strong base and at a high temperature. Indole
itself can be prepared by this method. This reaction has been applied

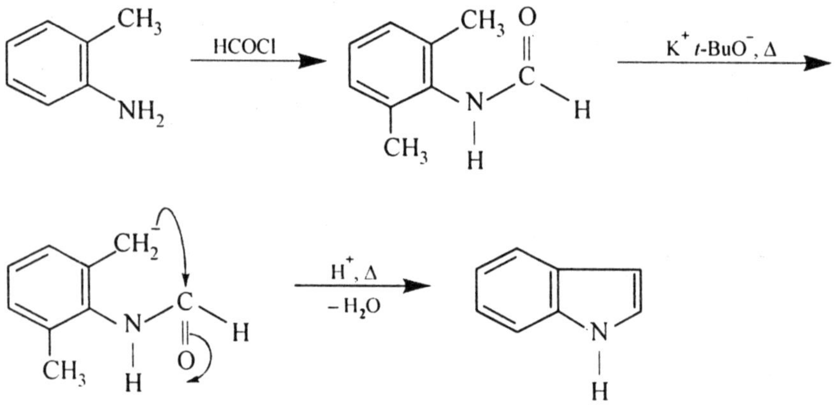

successfully to a number of acyl derivatives of *o*-toluidine which can be
prepared from *o*-toluidine and an appropriate carboxylic acid. This reaction is
employed particularly for the synthesis of simple 2-alkyl and 2-aryl-indoles
as well as for 2, 3-unsubstituted indoles carrying substituents instead on the
benzene ring. The mechanism probably involves the following sequence of
reactions.[15,17]

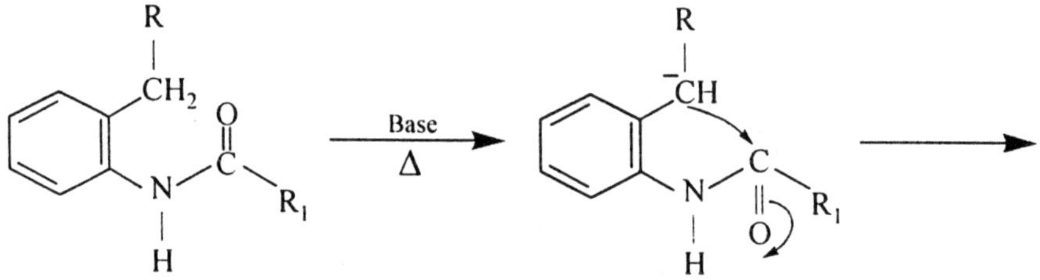

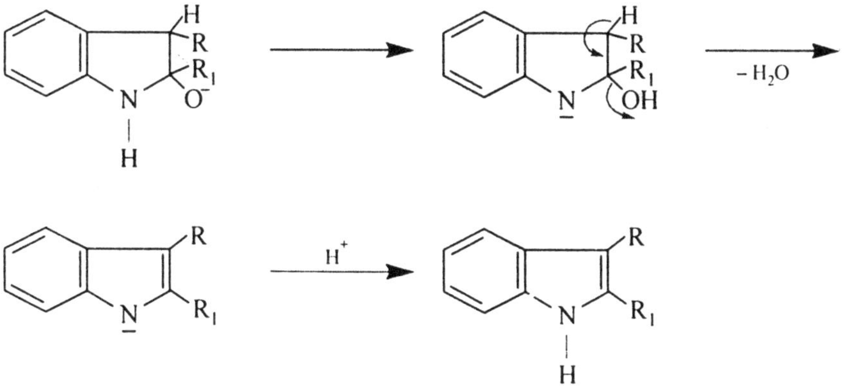

In recent years the Madelung synthesis of indoles has undergone several modifications.[18] Bartoli and coworkers[19] have proposed a synthesis of indoles based on lithium diisopropylamide promoted cyclization of *ortho* (trimethylsilylmethyl) N-methylanilides as illustrated by the following example, equation (7.3).

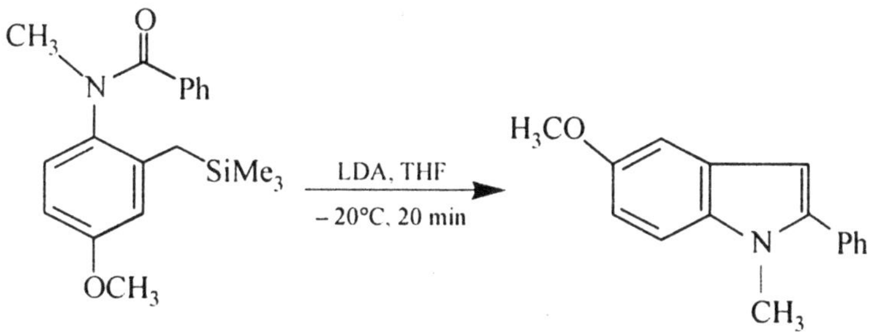

$$\tag{7.3}$$

3.   *The Nenitzescu Synthesis :* This reaction provides a direct route to 5-hydroxyindoles. It consists of a condensation of *p*-benzoquinone with 3-aminocrotonate usually at the refluxing temperatures of the solvent such as acetone[20], equation (7.4).

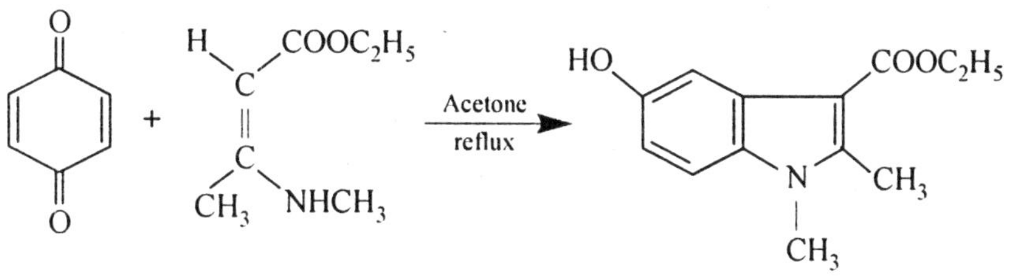

$$\tag{7.4}$$

Monosubstituted *p*-benzoquinones yield a mixture of both 6-and 7-substituted 5-hydroxyindoles (21) and (22) equation (7.5). 2-Carbomethoxy *p*-benzoquinone is found to give rather a 4-substituted 5-hydroxy product,[21] equation (7.6). The final esters can be hydrolyzed and decarboxylated by refluxing in 20% hydrochloric acid.

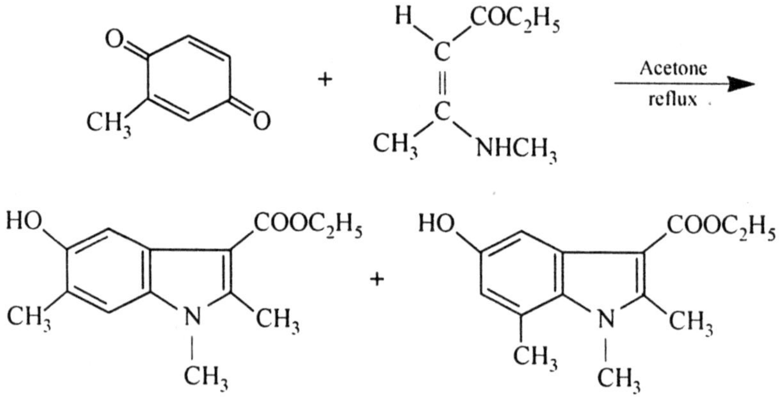

$$(7.5)$$

(21)   (22)

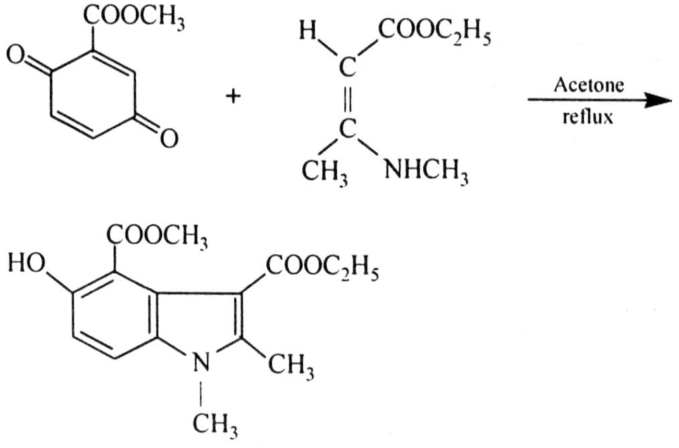

$$(7.6)$$

The mechanism of the reaction involves the following accepted pathway.

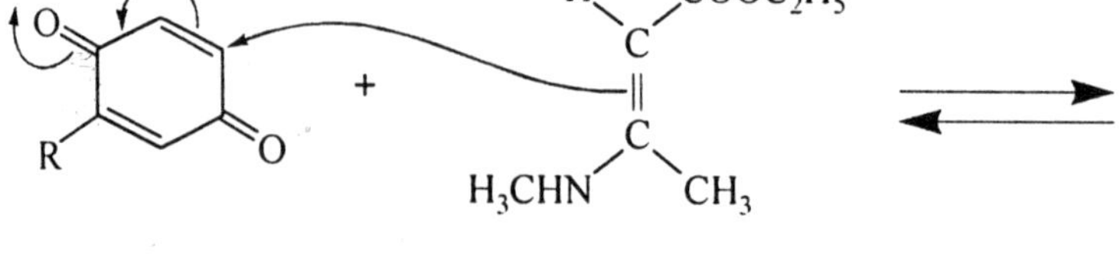

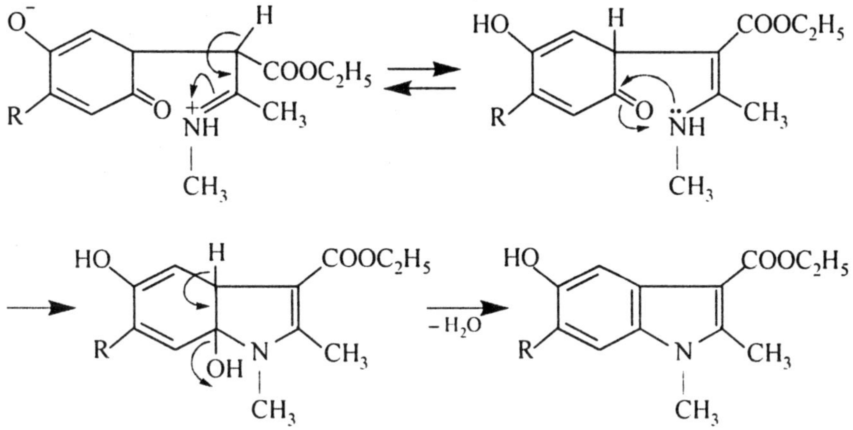

The formation of the 6-and 7-isomers presumably is a manifestation of the steric influence during the final cyclization step in which the heterocyclic ring is formed.

4.   *The Bischler Synthesis:* This is a reaction of an α-arylaminoketone or aldehyde, prepared from α-halo precursor and an arylamine which on heating with zinc chloride or an acid forms an indole derivative,[22] equation (7.7).

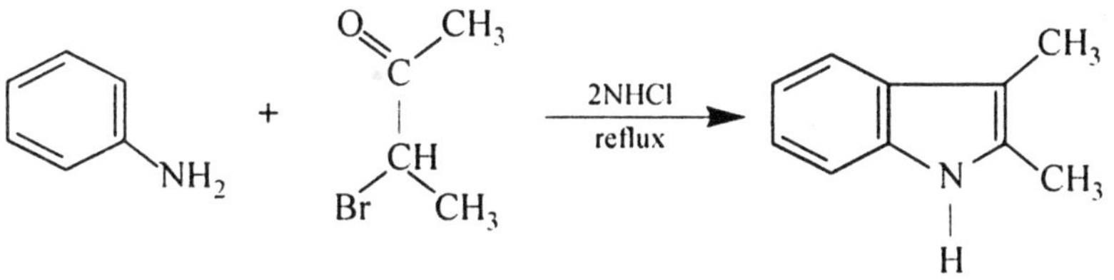

$$(7.7)$$

The mechanism of this reaction probably involves the following steps:

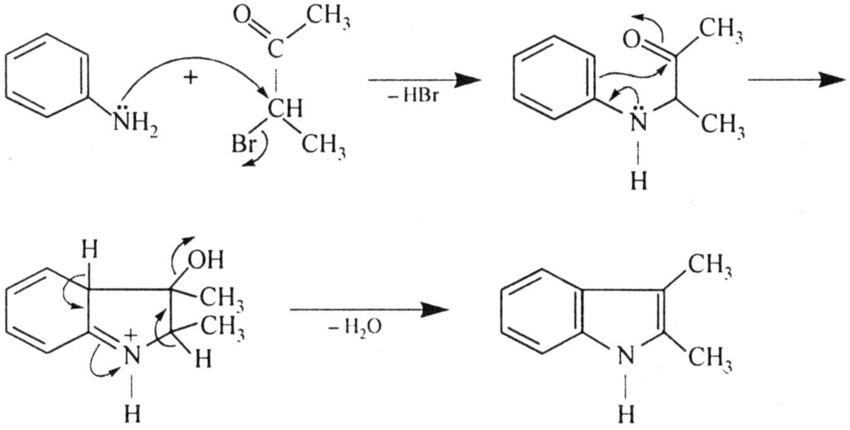

If an unsymmetrical ketone is employed like the one shown in (23), then a complication arises due to formation of two isomeric indoles[23], (24) and (25).

(23)

(24)   (25)

32%   18%

5.   *The Reissert Synthesis:* This procedure also offers a convenient method for the preparation of indole and its derivatives. It involves[24] a base-catalyzed condensation of *o*-nitrotoluene with oxalic acid ester (diethyl oxalate) in the presence of sodium ethoxide. The resulting *o*-nitro-phenylpyruvate or the corresponding acid after hydrolysis functions as the starting material for this reaction. This on reductive (Zn/CH$_3$COOH) cyclization yields indole. This synthesis is also useful for indoles substituted in the aromatic ring.

6.   *Miscellaneous Methods:* Besides the methods discussed above, a number of additional synthesis have been reported.

(*a*) *Microwave Irradiation:* In a further modification of the Fischer indole synthesis, Brönsted acids have been replaced by montmorillonite clay which is acidic. Thus 2-phenylhydrazine and acetone on treatment with the clay and under microwave irradiation, equation (7.8) gives an excellent yield (86%) of 2-methylindole.[25]

$$\text{(7.8)}$$

Earlier this reaction was achieved[26] in the absence of microwave irradiation.

(*b*) *From Benzocyclobutenone:* A new and simple method[28] for the preparation of 2-indole derivatives involves the benzocyclobutenone as the starting material. This is treated with a Grignard reagent to afford the corresponding alcohol. The alcohol is followed by its treatment with hydrazoic acid and bromotrifluoro etherate to allow the formation of the azide. The azide on acid catalyzed rearrangement (Schmidt reaction) leads to 2-substituted indole.

(*c*) *From 2-Chloro-N-Methyl-N-Allylaniline:* Organo-metallic complexes have been used extensively in the synthesis of heterocyclic ring systems.[29] Aryl nickel complex, tetra *kis* (triphenylphosphine) nickel, for instance, reacts with the title compound on refluxing in ether to yield a indole.[31] The mechanism proceeds in the following steps:

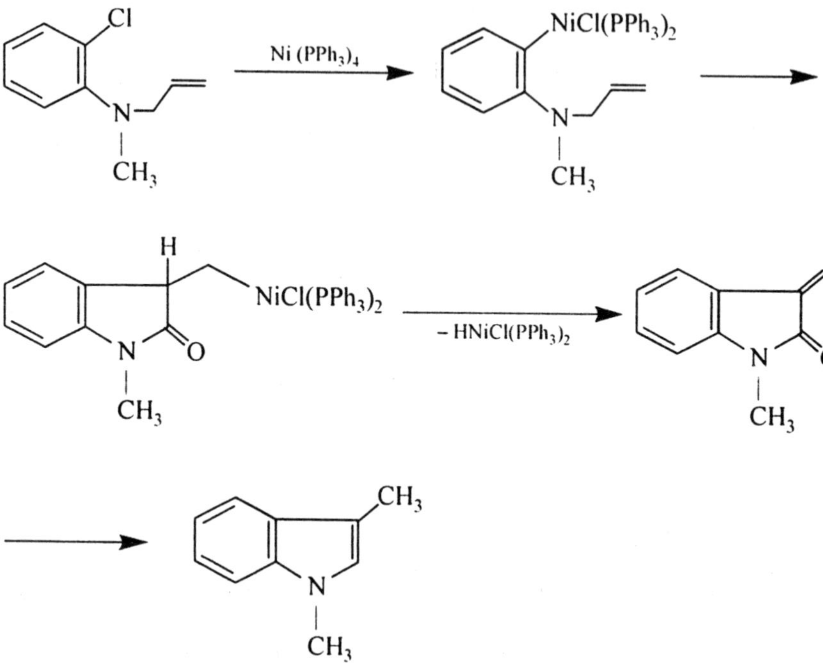

(45%)

(*d*) *From Azirine Derivatives[31]:* 2, 3-Diphenylazirine reacts with benzyne to give a 1:1 adduct (26) which gives rise to two indoles. A 1, 2-hydrogen shift to the nitrogen atom yields 2, 3-diphenylindole [26a]. However, a 1, 2-hydrogen shift to the carbon atom leads to a 3H-indole system which is converted by benzyne to 1, 2, 3-triphenylindole (26b).

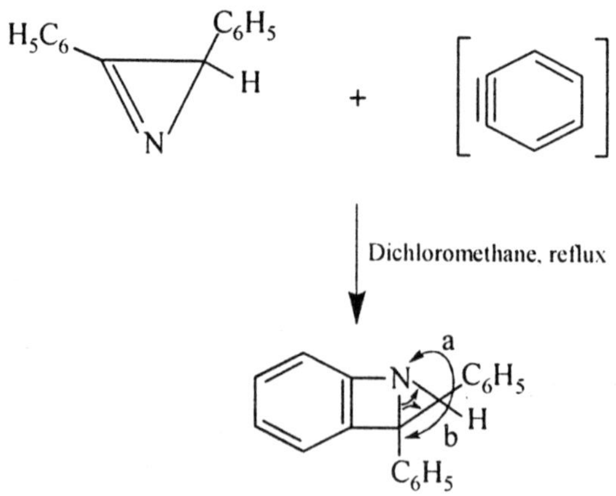

Dichloromethane, reflux

(26)

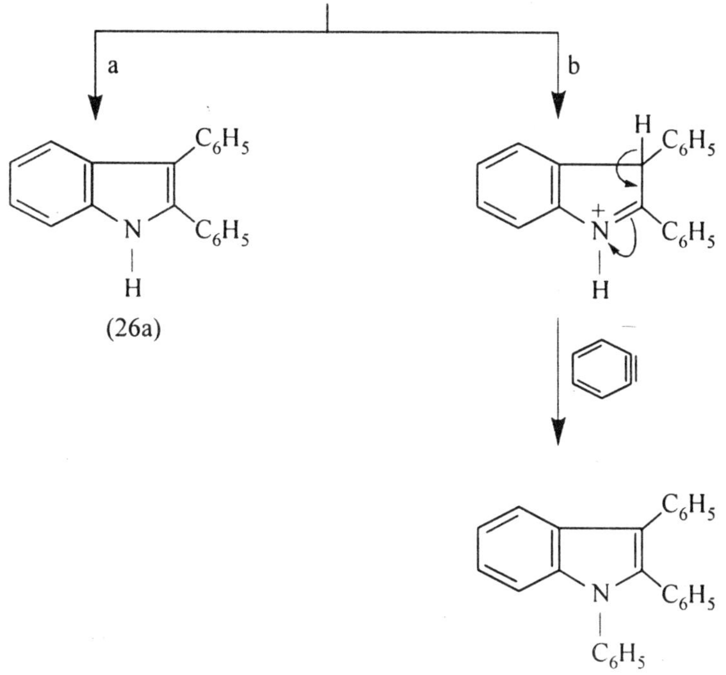

(26a)

(26 b)

(*e*) *The Gassman Synthesis:* This synthesis depends on sigmatropic rearrangement to generate[32] the ring carbon atom. In this case N-chloroaniline forms a sulfonium salt with β-keto sulfide, this salt is deprotonated to an ylide (27) which subsequently rearranges and cyclizes. A final desulfurization gives indole.

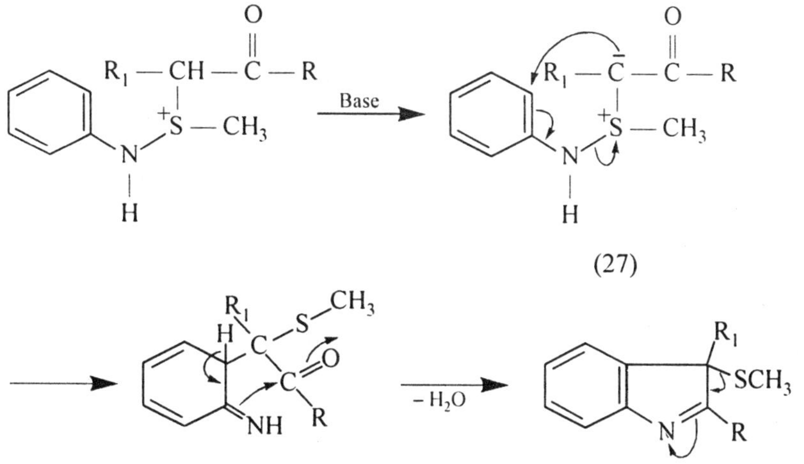

(27)

$$\xrightarrow{-\text{SCH}_3}$$

Indole derivatives have been prepared by some additional methods[33, 34] as well.

(f) Palladium Catalyzed Synthesis: Palladium acetate $[\text{Pd(OAc)}_2]$ can also be used as catalyst[35] in the synthesis of 2-phenylindole, equation (7.9).

$$\xrightarrow[\substack{\text{Pd(OAc)}_2,\ \text{toluene}\\100°C}]{\text{CH}_2=\text{CHCH}_2\text{OCOOCH}_3} \tag{7.9}$$

92%

Annelated indoles are synthesized[36] using a mixture of indole, tetra *kis* (triphenylphosphine) palladium [0], potassium acetate and an aryl bromide in N, N-dimethylacetamide on heating at 160°C for 6 hr.

Novel synthesis of indoles *via* palladium catalyzed reductive N-heterocyclization[37a] of *o*-nitrostyrene derivatives and using polymer bound hydrazines[37b] has also been accomplished.

7.1.3 Chemical Reactions

Indole undergoes a number of interesting chemical reactions similar to other heterocyclics.

1. Reaction with Acids: Indole like pyrrole is a very weak base and even less so than indolizine. In dilute acid, (5×10^{-3}M sulfuric acid) a β-protonated $3H$-indolium cation (28) is formed which has been supported by *u.v.* and *n.m.r.* data as well. The reason for its thermodynamic stability and hence its dominance in equilibrium is retention of benzene aromaticity and delocalization

(28)   (29)   (30)

of the positive charge over the nitrogen and the α-carbon. A proton can however, be added at both 1-and 2-positions in strong acid solution,[41] to form cations (29) and (30) respectively. It may be noted that 2-protonated species, *i.e.* 1*H*-indolium cation (29) also has the benzene ring intact but the positive charge is almost entirely localized on the nitrogen atom. Deuterium exchange in H_2SO_4 takes place at the 3-position only.

2.   *Alkylation:* Because of the involvement of the nitrogen lone-pair in indole with the aromatic sextet, it is not surprising that unlike secondary amines reaction with alkyl halides does not take place. Indole itself reacts with methyl iodide in dimethylformamide above 100°C forming 1, 2, 3, 3-tetramethyl-3 *H*-indolium iodide[39] (31). Initially attack takes place at the carbon atom followed by a series of alkylations and deprotonations and a C_3 to C_2 methyl shift to give eventually the indoleninium salt:

(31)

For the alkylation of the nucleus various basic reagents such as sodium amide or potassium hydroxide in dimethylsulfoxide (DMSO) have been used to generate the indolyl anion.

It has been demonstrated that alkylation can easily be carried out using a two phase system consisting of an aqueous solution of 50% soldium hydroxide and a benzene solution of the akylating agent, *i.e.*, phase transfer catalyst,[40] equation (7.10).

(7.10)

3.   *Electrophilic Substitution:* Indole is a π-excessive molecule similar to pyrrole, therefore, the electron density on its carbon atoms is greater than in the benzene molecule. Indole, therefore, reacts easily with an electrophile and the attack takes place in the heterocycilc ring. It is markedly less reactive than the ocrresponding monocyclic heterocycles. The reactivity of indole towards electrophiles is highly regiospecific for position-3, whereas in the case of pyrrole the attack takes place at position-2. It may be noted for comparison that an electrophile attacks at C-2 in benzofuran while at C-2 as well as C-3 in benzothiophene. The orientation of an electrophilic attack can be rationalized by considering the intermediate ions produced by attack at these two positions.

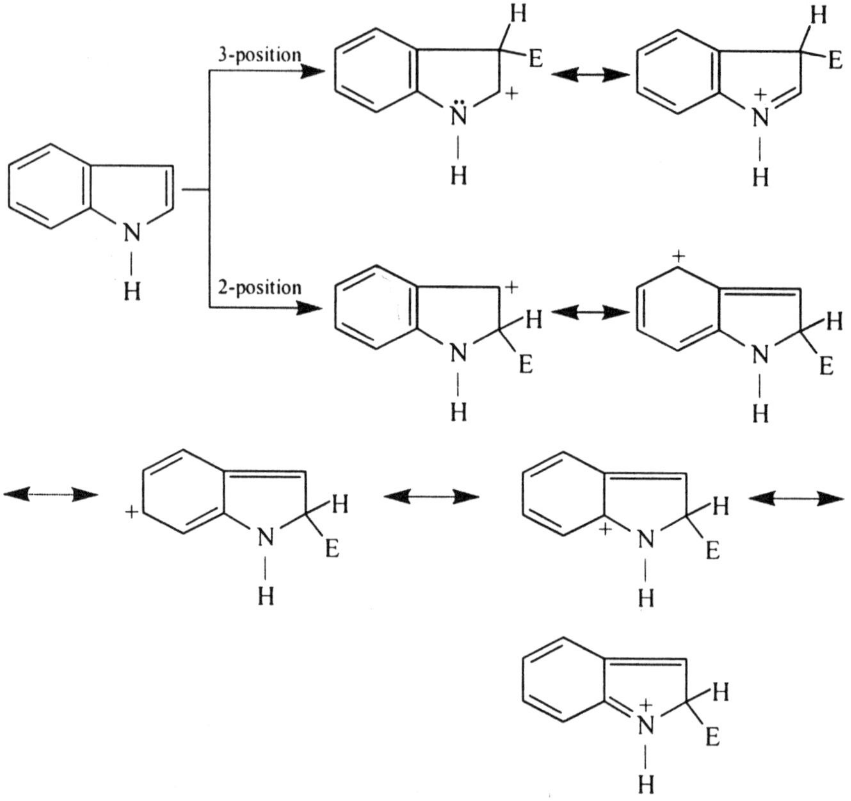

Attack of electrophile at C-2 generates a carbocation in which the charge is distributed on the benzene ring as well as on the hetero atom, but the structures with the charge on the hetero atom no longer possess a benzenoid ring. On the other hand, attack at C-3 does not permit effective distribution of charge around the ring but the electron pair on the hetero atom is used efficiently and the aromaticity of the benzene ring is retained. If this position is occupied then attack occurs at C-2 or in the benzene ring.

Halogenation: Milder reagents are used to effect halogenation of indole. Chlorination of indole has been widely studied and many reagents have been used for this purpose.[41, 42] Earlier workers[43] used sulfuryl chloride to obtain 3-chloroindole. Other reagents such as chlorine,[48] N, N-dichlorocarbonate,[44] phosphorous pentachloride[45] and *tert.* butyl hypochloride have been used. Bromination (Br_2/dioxane/0°C or Br_2/CH_3COOH or NBS) of indole gives 3-bromoindole.[46] If the 3-position is occupied by electron-withdrawing groups then attack occurs in the benzene ring,[47] equation (7.11).

$$(7.11)$$

Nitration: Indole is very reactive towards the normal nitrating agents HNO_3/H_2SO_4 and results in polymeric products. 3-Nitroindole can be obtained by nitration in non-acidic medium with benzoyl nitrate in CH_3CN solvent at 0°C.

Studies on nitration with 2-methylindole have shown that electrophilic substitution with nitronium ion, the reactivity order is 3 > 6 > 4. Thus in nitric acid 3-substitution is rapidly followed by C-4 (minor) and C-6 (major) nitration yields are generally poor due to competing oxidation reactions.

Sulfonation: Sulfonation of indole is carried out only under milder conditions using pyridine-sulfur trioxide complex in order to minimize problems associated with acidity, equation (7.12).

$$(7.12)$$

3-Indolesulfonic acid

Acylation: Indole itself does not respond to the Friedel-Crafts reaction, as acylation of the nitrogen atom takes place.[48] Acylation, however, of N-substituted indoles has been carried out successfully.[49, 50] Formylation of indoles by the Vilsmeier-Haack procedure is very efficient and yields 3-formylindole.[51] It is yet another example of an efficient electrophilic substitution in indole chemistry. Diazo coupling occurs very rapidly with pyrroles and indoles with benzene diazonium salts. Reaction is much more faster in alkaline medium when the species undergoing the reaction are the N-deprotonated heterocycles.

4.   *Reaction with Carbenes:* The most highly investigated reaction of a carbenoid attack on indole is that of a halocarbene.[52] A dichlorocarbene, for instance, adds to the 2, 3-C—C double bond of 3, 3-dimethylindole to give a mixture[53] of products, namely 3-chloro-2, 4-dimethylquinoline (32) and 3-dichloromethyl-2, 3-dimethyl-3*H*-indole (33). Indole itself reacts in a similar fashion to yield 3-formylindole and 3-chloroquinoline.

(32)   (33)

5.   *Reaction with Oxidizing Agents:* In the presence of air and light, indole itself is autoxidized to indoxyl which reacts further to give indigo. The C_2—C_3 double bond in substituted indoles can be readily cleaved by different reagents such as oxygen, ozone, sodium hypoiodate or perbenzoic acid, etc. It seems that a substituent at the 3-position is necessary for reaction with these reagents as indole itself gives only a resinous material, whereas 3-methylindole yields *o*-formaminoacetophenone.

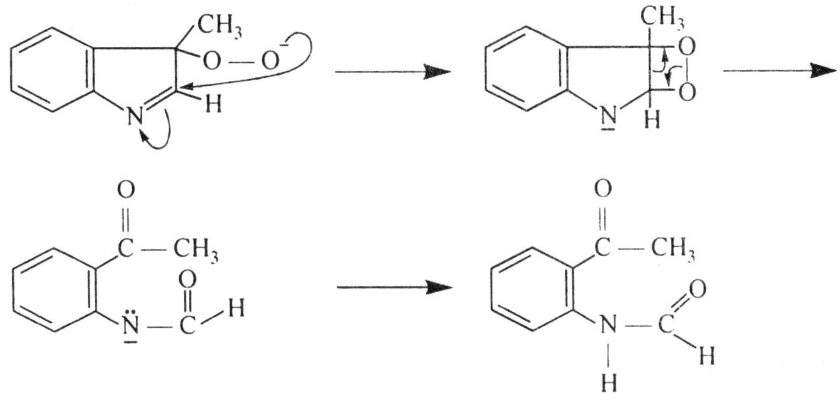

Similarly 2, 3-diphenylindole forms *o*-benzoylaminobenzophenone, equation (7.13). Autoxidation of 2-benzyl-3-phenylindole leads to a dioxindole

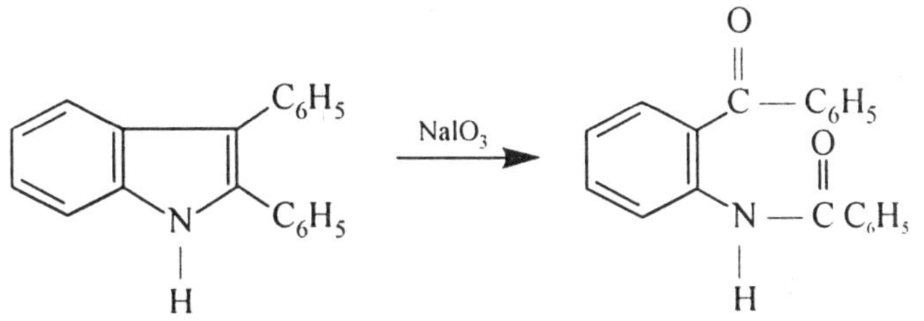

$$\text{(7.13)}$$

(34) but oxygen and a catalyst oxidizes the methylene group to give 2-benzoyl-3-phenylindole (35).[54] Electrochemical oxidation of 1-methylindole forms

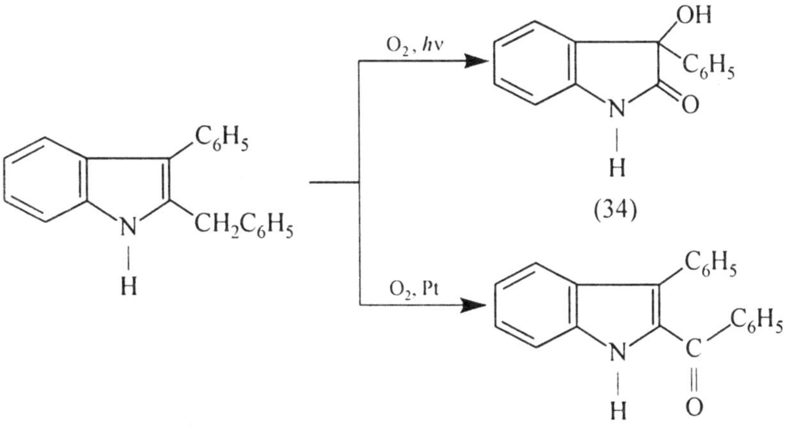

a mixture of products.[55] The indole ring can be reduced selectively in the carbocyclic or the heterocyclic ring. The neutral indole ring system is not reduced by nucleophilic reducing agents such as lithium aluminum hydride sodium borohydride or sodium in alcohol.[56-58] The heterocyclic ring, however, can be reduced in acidic reagents. Thus zinc and hydrochloric acid convert indole into a dihydro derivative called indoline, equation (7.14). A better reagent

$$\text{(7.14)}$$

Indoline

for reduction is sodium cyanoborohydride (NaCNBH$_3$/CH$_3$COOH) which yields indoline in about 90% yield.[59] A combinatin of lithium-ammonia-ethanol results in the reduction of the benzene ring in indole,[60] equation (7.15). Hydrogenation with more powerful reducing agents (Pt/CH$_3$COOH) yields

$$\text{(7.15)}$$

cis-octatetrahydroindole, equation (7.16). The stereochemistry of indole reduction has been studied,[61] addition of hydrogen takes place in a *cis* manner.

$$\text{(7.16)}$$

6.   *Reactions of N-Metallated Indoles:* The most acidic proton in indole is the N-hydrogen and can easily be converted into salts by strong bases such as sodium hydride, potassium *t*-butoxide as well as Grignard reagents. The indolyl anion so produced is mesomeric with the negative charge delocalized but mainly present on the nitrogen atom and β-carbon. In its chemical reactions, therefore this anion behaves as an ambident nucleophile and the ratio of N-to β-substitution by an electrophile depends on the nature of the associated metal cation, the substituent already present, the nature of the solvent and temperature of the reaction. As a general rule, sodio-or lithio-indole reacts predominantly at the nitrogen atom while alkyl magnesiylindole reacts mainly at the β-carbon atom to yield the acid.[61, 62]

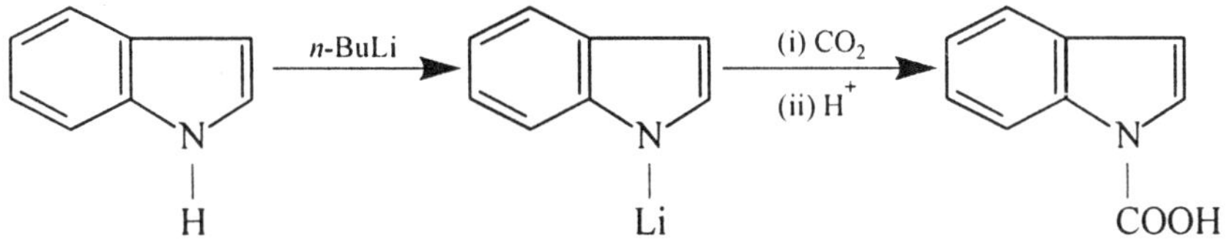

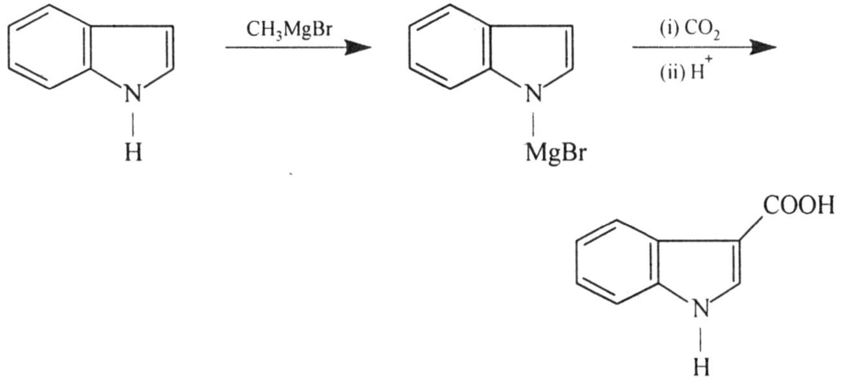

7. *Nucleophilic Substitution Reactions:* Nucleophilic substitution reactions of indole have been little studied. It is, however, known that direct displacement of nucleus bound halogen atom is difficult to achieve unless the halogen is activated by neighbouring electron-withdrawing groups. N-alkylindoles however, are deprotonated by nucleophilic reagents such as *n*-BuLi in the same manner as furan and thiophene.[62] The resulting salt serves as a useful intermediate.

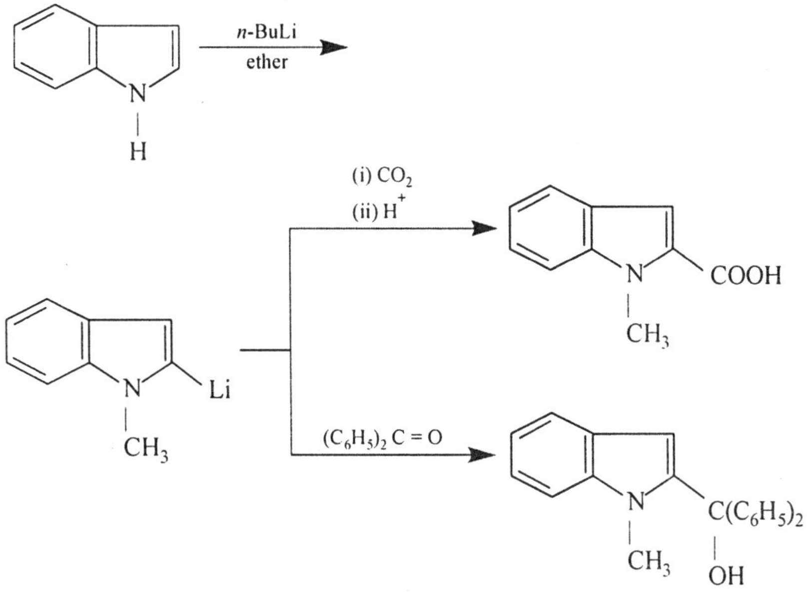

8.   *Photochemical Reactions:* Various indole derivatives have been known to undergo photochemical transformations.[63] 1-Ethoxyindole[64] and aryl esters[65] of indole undergo photochemical Fries rearrangement to give a mixture of products and thus the reaction has made available 3-, 4-, and 6-derivatives of indole which are obtained by rather cumbersome procedures. A large number of photosensitization reactions have also been observed[66–69] with a wide variety of indole derivatives. Similarly 3-(2-nitroprop-1-enyl) indole (36) rearranges to 3-(2-hydroxyiminopropylidene) 2-oxindole (37) according to the following steps:

(36)

(37)

The Diels-Alder reaction between 1-methylindole, an electron rich dienophile cyclohexa-1, 3-diene an electron-rich diene takes place by a photochemically induced catalyzed electron transfer reaction to give (38), equation (7.17).

(38)

(7.17)

The sensitizer employed was triaryl pyrylium tetrafluroborate. Surprisingly 2-methylindole does not react under these conditions, may be because of steric factors.

In contrast N-arylindoles undergo dimerization leading to cyclobutane adducts formed by bonding between the 2-position and 3-position of the indole ring.[72] The *anti* head to head regioisomer of the photodimers, equation (7.18) are formed predominantly though in low yields. Chemical yields of the

$$\text{(7.18)}$$

adducts are enhanced by the use of acetophenone as sensitizer. The analogous reaction does not take place with indole.[73]

7.1.4 Derivatives of Indole

Alkyl-and Arylindoles: Alkyl indoles have been prepared by the Fischer and the Reissert methods. 2-Phenylindole is prepared by the Fischer method according to equation (7.19). Another method for its preparation includes

$$\text{(7.19)}$$

starting from *o*-iodoaniline and cuprous phenylacetylide. 2-Methylindole on nitration (conc. HNO_3/r. t.) yields a mixture of polynitrated products whereas 2, 3-dimethylindole yields the 5-nitro derivative.[76] Skatole on bromination (NBS) gives 3-bromomethylskatole. 2-Methyl-or arylindoles undergo diazonium ion coupling at the β-position[77] to give a β-azo derivative, equation (7.20). 2-Methylindole condenses with methyl vinyl ketone to afford carbazole, equation (7.21).

$$\text{(7.20)}$$

$$(7.21)$$

Indole Aldehydes and Ketones: These derivatives have been prepared by a number of different methods. Thus indole reacts with chloroform and alkali in the familiar Reimer-Tiemman reaction to give indole 3-carbaldehyde although in a poor yield, 3-chloroquinoline being the major product. Formylation of indole can be accomplished by the Vilsmeier reaction. Indole 3-carbaldehyde responds to all the reactions of aromatic aldehydes. It forms oximes, phenylhydrazones semicarbazones and laso undergoes benzoin condensation and Cannizzaro reaction. On nitration indole yields 6-nitroindole.

Indole can also be converted to tryptophan. 3-Indolecarbaldehyde is also employed for the preparation of indigo. 3-Acylindole is obtained through the reaction of acid chloride with indolylmagnesium iodide, equation (7.22).

$$(7.22)$$

If the 3-position is occupied than the acyl group goes to the 2-position. The keto group can be reduced to an alkyl group with lithium aluminum hydride and diborane similar to pyrrole ketone to yield alkyl derivatives.

Indole Carboxylic Acid: The Fischer synthesis has been the most appropriate procedure to obtain various indole carboxylic acids. Indole 2-carboxylic acid is a crystalline compound, m.p., 206-208°C, the 3-isomer has a *m.p.* of 218°C. From the ionization constant values it is known that indole 2-carboxylic acid is stronger than the 3-isomer. Both these acids are decarboxylated rather easily but the decarboxylation of the 3-isomer is more facile.

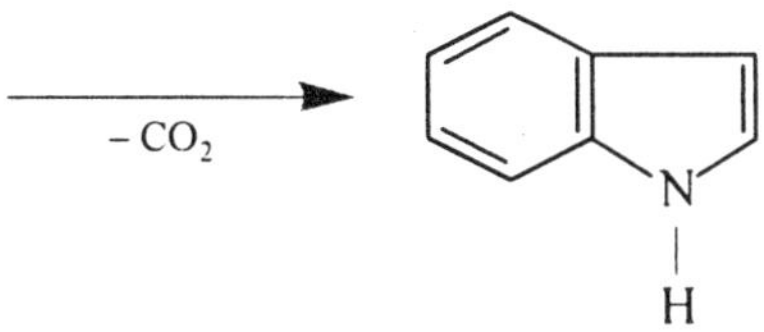

80%

Oxy-Indole Derivatives: Indole forms a number of important oxyderivatives namely, oxindole (39), indoxyl (40), isatin (41) and indigotin or indigo (42).

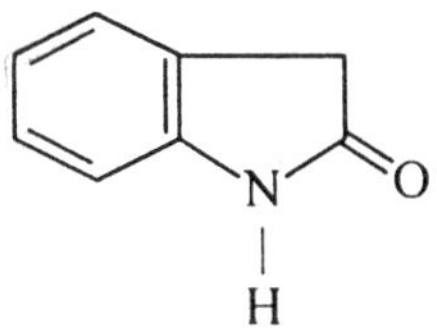

(39) 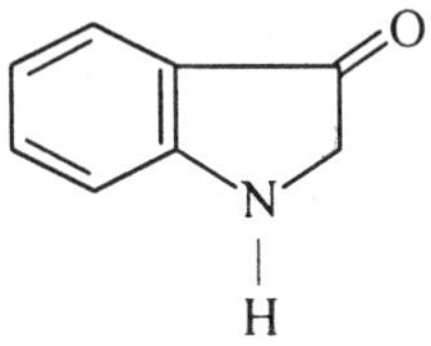   (40) 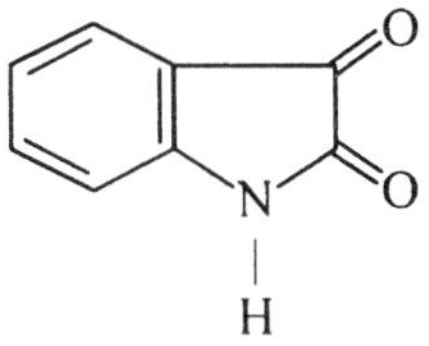   (41)

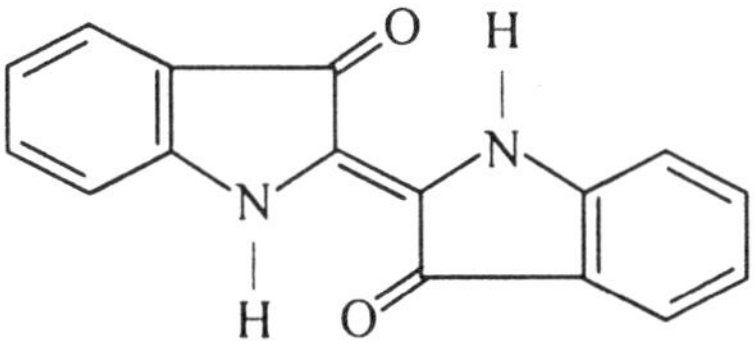

(42)

Oxindoles: Oxindole is best represented by structure (39) but can exist in two tautomers (43) and (44). Oxindole may be obtained by the direct reduction of

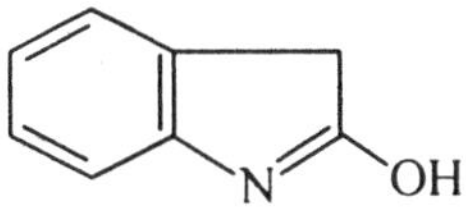

(43) 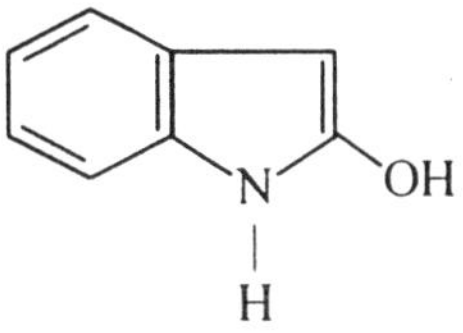   (44)

isatin in two steps. The first step uses sodium amalgam in an alkaline medium to give dioxindole which is further reduced (Sn/HCl) to oxindole. Alternatively direct Wolff-Kishner reduction of isatin also furnishes oxindole. The following ring closure method[78] utilizes an α-halogenated acid halide and an aromatic amine.

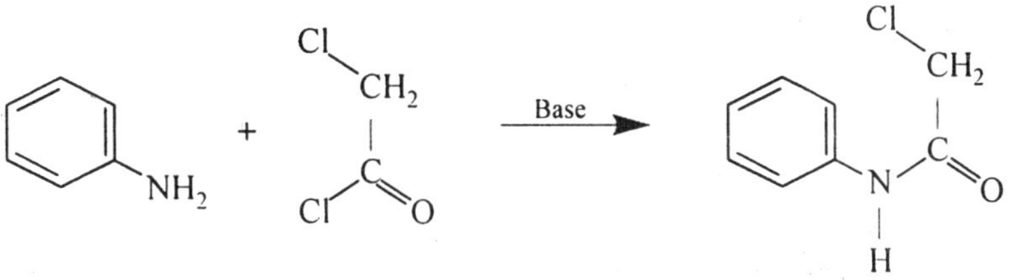

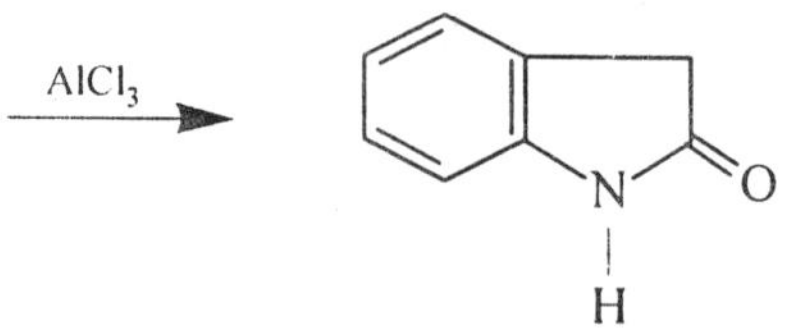

A more recent method involves an initial Diels-Alder reaction between N-acylpyrrole and 1, 3-dicarboethoxyallene leading to an adduct (45) followed by the treatment of this adduct with potassium hydride and ammonium chloride[79] to give, 4-carboethoxyoxindole. Formation of oxindole has also been observed in certain other reactions as well.[80]

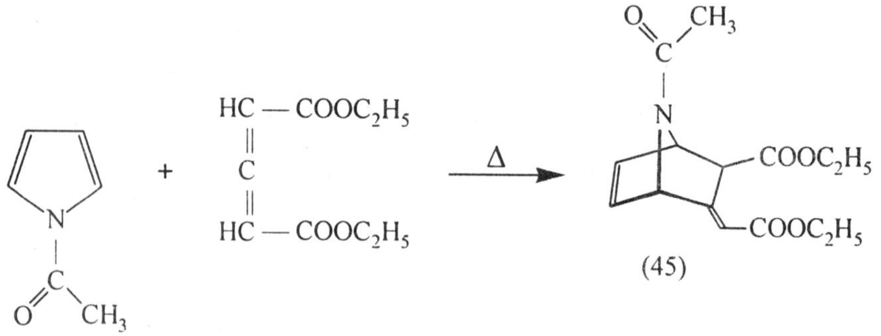

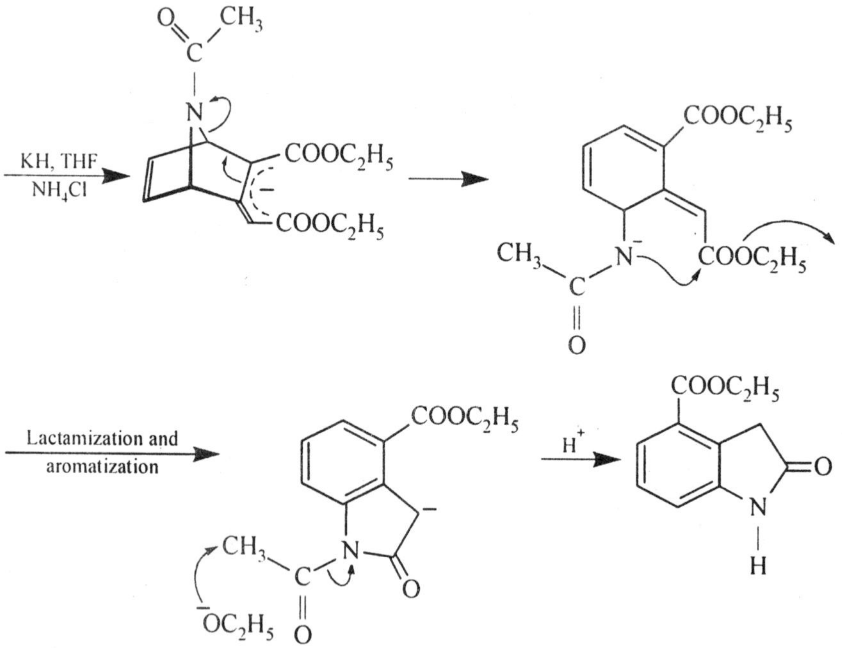

Oxindoles is a colorless solid m.p., 126-127°C and can be recrystallized from water. It is weakly basic and forms water insoluble hydroxides and hydrochlorides with HCl. Bromination (Br_2/H_2O) and nitration (HNO_3/H_2SO_4)

give rise to 5-substituted oxindoles. Nitric acid nitrates at the 3-position. Oxindole possesses an active methylene group at the 3-position as the anion can be stabilized with the carbonyl group. It, therefore, undergoes the usual condensation reactions of aldol and Claisen type. Oxindole forms sodium and silver salts with respective reagents which react with methyl iodide to give N- and C-alkyl derivatives. Oxindole behaves as a simple lactam, and undergoes conversion to 2-chloroindole with phosphorus pentachloride. The heterocyclic ring in oxindole is readily cleaved with barium hydroxide to give barium 2-aminophenyl acetate which recyclizes on acidification.

Indoxyls: Indoxyl (40) is the 3-keto tautomer of 3-hydroxyindole (46), though it exists almost entirely in the carbonyl form:

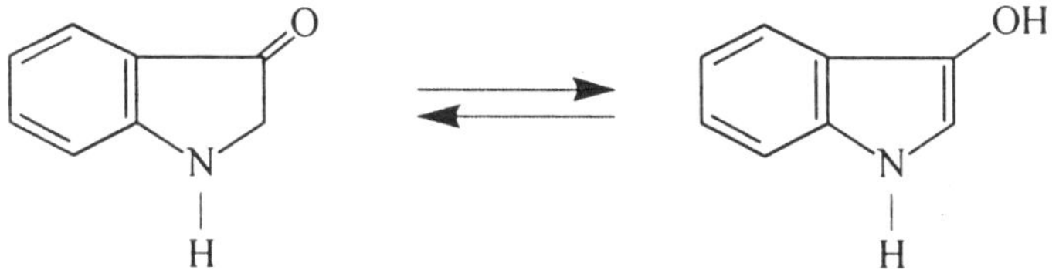

(39)   (46)

It is an important intermediate in the synthesis of indigo and is also of biochemical importance because of its accurrence in the form of its glucoside in certain plants. It may be prepared according to the following sequence commencing from aniline.

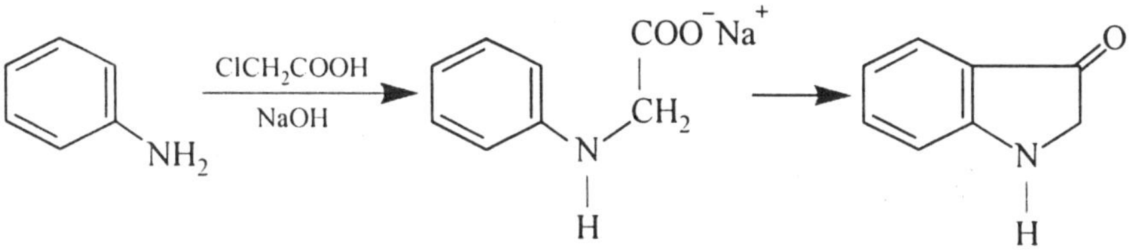

Indoxyl is a yellow crystalline solid, m.p., 85°C. It gives a strong unpleasant odor and is soluble in hot water. Indoxyl also deprotonates α to the carbonyl group like oxindole and thus condenses with aldehydes and ketones. With methyl iodide in sodium hydroxide, indoxyl is alkylated at the 2-position to yield 2, 2-dimethylindoxyl. In alkaline solution, indoxyl is converted to indigo. Indoxyl in the presence of Zn/HOAc is reduced to indole.

Isatin (Dioxindole): Isatin (41) is an indole derivative containing keto groups at 2-and 3-positions. Early studies on the action of oxidizing agents on indigo led to the discovery of an oxidation product to which the name isatin was given and could be obtained in 70-80% yields by the action of chromic acid on indigo. It has, however, been prepared by a number of other methods as well. The most general method is due to Sandmeyer[81] in which aniline and chloral hydrate first form iso-nitroacetanilide in the presence of hydroxylamine.

This subsequently cyclizes in the presence of sulfuric acid to isatin.

According to another synthesis 2-nitrobenzoyl cyanide, prepared from *o*-nitrobenzoyl chloride and silver cyanide, is hydrolyzed and then reduced with ferrous hydroxide to *o*-aminophenylaxalic acid. Acidification results in

isatin formation. N-substituted derivatives of isatin can be conveniently prepared from N-substituted anilines and oxalyl chloride followed by cyclization with aluminum chloride.

Isatin forms red needles, m.p., 200-201°C and can be recrystallized from hot water or alcohol. It is basic in nature and dissolves in conc. hydrochloric acid or sulfuric acid. With perchloric acid it forms stable perchlorates. The N—H hydrogen is sufficiently basic and forms salts with sodium or potassium metals. Halogenation results in the formation of a 5-haloi satin. If the 5-position is occupied then the electrophile takes the 7-position. Chromic acid or hydrogen peroxide oxidation results in isatoic anhydride which on subsequent hydrolysis and decarboxylation yields isatoic acid.

Isatin condenses with thiophene in conc. sulfuric acid to give a deep blue color due to the formation of indophenine (47).[82] This is used as a test for the detection of thiophene. The α-positions of the thiophene molecule are utilized in the reaction, therefore, α-substituted thiophene will not respond to this test.

(47)

The carbonyl group at the 2-position, on the other hand, is adjacent to the hetero atom and is stabilized by resonance. It thus behaves as a typical amide in its properties. The heterocyclic ring is opened by alkalies and the intermediate so obtained can condense with ketones to give quinoline derivatives.[83]

Indigo: Indigo (42) occurs in plants (indigofera tinctoria) as indican, the β-glucoside of which yields indoxyl and glucose on hydrolysis. Air oxidation of indoxyl then gives indigo which is used as a dye (vat dye). Indigo is nowadays produced synthetically and is reduced to *leuco* form (48) with sodium

Leuco form

(48)

hydrosulfite. It can be oxidized by exposure to air or more rapidly by an oxidizing agent such as sodium perborate. Indigo can be obtained by the Dakin process on 3-formylindole.[84] This consists in the treatment of 3-formylindole with hydrogen peroxide in the following steps. Indigo can also be obtained

starting from anthranilic acid. Indigo has also been obtained by a electrochemical method.[85] For dyeing purposes indigo is reduced to the yellow *leuco* indigo (48) by sodium hydrosulfite. The fabric is then treated with a solution of (48) in aqueous sodium bicarbonate and then indigo is regenerated by atmospheric oxidation.

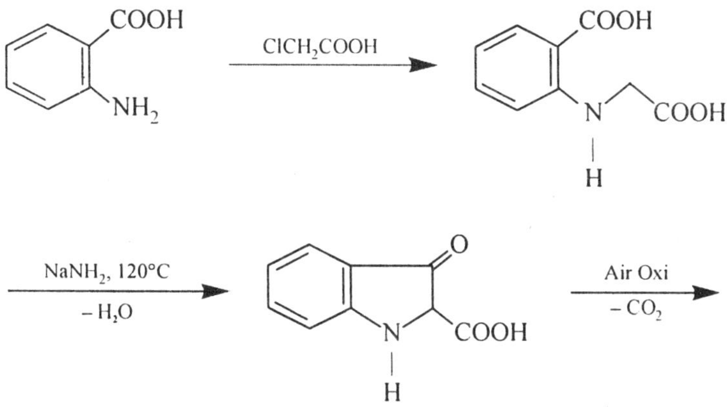

7.1.5 Naturally and Biologically Active Compounds

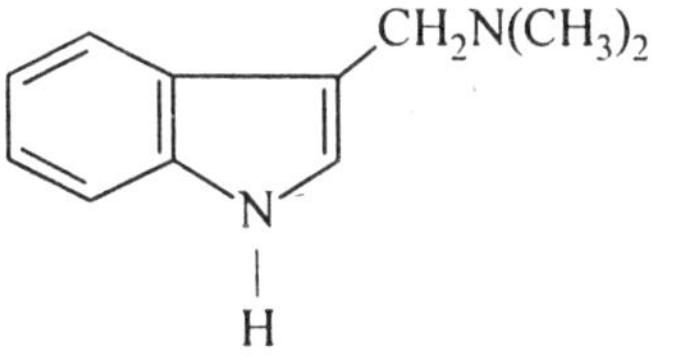

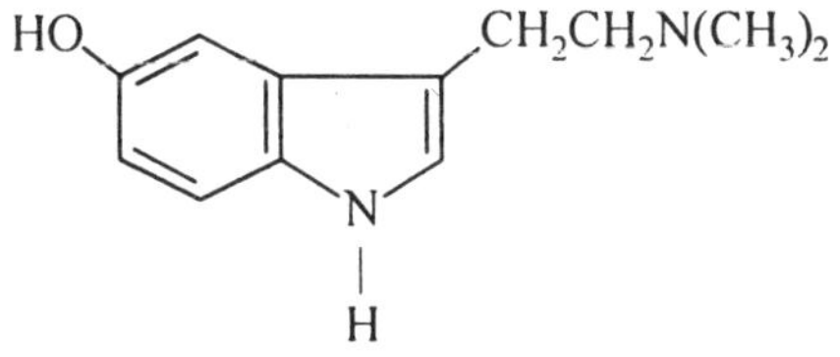

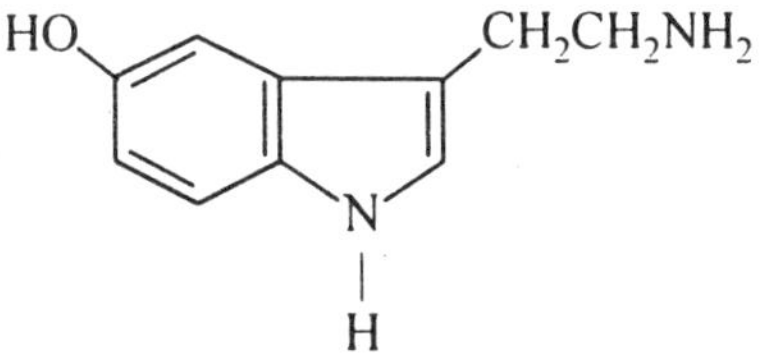

Indole derivatives find a large number of uses as coloring agents, essential amino acids, plant growth hormones and in blood sugar lowering agents. Of the large number of nitrogen bases that occur in nature, a considerable portion contain indole ring and majority of them are alkaloids which have been widely studied.[86] The large number of indole alkaloids are divided into five categories namely (a) the simple alkaloids (b) the ergot alkaloids (c) the yohimbe alkaloids (d) the harmala alkaloids and (e) the strychnos alkaloids. The first type of alkaloids are closely related structurally to tryptophan. Some examples are gramine (49), produced in sprouting barely, bufotenine (50) occurs in fungi (mushrooms) and has been shown to be the active principle of the narcotic *snuff*. Serotonin (51) is widely distributed in nature and is present in salivary glands and gastrointestinal tissue of mammals.

Ergot is a fungus which grows principally on cereal especially rye. All the six ergot alkaloids are derivatives of lyserigic acid (52). The most important alkaloid of this series is its diethylamide (does not occur naturally) known as

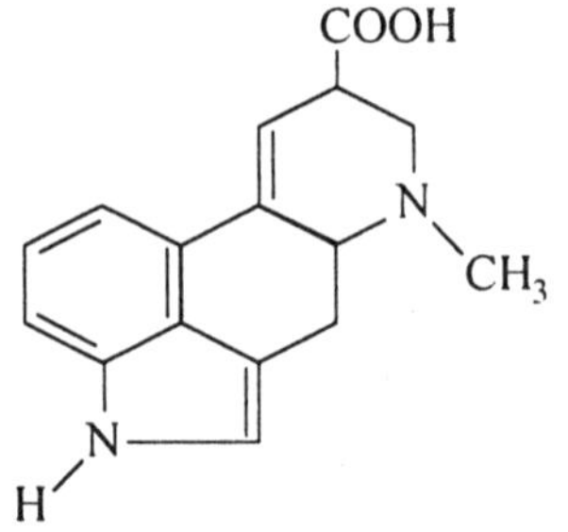

(52)

LSD. It is known to cause hallucinations in humans presumably by altering the serotonin level in the brain.

The most important member of yohimbe class of alkaloids is reserpine (53). This alkaloid is used considerably in the treatment of hypertension (high blood pressure) and is a major tranquillizer.

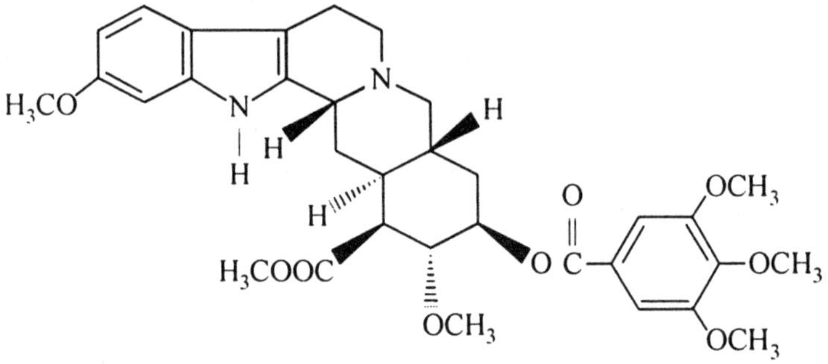

(53)

The harmala alkaloids include *harman* (54) and harmine (55).

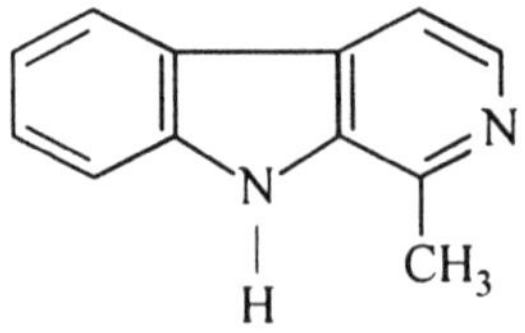

(54) 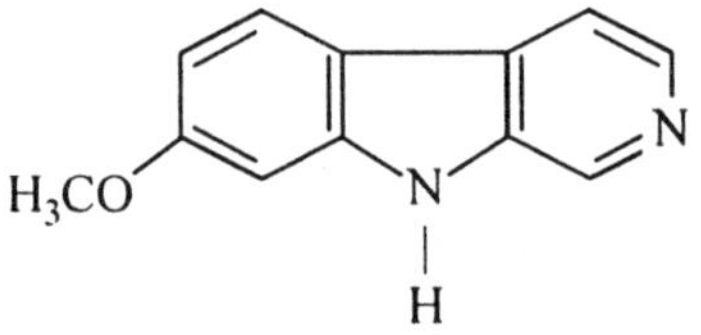   (55)

Among the strychnos alkaloids, the most popular examples are strychnine (56) and brucine (57) which are very complex.

(55) R = H

(56) R = OCH$_3$

A new alkaloid of potential pharmacological activity is sorelline (58).

(58)

7.2 ISOINDOLES[88]

In addition to indole, another benzopyrrole, *i.e.* isoindole (6) is possible by *ortho* condensation of benzene ring across the 4, 5-positions of the pyrrole nucleus. It is isomeric with indole. The ring system is numbered as shown.

(6)

Isoindole 2*H*-benzo [c] pyrrole or 1*H*-isoindole, in general, is quite reactive and thus eluded isolation[89] until 1951. Isoindole responds to Ehrlich test. The parent compound can potentially tautomerize to the 1*H*-isomer, *i.e.* isoindolenine or 1*H*-isoindole (6a), but exists largely as (6), in solution.

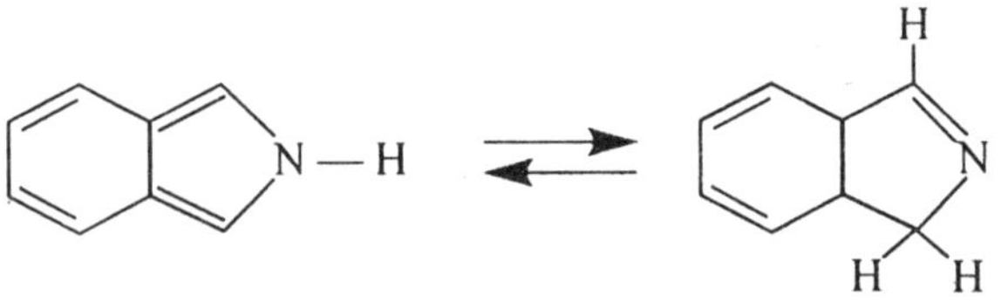

(6)   (6a)

The greater stability of the isoindole form (6) is ascribed to its lower π-electron energy, as it behaves as a 10 π-electron system.[90]

7.2.1 Physical Properties

Isoindole is white solid and can be isolated at low temperature but decomposes at room temperature. It is stable only as its Diels-Alder adduct. 2-Methylisoindole is a colorless compound, m.p. 90-91°C. Isoindole exists in the following canonical forms:

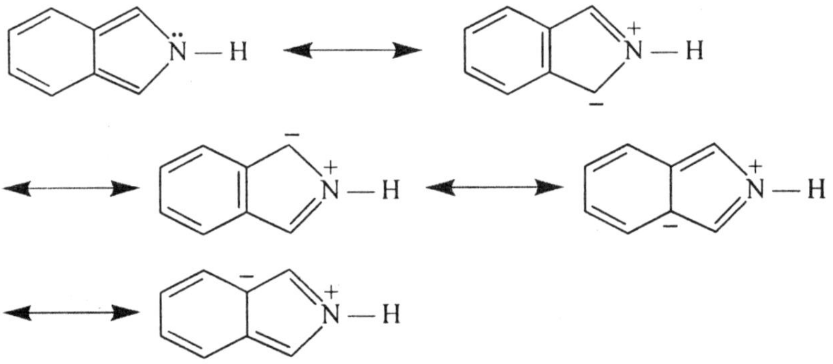

Isoindole behaves as a typical secondary amine, its pKa is not known.

7.2.2 Synthetic Methods

1. *From 2-(Methoxycarbonyloxy) Isoindole:* The elusive indole was synthesized[91] by the pyrolysis of the title compound in a silica tube at 500°C/0.01 mm Hg.

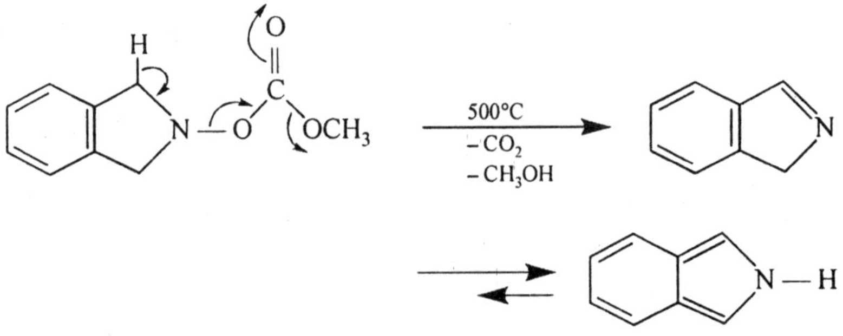

(6)

2. *From Isoindolinium Salts:* 2-Methylisoindole was prepared by treating the isoindolinium salt with phenyllithium and was isolated as its D.A. adduct with maleic anhydride. 2-Methylisoindole has also been obtained by the action of potassium amide/liq ammonia on N-2-chlorobenzyl-N-methylamminoacetonitrile in 80% yield.[92, 93]

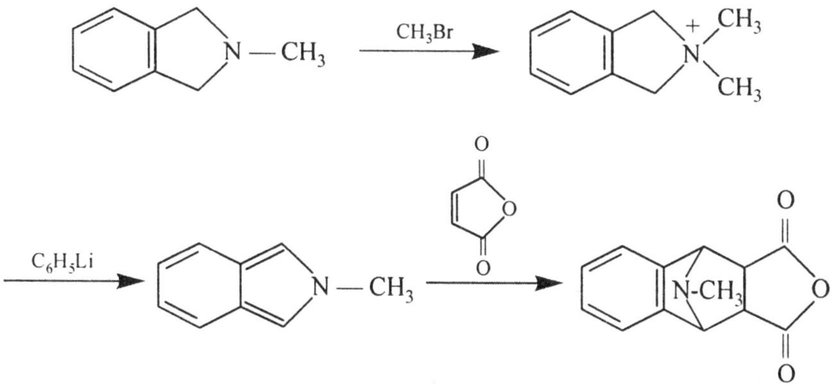

3. *From Cyclohexadiene:* An initial Diels-Alder adduct (59) formation between 1, 3-butadiene and dibenzoylacetylene on reaction with an amine yields[94] a substituted isoindole. This reaction utilizes the ability of phenyl group to participate in a 1, 5-sigmatropic rearrangement.

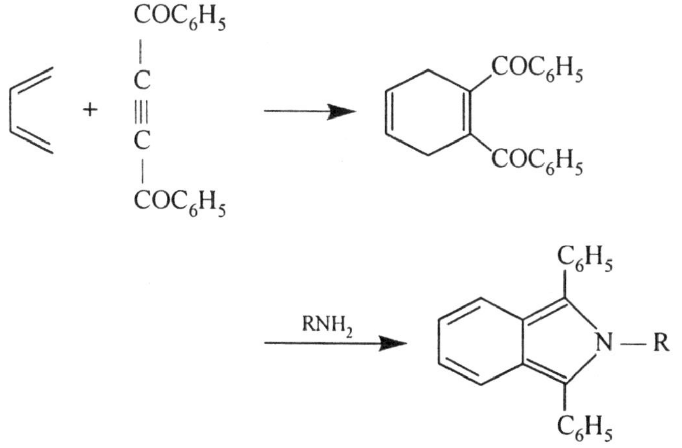

In spite of the π-electron bicyclic heterocyclic system and apparently aromatic structure, N-substituted and especially halogen substitution are required to increase the stability of isoindoles not conjugated with the unsaturated functional group.[95] As a result very few oxygen or nitrogen substituted isoindoles have been reported.[96]

7.2.3 Chemical Reactions

Isoindole forms isoindolinium salts (60) with acids. The π-electron density is highest at C-1. It undergoes electrophilic substitution at this position to yield a stable intermediate (61). Isoindole and 1,3-disubstituted isoindoles readily give cycloaddition products across the 1, 3-positions.

(60)   (61)

In this reaction isoindole behaves much like furan. Isoindoles are sensitive to atmospheric air and often form mixture of oxidation products or polymers. They are preferentially oxidized in the five-membered ring to give phthalic acid. Isoindoles can be reduced (Raney nickel, Zn/CH_3COOH) to the corresponding isoindolines.

Isoindole forms Diels-Alder adducts with several dienophiles such as maleic anhydride, phenylmaleic anhydride dimethylacetylene dicarboxylate, etc. equation (7.23).

$$(7.23)$$

7.3 INDOLIZINES

Since nitrogen is trivalent, an additional structure can be written called indolizine (62) which was earlier known as *pyrrocoline.* It is isoelectronic

(62)

with indole and is also a 10 π-electron bicyclic aromatic heterocyclic system. It does not exist as a tautomeric form but can be expressed into a number of charged mesomeric structures. Indolizine is a solid, m.p., 74°C and possesses an odor resembling that of naphthalene. Indolizine behaves as a weak base (pKa 3.9) but is stronger than indole (pKa-3.5).

7.3.1 Synthetic Methods[97]

Indolizine has been prepared by the following methods starting from pyridine derivatives:

1.   *From 2-Picoline:* This method consists of treating 2-picoline with α-bromoethylpyruvate and the intermediate so formed is cyclized in the presence of sodium bicarbonate.[98]

2.   An indolizine derivative namely 1-[(*cis*-2, 3-diphenylacryl) oxy]-2, 3-diphenylindolizine was prepared from 2, 3-diphenylcyclopropenone and pyridine by Breslow.[99]

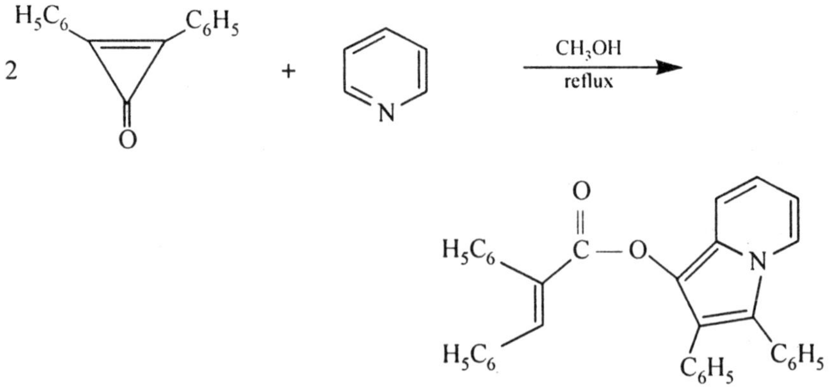

3.   One of the most useful preparation of indolizine consists of a reaction between 2-pyridyllithium and 2-chloromethyloxirane or epichlorohydrin. A salt is formed which on treatment with sodium hydroxide yields the parent compound indolizine.

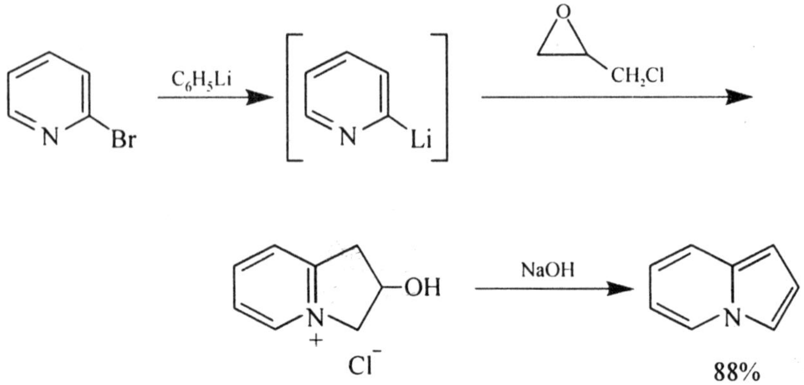

7.3.2 Chemical Reactions

Indolizine is a weak base and protonates mainly at position-3.[100] The electrophilic attack takes place at C-1 and C-3 as the cations (63) and (64)

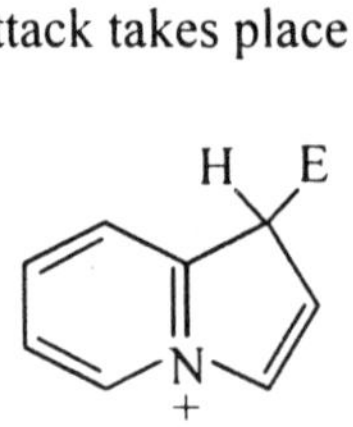

(63) 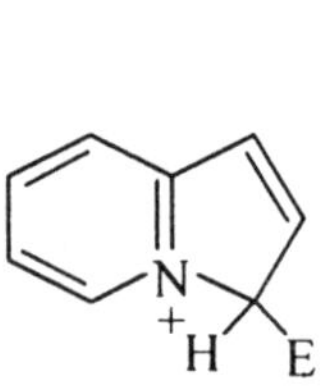   (64)

do not differ much in stability. Indolizines undergo nitration, Friedel-Crafts reaction Vilsmeier formylation, etc., equation (7.24) and (7.25). Indolizines are oxidized (H_2O_2) and as in the case of indoles, the five-membered ring is more susceptible to attack. On catalytic reduction, different products are obtained depending on the reagent employed. Indolizines are often used

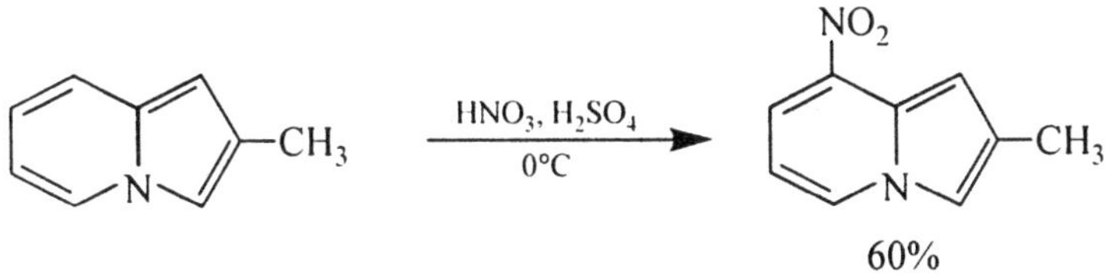

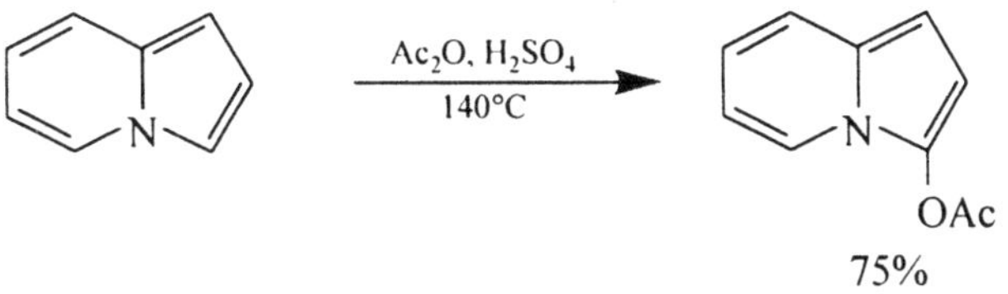

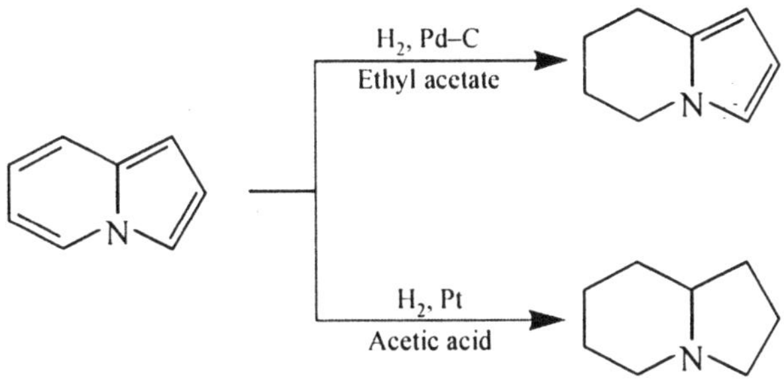

to synthesize cyclazines, i.e., tricyclic aromatic systems containing a central nitrogen atom. Thus tricyclo [3, 2, 2] azine (65) is obtained from 5-methylindolizine in the following manner.[99]

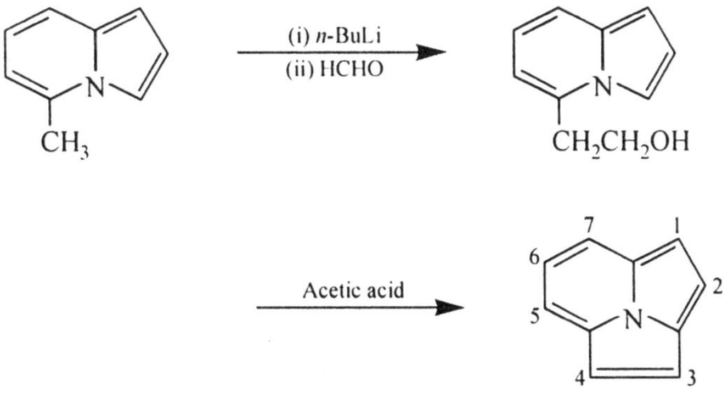

Indolizines are preferentially oxidized in the five-membered ring similar to isoindole to produce picolinic acid derivatives.

7.4 DIBENZOPYRROLES[101]

Dibenzopyrrole or carbazole (66) was discovered in 1827 in the crude anthracene fraction of coal tar. It responds to the pine splinter test, i.e.,

(66)

development of a red color when a pine splinter soaked in hydrochloric acid is held in the vapors of an alkali solution of carbazole. The numbering of the atoms in carbazole is shown above, the nitrogen atom is assigned position-9. It is aromatic and possesses a resonance energy of 74 Kcal/mole which is slightly greater than diphenyl (71 Kcal/mole). Carbazole is a solid m.p., 245-246°C and is soluble in most organic solvents.

7.4.1 Synthetic Methods

Carbazole and its derivatives can be obtained by several methods:

1.   *From Diphenylamine:* Palladium acetate has been used as a catalyst in various cyclization reactions.[102, 103] This catalyst has been successfully employed in the cyclization of diphenylamine and its derivatives to carbazoles,[104] equation (7.26)

$$\text{(7.26)}$$

2.   *The Graebe-Ullman Synthesis:* In this reaction the diazonium salt of 2-aminodiphenylamine cyclizes to benzotriazole (67) which on pyrolysis loses nitrogen to yield carbazole.

(67)

Ring closure of 4, 4′-di-*t*-butyl-2, 2-diaminodiphenyl has been achieved under milder conditions using Nafion H (a solid super acid) to obtain 2, 7-di-*t*-butyldibenzopyrrole, equation (7.27).

$$(7.27)$$

3. **From *p*-Benzoquinone:** Substituted carbazoles are obtained by reacting *p*-benzoquinone and *p*-nitroaniline in the presence of acetic acid. The method is exemplified for the preparation of 3-nitro-6-hydroxycarbazole.[106]

4.   *The Borsche-Synthesis:* According to this method phenylhydrazine and cyclohexanone react to give a hydrazone which cyclizes to tetrahydrocarbazole (68) in the presence of dil. sulfuric acid.[107] This is dehydrogenated to carbazole with lead dioxide.

(68)

5.   *From 2-Nitrodiphenyl:* The reaction[108] of 2-nitrodiphenyl with triethylphosphite yields a carbazole *via* a nitrene intermediate.

7.4.2 Chemical Reactions

The N – H bond in carbazole is more acidic than that in diphenylamine and it reacts with bases or Grignard reagents to form salts. Electrophilic attack in carbazole takes place at the 3-and 6-positions. Chlorination with sulfuryl chloride gives mixture of 3-and 3, 6-dichlorides. Bromination with N-bromosuccinimide produces 3-bromocarbazole. Nitration (HNO_3/H_2SO_4) takes place to give 3, 6-dinitrocarbazole.

Acyl halides and anhydrides react with carbazole to give N-acyl derivatives. 2-Aminocarbazole, a useful intermediate in the synthesis of dyes and pigments, stabilizer for polymers and in pesticides has been obtained from carbazole in the following manner.[109]

7.5 BENZOFURAN, ISOBENZOFURAN AND DIBENZOFURAN

In a manner similar to pyrrole, a benzene nucleus can condense at the 2, 3-and 3, 4-positions of furan to form benzofuran (4) and isobenzofuran (7) respectively. Condensation of a second molecule of benzene at the 2, 3-positions of benzofuran leads to dibenzofuran. These are all aromatic and respond to typical electrophilic substitution reactions. Isobensofuran is stable only at low temperatures.

(4)   (7)   (69)

7.5.1 Benzofuran

Benzofuran was earlier known as *coumarone* and was discovered in coaltar by Kraemer and Spikler.[110] It was subsequently prepared by the decarboxylation of benzofuran 2-carboxylic acid. The numbering of the ring system commences from the oxygen atom. This compound is also designated as benzo [b] furan.

(4)

A number of derivatives of benzofuran are present in plant products and have pharmacological applications. Benzofuran is highly aromatic and is a resonance hybrid of the following resonance structures.

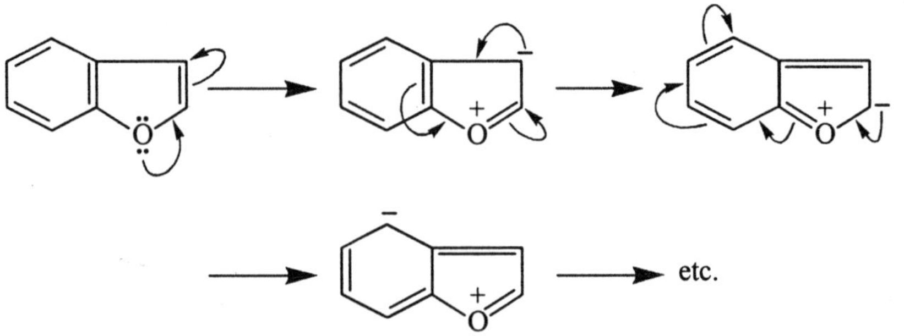

Benzofuran is a colorless liquid, b.p. 173°C.

7.5.1.1 Synthetic methods

The synthesis of benzofuran generally involve the ring closure of the furan ring.

1. ***From Coumarin:*** This method involves initially the bromination of coumarin to 3, 4-dibromocoumarin followed by treatment with alkali to coumarillic acid. The acid decarboxylates to benzofuran. If coumarin has a substituent at the 4-position, then according to the mechanism above, the substituent appears at the 3-position[111] in benzofuran.

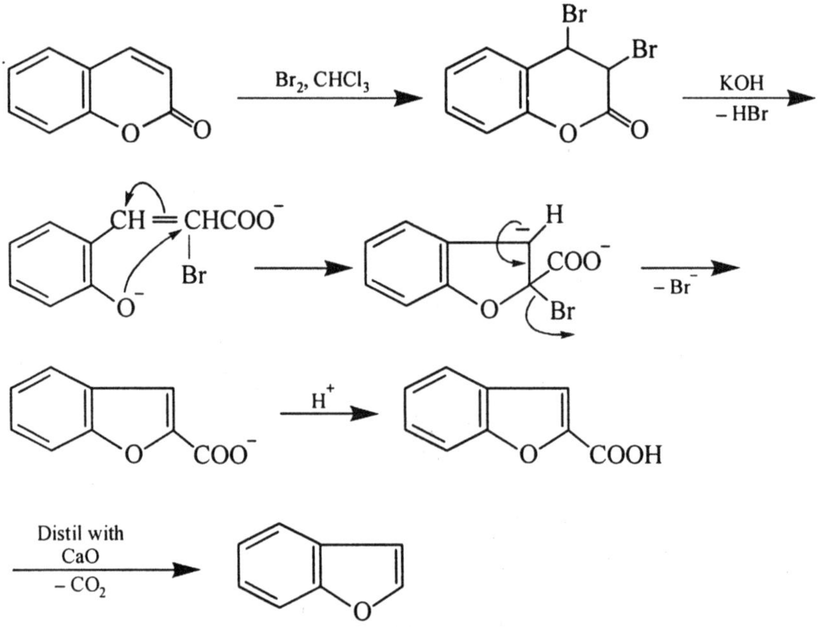

2. In another method 2-methoxyphenylmethyl 4-chlorophenyl ketone is first treated with HI (47%)/CH_3COOH to bring about an ether cleavage and subsequent cyclization of the intermediate leads to benzofuran.[112]

3. *From Aryloxyacetals:* Ring closing dehydration of aryloxyacetaldehydes or of their acetals on heating at 100°C in the presence of polyphosphoric acid produces the parent compound benzofuran.[120] Various additional benzofuran derivatives have also been prepared.[113]

4. *Claisen Rearrangement:* 2-Methylbenzofuran may be prepared by a Claisen rearrangement of aryl-2-chloroprop-2-en ethers.[114]

5. Palladium catalyzed hetero-annelation of acetylidienic compounds leads to benzofuran formation[115] as represented by the following example, equation (7.28).

$$(7.28)$$

7.5.1.2 Chemical reactions

The electrophilic attack on benzofuran proceeds under mild conditions and in contrast to pyrrole the attack takes places at the 2-and 3-positions. Chlorination (Cl_2/CS_2/-15°C) and bromination (Br_2/CS_2/-5°C) yield a mixture of 2-and 3-halobenzofurans. Nitration (HNO_3/CH_3COOH) and Vilsmeier formylation gives the 2-substituted derivative. Sulfonation (conc. H_2SO_4) yields resinous products. The Friedel-Crafts acylation (acetic anhydride/$SnCl_4$) yields 2-acylbenzofuran.

Reaction of benzofuran with dichlorocarbene gives ring expansion product[116] while nitrenes add to the 2, 3-carbon-carbon double bond.[117]

Benzofuran is easily reduced by hydrogen over palladium or sodium/ethanol to the corresponding 2, 3-dihydrobenzofuran known as coumaranes (equation 7.29).

$$Na, C_2H_5OH \tag{7.29}$$

70%

It is interesting to note that reduction of 2-nitrobenzofuran with Sn/HCl gives benzofuranone and not the expected 2-aminobenzofuran, (equation 7.30).

$$Sn, HCl \quad \Delta \tag{7.30}$$

Metallation of benzofuran with organolithium and subsequent carbonation yields benzofuran-2-carboxylic acid.[118]

The 3-lithio derivative, in contrast, opens to give 2-ethynylophenol (equation 7.31).

$$\tag{7.31}$$

Benzo [b] furans similar to indoles do not take-part in the Diels-Alder reaction.

2, 3-Dimethylbenzofuran undergoes epoxidation with dimethyl dioxirane in quantitative yield,[119] equation (7.32).

$$(7.32)$$

7.5.2 Isobenzofuran

The isobenzofuran system results by the *ortho* condensation of a benzene ring across the 3-and 4-positions of furan and is iso-electronic with naphthaline. It is also known as benzo [c] furan. It is interesting to note that it is not possible

to write a structure for isobenzofuran in which the Kekule structure of benzene is retained. Isobenzofuran is unknown while 1, 3-di-phenylisobenzofuran is stable and highly fluorescent compound. Most studies have been carried out with this derivative of isobenzofuran.

7.5.2.1 Synthetic Methods

The parent compound has been prepared by Warrener [120] from 3, 6-di (2′-pyridyl) *s*-tetrazine and 1, 4-dihydro-1, 4-*endo* naphthalene oxide, the initial yellow adduct (71) formed decomposes to isobenzofuran at room temperature (isobenzofuran is stable at –20°C).

(71)

The 1, 3-diphenylisobenzofuran can be prepared by the action of alkali on 2, 3-diphenylindenone (72) according to the following steps:

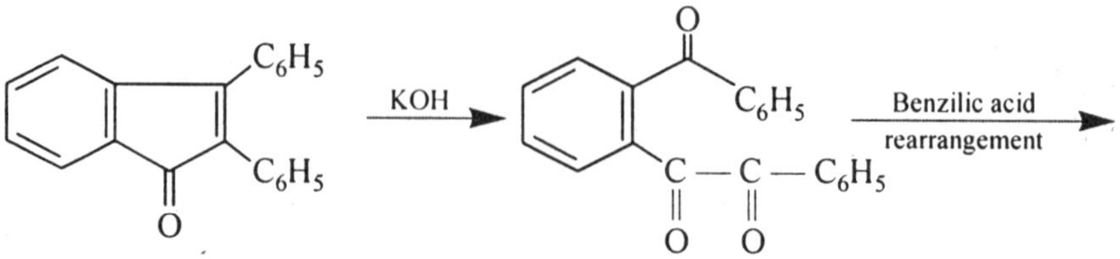

(72)

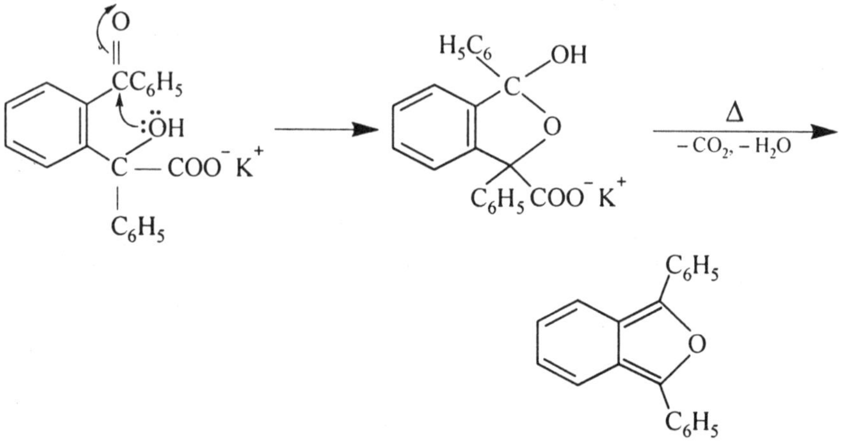

Isobenzofuran may also be prepared by heating the adduct obtained from
4-phenyloxazole and benzyne in a Diels-Alder reaction.[121]

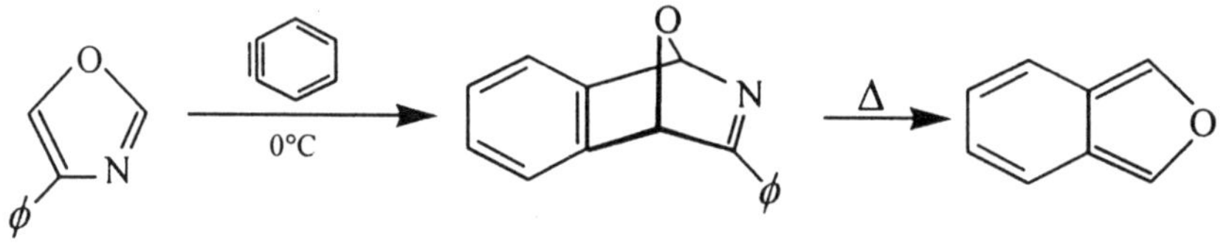

7.5.2.2 Chemical reactions

Isobenzoruran and its derivatives have been used as synthetically useful
intermediates. It takes part in the Diels-Alder reaction. [120, 121] 1, 3-
Diphenylisobenzofuran behaves as a typical and reactive diene and
consequently enters into Diels-Alder adduct formation with a number of
dienophiles.[123-127]

Nitrosobenzene also cycloadds to 1, 3-diphenylisobenzofuran to form the adduct. This adduct decomposes to a nitrone which on hydrolysis yields 1, 2-dibenzoylbenzene in quantitative yield.[122] 1, 3-Diphenylisobenzofuran can be easily oxidized and it gives *o*-dibenzoylbenzene which on reaction with Zn/CH_3COOH cyclizes to 1, 3-diphenylisobenzofuran.

7.5.3 Dibenzofuran

Dibenzofuran (72) is a fully aromatic compound and occurs in the anthracene oil fraction of coal tar. The ring system is numbered as shown. It can be

(72)

prepared[128] by the pyrolysis of diphenyl ether. Another method for its synthesis includes diazotization of *o*-aminodiphenyl ether and its subsequent cyclization in the presence of an acid.

Dibenzofuran is much more stable towards acids than benzofuran. The presence of extra benzene ring, compared to benzofuran, imparts protection to the furan ring. The electrophilic substitution on dibenzofuran takes place in an anamolous manner as the incoming electrophile takes a position which is determined by the nature of the reagent. Halogenation, for instance, occurs at the 2-position while nitration in acetic anhydride gives a mixture of 1-, 2-and 3-nitro derivatives. [129] Sulfonation with chlorosulfonic acid yields a single product, dibenzofuran 2-sulfonic acid in 80% yield. Reduction with a variety of reducing agents yields 1, 2, 3, 4-tetrahydrodibenzofuran.

Metallation with organolithium compounds takes place at the 4-position.

7.6 BENZOTHIOPHENE, ISOBENZOTHIOPHENE AND DIBENZOTHIOPHENE

The three title compounds *benzothiophene* (5), *isobenzothiophene* (8) and *dibenzothiophene* (73) are obtained by the condensation of thiophene with benzene nucleus.

(5)   (8)   (73)

These are aromatic and undergo electrophilic substitution reactions. Isobenzothiophene is stable only at a low temperature ($-30°C$).

7.6.1 Benzothiophene

Benzothiophene was first obtained in 1893. This compound and its derivatives occur in the naphthalene fraction of coal tar, crude petroleum and shark oil. It is a major sulfur containing impurity in technical naphthalene. It has also been

(5)

called as *thionaphthalene* or *thianaphthalene,* however, benzo [b] thiophene is the accepted name for this compound. The numbering of the ring atoms commences from the hetero atom and the other atoms are numbered anti-clockwise. Benzothiophene can be expressed as a resonance hybrid of the following canonical forms.

It possesses a dipole moment of 0.62 D which is similar to thiophene.

Benzothiophene is a colorless solid, m.p. 32°C and is stored in amber colored bottles.

7.6.1.1 Synthetic methods

Similar to many other heterocyclic compounds, the synthesis of benzothiophene also involves the building of the heterocyclic ring onto the benzene nucleus.

1.   *From o-Mercapto β-Chlorostyrene:* This compound on refluxing with an alcoholic solution of alkali cyclizes to benzothiophene (equation 7.33).

$$\text{(7.33)}$$

2.   *From Aryl Keto Sulfides:* These compounds can be cyclized to benzothiophene in the presence of polyphosphoric acid (PPA). Acetonylphenyl sulfide, for instance, produces 3-methylbenzothiophene equation (7.34). An extension of this method[130] is the cyclization of (arylthio) acetaldehyde dimethylacetal in the presence of PPA.

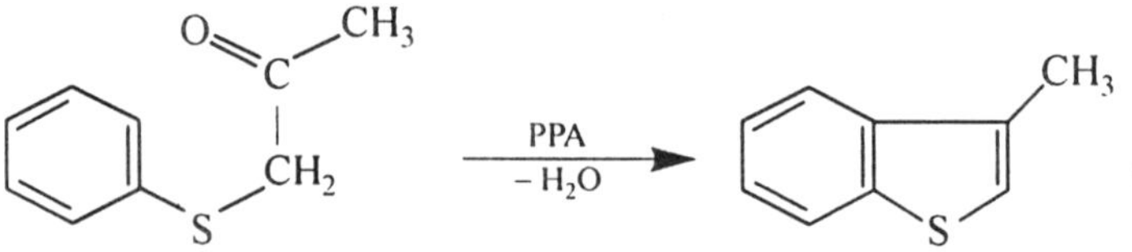

$$\text{(7.34)}$$

Benzothiophene has also been prepared by annelation (i.e., building one ring over the other) of disubstituted thiophene and an alkene. This interestingly builds the benzene ring[131] in contrast to the above methods.

3.   *From Diels-Alder Reaction:* The adduct (74) obtained from maleic anhydride and 2-vinylthiophene on decarboxylation and dehydrogenation yields benzothiophene.

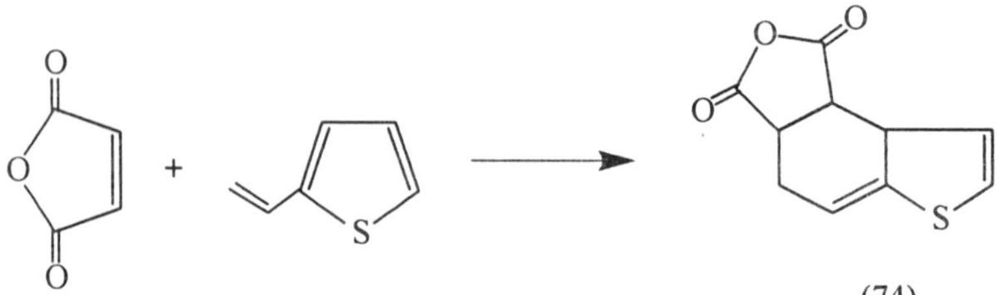

(74)

(i) Decarboxylatian
(ii) Dehydrogenation

4.   Benzo [b] thiophene can be synthesized by the direct pyrolysis of thiophene *via* the thiopyne intermediate. This intermediate undergoes (4 + 2) cycloaddition with another molecule of thiophene, followed by cheletropic elimination of sulfur to yield benzothiophene.

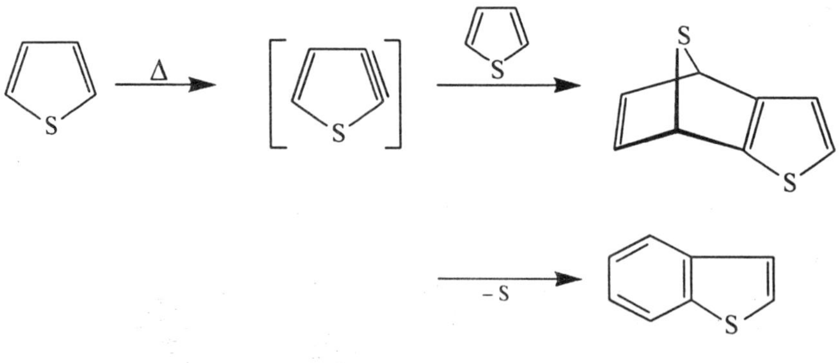

7.6.1.2 *Chemical reactions*

Benzothiophene readily polymerizes in the presence of acids.[132]

1. *Electrophilic Substitution:* Benzothiophene resembles naphthalene in its chemical properties and undergoes most electrophilic substitution reactions. In marked contrast to benzofuran and thiophene, the substitution predominantly takes place at the 3-position. This behavior is similar to indole. This difference may be ascribed to the weaker electronegativity of the sulfur atom relative to the oxygen atom. The electrophilic attack at the 3-position is further documented from the resonance structure in which the Kekule structure is present when C-3 has a negative charge. Bromination gives 3-bromobenzothiophene which also forms Grignard reagents. Iodination (I_2/HgI_2) also results in 3-substitution. Nitration (HNO_3/CH_3COOH) yields 3-nitrobenzothiophene contaminated with a small amount of the 2-isomer. Sulfonation of benzothiophene has been little studied. In the Friedel-Crafts reaction also the 3-isomer predominates[133] while the isopropylation surprisingly takes place at the 2-position.[134] Benzothiophene does not undergo formylation under the Vilsmeier conditions which occurs readily with thiophene.

2. *Metallation:* Direct metallation of benzothiophene with *n*-butyl lithium or sodium takes place at the 2-position and subsequent carbonation leads to benzothiophene 2-carboxylic acid.

3. *Reaction with Oxidizing Agents:* Benzothiophene on refluxing with 30% hydrogen peroxide in acetic acid yields benzothiophen-1, 1-dioxide or benzothiophene sulfone (75). This dioxide is much more stable than thiophene,

1, 1-dioxide and undergoes the self Diels-Alder reaction at 180°C. The adduct (76) loses SO_2 to form 9-thia-3, 4-benzofluoren-9, 9-dioxide (77). Thioindoxyl,

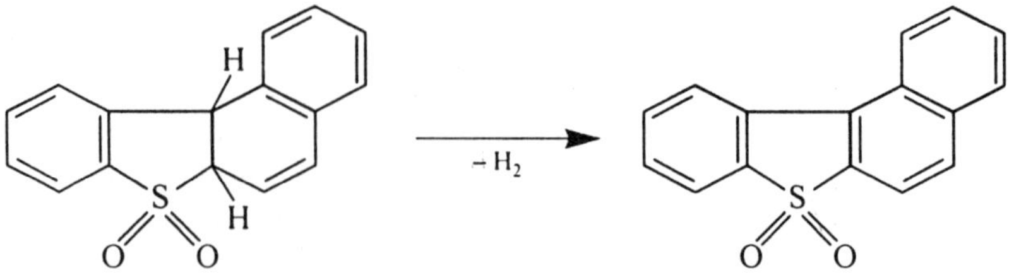

(77)

an important derivative of benzothiophene, on oxidation with alkaline potassium ferrocyanide yields thioindigo, the sulfur analog of indigo.

4.   *Desulfurization:* Both benzothiophene and dibenzothiophene are desulfurized by the action of nickel borohydride prepared from nickel chloride and sodium borohydride.[135] Thus benzothiophene yields 2-ethylbenzene while dibenzothiophene leads to biphenyl, equation (7.35) and (7.36).

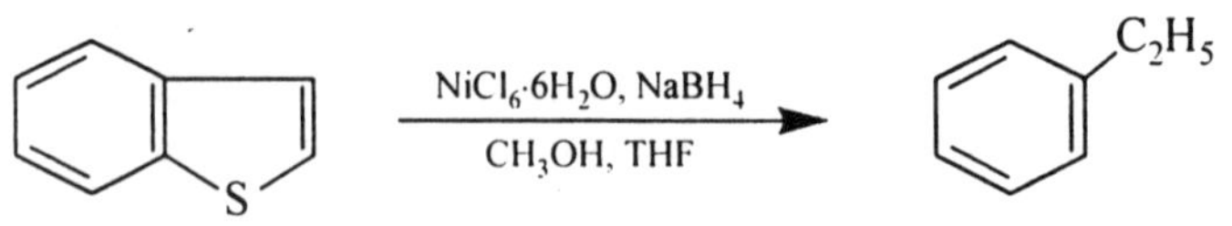

$$(7.35)$$

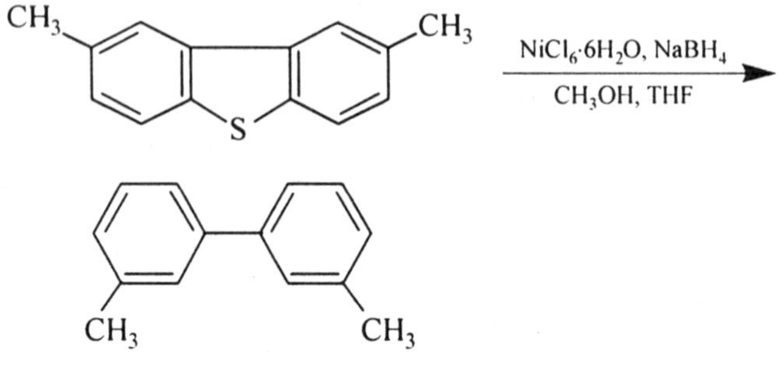

$$(7.36)$$

5.   *Photochemical Reactions:* The photochemistry of benzo [b] thiophene is different from that of thiophene. It does not undergo isomerization and is not converted to indole when irradiated in the presence of amines. It rather undergoes addition of amines at the 2, 3-position to give the addition product. With 3-propylamine, on irradiation, 3-propylamino-2, 3-dihydrobenzene [b] thiophene is obtained,[136, 137] equation (7.37). Both unrearranged (78) and

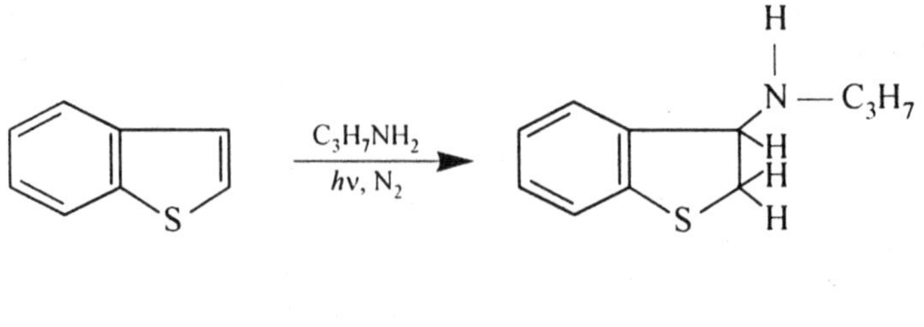

$$(7.37)$$

rearranged (79) adducts are obtained on irradiation of benzothiophene with dimethylacetylenedicarboxylate.

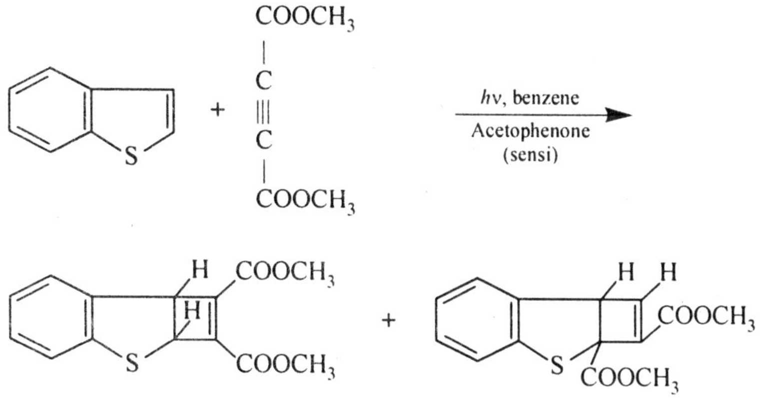

(78)   +   (79)

7.6.2 Isobenzothiophene

Isobenzothiophene (7) is an isomer of benzothiophene and is also designated as benzo [c] thiophene or *isothianaphthalene*. It is stable only at low temperature ($-30°C$) and has an odor resembling naphthalene.

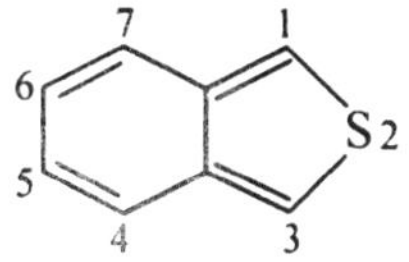

(7)

It is obtained[138] in 90% yield from 1, 3-dihydrobenzo [c] thiophene 2-oxide and aluminum oxide mixed in the ratio of 1:2 and heated to 100°C equation (7.38).

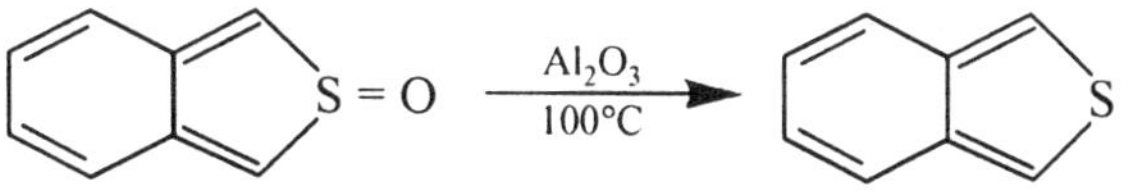

$$(7.38)$$

It is also often produced by dehydrogenation of hydrobenzo [c] thiophene, equation (7.39).

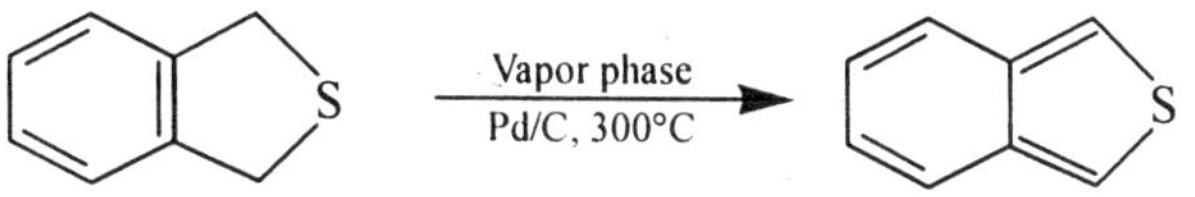

$$(7.39)$$

Isobenzothiophene undergoes the Diels-Alder reaction with maleic anhydride which on treatment with sodium hydroxide is desulfurized to naphthalene.[139] 1,3-Dihydrobenzo [c] thiophene-2, 2-dioxide can be alkylated using RX/KH–DME.[140]

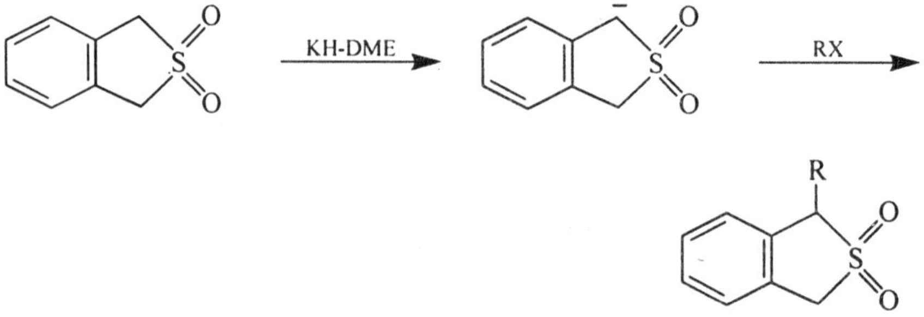

7.6.3 Dibenzothiophene

Dibenzothiophene (80) occurs in the phenanthrene fraction of coal tar. It is also obtained in small amounts during the pyrolysis[141] of spirotrithianes.

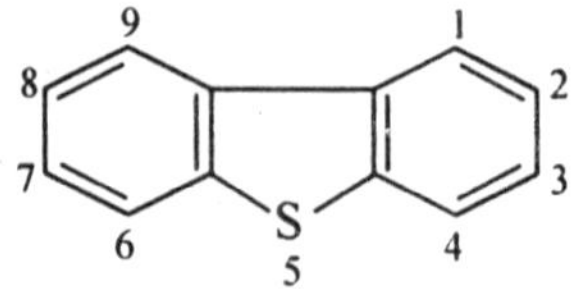

(80)

The best method for its preparation consists of the action of sulfur on biphenyl in the presence of aluminum chloride, equation (7.40).

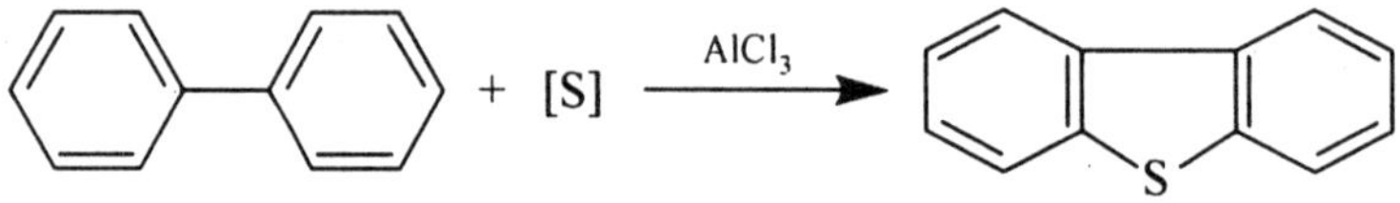

$$\text{biphenyl} + [S] \xrightarrow{\text{AlCl}_3} \text{dibenzothiophene} \qquad (7.40)$$

Electrophilic substitution in dibenzothiophene gives a mixture of 2-and 7-, and 8-disubstituted products. This behavior is evident from the different resonance structures of this molecule.

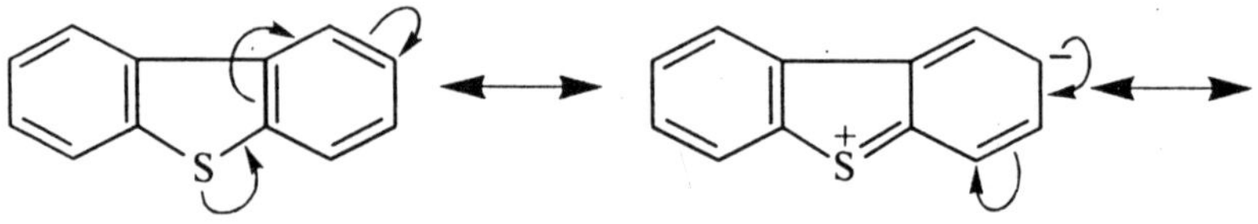

REFERENCES

1. R. J. Sundbrg, *The Chemistry of Indoles,* Academic Press, New York (1970); R. Nagarajan and P. T. Perumal, Tetrahedron, **58,** 1229 (2002).
2. B. Robinson, *Chem. Rev.,* **69,** 227 (1969); B. Robinson, *The Fischer Indole Synthesis,* John Wiley, New York (1982).
3. H. Fischer and F. Jourdan, *Ber.* **16,** 2241 (1983).
4. H. R. Snyder and C. W. Smith, *J. Am. Chem. Soc.,* **65,** 2452 (1943); K. C. Joshi *et al., J. Prakt. Chem.,* **320,** 701 (1978); F. Sannicolo, *Tet. letters,* 3101 (1984).
5. H. Posvic, R. Dombro, H. Ito and T. Tolinski, *J. Org, Chem.,* **39** 2575 (1974).
6. H. Illy and L. Funderburk, *J. Org. Chem.,* **33,** 4283 (1968).
7. G. S. Bajwa and R. K. Brown, *Canad. J. Chem.,* **48,** 2293 (1970).
8. C. F. H. Allen and C. V. Wilson, *J. Am. Chem. Soc.,* **65,** 611 (1943).
9. D. L. Hughes and D. Zhao, *J. Org. Chem.,* **58,** 228 (1993).
10. R. E. Lyle and L. Skarlos, *Chem, Comm.,* **644,** (1966).
11. R. B. Carlin, A. J. Magistro and G. J. Mains, *J. Am. Chem. Soc.,* **86,** 5300 (1964).
12. B. Miller and E. R. Matjeka, *J. Am. Chem. Soc.,* **102,** 4772 (1980).
13. R. R. Philips, *Org. Reactions,* **10,** 143 (1953).
14. R. A. Abramovitch and D. Shapiro, *J. Chem. Soc.,* **102,** 4772 (1980).
15. A. Galat and H. L. Friedman, *J. Am. Chem. Soc.,* **70,** 1280 (1948).
16. F. T. Tyson, *ibid.,* **72,** 2801 (1950).
17. R. L. Augustine, A. J. Gustavosen, S. F. Wanat, I. C. Pattison, K. S. Haughston and G. Koletar, *J. Org. Chem.,* **38,** 3004 (1973).
18. M. Le Corre, A. Hercouet and M. Le Baron, *Chem. Comm.* 14 (1981).
19. G. Bartoli, M. Bosco, R. Dalpozzo and P. E. Todesco, *Chem. Comm.,* 807 (1988).
20. S. A. Monti, *J. Org. Chem.,* **31,** 2669 (1966).
21. J. Bernier, J. Henichart, C. Vaccher and R. Houssin *J. Org. Chem.,* **45,** 1493 (1980).
22. G. R. Allen M. J. Wein, *J. Org. Chem.,* **33,** 198 (1968).
23. P. E. Verkade and J. Lieste, *Rec. Trav. Chim.,* **65,** 912 (1964).

24. P. L. Julian *et al., J. Am. Chem. Soc.*, **67,** 1203 (1945).

25. D. Villemin, B. Labiad, and Y. Ouhilal, *Chem. and Ind* (*London*), 607 (1989).

26. P. Bhattacharya and S. S. Jash, *Ind. J. Chem,* **26B,** 1177 (1987).

27. J. Bourdais and C. Germain, *Tet. letters,* 195, (1970).

28. T. Saegusa K. Kobayashi and Y. Ito., *J. Am. Chem. Soc.,* **100,** 5800 (1978).

29. G. Smolinsky and C. A. Pryde, *J. Org. Chem.,* **33,** 2411 (1968).

30. L. S. Hogedus, G. F. Allen, J. J. Bozell and E. L. Waterman *J. Am. Chem. Soc.,* **100,** 5800 (1978).

31. M. Mori and Y. Ban, *Tet. letters,* 1803 (1976).

32. V. Nair and K. H. Kim, *J. Org. Chem.,* **40,** 3784 (1975).

33. K. B. Wilberg and R. R. Squires, *J. Am. Chem. Soc.,* 101, 5512 (1979).

34. H. Suzuki S. V. Thiruvikarman and A. Osuka., *Synthesis,* 616 (1984).

35. T. Izumi and T. Yokota, *J. Heterocyclic Chem.,* **29,** 1085 (1992).

36. A. P. Kozikowski and D. Ma, *Tet. Letters,* 3317 (1991).

37a. M. Akazome, T. Kondo and Y. Watanabe, *Chem. Lett.,* 769 (1992), Also see, A. Kasahara, T. Izumi and L. X. Ping, *Chem. and Ind.* (London), 50 (1988).

37b. C. Rosenbaum, *Chem. Comm.,* 1822 (2003).

38. R. L. Hinman and C. P. Bauman, *J. Org. Chem.,* **29,** 2437 (1964).

39. R. L. Hinman and E. B. Whipple, *J. Am. Chem. Soc.,* **84,** 2534 (1962).

40. G. M. Rubottom and J. C. Chabala, *Synthesis,* 566 (1972).

41. E. V. Dehmlow, *Angew. Chem. Int. Edin.,* (Engl.), **13,** 170 (1974).

42. J. C. Powers in, "*Indoles*", W. J. Joulihan, (*Ed.*), Part II, Wiley Interscience. New York (1972), pp. 137-139; J. C. Powers, *J. Org. Chem.,* **31,** 2627 (1966).

43. M. Derosa and A. Triana, *J. org. Chem.,* **43,** 3639 (1978).

44. J. Bergmann, *Acta. Chem. Scand.,* **25,** 2865 (1971).

45. R. J. Bass, Tetrahedron, **27,** 3263 (1971); J. M. Muchowski, *Canad. J. Chem.,* **48,** 422 (1970).

46. G. M. Brooke, R. D. Chambers, W. K. R. Musgrare, R. A. Storey and J. Yeadon, *J. Chem. Soc. Perkin Trans.,* **1,** 162 (1976).

47. R. J. Owellen, *J. Org. Chem,* **39,** 69 (1974).

48. G. Berti A. Dasettimo and E. Nannipieri *J. Chem. Soc.,* 2145 (1968).

49. V. O. Illi, *Synthesis,* 378 (1979).

50. A. Shafife and S. Sattari, *Synthesis,* 389 (1981).

51. N. N. Suvorov and N. P. Sorokina, *J. Gen. Chem.,* U.S.S.R., **30,** 2036 (1960).

52. P. N. James and H. R. Synder, *Org. Synth. Coll. Vol.,* **4,** 539 (1963).

53. H. E. Dobbs, *Tetrahedron,* **24,** 491 (1968); C. W. Rees and C. E. Smithen, *Chem. and Ind.* (London), 1033 (1962); *J. Chem. Soc.,* 298 (1964).

54. L. J. Dolby and R. M. Rodia, *J. Org. Chem.*, **39**, 1493 (1970); *J. Heterocyclic Chem.*, **3**, 124 (1960).

55. F. Y. Chen and E. Leete, *Tet. letters*, 2013 (1963).

56. M. Janda *et al.*, *Coll. Czech. Chem. Comm.*, **46**, 3278 (1981).

57. A. E. Lanzillotti, R. Littell, W. J. Fanshawe, T. C. Mekenzie and F. M. Lovell, *J. Org. Chem.*, **44**, 4809 (1979).

58. J. G. Beyer, *Synthesis*, 508 (1974).

59. J. I. G. Cadogan and M. Cameron-Wood, *Pro. Chem. Soc.*, 361 (1962); also see R. K. Bansal, *J. Ind. Chem. Educ.*, **6**, 18 (1979).

60. G. W. Gribble and J. H. Hoffman, *Synthesis*, 859 (1977).

61. W. A. Remers, G. J. Gibs, C. Pidacks and M. J. Weiss *J. Org. Chem.*, **36**, 279 (1971); *J. Am. Chem. Soc.*, **89**, 5513 (1967).

62. B. E. Maryanoff, D. F. McComsey and S. O. Nortey *J. Org. Chem.*, **46**, 355 (1981).

63. D. A. Shivley and P. A. Roussel, *J. Am. Chem. Soc.*, **75**, 375 (1953).

64. H. Bader and W. Oroshnik, *ibid.*, **81**, 163 (1959).

65. M. Someh and M. Natsume, *Tet. letters*, 2451 (1973).

66. C. Thal *et al.*, *C. R. Acad. Sci. Paris*, **274**, 532 (1972).

67. G. Gauzzo and G. Jori, *J. Org. Chem.*, **37**, 1429 (1972).

68. G. Jori *et al.*, *Photochem. Photobiol*, **9**, 179 (1969).

69. O. DeSilva and V. Snieckus, *Synthesis*, 254 (1971).

70. J. Bergman, *Tetrahedron*, **26**, 3353 (1970).

71. A. Giesler, E. Steckhan, O. Wiest and F. Knock, *J. Org. Chem.*, **56**, 1405 (1991), also see, P. M. Jackson and C. J. Moody, *Tetrahedron*, **48**, 7447 (1992).

72. D. L. Oldroyd, N. C. Payne, J. J. Vittal, A. C. Weedon and B. Zhang, *Tet. letters* 1087 (1993).

73. D. R. Julian and R. Foster, *Chem. Comm.*, 311 (1973).

74. R. Hunt, S. T. Reid and K. T. Taylon *Tet. letters*, 2861 (1972); J. S. Cridland and S. T. Reid, *Chem. Comm.*, 125 (1969).

75. M. Ikeda, *J. Chem. Soc. Perkin Trans.*, **1**, 405 (1984).

76. S. Blechert, *Tet. letters*, 1547 (1984); A. Rykonski and M. Makosza, *Tetrahedron* 4795, (1984); W. Wierenga, J. Griffin and M. A. Warpehoski, *Tet. Letters*, 2437, (1983), *Tet. letters*, 2437 (1983); T. Shono, *ibid*, 1259 (1983); H. Maehr and J. M. smallhea, *J. Org. Chem.*, **46**, 1752 (1981); H. Suzuki *et al.*, *Synthesis*, 616 (1984).

77. C. E. Castro, E. J. Gaughan and D. C. Owseley. *J. Org. Chem.*, **31**, 4071 (1966).

78. B. C. Challis and H. S. Rzepa, *J. Chem. Soc., Perkin I*, 1209 (1975);

79. K. S. Pusztay and L. Sazabo, *Synthesis*, 276 (1975).

80. A. P. Kozikowski and M. P. Kuniak, *J. Org. Chem.*, **43**, 2083 (1978).

81. A. G. Schultz and W. R. Hagmann, *J. Org. Chem.*, **43**, 4239 (1978); also see R. S. J. Beer *et al.*, *J. Chem. Soc.*, 4139 (1954).

82. G. Reissenweber and D. Magnold, *Angew. Chem. Int. Edin.*, (Engl.), **20**, 882 (1981).

83. D. St. C. Black and L. C. H. Wong, *Chem. Comm.*, 200 (1980).

84. W. Reid and P. W. Weidmann, *Ber.*, **104**, 3741 (1971).

85. A. Chatterjee, G. K. Biswas and A. B. Kundu, *J. Ind. Chem., SOC.*, **46**, 429 (1969).

86. S. Torri, T. Yamaraka and H. Tanaka, *J. Org. Chem.*, **43**, 2882 (1978).

87. M. Dober, R. Beerli, W. K. Weissmahr and H. J. Borschberg, *Tet. Asymmetry*, **3**, 1411 (1992).

88. J. P. Kutney in, *"The Synthesis of Natural Products:,* Wiley Interscience, New York (1977), Vol. 3, pp. 273-438.

89. For a review see, F. S. Bahicherch *et al.*, *Russ. Chem. Rev.*, **50**, 1087 (1981).

90. J. D. White and M. E. Mann, *Adv. Heterocyclic Chem.*, **10**, 113 (1969).

91. E. Chacko J. Bornstein and D. J. Sardella, *Tet. letters*, 1095 (1977).

92. R. Bonnett and R. F. C. Brown., *Chem. Comm.*, 393 (1972); also see G. W. Priestley and R. W. Warrener, *Tet. letters*, 4295 (1972).

93. B. Jacques and R. G. Wallace, *Tetrahedron*, **33**, 581 (1977); also see, B. Jacques *et al., ibid.*, **33**, 2255 (1977).

94. R. Kreher and J. Senbert, Angew. *Chem. Int. Edin.*, (Engl.), **5**, 967 (1966).

95. W. Thielacker *et al., Angew. Chem.*, **673**, 96 (1964).

96. R. Krehr and H. Hennige, *ibid.*, 1911 (1973).

97. R. Krehr and H. Hennige, *ibid.*, 1911 (1973).

98. For a review see, T. Ucida, *Synthesis*, 209 (1976).

99. R. Breslow, *J. Am. Chem. Soc.*, **87**, 1320 (1965); Also see, E. Polijala, *Heterocyclic Chem.*, **15**, 955 (1978); D. H. Wadsworth S. L. Bender, D. L. Smith, H. R. Luss and C. H. Weidner *J. Org. Chem.*, **51**, 4639 (1986).

100. J. Mahon, L. K. Mehta, R. W. Middleton, J. Parrick and H. K. Rami, *J. Chem. Research*, 362 (1992).

101. V. Boekelhiede *et al., J. Am. Chem. Soc.*, **46**, 1951 (1963).

102. For a review see, J. A. Joule, *Adv. Heterocyclic Chem.*, **35**, 83 (1984).

103. J. M. Davidson and C. Triggs, *J. Chem. Soc.* (a), 1324 (1968).

104. H. Lakaki and H. Yoshimoto, *J. Org. Chem.*, **38**, 76 (1973); also see R. O. C. Norman, *J. Chem. Soc. Perkin Trans*, **1**, 1289 (1974).

105. G. Olah *et al., J. Org. Chem.*, **56**, 6248 (1991).

106. B. Akermark, L. Eberson, E. Jonsson and F. Petterson *J. Org. Chem.*, **40**, 1365 (1975).

107. J. Bernier, J. Herichart, C. Vaccher and R. Houssin, *J. Org. Chem.*, **45**, 1493 (1980).

108. L. S. Hegedus, G. F. Allen and E. L. Waterman, *J. Am. Chem. Soc.*, **98**, 2674 (1976).

109. J. I. G. Cadogan and M. Cameron-Wood, *Proc. Chem. Soc.,* **36,** (1962); also see R. K. Bansal, *J. Ind. Chem. Educ.,* **6,** 18 (1979).

110. J. B. Kyziol and A. Lyzniak, *Tetrahedron,* **36,** 3017 (1980).

111. A. Mustafa, *"Benzofurans"*, Wiley Interscience, New York (1974).

112. P. D. Seemuth and H. Zimmer, *J. Org. Chem.,* **43,** 3063 (1978).

113. P. Spagnolo, M. Tiecco, Tundo and g. Martelli, *J. Chem. Soc.,* 556 (1972).

114. M. Gill, *Tetrahedron,* **40,** 621 (1984); B. D. Dean and W. E. Truce, *J. Org. Chem.,* **46,** 3575 (1981); M. A. Makhlouf and B. Rickborn, *ibid.,* **46,** 2374 (1981).

115. N. G. Kundu, M. Pal, J. S. Mohanty and S. I. C. Dasgupta, *Chem. Comm.,* 41 (1992).

116. W. K. Anderson, *J. Chem. Soc. Perkin Trans. I,* **1,** (1976).

117. W. E. Parham, C. G. Fritz, R. W. Soreder and R. M. Dodson, *J. Org. Chem.,* **28,** 577 (1963).

118. D. W. Jones, *Chem. Comm.,* 113 (1971).

119. W. Adams, J. Bialas, L. Hadjiarapoglou, and M. Santer, *Chem. Ber.,* **125,** 231 (1992).

120. R. N. Warrener, *J. Am. Chem. Soc.,* **93,** 2346 (1971).

121. S. E. Whitney and B. Rickborn, *J. Org. Chem.,* **53,** 5595 (1988).

122. C. T. Lin, L. W. Din and S. L. Wang, *Heterocycles,* **29,** 263 (1989).

123. W. Friedrichsen, *Adv. Heterocyclic Chem.,* **26,** 235 (1980); J. G. Smith and N. G. Chu, *J. Org. Che.,* **46,** 4083 (1981) K. Saito *et al., Tet. letters,* 2573 (1984).

124. M. P. Cava and F. M. Scheel, *J. Am. Chem. Soc.,* **32,** 1304 (1967); also see T. Sasaki *et al., J. Am. Chem. Soc.,* **97,** 355 (1977).

125. G. Wittig and P. Fritze, *Angew, Chem. Int. Edin.,* (Engl.) **5,** 846 (1966).

126. A. Hassner and D. J. Anderson, *J. Org. Chem.,* **39,** 2031 (1974).

127. V. Nair, *ibid.,* **37,** 2508 (1972).

128. J. Natks, S. L. Crump and B. Rickborn, *J. Org. Chem.,* **51,** 1189 (1986).

129. T. Keumi *et al., J. Org. Chem.,* **56,** 4671 (1991).

130. B. D. Tilak and K. Rabindram, *Current Sci.,* **20,** 1205 (1951); B. D. Tilak, *Proc. Ind. Acad. Sci.,* **32A,** 390 (1950).

131. N. B. Chapman, K. Clarke and S. N. Sawhney, *J. Chem. Soc. (C),* 518 (1968).

132. J. W. Terpstra and A. M. CanLeusen, *J. Org. Chem.,* **51,** 230 (1986).

133. H. B. Hartough and S. L. Meisel in *"Compounds and condensed Thiophene Rings"* A. Weissberger, *Ed.,* Wiley, New York (1954); p. 29.

134. A. C. Cope and W. D. Burrous, *J. Org. Chem.,* **31,** 3093 (1966).

135. T. G. Back, K. Yang and H. R. Krouse, *J. Org. Chem.,* **57,** 1986 (1992).

136. S. F. Bedell, E. Spaeth and J. M. Bobbitt, *ibid.,* **27,** 2026 (1962).

137. C. Parkanyi *et al., Tetrahedron,* **29,** 651 (1973).

138. S. R. Ditto, P. Davis and D. C. Neckers, *Tet, letters,* 521 (1981).

139. M. P. Cava and N. M. Pollack, *J. Am. Chem. Soc.,* **88,** 4112 (1966); also see, R. Mayer, *Angew. Chem. Int. Edin.,* (Engl.) **1,** 115 (1962).

140. B. D. Tilak H. S. Desai and S. S. Gupte., *Tet. letters,* 1956 (1966).

141. K. C. Nicolaou, W. E. Barnette and P. Ma, *J. Org. Chem.,* **45,** 1463 (1980).

142. P. S. Fraser L. V. Robbins and W. S. Chilton, *J. Org. Chem.,* **39,** 2509 (1974).

REVIEW PROBLEMS

7. 1. Predict the major product of the following reactions:

(a)

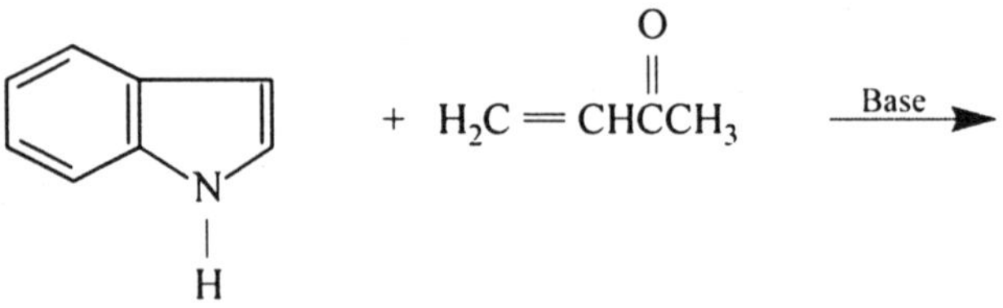

$$H_2C = CHCCH_3 \quad \xrightarrow{\text{Base}}$$

(b) $\xrightarrow{\text{HNO}_3,\ \text{Ac}_2\text{O}}$

(c) $+ \ H_2NNH_2 \longrightarrow$

(d) COOH $\xrightarrow[\Delta]{\text{PPA}}$

(e) $\xrightarrow{\text{Base}}$

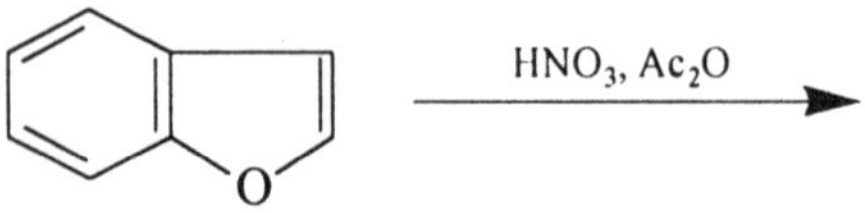

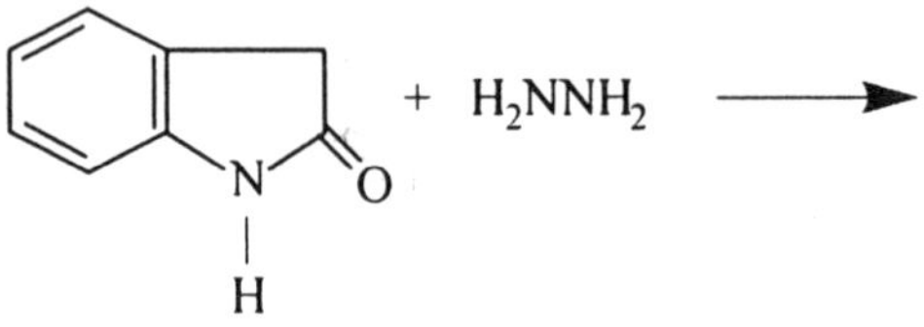

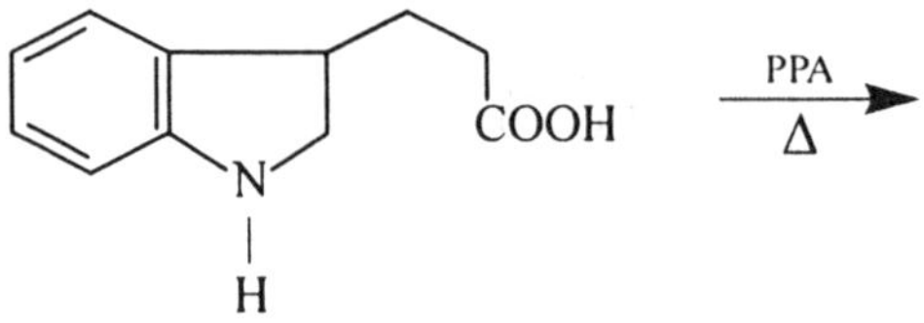

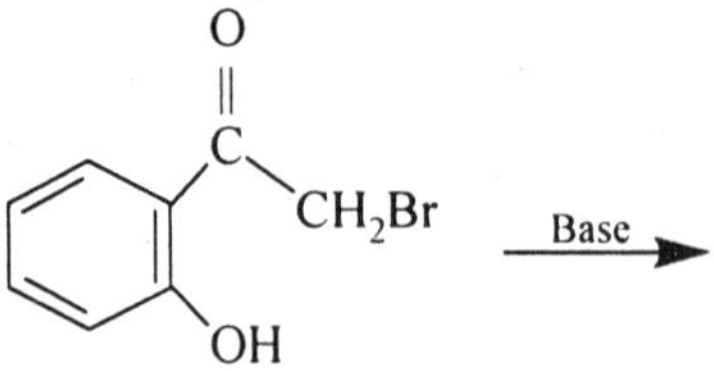

(f) $(CF_3CO)_2O \longrightarrow$

(g) $+$ OHC $\xrightarrow{\text{Base}}$

(h) R $\xrightarrow{H_2O_2,\ CH_3OH}$

(i) $\xrightarrow{\text{Base}}$

(j) $CH_2 = C - CH(CH_3)_2$ $\xrightarrow{ZnCl_2}$

(k) H_5C_6 $C=O$ / $C=O$ H_5C_6 $+$ NHCH$_3$ $\xrightarrow{\text{HCl, reflux}}$

(l) CH$_3$ $\xrightarrow{Br_2}$

(*m*) 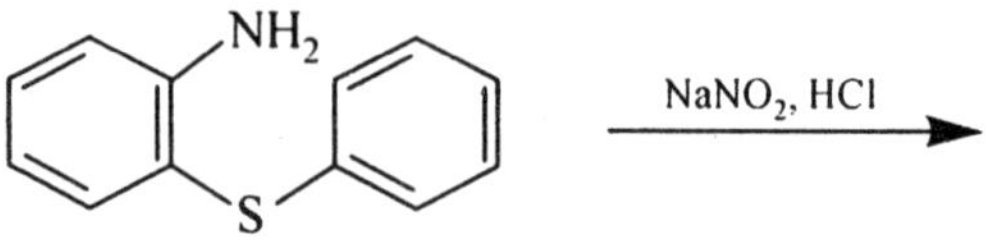
NH2
S
NaNO2, HCl

(*n*) 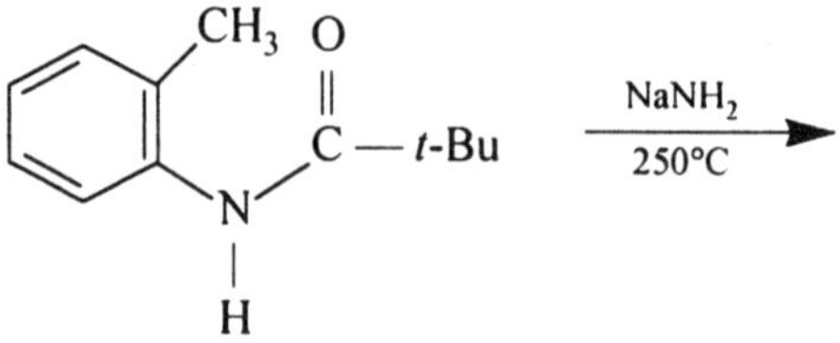
CH3 O
C — t-Bu
N
H
NaNH2
250°C

(*o*) 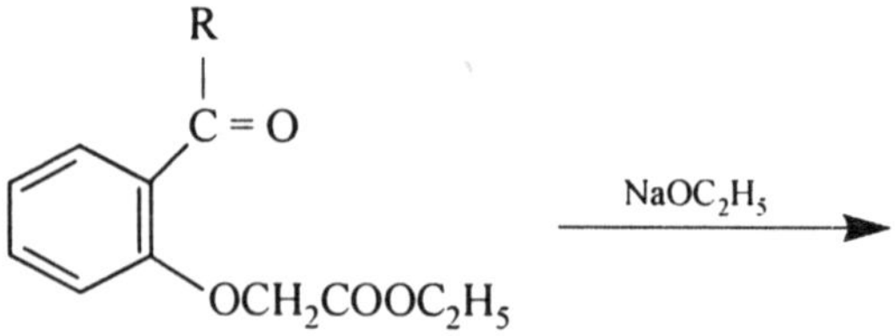
R
C = O
OCH2COOC2H5
NaOC2H5

(*p*) 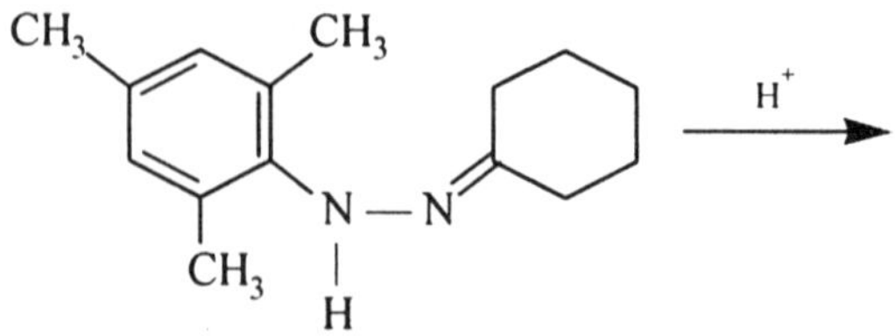
CH3
CH3
CH3
N — N
H
H+

(*q*) 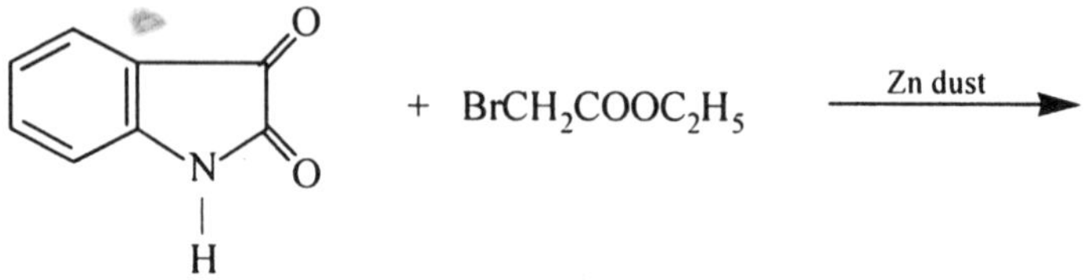
O
O
N
H
+ BrCH2COOC2H5
Zn dust

(*r*) 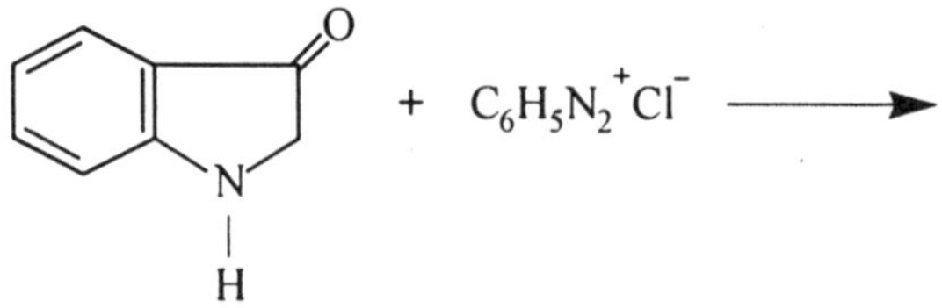
O
N
H
+ C6H5N2+Cl−

(*s*) 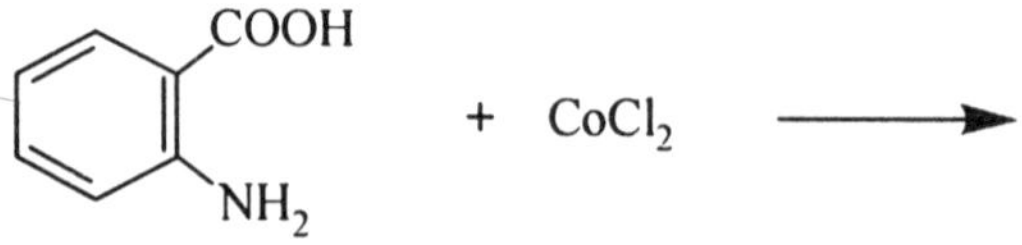
COOH
NH2
+ CoCl2

(*t*) [structure: 1-methylindoline] $\xrightarrow{\text{Br}_2 \atop \text{CCl}_4,\ \text{dark}}$

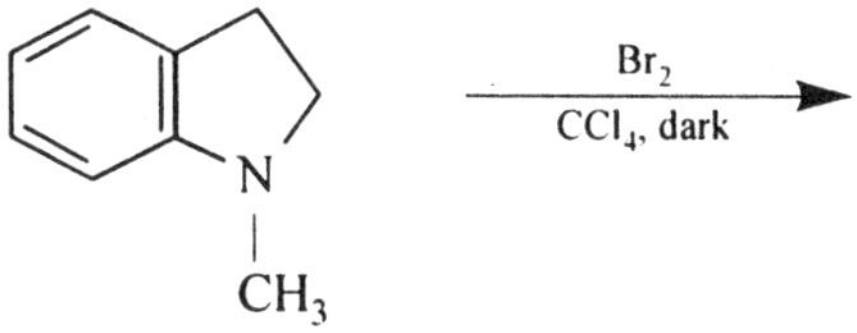

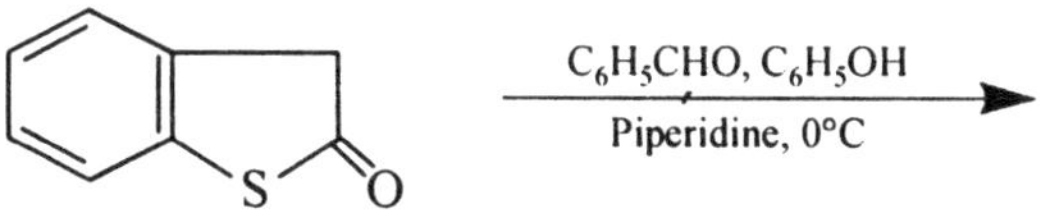

(*u*) [structure: benzothiophen-2(3H)-one] $\xrightarrow{\text{C}_6\text{H}_5\text{CHO, C}_6\text{H}_5\text{OH} \atop \text{Piperidine, 0°C}}$

(*v*) [structure: 1-(2-bromobenzoyl)oxindole] $\xrightarrow{\text{Pd(Ph}_3\text{P)}_4 \atop \text{KOAC, DMA} \atop 130°\text{C}}$

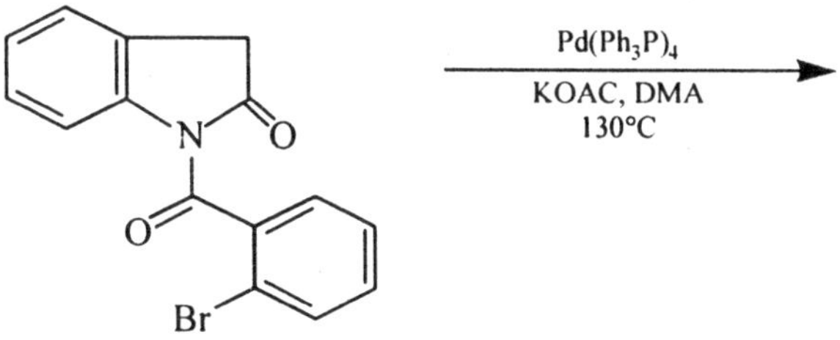

7. 2. Postulate a reasonable mechanism for each of the following reactions:

(*a*) [structure: 1,3-dimethylindole] $\xrightarrow{(\text{CH}_3)_2\text{C}=\text{CHCOCH}_3 \atop \text{aq. C}_2\text{H}_5\text{OH, HCl, } r.\ t.}$ [product structure]

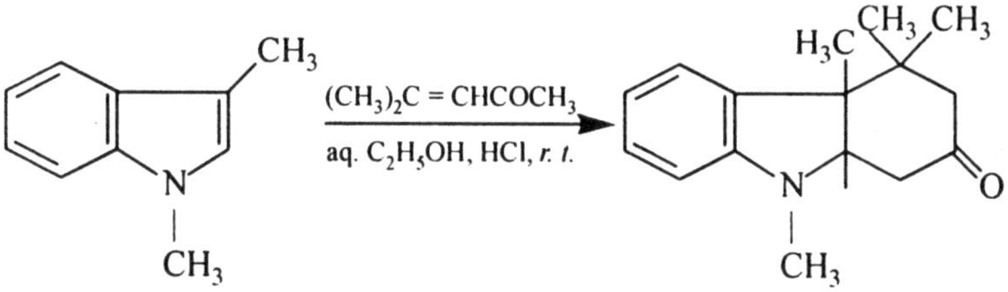

(*b*) [structure: 2,3-dimethylbenzothiophene] $\xrightarrow{\text{HNO}_3,\ (\text{CH}_3\text{CO})_2\text{O} \atop 0°\text{C}}$ [product structure with CH$_3$ and CH$_2$NO$_2$]

(*c*) [structure: 4-ethylcoumarin] $\xrightarrow{\text{(i)Br}_2,\ \text{CHCl}_3 \atop \text{(ii) Base} \atop \text{(iii) Decarboxylation}}$ [3-ethylbenzofuran structure]

(*d*) [structure: 2,3-dimethyl-2H-azirine] + [benzyne] $\longrightarrow$ [1,2,3-trimethylindole... 2,3-dimethylindole with CH$_3$ groups]

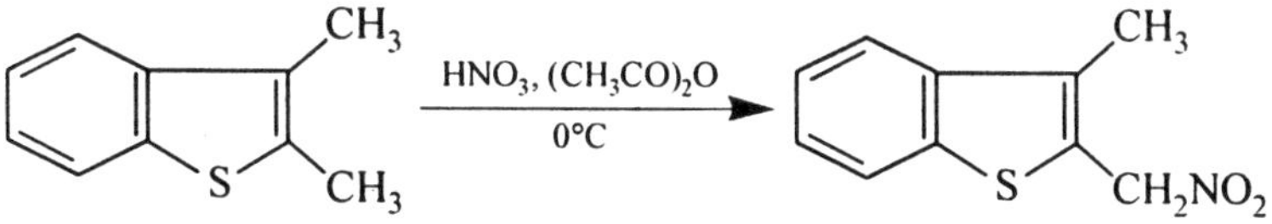

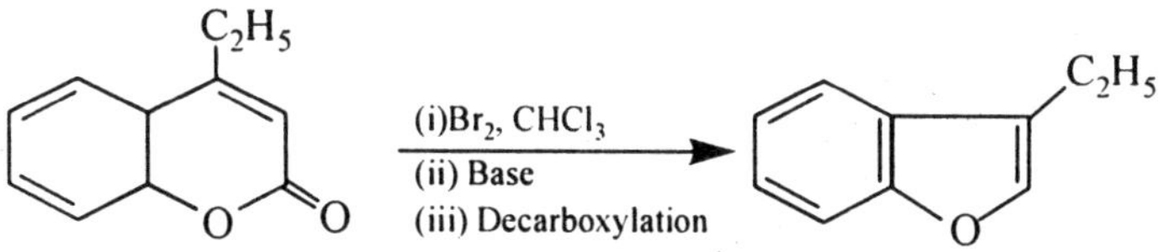

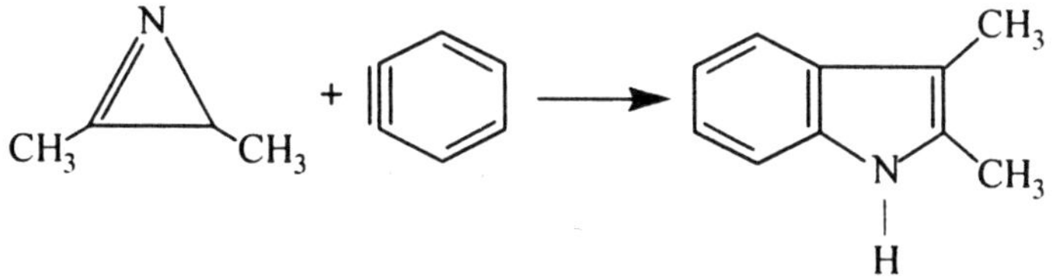

(e) [indole with N–CO–CH$_3$ substituent] →[Alc. KOH, r. t.] [indole, N–H]

(f) [2,3-dimethylindole] →[O$_2$, t-BuO$^-$K$^+$] [2-acetyl-N-(acetyl)aminobenzene product]

(g) [2-chlorophenyl hydrazone of acetone-type, N–N=C(CH$_3$)CH$_3$] →[ZnCl$_2$, Δ] [5,7-dichloro-2-methylindole]

(h) [2-methylphenyl N=CHN(CH$_3$)$_2$] →[(i) C$_6$H$_5$N$^-$Na$^+$ (CH$_3$); (ii) H$^+$] [indole, N–H]

7. 3. Discuss and compare the electrophilic substitution reactions of indole, benzofuran and benzothiophene.

7. 4. Offer explanation for the following observations:

    (*a*) Isoindole is more reactive than indole itself.

    (*b*) Oxindole readily undergoes aldol and Claisen condensation.

    (*c*) The electrophilic attack in pyrrole takes place at the 2-position but at position-3 in indole.

    (*d*) Benzofuran is more stable than furan towards acids.

Chapter 8

Bicyclic Ring Systems Derived from Pyridine

In the previous chapter we have seen that five-membered heterocyclic rings can be fused with benzene to give rise to new heterocyclic compounds which display very interesting physical and chemical properties. The pyridine ring in the same fashion may be fused with the benzene nucleus in different ways with the resultant formation of *quinoline* (1), *isoquinoline* (2) and *quinolizinium salts* (3). Only the first two have received considerable attention. These heterocycles have been obtained from coal tar. Their derivatives occur naturally,

(1)   (2)   (3)

particularly as constituent of alkaloids. They possess 10 π-electrons and behave as typical aromatic compounds. They undergo electrophilic substitution more readily than pyridine but less so than naphthalene. They show similarity in chemical behavior to naphthalene but there are marked differences as well. Both quinoline and isoquinoline are weakly basic and similar to pyridine the electron pair on the nitrogen atom is not involved in conjugation with the ring. As a result, like pyridine both of these compounds also undergo nucleophilic substitution readily on the heterocyclic ring.

The quinolizinium cation (3) occurs naturally in the form of alkaloids. Bicyclic six-membered aromatic compounds with hetero atom in both the rings are called naphthyridines and all the six of them are known. These compounds

1, 5-Naphthyridine   1, 6-Naphthyridine

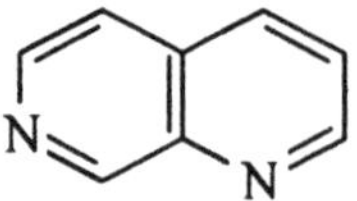

1, 7-Naphthyridine

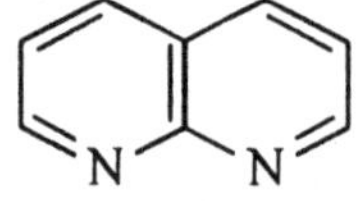

1, 8-Naphthyridine

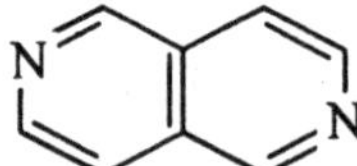

2, 6-Naphthyridine

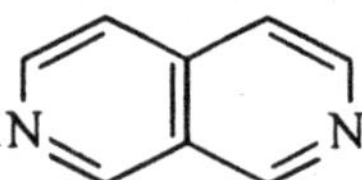

2, 7-Naphthyridine

are resistant to attack by electrophilic reagents, undergo quaternization and on oxidation with peracids yield N-oxides. They have been employed in the synthesis of products of chemotherapeutic and pharmacological importance.

8.1 QUINOLINES

Quinoline ring structure is obtained by *ortho*-condensation of benzene ring with pyridine. It is also called *l-azanaphthalene* or *benzo* [b] pyridine. It was first isolated by Runge in 1834, from coal tar bases and subsequently Gerhardt in 1842 obtained it from the alkaline pyrolysis of cinchonine, an alkaloid related to quinoline. The numbering in quinoline commences from the nitrogen atom which is assigned position-1. The dehydrogenation of quinoline with different types of reducing agents yields 1, 2-dihydroquinoline (4), 1,2,3,4-tetrahydroquinoline (5) and decahydroquinoline (6).

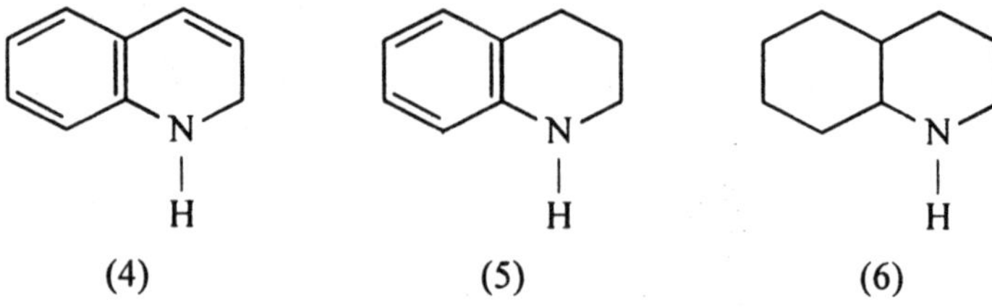

(4)   (5)   (6)

8.1.1 Physical and Spectroscopic Properties

Quinoline is a colorless hygroscopic liquid, b.p. 237°C and has a characteristic smell resembling that of pyridine. On exposure to air it develops a yellow color. Quinoline is miscible with organic solvents and is soluble in water to the extent of 0.7%.

Quinoline is highly aromatic and has a resonance energy of 47.3 kcal/mole and is considered a resonance hybrid of the following contributing structures:

(1a) (1b) (1c)

(1d) (1e)

As in the case of naphthalene, structures (1a), (1b) and (1c) are of low energy, however, additional charged structures are also possible because of the introduction of an electronegative nitrogen atom. The dipole moment of quinoline is 2.10 D which indicates charge separation in the ring.

Quinoline is weakly basic (*p*Ka 4.94) a value which is intermediate between aniline (*p*Ka 4.58) and pyridine (*p*Ka 5.17). Presence of electron-donating groups at the 2-and 4-positions of quinoline increase the basicity (2-methylquinoline pKa 5. 83). On the other hand, occurrence of hydrogen-bonding of group with the ring nitrogen atom decreases the basicity. For instance, 8-aminoquinoline (*p*Ka 3.93) is a much weaker base than 7-aminoquinoline (*p*Ka 6.5). The pyridine ring in quinoline is π-electron deficient, therefore, nucleophilic attack takes place at the 2-and 4-position. The π-electron densities have been calculated for quinoline by the molecular orbital method and show electron deficiency at these two positions. The electrophilic attack preferably takes place at the 5- and 8-positions. The close similarity of the absorption bands in the *u.v.* spectrum of quinoline to those of naphthalene further confirms similarities in their structures. The reduced double bond character of 2, 3- and 6, 7-bonds in quinoline is confirmed by *n.m.r.*.
Both quinoline and isoquinoline molecular ions lose HCN to yield radical cation (7) while loss of HCN from acridine gives structure (8).

(7) (8)

Methylquinoline radical ions (9) and (10), on the other hand, may involve azabenzotropylium (11) or pyridotropylium cations (12) during their fragmentation.

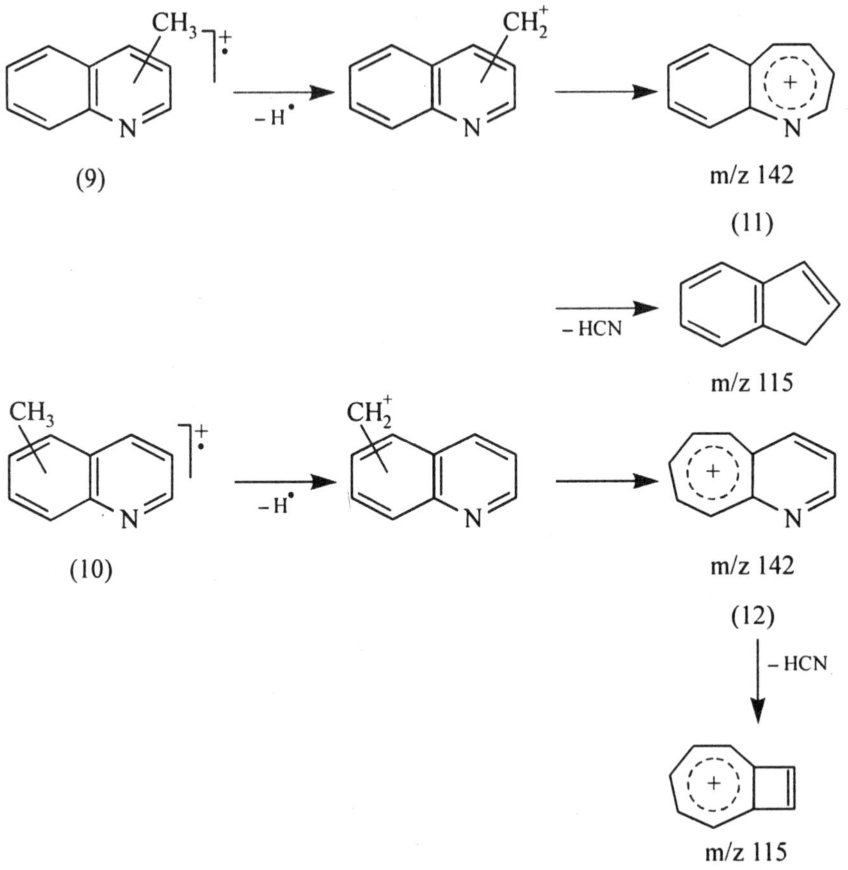

(9)   m/z 142

(11)

m/z 115

(10)   m/z 142

(12)

m/z 115

8.1.2 Synthetic Methods

Several well known methods have been used for the preparation of quinoline and its derivatives.[2,3]

1. The Skraup Synthesis: This is probably the most important preparation of quinoline. In this method[4] an aniline derivative having a vacant *ortho* position is heated with glycerol, conc. sulfuric acid and an oxidizing agent. In the simple case when the amine is aniline and oxidizing agent is nitrobenzene, the overall course of the reaction of refluxing in sulfuric acid is shown below:

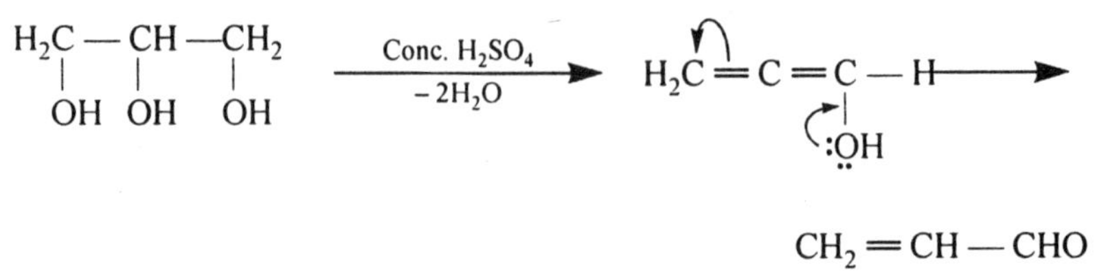

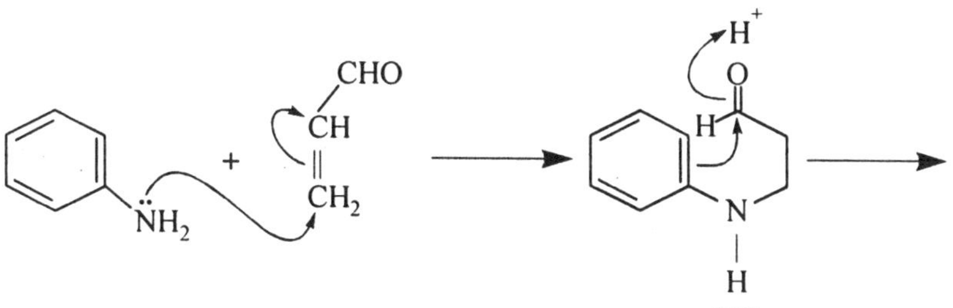

(13)

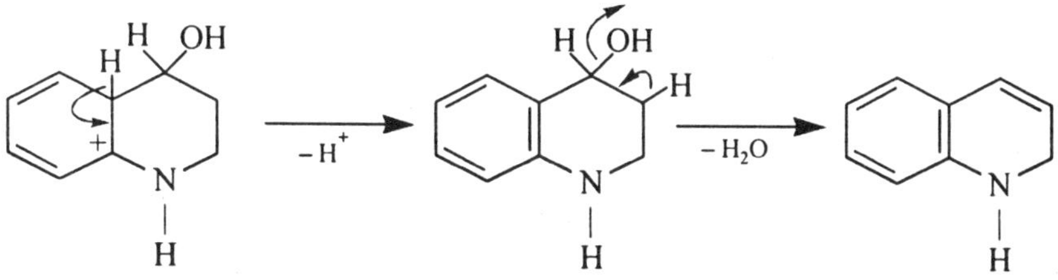

(14)

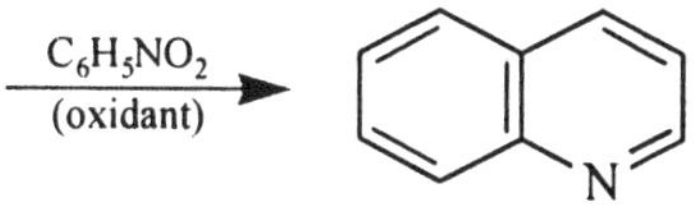

The reaction has been shown to proceed by dehydration of glycerol to α, β-unsaturated aldehyde, acrolein (produced *in situ*). Aniline adds to acrolein *via* the Michael addition across the vinyl group to give aniline propanal (13). This is followed by the electrophilic attack of protonated (13) on the aromatic ring resulting in ring closure. Dehydration leads to 1, 2-dihydroquinline (14) which on oxidation or dehydrogenation results in quinoline. The oxidant is included in the reaction mixture.

A nitro derivative of the corresponding aromatic amine is a convenient oxidizing agent, however, other oxidizing agents like iodine, ferric chloride have also been suggested. Substituted quinolines may be obtained by using substituted α, β- unsaturated carbonyl compounds[5], equation (8.1 and 8.2).

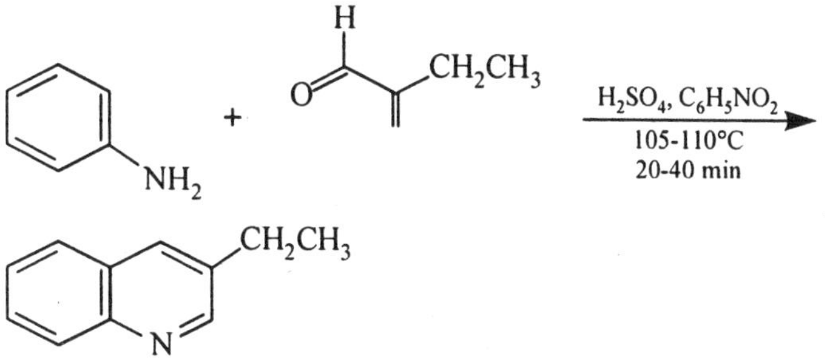

(8.1)

$$(8.2)$$

Using glycerol has the advantage of supplying a low concentration of acrolein which minimizes its polymerization.

Substituted anilines give quinoline derivatives, substituent appearing in the benzene ring. Anilines substituted in one *ortho*-position give 8-substituted quinolines. The symmetrical *para*-substituted anilines give 6-substituted derivative while *meta*-substituted anilines give rise to a mixture of 5- and 7- substituted quinolines. *m*-Chloroaniline, for instance, yields both 5-chloro and 7- chloroquinolines, equation (8.3).

$$(8.3)$$

48%   53%

2. **The Doebner-Miller Synthesis:** It is closely related to the Skraup synthesis as it also utilizes an aromatic amine and proceeds through the intermediate formation of a dihydroquinoline. The reagent which brings about dehydrogenation differs from the Skraup synthesis. In this synthesis an aromatic amine, aniline and two molecules of an aldehyde (acetaldehyde) are

heated in the presence of hydrochloric acid[6] to form schiff's base. Two molecules of this base self condense[7] to form quinoline nucleous.

Though crotonaldehyde does not participate in the reaction (as was earlier considered[8]) it is possibly formed during the reaction. A part of aniline also combines with crotonaldehyde and the resulting Schiff's base functions as the oxidizing agent and converts dihydroquinoline to quinoline. In general, the aldehyde R_1CH_2CHO will give quinoline (15).

(15)

If an aldehyde is replaced by a ketone, for instance, acetone, then 2,2, 4-trimethyl-1, 2-dihydroquinoline is initially obtained which eliminates methane to form 2, 4-dimethylquinoline.

$$C_6H_5N = C(CH_3)_2$$

3.   *The Friedlander Synthesis:* This is another generally useful method for obtaining quinoline. In this case *o*-aminobenzaldehyde or *o*-aminoacetophenone is condensed with an aldehyde or ketone containing an active methylene group in refluxing alcoholic sodium hydroxide solution to yield quinoline.[9] The first step is the formation of the Schiff's base followed

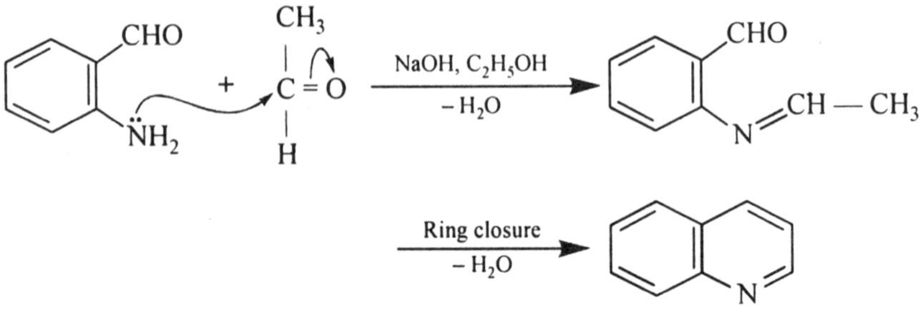

by ring closure to quinoline by a Knoevenagel condensation. More recently it has been observed[10,11] that acids are particularly effective catalysts in this reaction. The ring closure, therefore, depends on the conditions employed as is apparent from the following example:

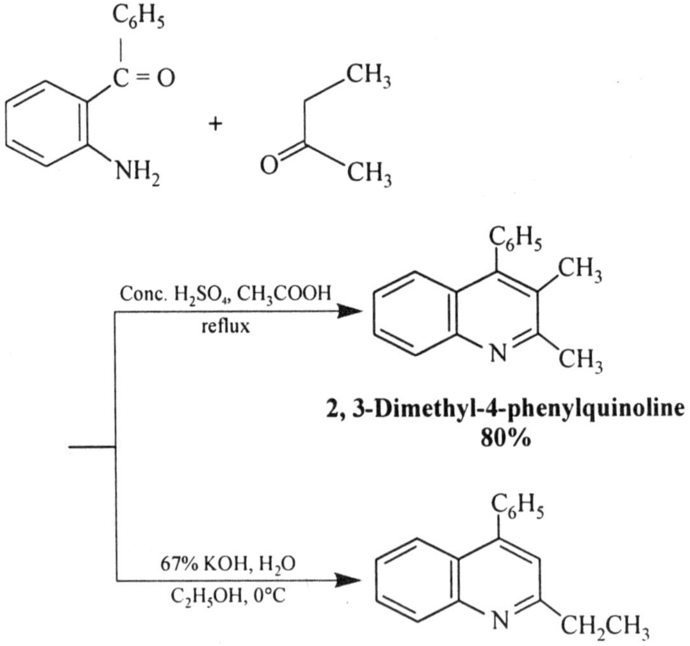

2, 3-Dimethyl-4-phenylquinoline
80%

2-Ethyl-4-phenylquinoline
71%

This method is particularly useful for the preparation of 2-substituted quinoline derivatives which are rather accessible with difficulty by other means.

4. ***The Pfitzinger Synthesis:*** This is a modification of the Friedlander method. The required o-aminobenzaldehydes in the Friedlander synthesis are difficult to obtain and are often unstable. The difficulty has been overcome in the Pfitzinger synthesis which instead uses isatin. Isatin in the presence of alkali is converted to isatoic acid which is condensed with a ketone to afford quinoline.

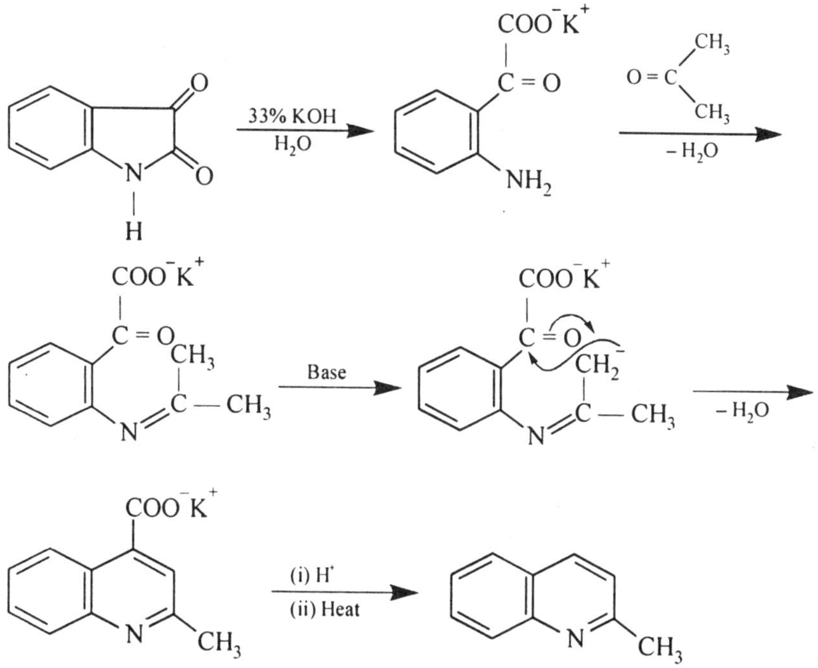

2-Methylquinoline

5. ***The Pictet Method:*** According to this procedure, N-ethylacetamide is heated with zinc chloride to yield 2-methylquinoline. Pictet found that N-ethylacetamide afforded a mixture of quinoline and toluidine.

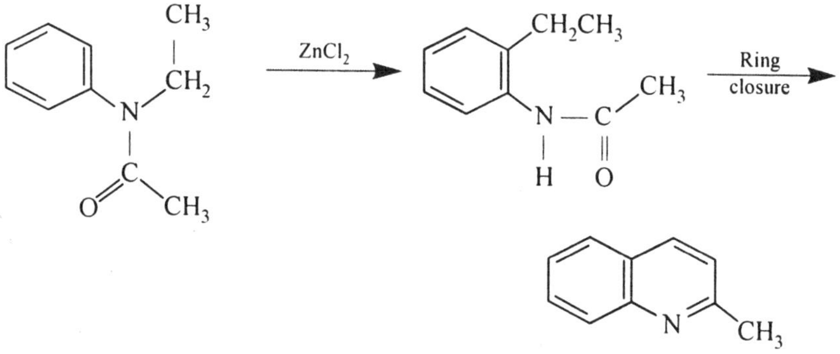

2-Methylquinoline

He, therefore, proposed that the ethyl group migrated to the *ortho* position first to produce *o*-ethylacetamide which then cyclized to methylquinoline. Liete[12] recently reinvestigated this reaction using (^{13}C -acetyl) -N-methylacetamide and heated it at 290°C for 6 hr and obtained quinoline in 6.5% yield in which quinoline was enriched at C-4. On this basis he suggested a modified mechanism for this reaction.

6.   *The Conrad-Limpach Synthesis:* In this reaction aromatic amines react with β-keto esters such as ethyl acetoacetate to give 2- and 4-quinolones. At

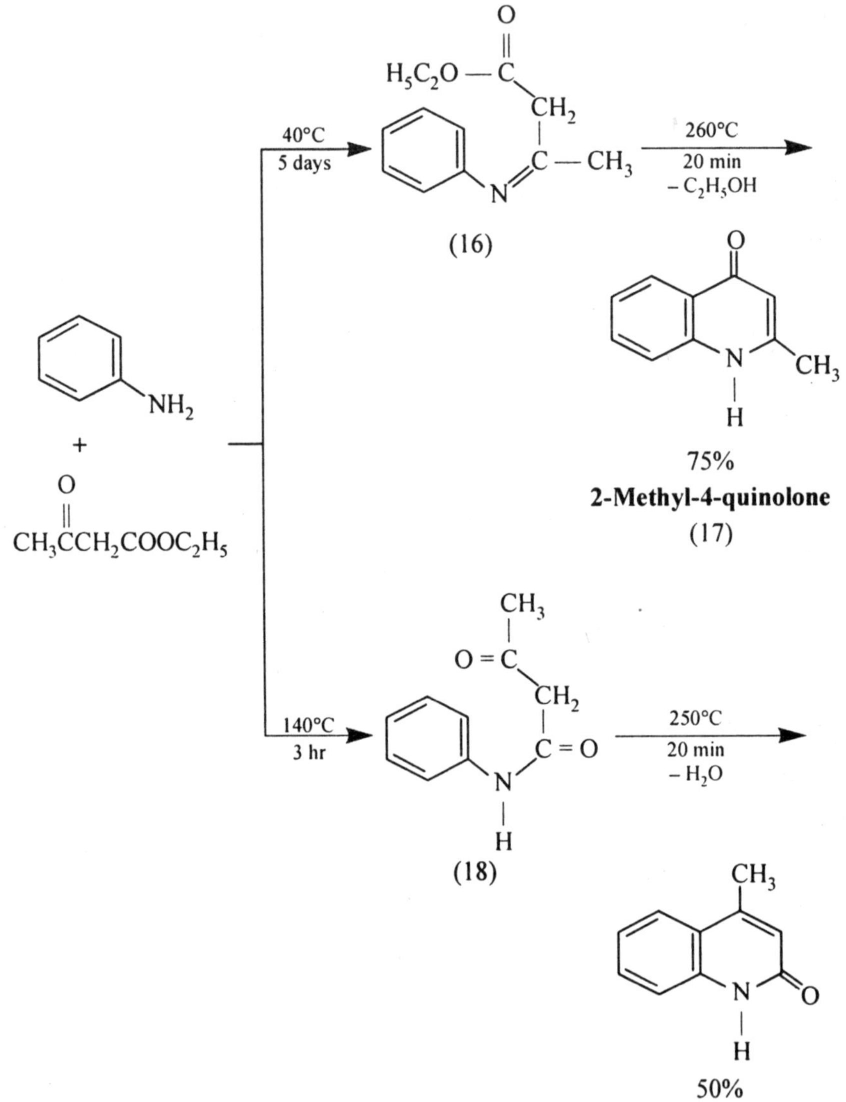

(16)

75%
2-Methyl-4-quinolone
(17)

(18)

50%
(19)
4-Methyl-2-quinolone

low temperature the amine condenses with the ester at the keto group to give an anil (16) (kinetically controlled) which cyclizes to 4-methylquinolone (17) on heating. On the other hand, at higher temperatures the initial product is an anilide (18) (thermodynamically controlled) by condensation of the amine at the ester function. This anilide undergoes ring closure on heating or in the presence of acid to 2-methylquinolone (19).

The Conrad-Limpach method is very versatile because aniline and the β-keto ester component may be altered structurally almost without limitations as exemplified in the following equations (8.4 and 8.5).

$$(8.4)$$

2-Carboethoxy-7-chloro-4-quinolone

$$(8.5)$$

3-Nitroquinoline

7.   *Miscellaneous Methods:* Substituted quinolines have been prepared recently by Rutherium-catalyzed oxidative coupling of alcohols. This is illustrated by the following example.

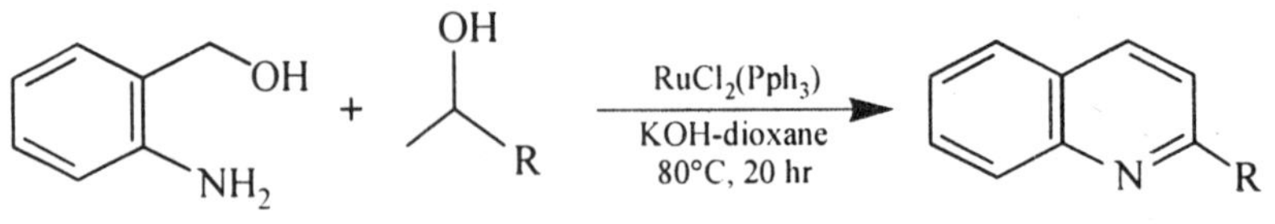

**2-Aminobenzyl
alcohol**

2, 3-Dimethylindole reacts in a familiar manner with chloroform and sodium hydroxide in the presence of tetralkylammonium salt to give 2, 4-dimethyl -3-chloroquinoline,[14] (equation 8.6).

$$(8.6)$$

Photocyclization of the following compound gives a thietane *via* an allowed $(2 + 2)$ cycloaddition process. A subsequent cleavage of thietane yields a thiol which eliminates hydrogen sulfide to give 2,3-dimethylquinoline.

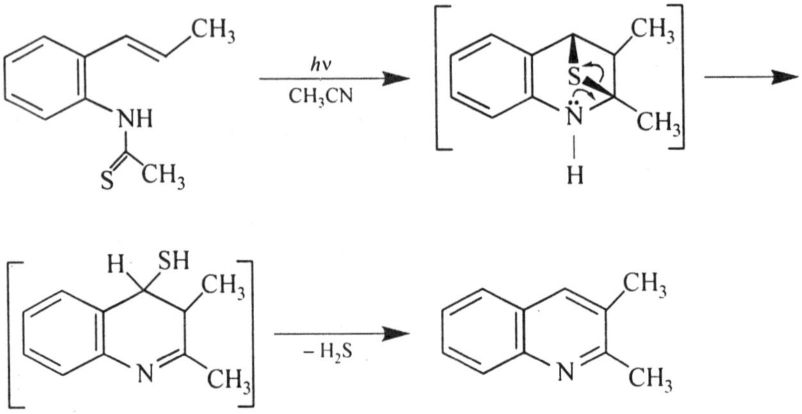

Quinolines have also been obtained by the photolysis[15,16] of o-vinylthioanilides.

A reaction of two compounds in which one or both contain an electrophilic and a nucleophilic center condense in the presence of base to give quinoline.[17] The Friedlander quinoline synthesis is an example of this type of reaction.

Substituted quinolines have also been prepared from an aromatic amine and an excess of 2-ethylacrolein in 70% H_2SO_4 at 110°C in the presence of catalytic amount of iodine.[18] The catalyst enables the oxidation of the intermediate dihydroquinoline by sulfuric acid used as solvent.

8.1.3 Chemical Reactions

Quinoline displays chemical properties associated with a tertiary amine. In addition because of the fusion of a benzene ring, properties of both benzenoid and pyridinoid compounds frequently manifest themselves.

1. *Reaction with Acids:* Quinoline is a week base and is thus protonated on the ring nitrogen with mineral acids to form water solube salts.

2. *Electrophilic Substitution:* The electron-rich nitrogen atom of quinoline is the main center for attack by electrophiles. The hetero atom has considerable deactivating effect on the ring towards electrophilic attack. In this respect the electrophilic reaction of quinoline may be compared with 1-nitronaphthalene just as that of pyridine with nitrobenzene. Therefore, electrophilic substitution on quinoline nucleus requires severe conditions though less than those in the case of pyridine. The attack in the protonated quinoline takes place in the carbocyclic ring at C-5 and C-8 positions because the corresponding intermediates (20) and (21) can be represented by resonance structures in which the aromaticity of the heterocyclic ring is still preserved.

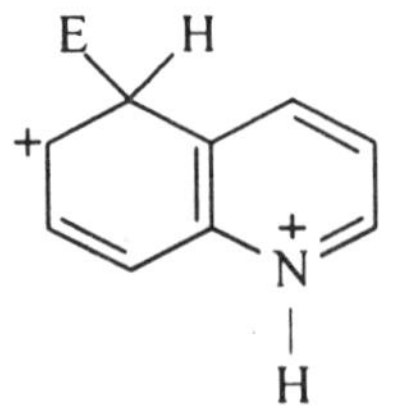

(20) 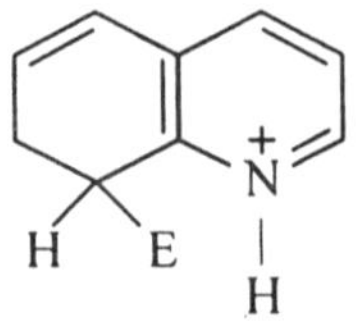   (21)

As a π-electron deficient heterocycle, quinoline is less reactive than benzene particularly in acid solutions.

Halogenation: Attack by a halogen depends on the nature of the reagent employed for halogenation. Chlorination (SO_2Cl_2) of quinoline for instance, yields 3-chloroquinoline while bromination (Br_2, CCl_4) yields 3-bromoquinoline. The product is obtained by the following addition – elimination mechanism. Orientation in the presence of strong acids (Br_2, Ag_2SO_4) proceeds to give a mixture of 5-and 8-bromoquinolines[19], the reacting species being quinolinium cation.

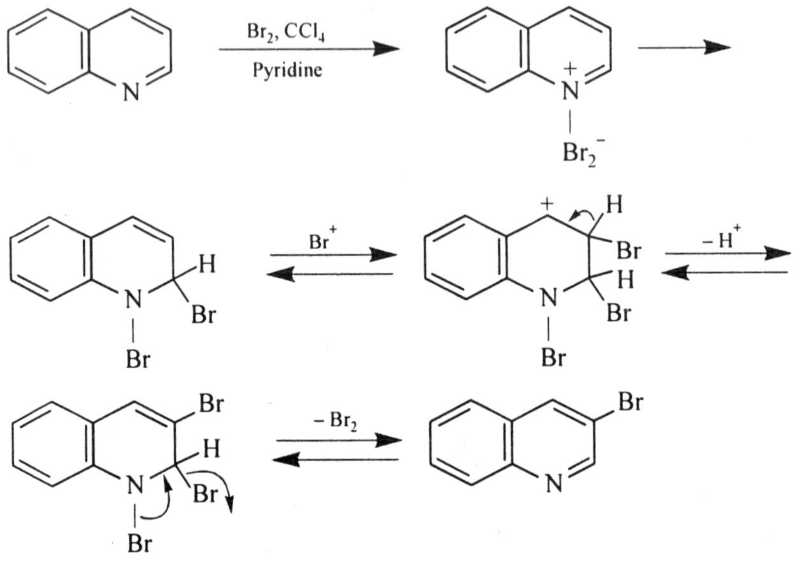

Nitration: The nitration and sulfonation of quinoline can be carried out under conditions which are less strenuous than those in the case of pyridine. The pyridine ring is already π-electron deficient and becomes more so by protonation of the ring nitrogen atom. Acidic electrophilic reagents thus show a strong preference for attack in the benzene ring. Nitration $(HNO_3, H_2SO_4$ 0°C) of quinoline gives a mixture of 5- and 8- nitroquinolines[20], in equal amounts. In acetic anhydride as solvent or with dinitrogen tetroxide as nitrating agent, the nitration proceeds in a totally different manner and very inefficiently.

After the reaction much quinoline is recovered unchanged and the main product is 3-nitroquinoline in a yield of 6% along with a small quantity of 6- and 8-nitroquinolines.[21,22]

Sulfonation: The products of sulfonation vary with the experimental conditions. In conc. sulfuric acid quinoline almost entirely is converted into quinolinium cation. Sulfonation at 220°C results in attack at the benzene ring to give largely quinoline 8-sulfonic acid. At a very high temperature (300°C), 6-quinolinesulfonic acid is obtained because the 5-and 8-isomers rearrange to the 6-isomer which is thermodynamically more stable.

Mercuration: With mercuric acetate, quinoline forms the quaternary N-mercuriacetate at room temperature. At 160°C further substitution occurs and treatment with sodium chloride, gives a mixture of 3- and 5-mercurichlorides, (equation 8.7).

$$(8.7)$$

Reactions of quinoline with free radicals are often unselective and form complex mixture of products. Thus phenylation of quinoline with benzoyl peroxide gives all the seven phenylquinolines.

The Friedel-Crafts Reaction: This reaction is rare in the quinoline series because of the deactivation of the ring by the nitrogen atom. However, the Friedel-Crafts acylation of 8-methoxyquinoline occurs at the 5-position but not position-7.

3.   *Reaction with Nucleophiles:* Attack by a nucleophile occurs in the pyridine ring of quinoline and position-2 is the preferred site for such an attack.[23] If this position is occupied then attack may take place at the 4-position. The Chichibabin reaction is an example of a normal attack at C-2, equation (8.8).

$$(8.8)$$

The addition of organolithium reagents to quinoline in a 1, 2-fashion is well established,[24] and the resultant 1, 2-dihydroquinoline can be manipulated.[25,26] Oxidation leads to 2-alkylquinoline while reduction with Na/C_2H_5OH or metal hydrides to 2-substituted 1, 2, 3, 4-tetrahydraquinoline.

2-Butylquinoline

2-Butyl -1,2,3,4-tetrahydro-quinoline

Quaternization at the nitrogen atom reinforces the ability to react with nucleophilic reagents. With sodium hydroxide, carbinolamine is formed but the actual product isolated is dimeric carbinolamine ether (22). Even weak nucleophiles attack readily on the 1-alkylquinolinium cations. This, behavior

(22)

is illustrated by the attack of a Grignard reagent and the resultant formation of 1-methyl-2-phenyl-1, 2-dihydroquinoline known as Reissert compounds,[27, 29] equation (8.9).

$$\text{(8.9)}$$

Halo derivatives of quinoline on reaction with potassium amide in liq ammonia give rise to product of amination *viz* the "*hetaryne*" type intermediate. For instance, 2-amino-3-bromoquinoline forms 2, 3-diaminoquinoline[30] under these conditions.

Besides halo derivatives of quinoline, isoquinoline and naphthalene undergo rapid nucleophilic substitution with nucleophiles like alkoxides and thiolate ions.[31] The reaction is accomplished using microwave irradiation.

2-Methoxyquinoline

4. *Reaction with Reducing Agents:* Quinoline is more readily reduced than naphthalene just as pyridine is easily reduced than benzene. Therefore, many types of reagents reduce the heterocyclic ring. Reduction of quinoline to 1, 2, 3, 4-tetrahydroquinoline is best achieved with Raney nickel or PtO_2, H_2.[32a] or ammonium carbonate,[32b] Pd/C. In acid media the benzo group can be selectively reduced.[33] Hydrogenation to decahydroquinoline is difficult,[34] ring rupture takes place when quinoline is reduced over nickel at 260-380°C.

1,2-Dihydroquinoline   **1,2,3,4-Tetrahydro quinoline**

5. ***Reaction with Oxidizing Agents:*** The course of degradative oxidation of quinoline and is derivatives complex and either ring may be opened. The pyridine ring is π-deficient and oxidation is independent of electron availability. The pyridine ring thus remains intact while the benzene ring is destroyed on treatment with alkaline potassium permanganate.[35] Lipidine (4-methylquinoline) gives 4-methylquinolinic acid. Quinolines with alkyl or aryl substituents a position-2 suffer ring opening of the heterocyclic ring on oxidation in acidic potassium permanganate to yield acyl- or aroyl-anthranilic acid.

Quinolinic acid

Nicotinic acid (Niacin)

Ozonolysis of quinoline gives glyoxal and pyridine-2, 3-dicarboxaldehyde, equation (8.11).

$$(8.11)$$

Oxidation of quinolines with peracids leads to quinoline N-oxides in excellent yields. In substituted quinolines the yield depends on the nature and location of the substituents.[37] The ring activation by N-oxide group of quinoline is analogous to that or pyridine N-oxide. The electrophilic attack in quinoline N-oxide takes place at C-4 to give 4-nitroquinoline N-oxide while at low temperature 5-and 8-isomers are also obtained.[38]

8.1.4 Derivatives of Quinoline

Alkyl-and Aryl-Quinolines: The 2- and 4-methylquinolines occur in coal tar. The isomeric 2-, 3- and 4-alkylquinolines have been obtained by the Skraup, Doebner-Miller and Combes methods among others. 2-Methylquinoline has

also been obtained from aniline and ethylene in the presence of rhodium chloride.[39] 2-Phenylquinoline has been prepared[40,41] among other products, by irradiating 3, 5-diphenylisoxazole.

The 2- and 4-methylquinolines display chemical behavior different from the 3-isomer. This is again similar to the case of 2-and 4-picolines and 3-picoline. The hydrogen in the 2- and 4-methylquinolines are active and can react with esters, aldehydes and ketones in basic medium to yield condensation products. With 2-methylquinoline methiodide, the positive charge further leads to activation of the 2-methyl group and thus facilitates condensation.

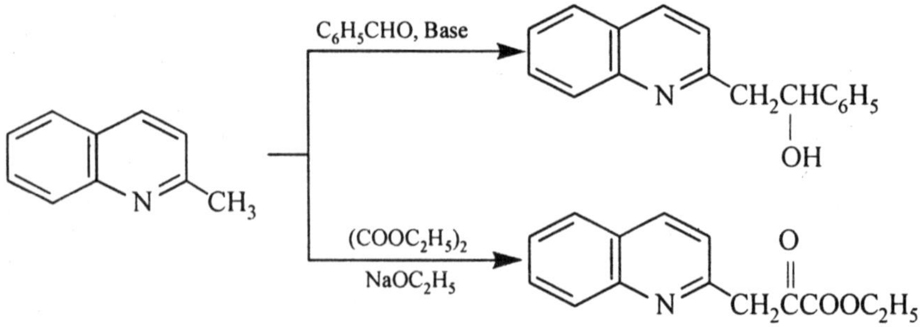

In contrast 3-methylquinoline fails to undergo condensation.

Halo Quinolines: Direct halogenation as discussed earlier can lead to halo quinolines in moderate yields. However, the most convenient method of preparing 2- and 4-chloroquinolines is from the corresponding quinolones by reaction with phosphorous pentachloride as shown below (equation 8.12).

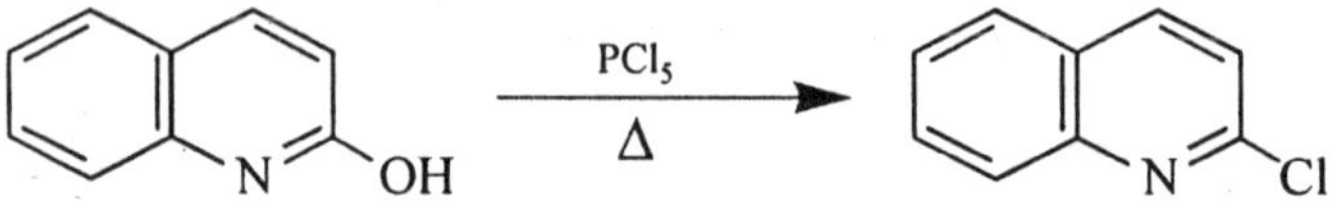

$$(8.12)$$

The 4-chloroquinoline has been obtained from quinoline N – oxide by treating with sulfuryl dichloride (SO_2Cl_2).

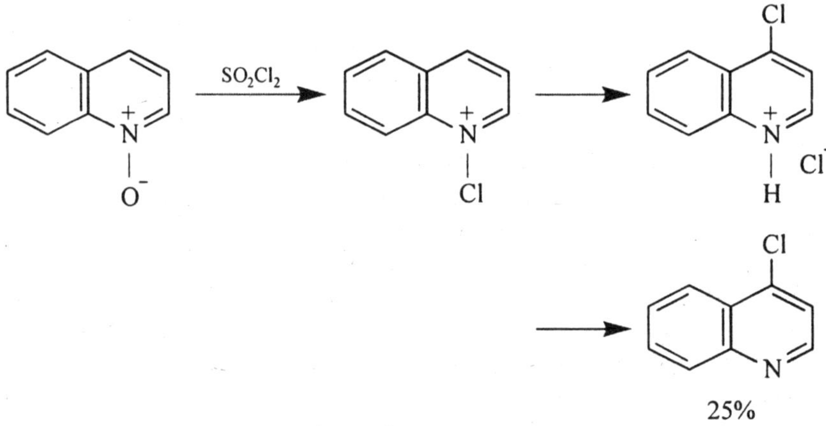

25%

The chloro group in 2-and 4-chloroquinoline can readily be displaced by nucleophiles such as H_2O, NH_3, $NaOC_2H_5$, etc. Though the 3-isomer is unreactive, but it forms 3-aminoquinoline on treating with ammonia in the presence of copper.

Hydroxyquinolines or Quinolinols: All the hydroxyquinolines are known to exist in tautomeric equilibrium with the corresponding quinolones. The behavior of 2-and 4-hydroxyquinolines is similar to 2- and 4-hydroxypyridines.[42] The quinolones have the following mesomeric structures:

The 2- and 4-quinolinols can usually be prepared by the Conrad Limpach or Knorr synthesis while the 3-isomer is most conveniently obtained by diazotizing 3-aminoquinolines followed by hydrolysis.

2-Hydroxyquinoline (carbostyril) is a colorless compound, m.p., 199°C. It is oxidized to isatin by potassium permanganate. The quinolinols behave as typical phenols and when the – OH group is present on the benzene ring they show all the reactions of the corresponding naphthols, i.e. they give violet color with ferric chloride, undergo Reimer-Tiemann reaction and couple with diazonium cations.

8-Hydroxyquinoline or oxine is the best known compound of this group. It can be prepared by fusing quinoline 8-sulfonic acid with potassium hydroxide. This compound is steam volatile and the – OH stretching frequency appears at 3416 cm^{-1} less than the normal – OH group. This is ascribed to the intramolecular hydrogen bond formation in 8-hydroxyquinoline. It has been employed in analysis as chelating agent for Zn^{II}, Mg^{II}, Bi^{II}, Mn^{II}, Al^{III}, etc. The chelate of type shown below is insoluble in water and is thus a very valuable reagent for the estimation of metallic ions.

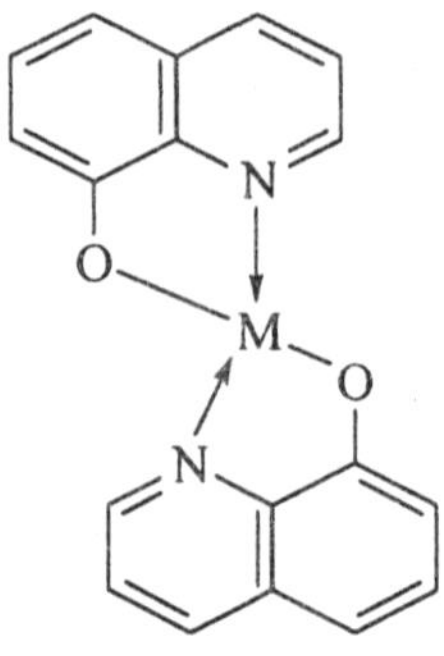

Quinoline Aldehydes and Ketones: Quinoline aldehydes are known in which the formyl group is found in each of the seven positions. Oxidation of the methyl derivatives with selenium dioxide is the most appropriate method for their preparation.[43] They can also be prepared by the hydrolysis of bromomethyl quinolines and by the Reimer-Tiemann reaction. The quinoline aldehydes participate generally in the reactions of aromatic aldehydes lacking an α-hydrogen. They, for instance, readily undergo benzoin condensation, the Cannizzaro reaction and the Perkin reaction.

A number of quinolyl ketones or acyl ketones are known. They have been prepared by a number of methods. Although quinolyl halides do not form Grignard reagents, esters of quinoline carboxylic acids can usually be converted

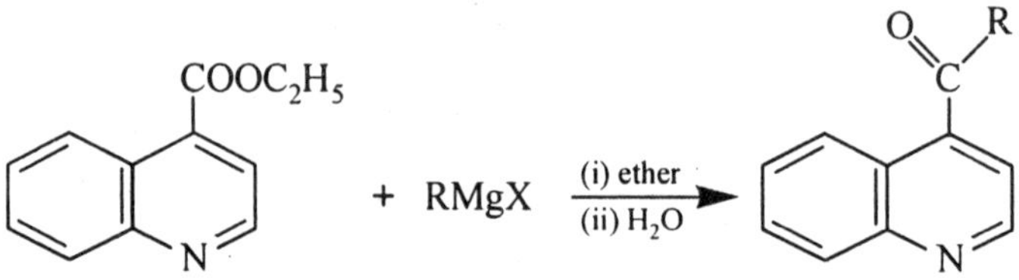

$$\text{(8.13)}$$

to ketones in the manner shown equation (8.13). They acyl halides of quinoline react with benzene in the Friedel-Crafts reaction to afford quinolyl aryl ketones.

Quinolyl ketones participate in the Beckman rearrangement and undergo many other reaction typical of aromatic ketones.

Quinoline Carboxylic Acids: The quinoline carboxylic acid may be prepared by several general methods. 2-Quinoline carboxylic acid (quinaldinic acid) can be obtained from 1-benzoyl -2-cyano-1, 2-dihydroquinoline on heating with hydrochloric acid (equation 8.14).

$$(8.14)$$

Oxidation of 3-methylquinoline with chromic acid yields the 3-isomer. Quinoline 4-carboxylic acid (cinchoninic acid) is obtained by the oxidation of lipidine or by the following sequence from 4-cyanoquinoline.

Quinoline 6-carboxylic acid was obtained by a modified Skraup synthesis on p-aminobenzoic acid.

Quinoline carboxylic acids can be decarboxylated readily,[44] the 2-isomer particularly decarboxylates at 160°C. The mechanism of decarboxylation involves the zwiterionic from of the acid. The 4-isomer although stable to fused aqueous potassium hydroxide which converts it to the 4-hydroxycarbonyl-2-quinolone can be decarboxylated by heating with copper power.

50%

8.1.5 Naturally and Biologically Active Compounds

Quinoline is not found in nature although a number of alkaloids[45] containing this nucleus are are well known and many of them function as therapeutic agents. Cinchona alkaloids[46] are isolated from the bark of cinchona trees and are quinoline bases. The structure of quinine (23) can be represented as follows:

(23)

Quinine has been a standard drug for the treatment of malaria for centuries, it has, however, been superseded by a number of more potent synthetic drugs such as chloroquin (24), plasmoquin (25) atebrin (26), etc.

(24)

(25)

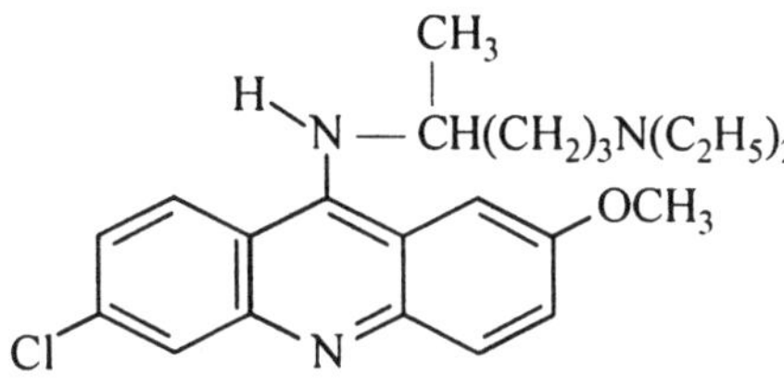

(26)

Many commercially available dyes[47] also contain the quinoline nucleus. The cyanin dyes used for improving the color sensitivity of photographic emulsions have been reviewed by Hamer.[48] The dye cyanine (27) a green sensitizer is formed by base-catalyzed condensation from equimolar quantities of the quaternary ethiodides of 3-ethylquinoline and 4-methylquinoline. The dye pinacyanol (28) a red sensitizer is obtained similarly from two moles of 1-ethyl-2-methylquinolinium iodide equation (8.15) and ethylorthoformate, equation (8.16).

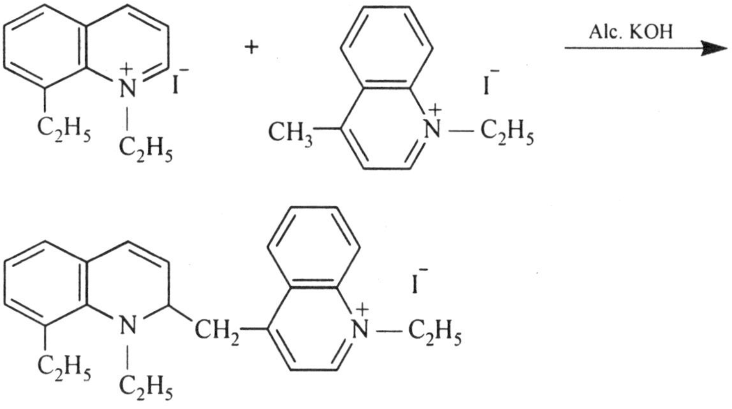

$$(8.15)$$

(27)

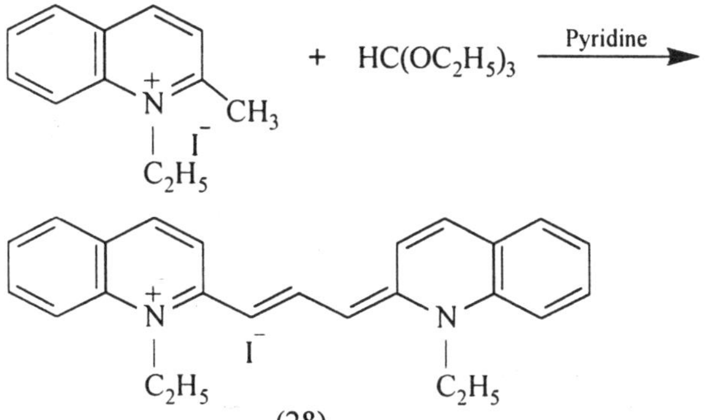

$$(8.16)$$

(28)

Since these dyes are sensitive to light, they cannot be employed for dyeing fabric.

8.2 ISOQUINOLINES

Isoquinoline (2), ring system is also obtained by fusion of pyridine with a benzene ring. It was first isolated by Hoogewerff and Dorp from the quinoline

(2)

fraction of coal tar in 1885. Many of its derivatives also occur in coal tar. Isoquinoline does not occur free in nature but abundantly in several alkaloids. It is called 2-*azanaphthalene* or *benzo* [b] *pyridine* and is numbered in the same manner as quinoline, but the nitrogen atom is assigned position-2. Because of close similarities in the structure of quinoline and isoquinoline they show a close relationship in their physical and chemical properties.

The reduced isoquinolines namely, 1, 2-dihydroisoquinoline (29), 1,2,3, 4-tetrahydroisoquinoline (30) and decahydroisoquinoline (31) are particularly important because of their occurrence in alkaloids.

(29)   (30)   (31)

8.2.1 Physical and Chemical Properties

Isoquinoline is a colorless solid, m.p., 243°C. It has a smell resembling that of benzaldehyde. It is steam volatile and sparingly soluble in water but soluble in most organic solvents. Isoquinoline turns yellow on keeping.

Isoquinoline is highly aromatic and may be considered as a resonance hybrid of the following structures:

Similar to pyridine, the pair of *2p* electrons on the nitrogen atom is not conjugated with the ring and therefore, this substance functions as a weak base (*p*Ka 5.14) compared with quinoline (*p*Ka 4.94). The dipole moment is 2.6 D which is slightly larger than pyridine and quinoline and gives support to charged resonance structures. The π-electron densities for isoquinoline have been calculated and show that electrophilic attack occurs preferentially at the C-5 and C-8 positions.

8.2.2 Synthetic Method[49]

Isoquinoline and most of its derivatives can be prepared by the following general methods. Most routes result in partially reduced heteroaromatic compounds. Aromatization is accomplished in a separate dehydrogenation step.

1. *The Bischler-Napieralski Synthesis:* This synthesis was first suggested by Bischler and Napieralski[50] and has been subjected to a number of improvements. This method involves a cyclodehydration of an acyl derivative of β-phenylethylamine to give 3, 4-dihydroisoquinoline, in the presence of

polyphosphoric acid, zinc chloride or phosphorous pentoxide. This is dehydrogenated (Pd, C, 160°C) to isoquinoline. The yields of the products are excellent if electron-donating groups are present on the benzene ring but poor if electron-withdrawing groups are present. The observation is supported in a lower yield (50%) of 7-nitro-3, 4-dihydroisoquinoline formation. This is in accord with the electro philic ring-closure nature of the ring.[51]

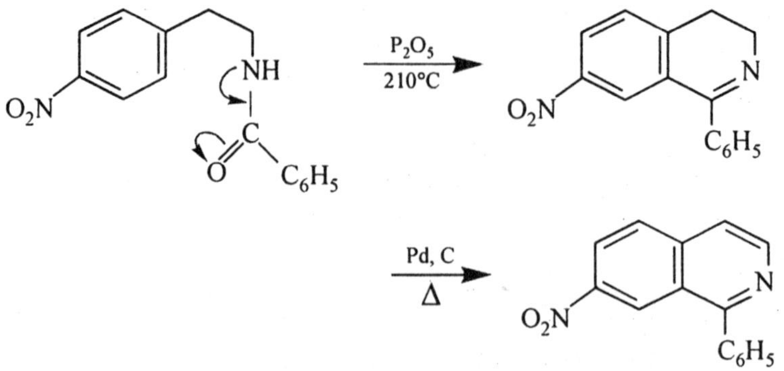

Presence of groups on the benzene ring also exert the usual directive effect which are normally operative under such conditions. Thus ring closure of structure (32) gives rise to 6-methoxy-1-phenyl-3, 4-dihydroisoquinoline (33) and not 8-methoxy-1-phenyl-3, 4-dihydroisoquinoline (34) because the ring closes *para* to the methoxy group[52], an electron-donating group.

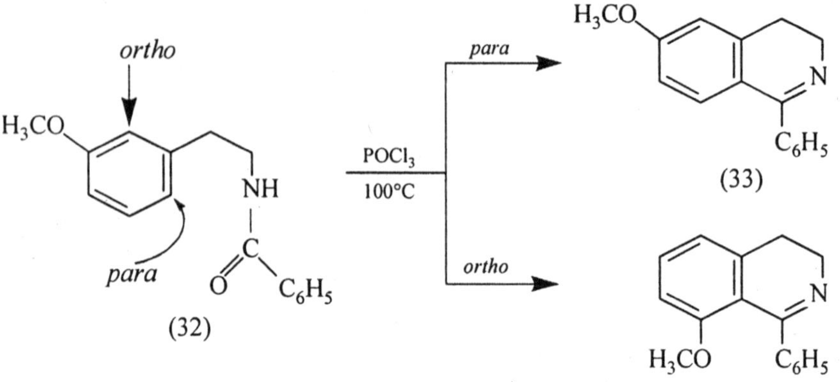

2. *The Picte-Gams Synthesis:* This is a modification of the above synthesis. In this method β-phenylethylamine is replaced by β-hydroxy-β-phenylethylamine. Such substances undergo a facile dehydration under above conditions and the styrylamide so obtained cyclizes to an isoquinoline derivative directly. In this procedure the final dehydrogenation step is eliminated.

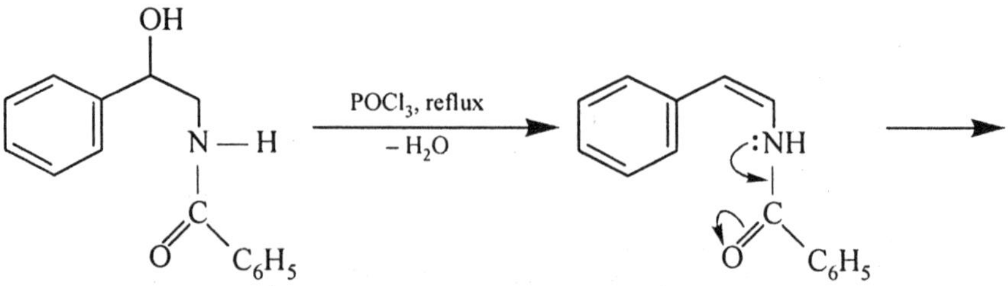

1-Phenylisoquinoline

3. ***The Pomeranz-Fritsch Synthesis*:** In this synthesis an aromatic aldehyde or a substituted benzaldehyde is condensed with aminoacetal to Schiff's base (aldimine) which is then cyclized in the presence of sulfuric acid or phosphorous pentoxide. The last step is similar to the Skraup synthesis for quinoline in that the acid causes the elimination of one mole of ethanol and the resultant species attacks the ring as an electrophile. The final product is obtained by elimination of a second mole of ethanol.

As is obvious from the mechanism of the reaction that the electron-donating groups accelerate the rate of the reaction while electron-withdrawing groups retard the rate. Furthermore, the electron-releasing groups accelerate

(35)

the reaction when present *meta* to the aldehyde carbon. The *meta* orientation permits the resonance electron to the center of electrophilic attack. The ring closure takes place invariably *para* to the activating substituents.

The yields with methoxy as substituent are good while with nitro group, oxazoles are produced as by-product[53]. In a detailed study Brown[54] has shown that *o-, m-*and *p*-substituted benzaldehydes form exclusively the oxazoles (35) and no trace of the corresponding isoquinolines was detected.

4.　*From Indene:* The parent compound isoquinoline has been prepared[55] in a high yield from indene by treating with ozone at-70°C followed by reduction of the dialdehyde with dimethylsulfide in the presence of NH_4OH.

8.2.3 Chemical Reactions

Isoquinoline resembles in many of its properties to quinoline.

1.　*Electrophilic Substitution:* Electrophilic attack in strongly acidic conditions on isoquinoline (i.e. on isoquinolinium cation) takes place under rather drastic conditions than in pyridine. The 5-and 8-positions are most suscepti-ble to attack and occurs readily than in naphthalene.

Halogenation: Chlorination (Cl_2, $AlCl_3$) gives 5-chloroisoquinoline in a low yield. Bromination occurs depending on the conditions of the reaction and appears to be anamolous. Bromination (Br_2, $AlCl_3$) takes place at 5-position but heating isoquinoline hydrobromide with bromine in sulfur monochloride, the 4-isomer is obtained.

Nitration: Nitration (HNO_3, H_2SO_4, 0°C) of isoquinoline gives mainly 5-nitroisoquinoline (72%) with a small amount of 8-nitroisoquinoline (8%).[56] At higher temperature the proportion of the 8-isomer is slightly increased.

Sulfonation: Isoquinoline undergoes sulfonation with lesser difficulty than quinoline and gives isoquinoline 7-sulfuric acid as the major product.

Amino group present at the 3-position of isoquinoline undergoes diazotization under normal conditions.

2.   *Reaction with Oxirane:* Isoquinoline reacts with oxirane in acetic acid to form oxazolidine[57,58]

3.   *Reaction with Nucleophiles:* The attack of a nucleophile takes place at position-1, if this position is occupied then the attack occurs at the 3-position. The Chichibabin reaction (equation 8. 17) is an example of a normal attack at C-1. The alkyl-or aryl-lithiums also attack at position-1.

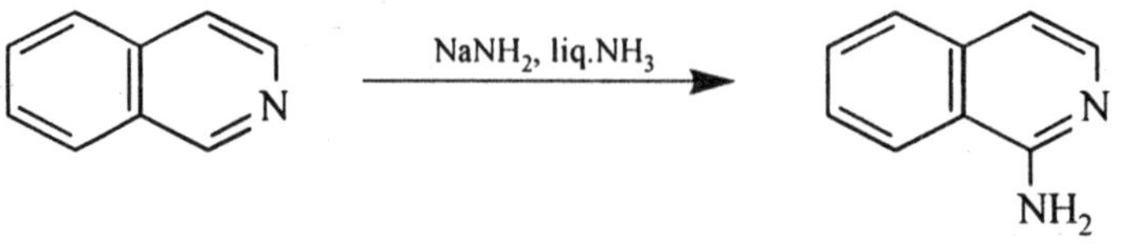

$$(8.17)$$

1-Aminoisoquinoline

The nitrogen atom in the ring reacts with alkylating agents to give quaternary salts and quaternization reinforces the ability of the pyridine ring in isoquinoline towards nucleophilic attack. This behavior is reflected in the formation of Reissert compounds.[59, 60]

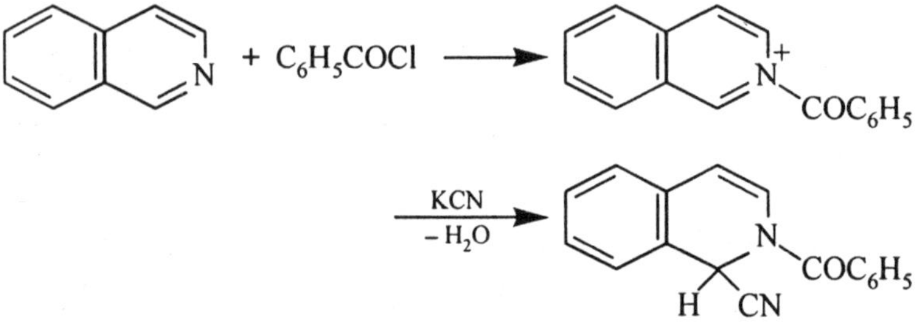

4.   *Reaction with Reducing Agents:* Isoquinoline is attacked by different reducing agents to furnish different reduced products.

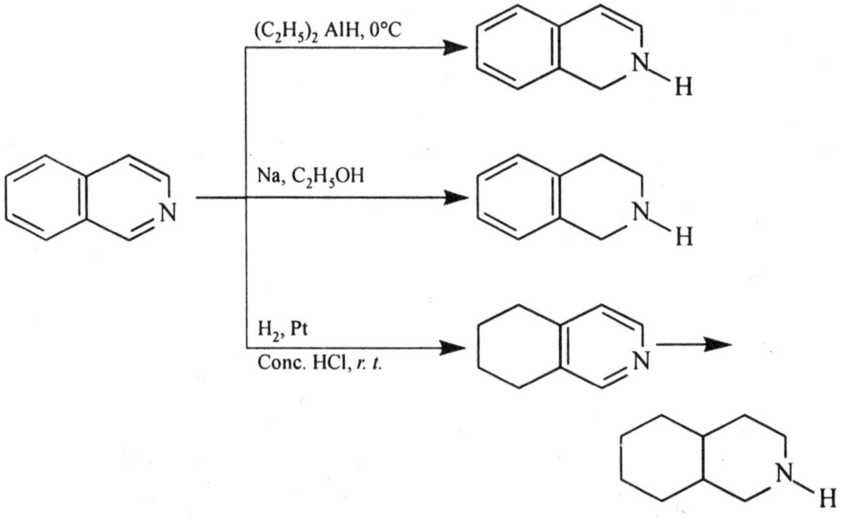

5.   *Reaction with Oxidizing Agents:* Isoquinoline is largely resistant to the action of most oxidizing agents but is attacked by alkaline potassium premanganate. Neutral potassium permanganate converts isoquinoline largely to phthalimide, equation (8.18), while alkaline $KMnO_4$ to phthalic acid and pyridine 3, 4-dicarboxylic acid, equation (8.19).

$$\text{(8.18)}$$

$$\text{(8.19)}$$

The action of perbenzoic acid converts isoquinoline to isoquinoline N – oxide which activates the pyridine ring toward electrophilic attack. Nitration (HNO_3, H_2SO_4) gives rise to 5-and 8-nitroisoquinoline N-oxides.[61]

Photolysis of isoquinoline N-oxide gives back isoquinoline.[62]

6. *Photochemical Reactions:* The photochemistry of isoquinoline is very much similar to that of quinoline. For instance, isoquinoline gives 1-alkylated isoquinoline on irradiation with aliphatic carboxylic acids.[63]

8.2.4 Derivatives of Isoquinoline

Alkyl-and Aryl Derivatives: The alkyl-and aryl-derivatives of isoquinoline can be prepared by some of the methods described earlier. The alkyl group at 1-position has reactivity similar to 2-alkyl substituted quinoline. 3-Methylisoquinoline undergoes bromination with N-bromosuccinimide.

Haloisoquinolines: 1-Chloroisoquinoline is obtained from isocarbostyril and phosphorous oxychloride, equation (8.20). Many halo-derivatives can

$$\text{(8.20)}$$

also be obtained from the corresponding amino compounds *via* diazonium salts. The halogen at 1-position is very reactive and can be readily replaced by amines and alkoxides. This reactivity is similar to a chloro group present at position-4 but approximately 50,000 more reactive than at the 3-position.[64] The nucleophilic substitution using microwave irradiation takes place rather more efficiently.

4-Methoxyisoquinoline

Hydroxyisoquinolines: The hydroxyisoquinolines may be obtained by the Pictet-Spengler and Pomeranz procedures. The diazotization of 1-aminoisoquinoline results in the formation of isocarbostyril [1 (2*H*)-isoquinolone]. The hydroxyisoquinolines exist in solution as equilibrium mixtures of the isoquinolinol and isoquinoline tautomers. The 4, 5-, 6-, 7-and 8-isoquinolinols have typical phenolic properties.

Isoquinoline Aldehydes and Ketones: 1-Methylisoquinoline, on oxidation with selenium dioxide furnishes isoquinoline-1-carbaldehyde and

Isoquinoline 3-carboxylic acid

3-methylisoquinoline furnishes isoquinoline-3-carbaldehyde.[65] Isoquinoline-1-carbaldehyde readily condense with compounds containing active methylene groups.

The Claisen ester condensation of isoquinoline-1 and 4-carboxylic esters with ethyl acetate followed by ketonic cleavage of the β-keto ester constitutes a satisfactory synthesis of 1-and 4-isoquinoline ketones.[66]

Isoquinoline Carboxylic Acids: Isoquinoline 1-carboxylic acid can be obtained from the Reissert derivative by treatment with a strong acid. The 3-isomer is prepared by SeO_2 oxidation of 3-methylisoquinoline as shown above. The remaining isomeric acids are obtained by hydrolysis of the corresponding isoquinoline nitriles.

8.2.5 Naturally and Biologically Active Compounds

The isoquinoline ring system is widely distributed in nature in the form of alkaloids.[67,68] The family of isoquinoline alkaloids is divided into four types namely papaverine, narcotine, berberine and emetine. Papaverine (36) known as the opium alkaloid occurs in the opium poppy *papaves somniferum*. Three types of alkaloids occur in poppy seeds namely papaverine, landanosine, and morpholine alkaloids (morphine, codeine, thebaine). Opium alkaloids have been used principally for the relief of pain. Narcotine (37) is used as a

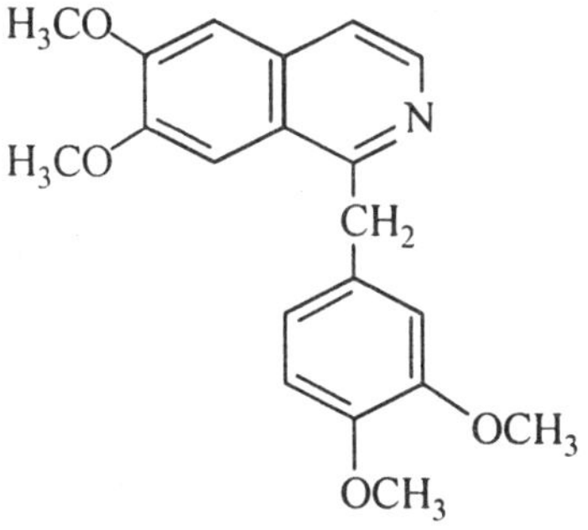

(36)

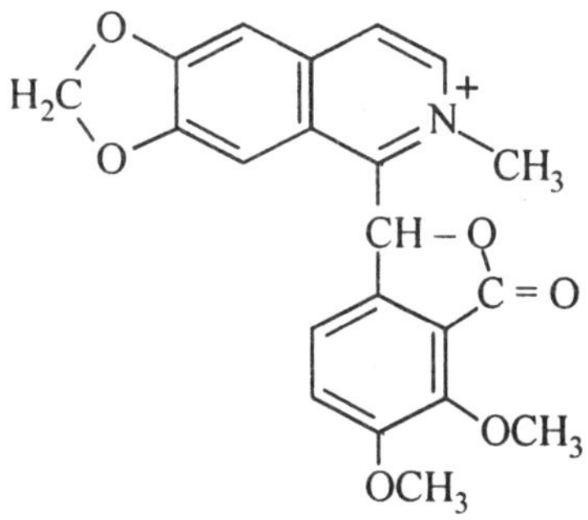

(37)

depressant. Beriberine (38) occurs in the berberi species. Emetine (39) is a member of the epicacauanha group of alkaloids.

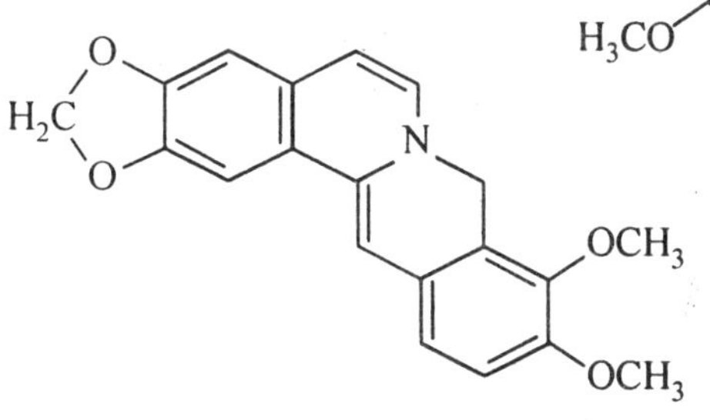

(38)

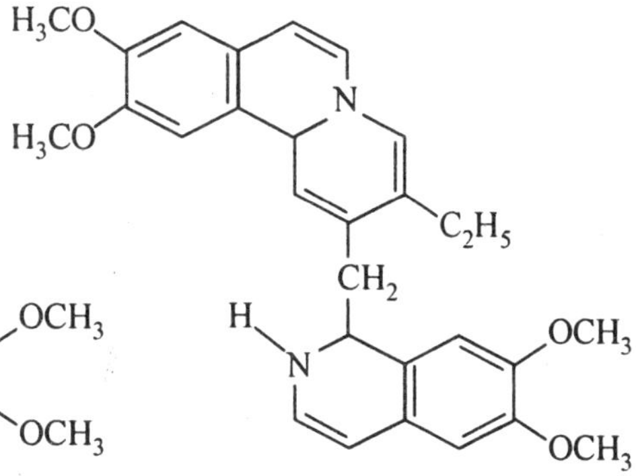

(39)

8.3 QUINOLIZINIUM SALTS

The quinolizinium cation (3) is the parent compound of the aromatic quinolizines. It is aromatic and isoelectronic with naphthalene, quinoline and

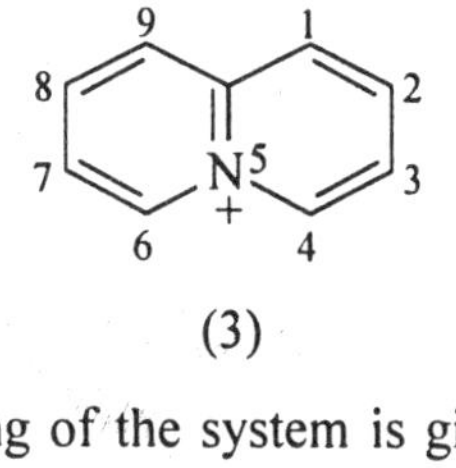

(3)

isoquinoline. The numbering of the system is given, the positively charged nitrogen atom is assigned position-5. The three possible benzo derivatives of quinolizinium are each well known, the benzo [a]-(40), benzo [b]-(acridizinium) (41) and benzo [c]-quinolizinium (42) ions.

There are three possible structures for the non-aromatic quinolizines depending on the location of the additional hydrogen atoms, namely 2*H*-quinolizine (43), 4*H*-quinolizine (44) and 9a *H*-quinolizines (45). None of these have been isolated in stable form. The fully saturated cycloaliphatic tertiary amine quinolizidine (46) is stable.

The quinolizinium salts are water soluble crystalline solids and melt without decomposition. The quinolizine ring system occurs in a number of alkaloids. The fully reduced quinolizine form occurs in lupins as for instance, lupinine (47), other forms occur in cytisine (48) a tricyclic poisonous alkaloid and anagyrine (49).

8.3.1 Synthetic Methods

Quinolizinium ion and its derivatives have been prepared by the following methods.

1. *Condensation Reactions:* Quinolizinium ion is synthesized from 2 – picolyllithium and β-ethoxypropionaldehyde in the following steps:

For the preparation of both alkyl-and aryl-quinolizinium salts Glover and Jones[70] allowed 3, 3-diethoxypropylmagnesium chloride to react with 2-acetylpyridine to produce an intermediate (50). which in the following steps yields 1-methylquinolizinium ion.

The preparation of benzo [a] quinolizinium ion and its congeners requires the conversion of 1-phenylpyridine into its quaternary salt by the action of α-halo Ketones or α-halo aldehyde derivatives.

This is followed by acid catalyzed cyclization of the salt to obtain the benzo [a] quinolizinium system.

Another method for the preparation of quinolizinium ion involves the reaction of 3-ethoxy propylmagnesium bromide with 2-cyanopyridine and the resulting imine is hydrolyzed to a ketone. The ether is cleaved with hydrobromic acid to form the bromide (51), the salt is neutralized and heating results in the formation of a, 1, 2, 3, 4-tetrahydroquinolizinium salt. This salt is refluxed in acetic anhydride to give the quinolinium ion.

For the synthesis of acridizinium salt, *o*-lithiobenzyl chloride is treated with 6-methyl-2-pyridinecarboxyaldehyde in ethanol at-100°C and the resulting alcohol is cyclized on heating.

(52)

4-Methylacridizinium bromide (52) is obtained on aromatization in the presence of hydrobromic acid.

8.3.2 Chemical Reactions

The quinolizinium ions behave in a similar manner to pyridinium salts in their chemical reactions.

1. *Electrophilic Substitution:* The electron localization energies calculated by Acheson and Goodall[71] indicate that these salts are unreactive towards electrophilic attack. The 1-and 3-hydroxyquinolizinium salts, on the other hand, rapidly brominate at the 2-and 4-positions respectively. They behave as typical phenols. 2-Hydroxyquinolizinium salt behaves as a strong acid and loses its proton to give 2-quinolizone.

It has been observed that an amino or hydroxy group present in position-1 directs the incoming electrophile to the 2-position while those in position-2 direct to 1-position, as shown in equations (8.21) and (8.22).

$$(8.21)$$

**2-Bromo-1-hydroxy
quinolizinium bromide**

$$(8.22)$$

**1-Bromo-2-hydroxy
quinolizinium bromide**

2. **Nucleophilic Substitution:** Quinolizinium salts react readily with nucleophilic reagents usually at position-4 with simultaneous opening of the ring to form derivatives of 2-butadienylpyridine[72], as illustrated by the following examples:

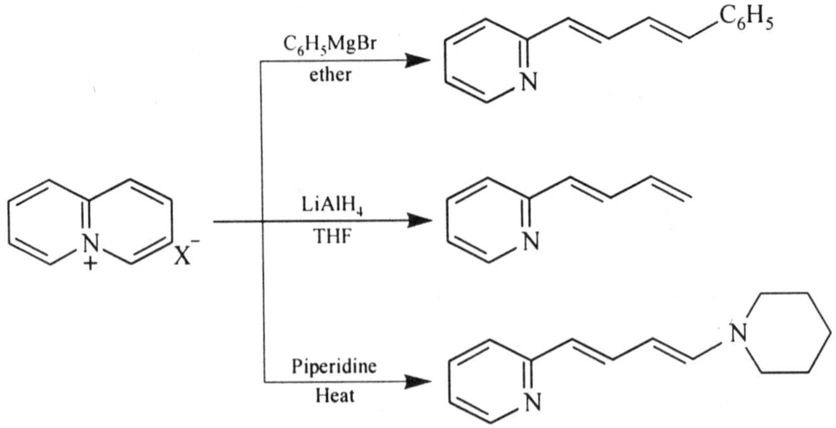

3.   *Condensation and Coupling:* The 2-and 4-methylquinolizinium salts are active and condense with p-N, N-dimethylaminobenzaldehyde to yield the corresponding styryl derivative (53), a Schiff's base equation (8.23).[75] 1-Hydroxyquinolizinium bromide couples with benzenediazonium chloride to give (54), equation (8.24).

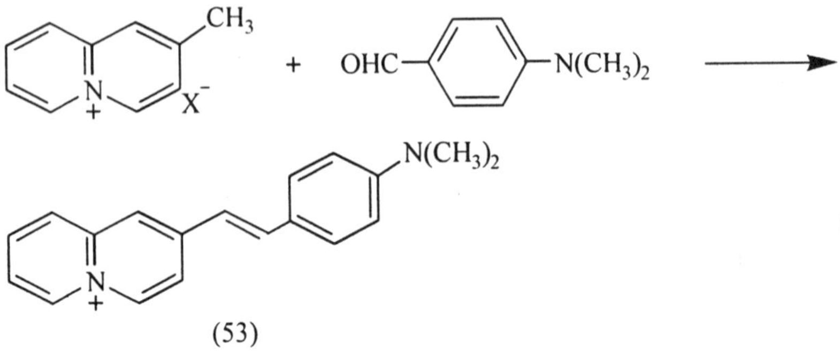

$$(8.23)$$

(53)

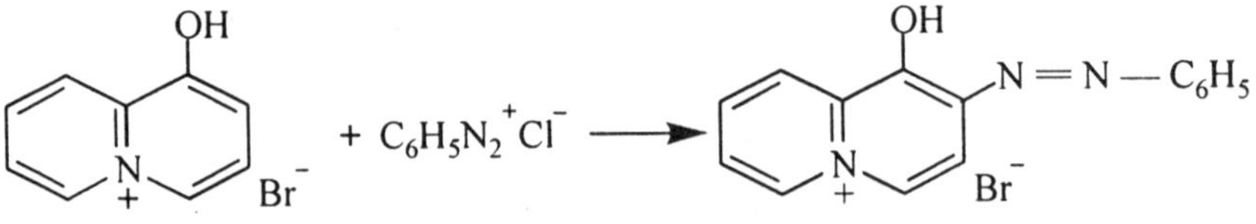

$$(8.24)$$

(54)

Quinolizinium salts do not take part in cycloaddition reactions even with strong nucleophiles. However, the only quinolizinium derivative to undergo cycloaddition is the anhydride (55) which reacts with styrene to give the adduct (56), equation (8.25).

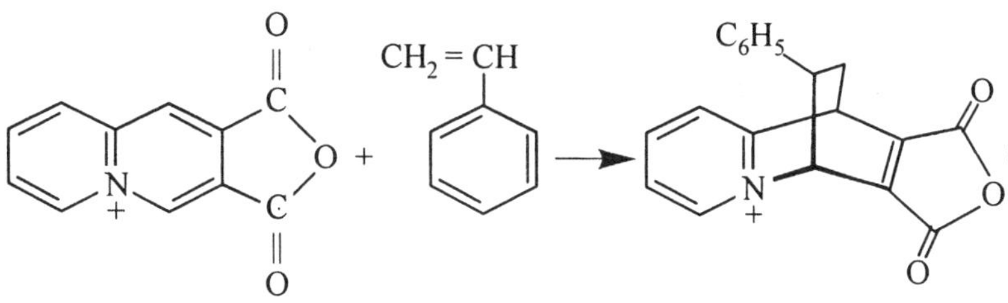

$$(8.25)$$

(55)   (56)

However, acridizinium ion undergoes cycloaddition rather easily with benzyne to afford the adduct (57), equation (8.26).

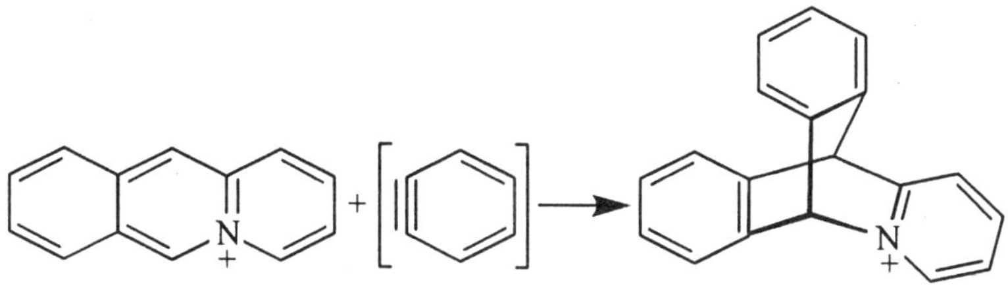

$$(8.26)$$

93%

(57)

This adduct on reduction with sodium borohydride followed by hydrolysis affords anthracene. This method has found application in the preparation of anthracene and related hydrocarbons.

8.4 ACRIDINES

Acridine (58) is the aza derivative of anthracene and was first discovered by Grache and Caro in 1871 in the anthracene fraction of coal tar[76]. This name was given to this compound because of its rather acid and irritating action on the skin. The numbering is shown in the above structure and the nitrogen atom is assigned position-10. Acridine and its derivatives constitute an important series of chemotherapeutic drugs and dyestuffs.

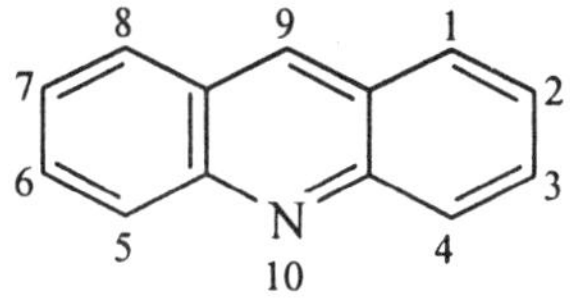

(58)

It is an aromatic compound with a resonance energy of the order of 105 Kcal/mole. It is weakly basic and its basicity does not differ much from quinoline and isoquinoline as is evident from the *p*Ka values. These values indicate that a fused benzene ring does not affect the basicity very much but the trend is on the basis of electron donation by the phenyl group as in the case of phenyl substituted pyridines.

*p*Ka   5.17   4.9   5.1

5.6

Acridine is a pale yellow crystalline solid m.p., 114°C. Acridine and many of its derivatives give highly fluorescent solution. It has a dipole moment of 2.10 D.

8.4.1 Synthetic Methods

Several synthesis of acridines involve formation of acridinones as intermediates.

1. *From Diphenylamine-2-Carboxylic Acid:* *o*-Chlorobenzoic acid is first condensed with aniline in the presence of a base to yield diphenylamine-2-carboxylic acid. The acid is cyclized in the presence of sulfuric acid to acridinone (59) whereas reaction with phosphorous oxychloride gives 9-chloroacridine (60). Reaction of (59) and (60) with appropriate reagents provide acridan or 9, 10-dihydroacridine (61). This on oxidation by air or ferric chloride leads to acridine.[77]

(59)

(60)

(61)

2. *From the Pyrolysis of o-Hydroxymethyldiphenylamines:* The title compound produces[78] acridine in one step during the pyrolysis of *o*-hydroxymethyldiphenylamine:

Mayo and coworkers[79] also obtained acridine from the pyrolysis of certain thioamides.

8.4.2 Chemical Reactions

Acridine possesses many properties which can be derived by extrapolation from the chemistry of pyridine and quinoline.

1. *Reaction with Acids:* Acridine is a weak base but forms soluble salts with mineral acids. Aminoacridines, on the other hand, are strongly basic because of the resonance that is possible in the ion and not in the unionized form. It forms a disalt and the first proton is accepted by the ring nitrogen atom.

2. *Electrophilic Substitution:* The electrophilic substitution as expected takes place in the benzenoid ring. Halogenation gives a mixture of addition and substitution products. Bromination of acridine gives 2-and 2, 7-dibromo products, equation (8.27).

$$(8.27)$$

Nitration yields mixed isomers. Acridine N-oxide, however, undergoes nitration (HNO_3, H_2SO_4, $0°C$) to produce 9-nitroacridine-N – oxide.

3. *Reaction with Nucleophilic Reagents:* The most outstanding chemical property of acridine is that it undergoes nucleophilic attack at the 9-position. This is in accordance with the electron density decrease at this position in comparison to at 1-, 2-, 3-, and 4-positions. Thus reaction of acridine with sodium amide in liquid ammonia gives 9-aminoacridine. This compound is a strong base and its hydrochloride is used as an anti-bacterial agent. The quaternary salts of acridine are also more reactive towards nucleophilic reagents.

4. *Reaction with Oxidizing Agents:* Acridine is a very stable ring system towards the action of oxidizing agents. It is thus not easily oxidized but is converted to acridine N-oxide with peracids. This oxide is a yellow solid, m.p. 169°C and in ethanol solution gives a green fluorescein. It brominates and nitrates at the 9-position.

5. *Reaction with Reducing Agents:* 9-Substituted acridines can be reduced by sodium in ether to the corresponding acridanes. The benzene ring of acridine can be selectively reduced by Pd/H_2 or Rh/C in hydrochloric acid, platinium oxide in trifluoroacetic acid to produce dioctahydroacridine, equation (8.28).

$$\text{(8.28)}$$

1, 4, 5, 8-Tetra hydroacridine is obtained by reduction over lithium in liquid ammonia, equation (8.29).

$$\text{(8.29)}$$

6.   *Photochemical Reactions:* Photochemical behavior of acridine has been investigated.[80] Sasaki and coworkers[81] observed that irradiation of acridine in the presence of quadricyclane or norbornadine yielded a mixture of 1:1 Diels-Alder adducts (62) and (63).

Acridine is reduced photochemically[82,84] in 2-propanol and the reaction proceeds by the abstraction of the most reactive atom from the substrate i.e., *via* a acridanyl radical (64) to give a mixture of acridane (65) (9, 10-dihydroacridane) and 9, 9-diacridane (66) as well as (67) obtained by the addition of 2-propanol.[84-86] Acridine has also been known to undergo thermal Diels-Alder reaction.

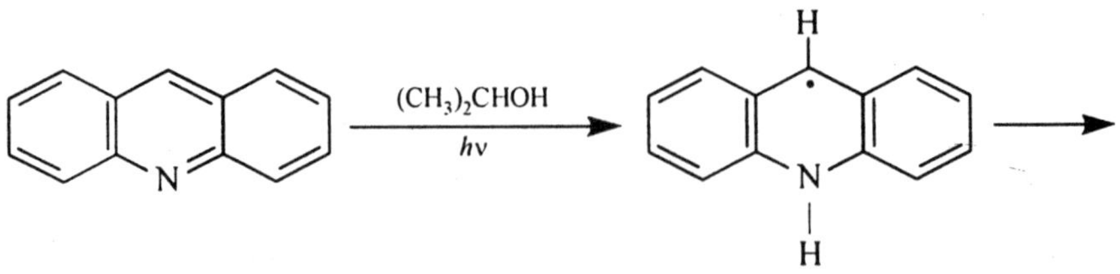

(64)

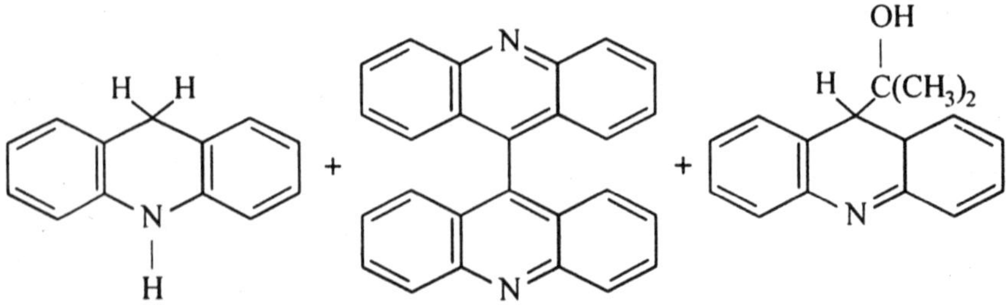

2.6%   46%

(65)   (66)   (67)

Ciganek[87] reported an interesting example of an intramolecular Diels-Alder reaction in which N-methyl-N-propargyl-9-acridine carboxamide (68) gives the Diels-Alder adduct (69).

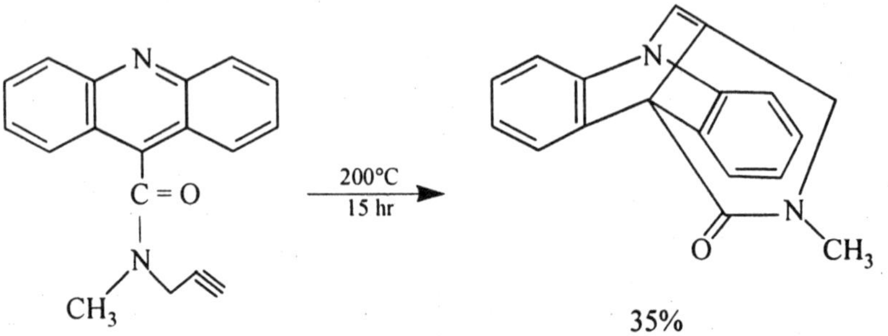

(68)   (69)

8.4.3 Naturally and Biologically Active Compounds

The acridone containing alkaloids have been isolated from the bark and leaves of several trees, for instance, *Melicope fareana, F. Muell and Evodia xanthoxyloids*. These trees are found in the forests of Queensland, (Australia), and the important alkaloids are the following (70-72).

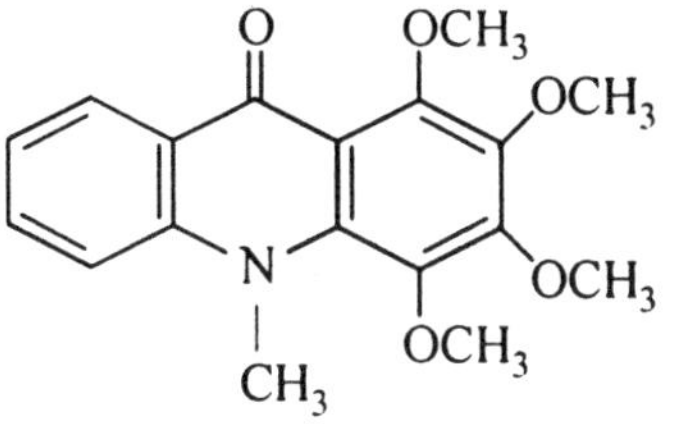

(70)

Melicopicine

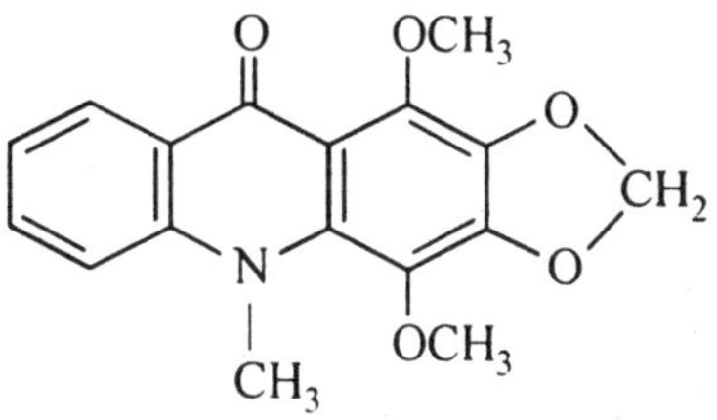

(71)

Melicopidine

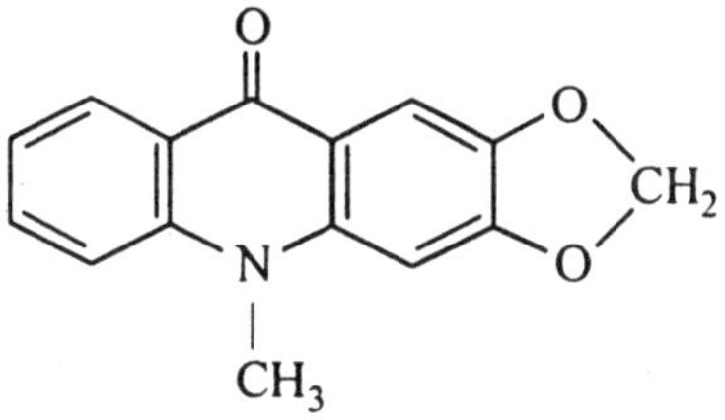

(72)

Eroxanthine

Many of the compounds containing the acridine ring system are employed as chemotherapeutic agents such as stebrin (73), acriflavin and proflavin.

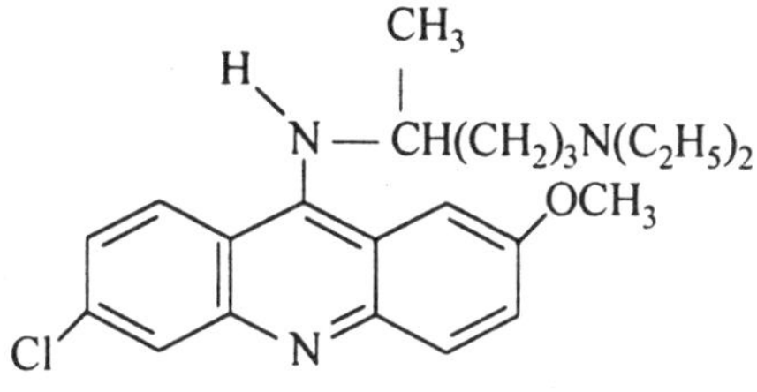

Stebrin

(73)

8.5 PHENANTHRIDINES

The benzoquinolines form two distinctly defined groups, those (74) in which the benzene ring is fused to the benzene nucleus of quinoline and (75) in which the fusion is to the heterocyclic ring. The first group resembles quinoline in its properties while the second resembles acridine.

(74)   (75)

Phenanthridine is also known as *benzo* [c] *quinoline* and is isomeric with acridine. It was first isolated in 1889 from coal tar. Phenanthridine is a colorless m.p., 108°C and gives a weak blue fluorescein in dilute ethanol.

8.5.1 Synthetic Method

A direct synthesis of phenanthridine involves the cyclodehydration of an acyl *o*-aminodiphenyl in the presence of phosphorous oxychloride.

R = H or Alkyl

8.5.2 Chemical Reactions

Phenanthridine is also an aza derivative of phenanthrene like acridine and is thus also has several properties that can be expected from the chemistry of pyridine and quinoline.

1.   *Reaction with Acids:* Phenanthridine and its derivatives behave as tertiary bases and form water soluble monoacid salts.

2.   *Electrophilic Substitution:* The electrophilic substitution reactions, in general, of phenanthridine have not been studied in details, however, such reagents yield a mixture of products. The π-electron density calculations show that position-6 is not attacked by the electrophilic reagents. In contrast to

acridine, all the mono-positions in phenanthridine are non-equivalent. Nitration (HNO_3, H_2SO_4) affords a mixture of 1-and 10-nitrophenanthridines with minor amounts of 2-and 2-, 4-, and 8-isomers as well.

3. *Nucleophilic Substitution:* The familiar Chichibabin reaction with sodium amide in liquid ammonia proceeds rapidly to give 6-aminophenanthridine.

4. *Reaction with Oxidizing Agents:* Phenanthridine is resistant to the action of oxidizing agents but with peracids, phenanthridine N – oxide is formed. Photolysis of 6-cyanophenanthridine N – oxide in ethanol at room temperature results in the formation of 5-ethoxyphenanthridinone (76). The mechanism is thought to proceed *via* an oxaziridine intermediate.[88]

(76)

5. *Reaction with Reducing Agents:* Phenanthridine is easily reduced with tin and hydrochloric acid to give 5, 6-dihydrophenanthridine. The same product can also be obtained by hydrogenation over Raney nickel.[89]

REFERENCES

1. H. C. Longuet-Higgins and C. A. Coulson. *Trans.* Farraday Soc., **43**, 87 (1947).
2. For Synthesis of quinoline, see, G. Jones, *"The Chemistry of Heterocyclic Compounds"*, Vol. 32, Part I, Wiley Interscience, New York, (1977), pp. 93-318, Part II (1985).
3. *"Physical Methods in Heterocyclic Chemistry"*, (Ed.) A. R. Katritzky, Vol. I and II, Academic Press London (1963).
4. N. Campbell in *"Rodd's Chemistry of Carbon Compounds"*, S. Coffey, (Ed.), Vol. 4, Elsevier, Amsterdam, 2nd ed., (1976), p. 231.
5. M. H. Palmer, *J. Chem. Soc.,* 3695 (1962).
6. G. W. H. Cheeseman, *J. Chem. Soc.,* 242 (1960).

7. T. P. Forrest, G. A. Dauphines and W. F. Miles, *Canad. J. Chem.*, **47**, 2121 (1969). et al.
8. T. J. Curphey, *J. Am. Chem. Soc.*, **87**, 2063 (1965).
9. R. H. Manske, *Chem. Rev.*, **30** 113 (1942).
10. G. Kempter and G. S. Hirschbeg, *Chem. Ber.*, **98**, 419 (1965).
11. E. A. Fehnel, *J. Org. Chem.*, **31**, 2899 (1966).
12. E. Liete, *Canad. J. Chem.*, **58** 1806 (1980).
13. C. S. Cho. *et al.*, *Tetrahedron*, **59**, 7997 (2003).
14. A. Cambacorta R. Nicoletti and M. L. Forcellese, *Tetrahedron*, **27**, 985 (1971). et al.
15. R. P. Thummel and D. K. Kohli, *J. Org. Chem.*, **42**, 2742 (1977).
16. P. de Mayo, L. K. Sydnes and G. Wenska *J. Org. Chem.*, **45**, 1549 (1980).
17. M. G. Bures and W. L. Jorgensen, *J. org. Chem.*, **53**, 2504 (1988).
18. C. O' Murchu, *Synthesis,* 880 (1989).
19. P. B. D. de la Mare *J. Chem. Soc.*, **561** (1960).
20. M. J. S. Dewar and P. Mailthis, *J. Chem. Soc.*, **944**, (1957).
21. M. J. S. Dewar and P. Mailthis, *ibid.*, 944 (1957).
22. J. H. Ridd, *Methods in Heterocyclic Chem.*, **1**, 109 (1963).
23. M. T. Leffler, *Org. Reactions*, **1**, 91 (1942).
24. *"The Chemistry of Hetrocyclic Compounds"*, G. Jones (Ed.), John Wiley, Bristol, Vol. 32. p. 2.
25. S. W. Goldstein and P. J. Dambek, *Synthesis* 221 (1989).
26. J. W. Bunting and D. J. Norris, *J. Am. chem. Soc.*, **99**, 1189 (1977).
27. A. McCoubrey and D. W. Mathieson, *J. Chem. Soc.*, 696 (1949).
28. For a review see, F. D. Popp. *Adv. Heterocyclic Chem.*, **24**, 187 (1979).
29. F. D. Popp, W. Blount and P. Melvin, *J. Org. Chem.*, **26**, 4930 (1961).
30. H. J. Den Hertog and D. J. Burman, *Rec, Trav. Chem.*, **91**, 841 (1972).
31. Y. Cherng, *Tetrahedron,* 58, 1125 (2002).
32a. F. W. Vierhapper and E. L. Eliel. *J. Org. Chem.*,**40**, 7229 (1975).
32b. P. Blaczewski and J. A. Joule, *Synthetic Comm.*, **20**, 2815 (1990).
33. V. I. Stenberg, J. Wang, R. J. Baltisberger, R. Van Buren and N. F. Woolsey, *J. Org. Chem., Soc.*, **43**, 2991 (1979).
34. F. W. Vierhapper and E. L. Eliel, *J. Am. Chem. Soc.*, **96**, 2256 (1974).
35. G. Jones in *"Quinolines"*, Part I, G. Jones (Ed.), John Wiley and sons, London (1982). p. 61.
36. E. Ochai, *J. Org. Chem.*, **18**, 534 (1953).
37. F. R. Stermitz C. C. Wei and C. M., *Chem. Comm.*, 482, (1968); *J. Am. Chem. Soc.*, **92**, 2745 (1970)
38. G. Jones and D. J. Baty, *"Quinolines"*, Part II, (Ed.), G. Jones, John Wiley and Sons, New York (1982), p. 379; A. R. Katritzky and J. M. Lagowski, *"Chemistry of Heterocyclic* N-oxides", Academic Press, London (1971); T. Kaiya and Y. kanazoe, *Tel. Letters*, 511 (1985).

39. S. E. Diamond A. Szal Kiewicz and F. Mares, *J. Am. Chem. Soc.*, **101**, 490 (1979); Also see U. Berger, T. Burgemeister, G. Dannhardt, K. K. Mayer and W. Wiegrebe, Tetrahedron, **40**, 220 (1984); K. Akiba, T. Kasai and M. Wadar, *Tet. Letters*, 1709 (1982).

40. H. Giezendanner *et al.*, *Pure Appl. Chem.*, **33**, 339 (1973).

41. H. Giezendanner, H. J. Rosenkranz, H. Hansen and H. Schmid, *Helv. Chim. Acta.*, **56**. 2588 (1973).

42. J. A. Gibson, W. Kynaston and A. S. Lindsey, *J. Chem. Soc.*, 4340 (1955)

43. M. Seyhan, *Chem. Ber.*, **85**, 425 (1952).

44. R. D. Brown, *Quart. Rev.*, **5**, 131 (1951).

45. K. W. Bentley. *"The Alkaloids"*, Interscience, New York)1957).

46. W. Soloman in, *"Chemistry of the Alkaloids"*, S. W. Pelletier, (Ed.), Van Nostrand Reinhold, New York (1970), Chapter 11.

47. R. Venkataraman, *"Synthetic Dyes"*, Academic Press, New York (1952)

48. E. M. Hamer, *"The Cyanin Dyes and Related Compounds"*, Interscience New York (1963).

49. S. F. Dyke, in Rodd's, *"Chemistry of Carbon Compounds"*, S. Coffey, (Ed.), Vol. IV, Elsevier, Amerstdam (1976), Chapter 27; also see. D. L. Boger, *et al.*, *Tetrahedron*, **37**, 3977 (1981).

50. W. M. Whaley and T. R. Govindachari, *Org. Reactions*, **6**, 151 (1952); also see H. Ishii, T. Ishikawa, T. Watanabe, Y. Ichikawa and E. Kawanabe, *J. Chem. Soc. Perkin Trans.*, **1**, 22839 (1984).

51. N. Itih and S. Suganawa, *Tetrahedron*, **1**, 45 (1957).

52. G. A. Reynolds and C. R. Nauser. *Org. Syn. Coll.* Vol., **3** 593 (1955).

53. W. E. Cass. *J. Am. Chem Soc.*, **64** 785 (1942).

54. E. V. Brown. *J. Org. Chem.*, **42** 3208 (1977).

55. R. B. Miller and J. M. Frincker. *J. Org. Chem.*, **45** 5312 (1980).

56. M. J. S. Dewar and P. M. Maitlis. *J. Chem. Soc.*, 2521 (1957).

57. W. Schneider and B. Miller. *Arch. Pharm.*, (Germany). **294**. 360 (1961).

58. C. N. Filler,. *J. Org. Chem.*, **43** 572 (1978).

59. F. D. Popp and W. Blount, *J. Org. Chem.*, **27**, 297 (1962).

60. M. D. Rozwadowska and D. Bronzda. *Canad. J. Chem.* **58**. 1239 (1980).

61. J. Gleghorn, R. B. Moodieand, K. Schofield, *J. Chem. Soc., Perkin Trans.*, **2** 229 (1972).

62. C. Lohse. *J. Chem. Soc., Perkin Trans.*, **2** 229 (1972).

63. A. Padwa and K. Crosby, *J. Org. Chem.*, **39**, 2651 (1974).

64. R. Noyori. M. Kato. M. Kawanisi and H. Nozaki, *Tetrahedron*, **25**, 1125 (1969).

65. G. Illuminati. *Adv. Heterocyclic Chem.*, **16**, 123 (1974).

66. C. E. Teaque and A. Roe, *J. Am. Chem. Soc.*, **73** 688 (1951).

67. J. J. Padbury and H. G. Lindwall, *ibid.*, **67**, 1268 (1945).

68. M. Shama in *"Chemistry of the Alkaloids"*, S. W. Pelletier, (*Ed.*), Van Nostrand Reinhold, New York (1970), p. 31.

69. R. K. Hill, in *"Chemistry of the Alkaloids"*, S. W. Pelletier, (*Ed.*), Van Nostrand Reinhold, New York (1970), p. 414.

70. E. E. Glover and G. Jones, *J. Chem. Soc.*, 3021 (1958); *Chem. and Ind.*, 1456 (1956).

71. R. M. Acheson and D. M. Goodall, *J. Chem. Soc.*, 3225 (1964).

72. P. A. Duke, A. Fozard and G. Jones, *J. Org. Chem.*, **30**, 526 (1965).

73. D. L. Fields T. H. Regan and J. C. Dignan, *J. Org. Chem.*, **33**, 390 (1968).

74. D. L. Fields, *idid.*, **36**, 3002 (1971).

75. T. Miyadera and Y. Kishida, *Tet. letters*, 905 (9165); T. Miyodera, *Chem. Pharm. Bull. (Japan)*. **13** 503 (1965).

76. A. Richards and T. S. Stevens, *J. Chem. Soc.*, 3067 (1958).

77. R. M. Acheson, *"Acridines"*, Interscience, New York (1956).

78. A. Albert and J. B. Wills, *Chem. and Ind.*, **65**, 26 (1946).

79. Y. Mao and V. Boekelheide. *J. Org. Chem.*, **45** 1547 (1980).

80. P. de Mayo, L. K. Syndes and G. Wenskar, *ibid.*, **45** 1549 (1980).

81. T. Sasaki, K. Kanematsu, I. Ando and O. Yamashita, *J. Am. Chem Soc.*, **99**, 871 (1977); *ibid.*, **98**, 2687 (1976).

82. M. Koizume, Y. Ikeda and T. Iwaoka, *J. Chem. Phys.*, **48**, 1869 (1968).

83. E. Vander-Donckt and G. Porter, *ibid.*, **46** 1173 (1967).

84. A. Kira, Y. Ikeda, and M. Koizumi, *Bull. Chem. Soc. Japan*, **39**, 1673 (1966).

85. V. Zanker and H. Linschitz, *J. Am. Chem. Soc.*,**85**, 58 (1963).

86. P. Geruthi and H. Goth, *Angew. Chem. Int. Edin.*, (Engl), **3**, 744 (1963).

87. R. M. Acheson and J. Woollard, *J. Chem. Soc., Perkin Trans.*, **1**, 438 (1975).

88. E. Ciganek, *J. Org. Chem.*, **45**, 1497 (1980).

89. K. Tokumura, H. Goto, H. Kasliwabara, C. Kaneko and M. Itoh, *J. Am. Chem. Soc.*, 102, 5643 (1980); *Tet. Letters*, 2027 (1979); *Chem. Pharm. Bull (Japan)*, **26**, 2508 (1978).

REVIEW PROBLEMS

8. 1. Predict the major product of the following reactions:

(a) (indole with 3-substituted CH$_2$CH(NH$_2$)COOH side chain) + 20% HCHO, 15 hr

(b) — reaction of 3-nitroaniline (O_2N–C$_6$H$_4$–NH_2) with glycerol (CH_2OH–$CHOH$–CH_2OH) under Conc. H_2SO_4, Δ, $C_6H_5NO_2$ →

(c) — reaction of 2-aminobenzaldehyde (CHO, NH_2) with pyruvic acid (CH_3–CO–COOH) under (i) Friedlanders conditions (ii) CH_3OH, HCl →

(d) — reaction of the isoquinolinium–piperidinium quaternary salt under (i) OH^- (ii) Heat →

(e) — reaction of 2-nitrobenzaldehyde (CHO, NO_2) with 1,3-dihydroisobenzofuran under NaOH, reflux →

(f) — reaction of 3-acetylindole (COCH$_3$, N–H) with quinoline under I_2, 100°C, 45 min →

(g) — reaction of 2-aminoacetophenone (CH_3–C=O, NH_2) with ethyl cyanoacetate (CH_2CN–$COOC_2H_5$) under H_2SO_4 →

(h) — reaction of 3-bromo-2-chloropyridine (Br, Cl) with Li, Hg then furan then Li, Hg →

8. 2. Suggest a plausible mechanism for each of the following transformations:

(*a*)

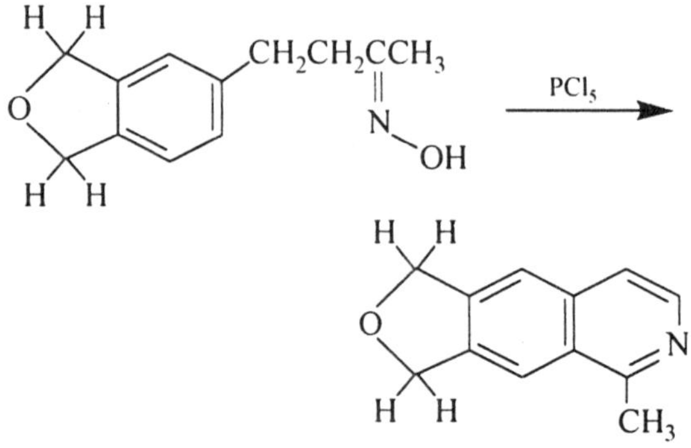

(*b*)

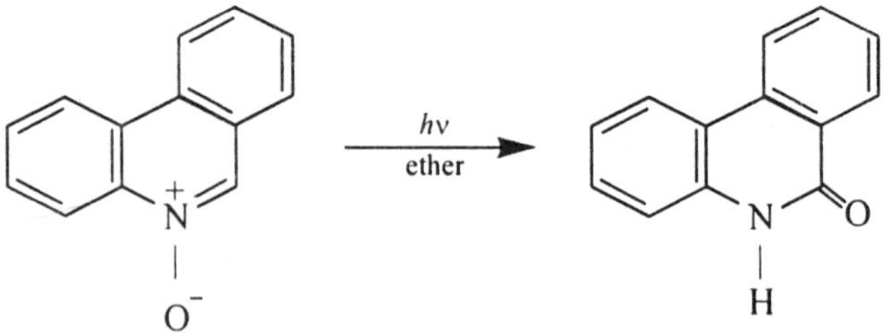

(*c*)

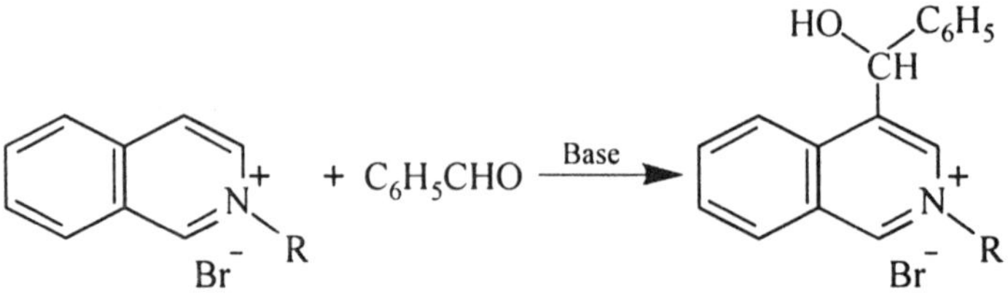

(*d*)

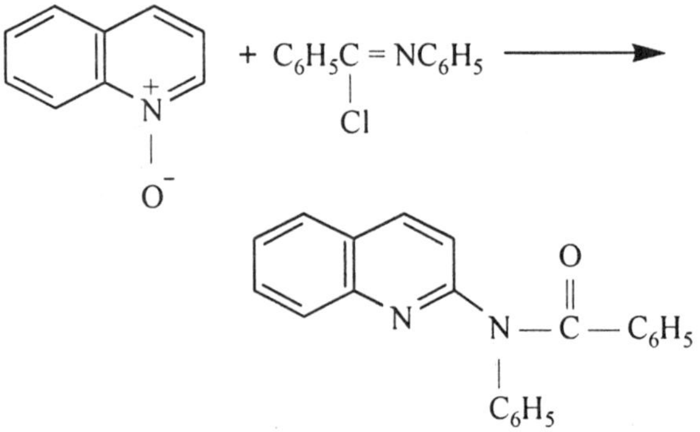

(*e*)

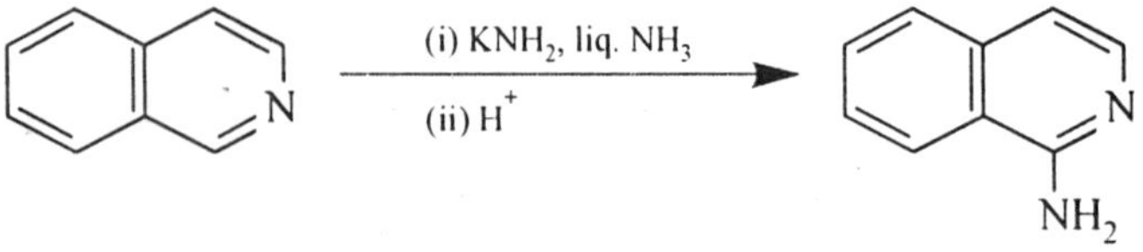

(*f*)

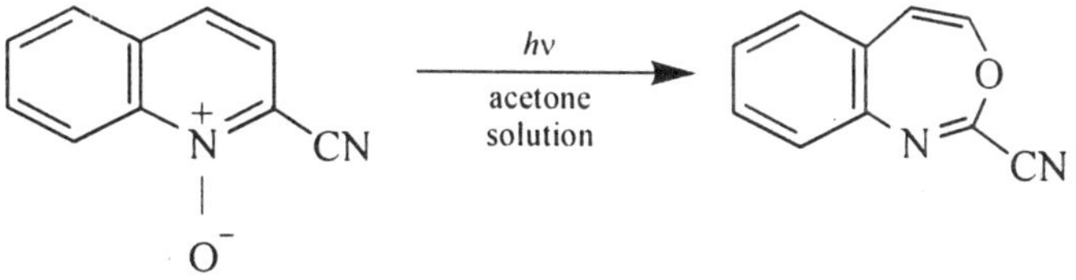
hν
acetone solution
CN
O
N

8. 3. Complete the following reactions:

(*a*)

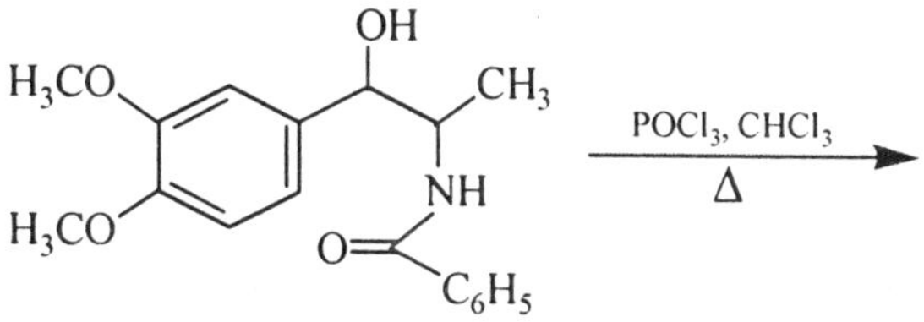
POCl3, CHCl3
Δ

(*b*)

+ CH3COCH2COOC2H5
(i) 110°C
(ii) H2SO4

(*c*)

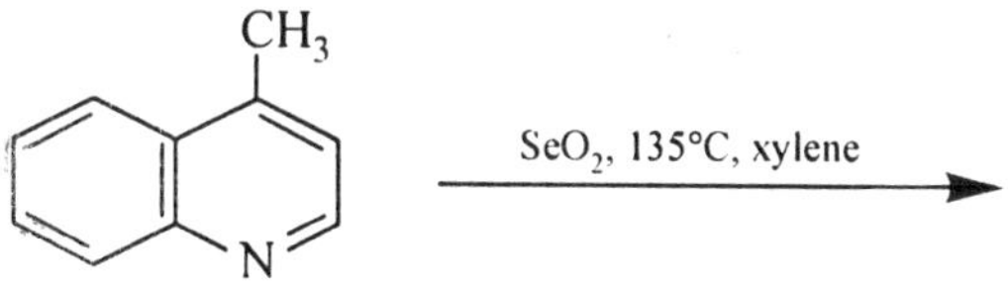
CH3
SeO2, 135°C, xylene

(*d*)

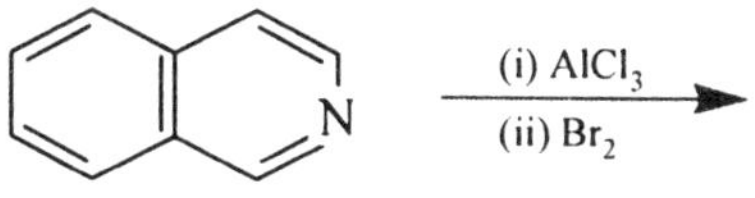
(i) AlCl3
(ii) Br2

(*e*)

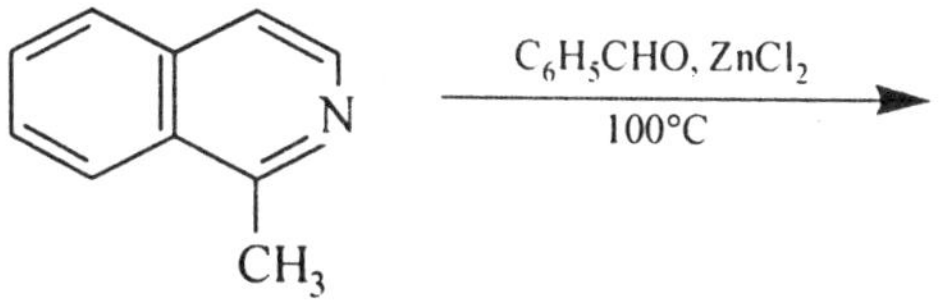
C6H5CHO, ZnCl2
100°C

(*f*)

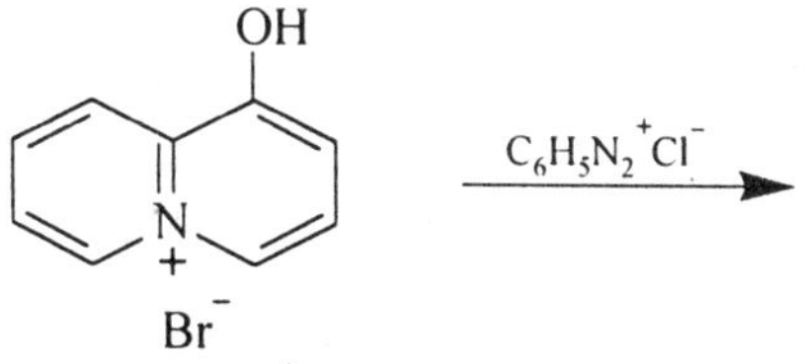
OH
Br
C6H5N2+Cl−

(*g*)

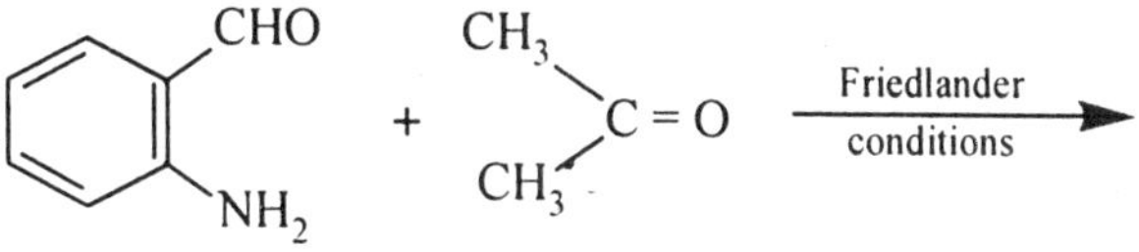
CHO
NH2
Friedlander conditions

(*h*) 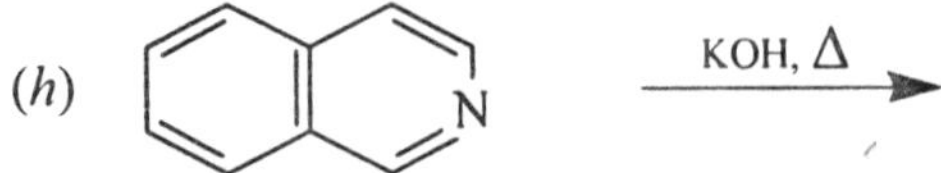

8. 4. Offer suitable explanation for each of the following observations:

    (*a*) 4-Aminoacridine (*p*Ka 8. 04) is more basic than the 3-isomer (*p*Ka 4. 40).

    (*b*) Acrolein itself is not used in the Skraup synthesis of quinoline.

    (c) The Bischler-Napieralski synthesis works best if electron-donating groups are present on the ring.

    (*d*) Quinoline is more easily reduced than naphthalene.

    (*e*) Quinolizinium ion is aromatic but does not undergo electrophilic substitution easily.

    (*f*) 7-Aminoquinoline (*p*Ka 6.5) is more basic than 8-aminoquinoline (*p*Ka 3. 43).

    (*g*) In the Doebner von Miller synthesis, aldehydes R_1CH_2CHO and R_2CH_2CHO give four products.

8. 5. Show by drawing resonance structures of the intermediate that the electrophilic attack in quinoline occurs at C-5 and not at C-6 under acidic conditions.

8. 6. Bring out the similarities and differences between Fischer and Doebner-Miller quinoline synthses.

8. 7. Write structure of the Diels-Alder adduct between acridine and

    (*a*) Quadricyclane and

    (*b*) Norbornadiene

8. 8. Discuss the Pomeranz-Fritsch and Pictet Spangler synthetic methods of isoquinoline.

8. 9. Illustrate with examples how quinoline compares with 1-nitronaphthalene in electrophilic substitution.

8. 10. Using Skraup synthesis prepare:

    (*a*) 8-Methylquinoline

    (*b*) 6-Ethylquinoline

    (*c*) 4-Ethylquinoline

8. 11. Discuss the preparation and properties of quinolizinium salts.

8. 12. Discuss two methods for the synthesis of quinoline.

8. 13. How is quinoline 2-carbaldehyde prepared? Describe its any four chemical properties.

8. 14. What reactants are required in a Skraup synthesis of 6-chloro-3, 4-dimethylquinoline?

8. 15. Write four chemical properties of 1-hydroxyquinoline.

8. 16. Discuss the preparation and chemical reactions of acridines.

Chapter 9

Seven Membered Heterocylic Compounds

The seven-membered heterocyclic rings containing one hetero atom are the heterocyclic analogues of 1, 3, 5-cycloheptatrien – these are *azepine* (1), *oxepin* (2) and *theipin* (3) The interest in the chemistry of these compounds is of recent origin. X-ray crystallography has shown that these systems are not

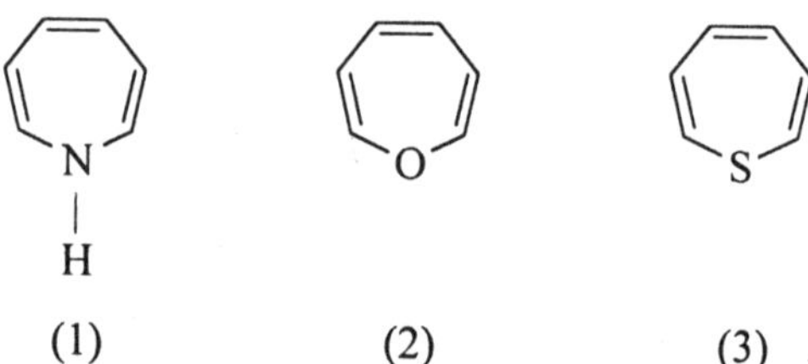

(1)   (2)   (3)

planar and hence bahave as polyenes. They do not comply with Hückel's rule of (4n + 2) π electrons and are thus non-aromatic and possess a high reactivity. They are potential 8π electron systems and are isoelectronic with the unstable cycloheptatrienyl anion.[1]

9.1 AZEPINES[2-4]

Four tautomeric forms designated as 1*H*, 2*H*-, 3*H*-and 4*H*-azepine may be drawn for azacycloheptatriene. The numbering commences from the nitrogen atom. Except 3*H*-azepine none of the others have been isolated.[5]

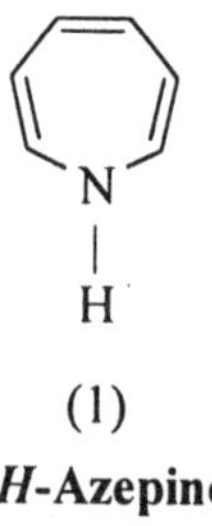

(1)

1*H*-Azepine

1*H*-azepine and *N*-substituted 1*H*-azepine are very unstable and readily rearrange to the tautomeric 3*H*-azepine. The 4*H*-isomer[6] can be isolated and stored but in basic solution it also isomerizes to (3); thus the stabilities of azepine tautomers decrease in the order $3H > 4H > 1H$. The benzo derivatives have been predicted to be more stable provided that annelation does not disturb the aromaticity of the benzene ring. In fact, the $N - H$ tautomer, 2, 3-benzazepin (7) is stable. In $1H$-azepine, there is absence of electron delocalization with

(4)   (5)   (6)   (7)

2H-Azepine **3H-Azepine** **4H-Azepine**

the ring and recent X-ray studies of its N-bromobenzenesulfonyl derivatives have confirmed that the molecule exists in a boat conformation.

9.1.1 Synthetic Methods

Azepines and their derivatives have been synthesized by several methods.

1. *Valence-Bond Isomerization:* This approach has been employed in recent years for the preparation of a large variety of compounds. This reaction differs mechanistically from the majority of the common molecular rearrangements. It is essential for the valence bond migration to take place that the atoms involved in bond reorganization be apparently in a parallel plane so that small adjustments in bond distances, angles and electron distribution can be achieved. This reaction involves a reorganization of the σ-and π-electrons and does not involve a migration of any atom or group. Such organizations which are

(8)

3H-azepine

accompanied by corresponding changes in atomic distances and bond angles are known as *valence bond-isomerization*. A relevant example is the preparation of N-ethoxycarbonyl-1*H*-azepine[8] (8) from benzene and ethoxy-carbonylnitrene (:NCOOC$_2$H$_2$).

Structure (8) is stable because the presence of electron-withdrawing group on the nitrogen atom stabilizes the ring. The parent compound on hydrolysis, however, rearranges to 3*H*-azepine. Paquette and coworkers[9] have reported a more versatile synthesis which permits the preparation of specifically substituted 1*H*-azepines.

2.   *From Azirines:* Azirines are potentially useful percursors for the synthesis of heterocyclic compounds. An azirine on addition to cyclopentadienones thermally affords 3*H*-azepines in excellent yields.[10]
A cycloaddition reaction between 2-phenylazirine and 2, 5-dimethyl-3, 4-diphenylcyclopentadienone in refluxing benzene forms the intermediate (9). This loses carbon monoxide to give (10) which isomerizes to the more stable 3*H*-azepine tautomer.[10]

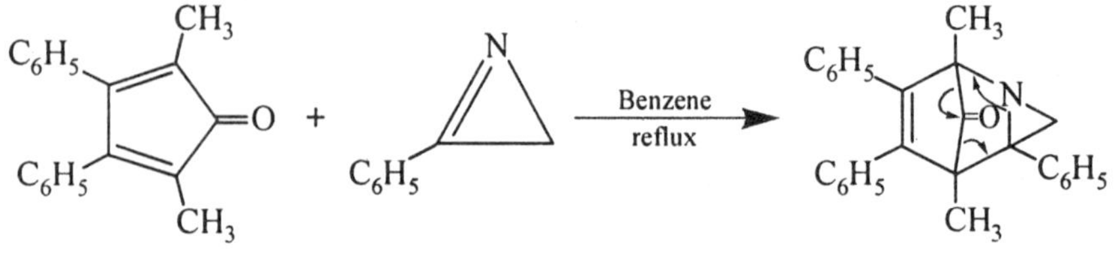

(9)

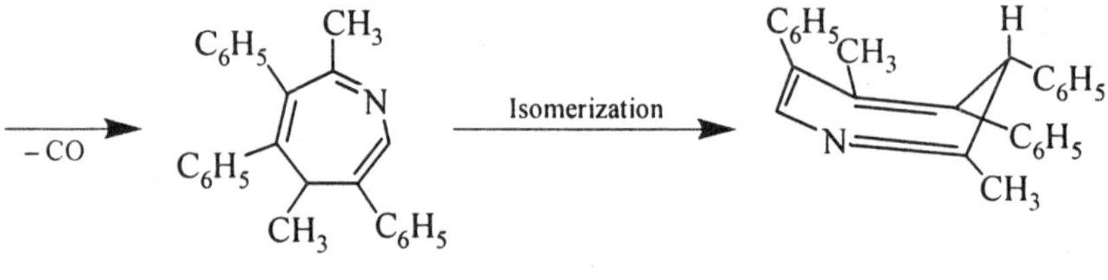

(10)

3.   *From Phenylazide:* A phenylazide on decomposition in boiling primary or secondary amine undergoes ring expansion to yield 3*H*-azepine derivative[11,12] *via* benzazirine (11). The 2-diethylamino-1*H*-azepine (12) tautomer is probably the intermediate[13] in this reaction.

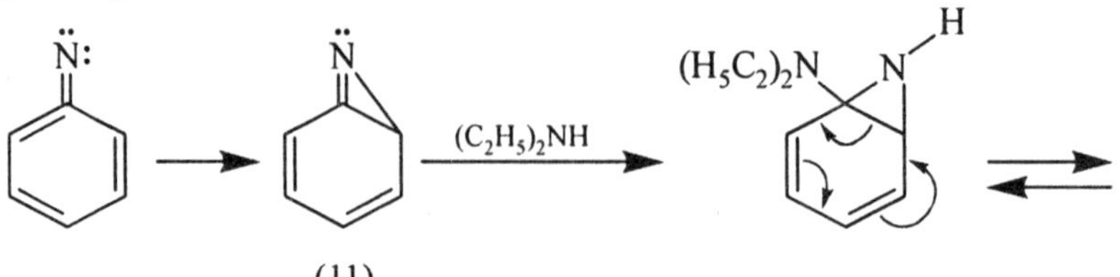

(11)

(12)

4. 2-Phenyl-1, 4-dicyano-1-butene having *syn* nitrile groups on treatment with hydrogen bromide in an anhydrous medium leads to 2-amino-5-aryl-7-bromo-3*H*-azepine.[14]

5. *From Nitrobenzene:* An interesting synthesis of 2-alkoxy-3*H*-azepines has been accomplished from nitrobenzene by its deoxygenation with tributyl-phosphine.[15] The reaction required a primary or secondary alcohol but fails in the presence of phenol. The mechanism involves an initial formation of arylnitrene.

6. N-methyladipimide cyclizes in the presence of thionyl chloride, the intermediate so obtained on further reaction with cupric bromide and triethylamine forms N-methylazepine-2, 7-dione.[16]

Azepine-2, 5-dione may be obtained starting from *p*-benzoquinone and reacting it with hydrazoic acid in the presence of sulfuric acid at 0°C.[17,18] This synthesis is essentially the Schmidt reaction with *p*-quinone instead of the more commonly used carbonyl compounds.

7. Substituted azepines have been prepared by the thermal rearrangement of 7-azaquadricyclanes.[19] The quadricyclanes are obtained from N-substituted pyrroles and dimethylacetylenedicarboxylate (DMAD) in the presence of aluminum chloride by the Diels-Alder reaction. The sequence of reactions is shown below for the preparation of 1, 4, 5-tricarbomethoxyazepine.

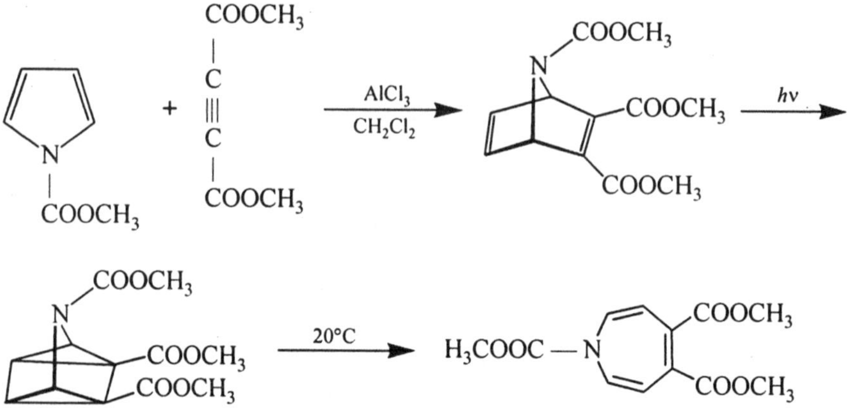

8. The more rare $4H$-azepines have been obtained by $(6 + 2)$ cycloaddition of cyclopropenes with 1, 2, 4-triazines.

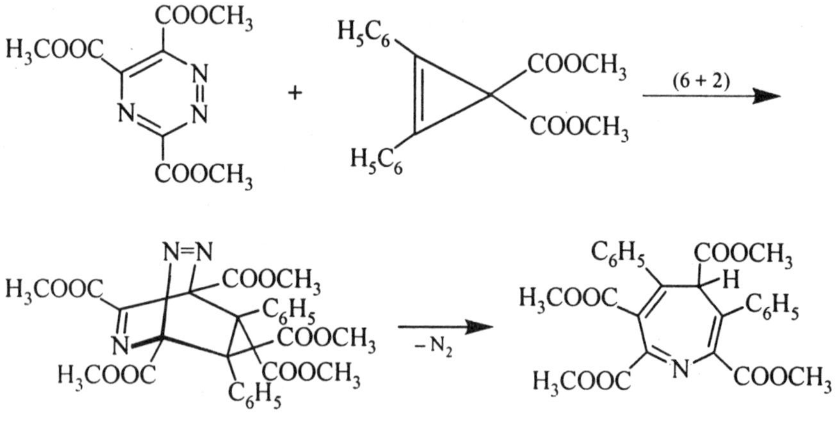

Base or heat treatment (above 180°C) of the metastable $4H$-azepine brings about rearrangement to a mixture of isomeric 3, 5-diphenyl-and 4, 6-diphenyl-$3H$-azepine tetraesters.

9.1.2 Chemical Reactions

Azepines and their derivatives undergo interesting chemical reactions.

1. *Thermal Reaction:* N-carboethoxy-$3H$-azepine on heating results in aromatization to give N-phenylcarbamate, similarly on heating, an aniline derivative 2-methyl-N-carboethoxyaniline results. In contrast N-carboethoxy-

1-*H*-azepine gives only a resinous product in the presence of an acid.[21,22] Substituted 1*H*-azepines give a mixture of products with nitrenes.[23a]

2. **Ring Contractions:** Azepines undergo photo-ring contraction to bicyclic valence tautomers of the type shown below. These transformations are thermally reversible and occur by an orbital symmetry controlled disrotatory electrocyclic process.[23] 2-Methyl-N-carbomethoxyazepine, for instance, yields 2-carbomethoxy-3-methyl-2-azabicyclo [3.2.0] hepta-3, 6-diene in 93% yield on photolysis.[23,24]

$$\text{(9.1)}$$

3. **Reaction with Oxidizing and Reducing Agents:** 2-Dimethylamino-3*H*-azepine may be reduced in the presence of 5% Pd/C and H_2 to give a mixture of 4, 5-dihydro-and 4, 5, 6, 7-tetrahydro-3*H*-azepines.[25] 3*H*-Azepin-2-ones are reduced to caprolactams with a platinum catalyst.

Information on the behavior of azepines towards oxidizing agents is scarce. However, recent work of Sundberg and coworkers[26] has shown that 3-alkyl-2-dialkylamino-1*H*-azepines undergo aerial oxidation to 1*H*-azepin-2-ones (2-azatropones).

4. **The Diels-Alder Reaction:** N-carboethoxyazepine cycloadds to 1, 5-dicarbomethoxy-3, 4-diphenylcyclopentadienone to yield a (6 + 4)π cycloadduct.

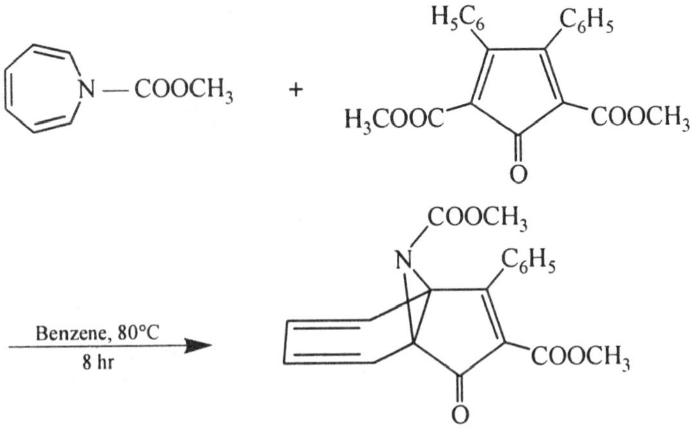

5. *Hydrogen Shifts:* Hydrogen shifts are common in large unsaturated rings such as azepines. A series of 1, 5-hydrogen shifts thermally allowed, connect the $1H$, $2H$, $3H$, and $4H$ isomers of unsubstituted azepines, oxepins and thiepins. Though sigmatropic mechanism permits the interconversions of the isomers, base-catalysis has been observed in some cases. As a result ionic mechanism must be considered.

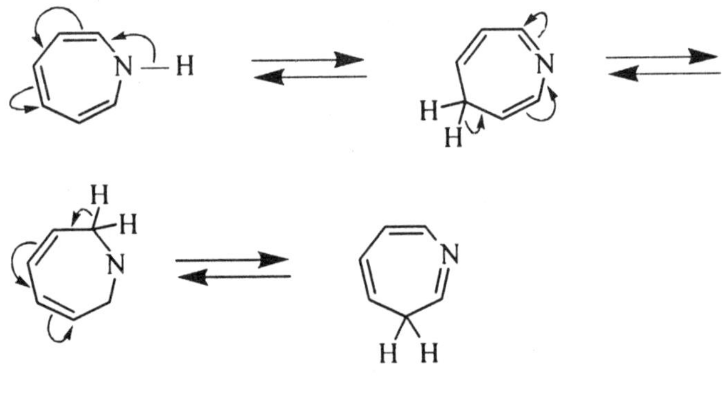

9.2 OXEPINS

Oxepin (2) containing one oxygen atom in the ring was first synthesized by Vogel and coworkers.[28] Subsequently it was demonstrated that benzene oxide (13) interconverts to its valence isomer oxepin by a thermally allowed disrotatory electrocyclic process.[29] Benzene oxide is obtained by the epoxidation of planar benzene.[30, 31]

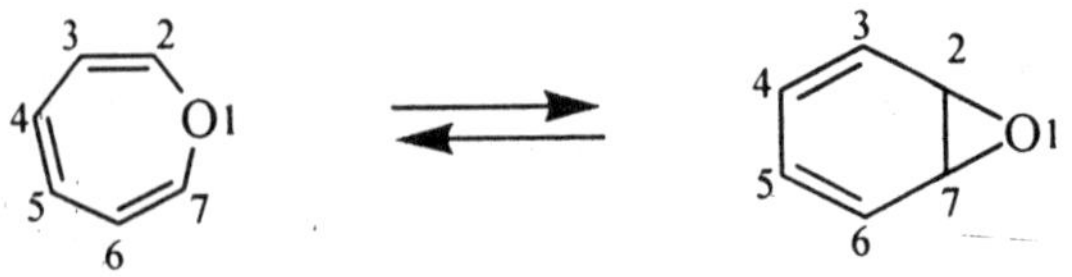

(2)   (13)

Oxepin is an orange liquid, b.p., 380°C/30 mm. Spectroscopic studies have played a crucial role in the investigation of the oxepin-benzene oxide equilibrium. The *n.m.r.* spectrum[32,33] of the liquid shows it to be a mixture of (2) and (13).

The ^{13}C *n.m.r.* spectrum of oxepin has been reported.[31] Substitution at position-2 in (13) favors oxepin while substitution at position-3 favors benzene oxide. The *u.v.* spectrum of the orange compound in isooctane solution shows λ max at 271 nm assigned to oxide with a shoulder at 305 nm due to oxepin.

Naphthalene 1, 2-oxide (14) does not equilibrate with its formal oxepin valence tautomer (15), presumably because conversion of (14) to (15) would involve loss of aromaticity.

(14)   (15)

9.2.1 Synthetic Methods

1.   Oxepin was first prepared by the treatment of 1, 2-epoxy-4, 5-dibromo-cyclohexane with a mild dehydrohalogenating agent (CH_3O^- Na^+/ether).

80%

Substituted oxepins can be obtained in the same manner from appropriate cyclohexa-1, 4-diene derivatives.[34-36] This is illustrated by the following examples:

2-Methyloxepin

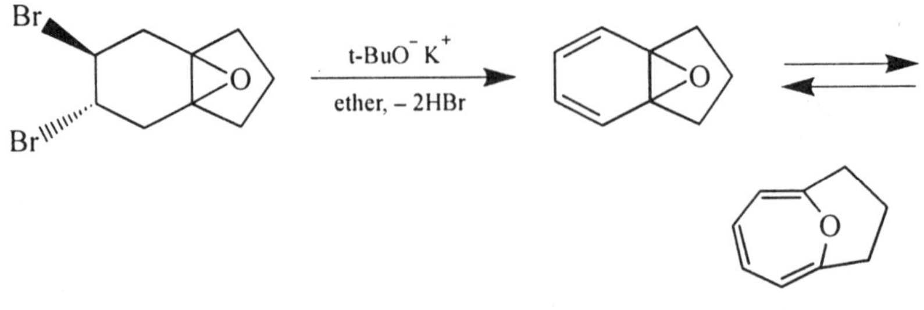

8, 9-Indan oxide

2.   *Valence-Bond Isomerization:* The compound 1, 2-divinylethylene carbonate (16) on pyrolysis in the presence of lithium chloride as catalyst transforms into a mixture of *cis*-and *trans*-divinylethylene oxides, (17) and (18) respectively. Both[37] rearrange to the valence bond tautomer, 4, 5-dihydroxepin.

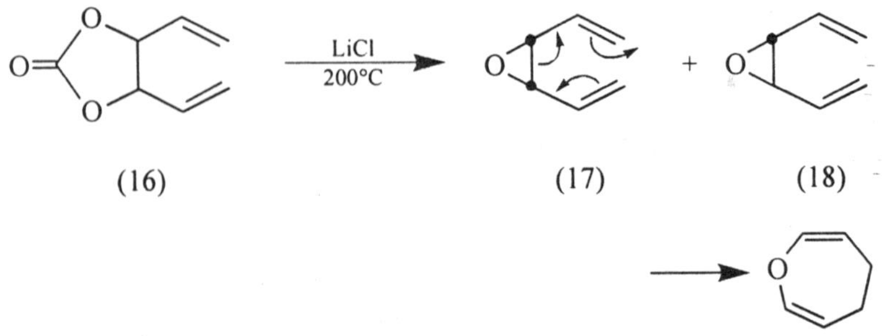

(16)   (17)   (18)

3.   Substituted oxepins (19) have been prepared by thermally induced rearrangement of 7-oxaquadricyclanes.[38,39] The oxaquadricyclane derivatives were obtained by photolysis of the Diels-Alder adducts of substituted furans with acetylene derivatives. The photolytic reaction is greatly promoted in the presence of iodine crystals.[38] Transition metal complexes in aprotic media also promote this isomerization.[40]

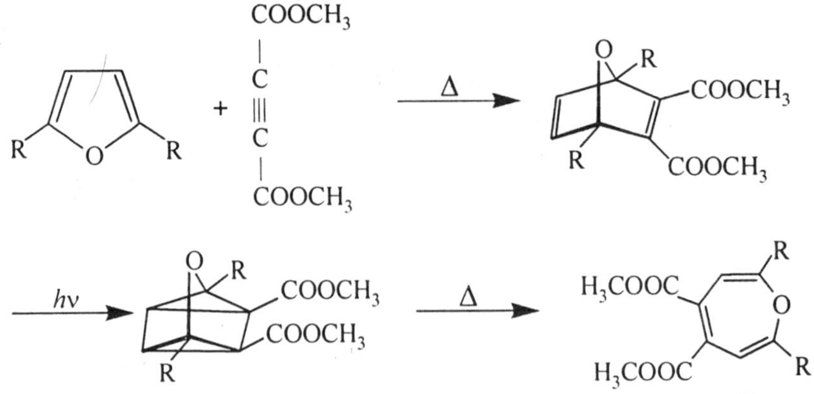

(19)

4. 1, 6-Oxido [10] annulene (20) in a non-polar solvent on silica gel rearranges[41] to 1-benzoxepin (21), equation (9.2).

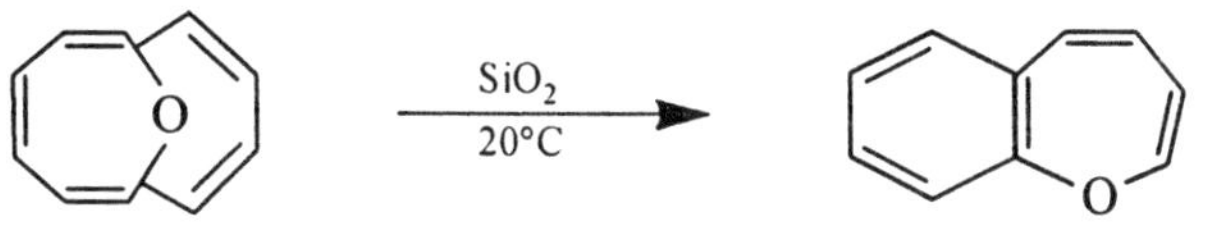

$$\text{(20)} \xrightarrow[20^{\circ}C]{SiO_2} \text{(21)} \qquad (9.2)$$

5. Another interesting method for the preparation of oxepin involves the following sequence of reactions starting from 1-carbomethoxycyclohexa-1, 4-diene.[42]

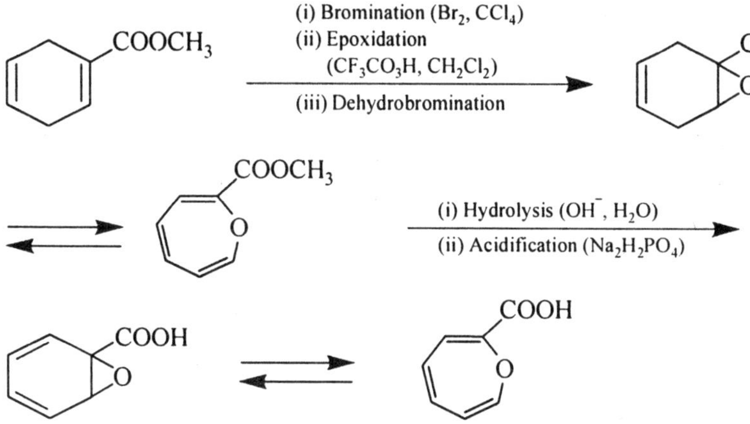

(i) Bromination (Br_2, CCl_4)
(ii) Epoxidation (CF_3CO_3H, CH_2Cl_2)
(iii) Dehydrobromination

(i) Hydrolysis (OH^-, H_2O)
(ii) Acidification ($Na_2H_2PO_4$)

6. Schweizer and coworkers[43] have reported the preparation of 1-benzoxepin from the sodium salt of salicyladehyde and epichlorohydrin in the following steps. In the key step, base treatment of the phosphonium salt (22) induced an intramolecular Wittig reaction together with dehydration to give 1-benzoxepin though in a poor yield (5%).

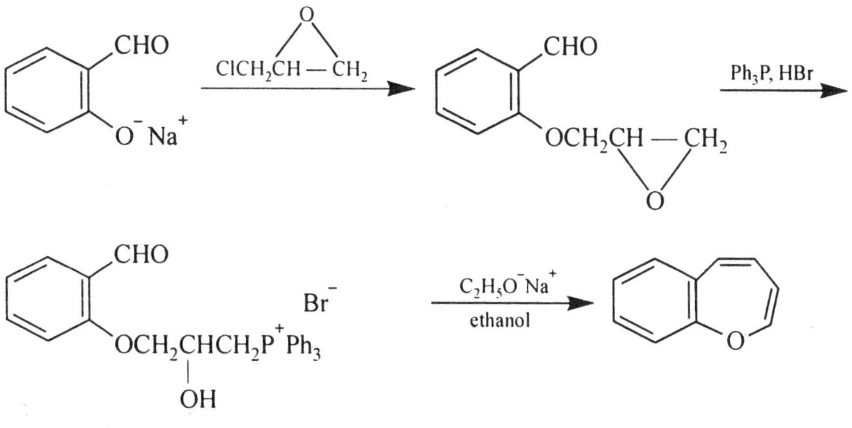

$\xrightarrow{ClCH_2CH-CH_2}$

$\xrightarrow{Ph_3P, HBr}$

$\xrightarrow[ethanol]{C_2H_5O^-Na^+}$

(22)

7.   Epoxidation of Dewar benzene with *m*-chloroperbenzoic acid, and subsequent photolysis or pyrolysis of the epoxide yields the parent oxepin.

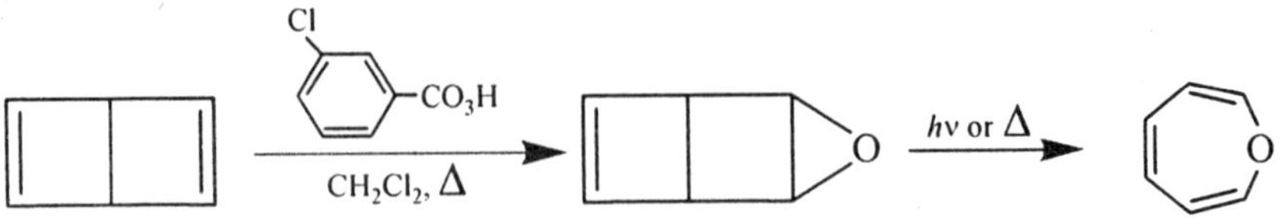

9.2.2 Chemical Reactions

1.   ***Reaction with Acids:*** Oxepin and its derivatives undergo facile thermal or acid catalyzed rearrangement to phenols *via* their arene oxide valence tautomers, equation (9.3). The use of isotopic labelling has revealed that these

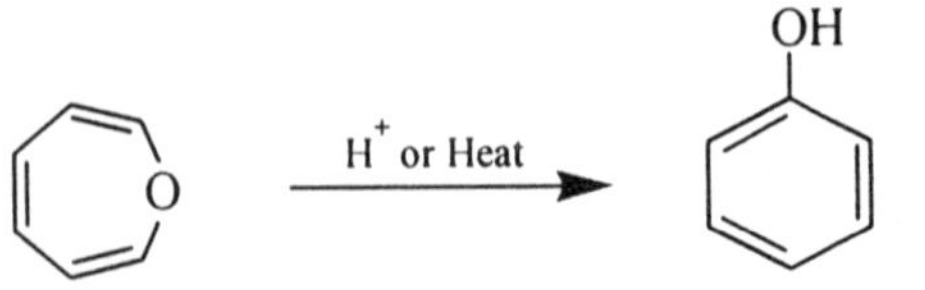

$$\text{(9.3)}$$

reactions are generally accompanied by 1, 2-hydride migration as demonstrated by the isomerization of deuterium labelled 4-methyloxepin (23) to *p*-cresol (24).[44]

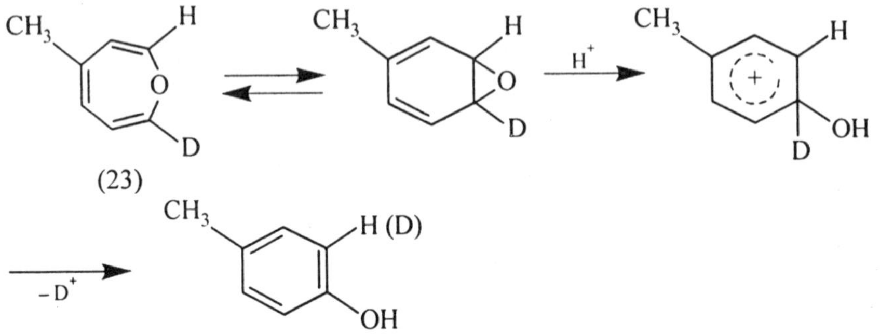

(23)

(24)

Naphthalene 1, 2-oxide similarly undergoes aromatization in aqueous solution catalyzed by acid.[45]

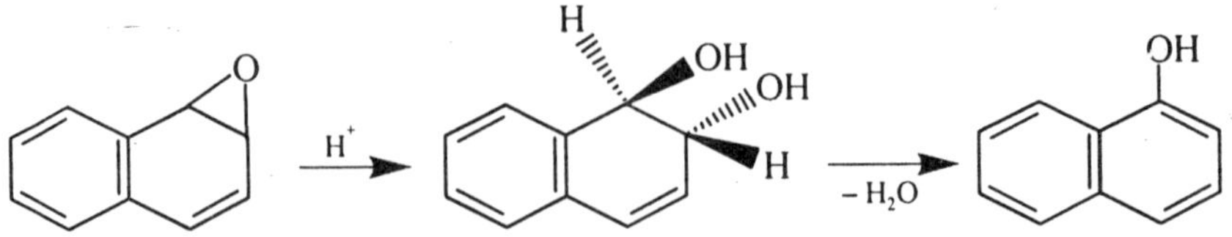

8, 9-Indane oxide behaves similarly towards Brönsted acid. In ether and acid, aromatization takes place to 4-indanol (25). In aqueous solution, on the other hand, 6-indanol (26) is the exclusive product.

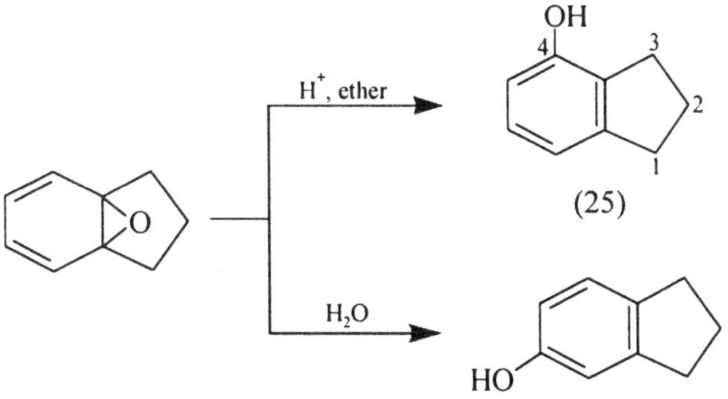

(25)

(26)

2. *The Diels-Alder Reaction:* Oxepin reacts *via* its arene oxide valence tautomer[30] with maleic anhydride to yield the adduct (27). For instance,

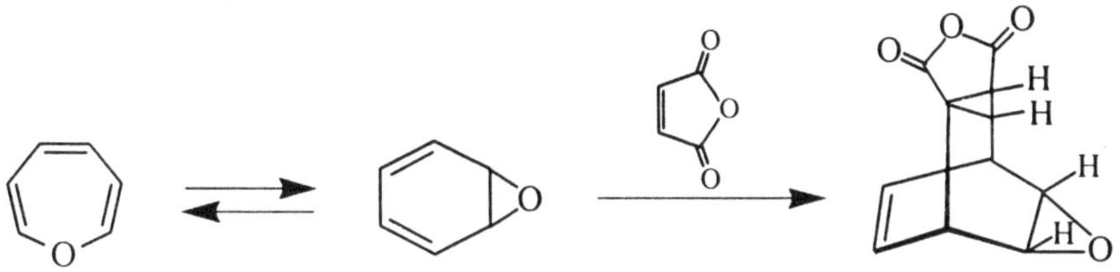

(27)

4-phenyl-1-benzoxepin gives the Diels-Alder adduct (28) in quantitative yield on reaction with tetracyanoethylene, a powerful dienoplile,[46] equation (9.4).

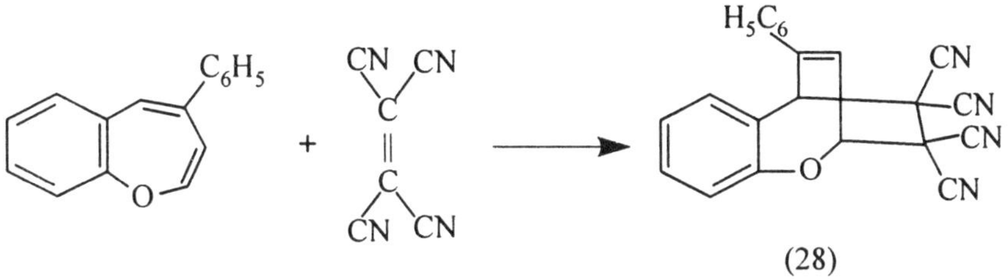

(28)

(9.4)

3. *Photochemical Reactions:* Photochemically oxepin is converted to 2-oxabicyclo [3.2.0] hepta-3, 6-diene[47], equation (9.5). Photolysis of

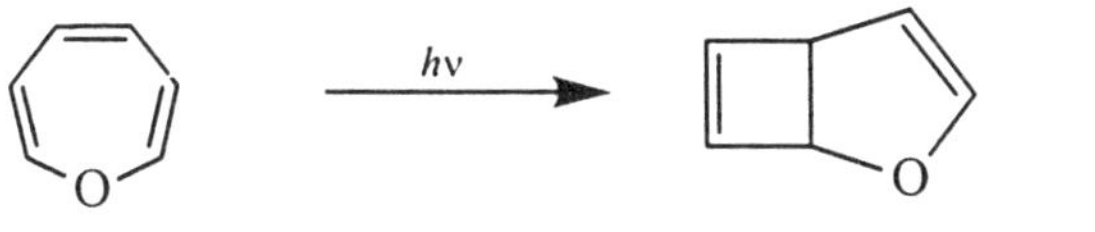

(9.5)

cyclooctatetraene monoperoxide on the other hand, results in oxonin.[48] Oxonin is stable only at low temperature. On heating (30°C) it rearranges to a bicyclic compound (29).

(29)

9.3 THIEPINS[49−51]

Thiepin (3) contains one sulfur atom in a seven-membered ring. It is an isotere of cyclooctatetraene. It contains 8π-electrons, is non-planar and

(3)

anti-aromatic.[52] An electrocyclic ring closure to benzene episulfide (30) followed by sulfur extrusion is a general reaction of all thiepins.[53] Not much is known about simple thiepins but their derivatives are known compounds. Parent thiepin has defied preparation todate.

(30)

9.3.1 Synthetic Methods

Most attempted syntheses of thiepins give benzene derivatives, however, highly substituted derivatives have been obtained in a stable form.

1.   Wynberg and Helder[54] reported that 2, 3, 4, 5-tetramethylthiophene and dicyanoacetylene in the presence of aluminum chloride yielded thiepin derivative (31).

(31)

2. Murata and coworkers[55] developed a novel approach in the isomerization of 4, 5-benzo-3-thiotricyclo [4.1.0.0] heptene[56] to 1-benzothiepin in the following steps.

3. Sterically hindered and stable thiepin has been prepared according to the following sequence of reactions.[57] The furan derivative (32) is oxidized to

(32)   (33)

(34)

dialdehyde (33) which is condensed in diglyme with dimethyl ester of thioglycolic acid and the resulting thiepin derivative (34) is esterified which is more stable.

4. Thiepin 1, 1-dioxide has been prepared by a reaction sequence utilizing a 1, 6-adduct of sulfur dioxide with *cis*-hexatriene.[58]

9.3.2 Chemical Reactions

The chemical reactions of thiepin derivatives have not been investigated to any great extent. The thiepin[54] derivatives (35) is known to give the adduct (36) on warming while on pyrolysis it yields tetramethylphathalonitrile (37). Thiepins behave similarly to oxepins in the cycloaddition reactions.

REFERENCES

1. W. von E. Doering and P. P. Gaspar, *J. Am. Chem. Soc.,* **85**, 3043 (1963)
2. For a review on azepines, see, L. A. Paquette, in *"Non-benzenoid Aromatics"*, J. P. Snyder, *Ed., Academic Press,* New York (1969), p. 249.
3. J. Streith and V. Snieckus, *Acc. Chem. Res.,* **14**, 348 (1981).
4. D. L. Lloyd and H. Mcnab. *Heterocycles,* **11**, 549 (1978).
5. G. Schaden, *Chem. Ber.,* **106**, 2084 (1973).
6. K. Hafner and J. Mondt, *Angew. Chem. Int. Edin.* (Engl.) **5**, 839 (1966).
7. J. C. Paul, S. M. Johnson, L. A. Paquette, J. H. Barrett and R. J. Haluska., *J. Am. Chem. Soc.,* **90**, 5023 (1968).
8. L. A. Paquette and D. E. Kuhla, *Tet. letters,* 4517 (1967); W. Lwowski, *Angew. Chem. Int. Edin.* (Engl.), **6**, 897 (1967). Also see (1980) U. Göckel, U. Hartmannsgruber, A. Steigel and J. Sauer, *Tet. letters,* 595, 599 (1980).
9. L. A. Paquette, D. E. Kuhla, J. H. Barrett and R. J. Haluska, *J. Org. Chem.,* **34** 2866 (1969).
10. D. J. Anderson and A. H. Hassner, *J. Am. Chem. Soc.,* **93** 4339 (1971) also see, A. Hassner and D. J. Anderson, *ibid.,* **94**, 8255 (1972); V. Nair, *J. Org. Chem.,* **33**, 802 (1972).
11. W. von E. Doering and R. A. Odum, *Tetrahedron,* **22**, 81 (1966).
12. R. Huisgen and M. Appl., *Chem. Ber.,* **91**, 12 (1958).
13. G. Maier, *Angew. Chem. Int. Edin.* (Engl.) **6**, 402 (1967).
14. W. A. Nasutavicus, S. W. Tobey and F. Johnson., *J. Org. Chem.,* **32**, 3325 (1967).
15. M. Masaki, K. Fukui and J. Kita., *Bull. Chem. Soc. Japan,* **50**, 2013 (1977).
16. R. Shapiro and S. Nesnow, *J. Org. Chem.,* **34**, 1965 (1969).
17. D. Misiti, H. W. Moore and K. Folkers, *Tet. letters,* 1071 (1965); also see R. W. Richards and R. M. Smith, *ibid.,* 2361 (1966).
18. C. G. Hughes, E. G. Lewars and A. H. Rees, *Canad. J. Chem.,* **52**, 3327 (1974).
19. R. C. Bansal, A. W. McCulloch and A. G. McInnes, *Canad. J. Chem.,* **47**, 2391 (1969).
20. E. Carstensen-Oeser, *Chem. Ber.,* **105**, 982 (1972).
21. L. A. Paquette D. E. Kuhla and J. H. Barrett *J. Org. Chem.,* **34**, 2879, (1969).
22. K. Hafner, *Angew. Chem. Int. Edin.* (Engl.), **3**, 165 (1964).
22a. T. Kumagai, K. Satake, K. Kidoura and T. Mukai, *Tet. letters,* 2275 (1983); *Heterocycles,* **5**, 1569 (1981).
23. G. Jones and L. J. Turbini, *J. Org. Chem.,* **41**, 2362 (1976).

24. L. A. Paquette and D. E. Kuhla, *J. Org. Chem.,* **34**, 2885 (1969) also see, J. S. Swenton, K. R. Burdett, D. M. Madigan and P. D. Rasso, *J. Org. Chem.,* **40**, 1280 (1975).

25. R. A. Odum *et al. Chem. Comm.,* 327, (1973).

26. E. Lerner, R. A. Odum and B. Schmall, *J. Am. Chem. Soc.,* **94**, 513 (1972).

27. R. J. Saunders, S. R. Soter and M. Brenner, *J. Org. Chem.,* **30**, 1887 (1965).

28. E. Vogel, R. Schubart and W. A. Böll, also see E. Vogel and H. Gunther, *ihid.,* **6**, 385 (1967), *Angew. Chem. Int. Edin.* (Engl.), 3510 (1964), Also see D. R. Boyd and M. E. Stubbs, *J. Am. chem. Soc.,* **105**, 2554 (1983); H. S. I. Chao and G. A. Berchtold, *J. Org. Chem.,* **46**, 813 (1981).

29. W. B. Jennings, M. Rutherford, D. R. Boyd, S. K. Agarwal and W. D. Sharma, *Tetrahedron,* **44**, 7551 (1988).

30. T. Nozoe in *"Non-Benzenoid Aromatic Compounds,"* D. Ginsberg, (Ed.,) Interscience, New York (1959);p. 339.

31. D. M. Jerina *et al., Heterocycles,* **1**, 267 (1973).

32. D. M. Hayes, S. D. Nelson, W. A. Garland and P. A. Kollman, *J. Am. Chem. Soc.,* **102**, 1255 (1980).

33. H. Gunther, *Tel. letters,* 4085 (1965).

34. D. M. Jerina, J. W. Daly and B. Witkop, *J. Am. Chem. Soc.,* **90**, 6523 (1968).

35. E. A. Fehnel, *ibid.,* **94**, 3961 (1972).

36. D. R. Boyd, G. A. Brechtold, M. J. McMarius, D. A. Kennedy and J. F. Malone, *J. Org. Chem.,* **51**, 2784 (1986).

37. R. A. Braun, *J. Org. Chem.,* **28**, 1383 (1963); E. L. Stogryn, M. H. Gianni, and A. J. Passannanic, *ibid.,* **29**, 1275 (1964).

38. R. K. Bansal, A. W. McCulloch, P. W. Rasmussen and A. G. McInnes, *Canad. J. Chem.,* **53**, 138 (1975).

39. H. Prinzbach and H. Babsch, *Angew. Chem. Int. Edin.* (Engl.), **14**, 753 (1975).

40. R. Roulet, J. Wenger, M. Hardy and P. vogel, *Tet. letters,* 1479 (1974).

41. E. Vogel *et at., Angew. Chem.,* **76** 785 (1964).

42. D. R. Boyd and G. A. Brechtold, *J. Am. Chem. Soc.,* **100**. 3958 (1978).

43. E. E. Schweizer, M. S. El. Bakoush, K. K. Light and K. H. Oberle, *J. Org. Chem.,* **33**, 2590 (1968).

44. D. M. Jerina, *et al., J. Am. Chem. Soc.,* **90**, 6523 (1968).

45. G. J. Kasperck and T. C. Bruice, *J. Am. Chem. Soc.,* **94**, 198 (1972).

46. H. Hofmann and P. Hofman, *Ann.,* 1571 (1975).

47. A. G. Anastassion and R. P. Cellura, *Chem. Soc.,* **88**, 1718 (1966).

48. A. G. Anastassion and R. P. Cellura, *Chem. Comm.,* 1521 (1969).

49. For a review see, A. Rosowsky, *"Seven-membered Heterocyclic Compounds containing Oxygen and Sulfur"*, Wiley Interscience, New York (1972), p. 573.

50. I. Murata and K. Nakasuji, *Topics Curr. Chem.*, **97**, 33 (1981).

51. D. N. Reinhoudt, *Rec. Trav. Chem.*, **101**, 277 (1982).

52. M. J. S. Dewar and N. Trinajstic, *J. Am. Chem. Soc.*, **92**, 1453 (1970).

53. T. J. Borton, M. D. Martz and F. G. Zike, *J. Org. Chem.*, **37**, 552 (1972).

54. H. Wynberg and R. Helder, *Tet. letters*, 3647 (1972).

55. T. Murata and T. Tatsuoka, *Tet. letters*, 2697 (1975).

56. I. Murata, T. Tatsuoka and Y. Sugihara., *Angew. Chem. Int. Edin.*, (Engl.), **13**, 142 (1974), *Angew. Chem. Int. Edin.* (Engl.), **13**, 142 (1974).

57. J. M. Hoffman, *Jr.*, and R. H. Schlessinger, *J. Am. Chem. Soc.*, **92**, 5263 (1970).

58. R. A. Aitken, *et. al.*, *Chem. Comm.*, 1164 (1982).

REVIEW PROBLEMS

9.1 Predict the products of the following reactions:

(a) [chemical structures] + [chemical structure] $\xrightarrow{\text{Toluene, }\Delta}$

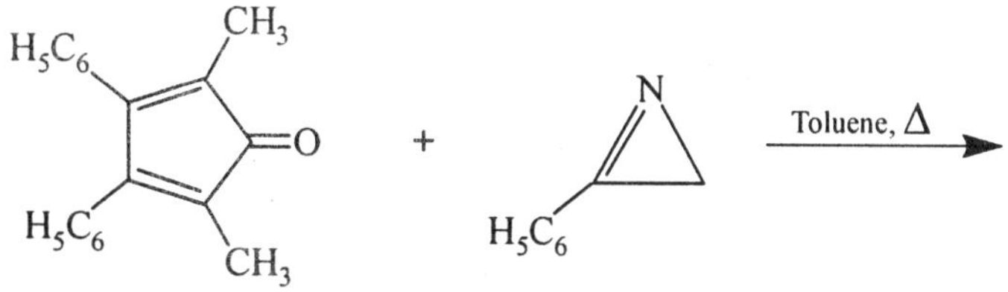

(b) [chemical structure] $\xrightarrow{h\nu}$

(c) [chemical structure] $\xrightarrow{CH_3O^- Na^+}$

(d) [chemical structure] $\xrightarrow{K^+ \ t\text{-BuO}^-}$

(e) [chemical structure] $\xrightarrow{\text{Cope rearrangement}}$

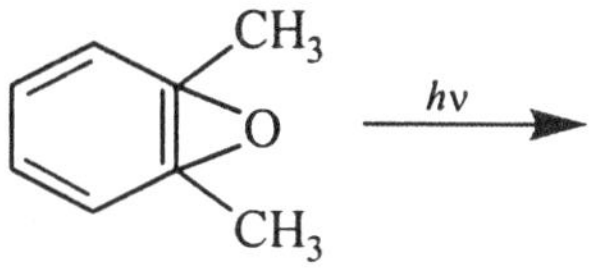

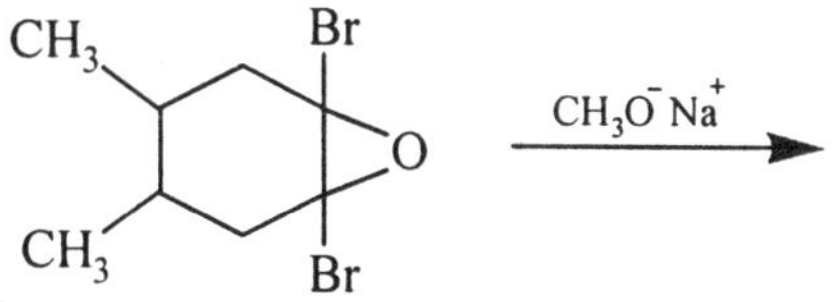

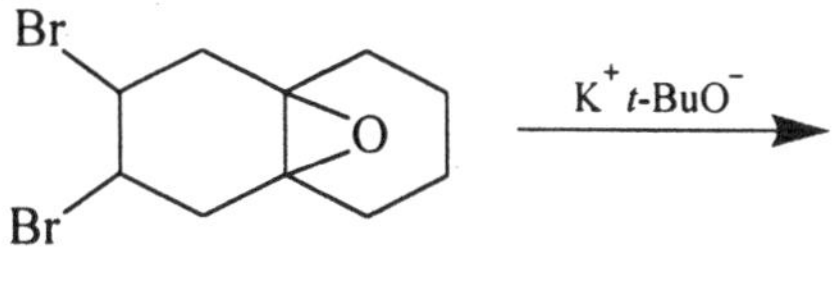

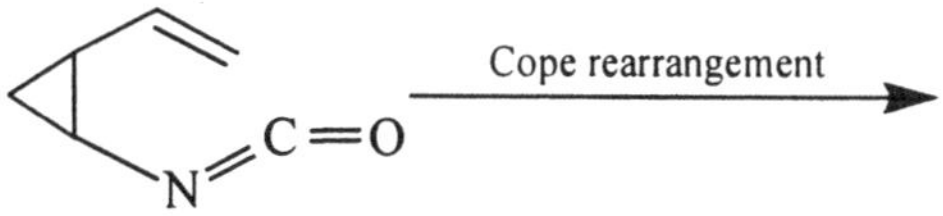

(f) $\Delta \longrightarrow$

(g) $\xrightarrow{h\nu}$

(h) $\Delta \longrightarrow$

9. 2 Describe different methods for the preparation of oxepins.

9. 3 Write a short note on valence-bond isomerization.

9. 4 Predict the mechanism for the following reactions:

(a) $\xrightarrow{H^+}$

(b) $\xrightarrow{h\nu}$

9. 5 Suggest a synthesis for 2, 7-diphenyloxepin.

9. 6 Write two methods for the synthesis of 3*H*-azepines from reagents involving nitrene intermediates.

Chapter 10

Five Membered Heterocyclic Compounds with Two Hetero Atoms

Replacement of a methylene group ($- CH_2$) in 1, 3-cyclopentadiene by N, O or S results in the resultant formation of pyrrole, furan and thiophene respectively. Since nitrogen is trivalent it can be substituted only for a methine ($- CH=$) group in a five-membered heterocyclic ring. Five membered heterocyclic compounds with an additional hetero atom are termed *azoles*. Thus *azoles* containing two nitrogen atoms; one oxygen and one nitrogen atom; one sulfur and one nitrogen atom in the 1, 2-position are designated as *pyrazole* (1), *isoxazole* (2) and *isothiazole* (3) respectively. When both the hetero atoms are present in a 1, 3-relationship then they are referred to as *imidazole* (4), *oxazole* (5) and *thiazole* (6). The numbering in these heterocyclic compounds,

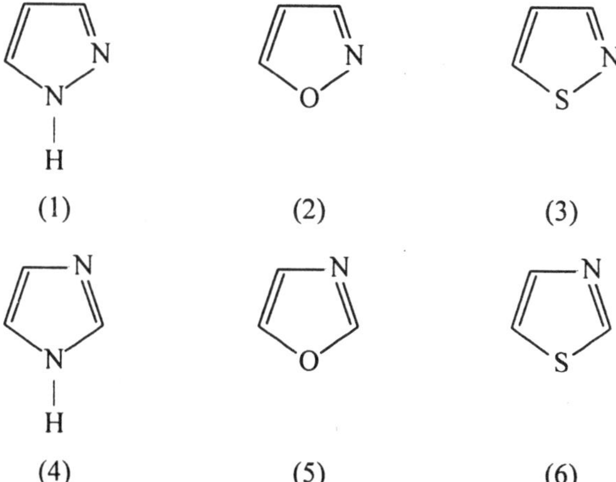

by convention, commences from the hetero atom. But when two hetero atoms are present in the ring then a choice is necessitated and the hetero atom which is in the periodic table and the element of least atomic weight in that group is often given the preference. This implies that among the three common hetero atoms, the order of preference is oxygen, sulfur and nitrogen. In other words, oxygen is assigned position-1 in preference to others.

Condensed ring systems of these heterocycles are also known and they are named as derivatives of the parent azoles.

Many of the azoles comprise the ring system of several natural and synthetic compounds which are important for the living systems and also as important drugs, dyes and agriculture chemicals.

The direct linking of two hetero atoms in 1, 2-azoles has a profound effect on basicity. This results in weakening of the basicity as in hydroxylamine (*p*Ka 5.8) and hydrazine (*p*Ka 7.9) as compared to ammonia (*p*Ka 9.2). Pyrazole (*p*Ka 2.5) is thus weaker than imidazole (*p*Ka 7.2) and isoxazole (*p*Ka 2.03) is weaker than oxazole (*p*Ka 0.8). The 1, 3-azoles, on the other hand, are basic compounds as the azomethine nitrogen has two electrons available for donation.

All the azoles are aromatic though resonance energies for many of them have not been measured, and their electronic structures follow from their relationship to pyrrole, furan and thiophene. The lone-pair of electrons on the hetero atom contributes towards the aromatic sextet. For the construction of the molecular orbital structure, each carbon atom contributes one p_z electron, the nitrogen atom gives the fourth electron and the second hetero atom (X), i.e. nitrogen in pyrazole or oxygen in oxazole, gives two electrons to complete the aromatic sextet. From the diagrams it is obvious that the azoles nitrogen

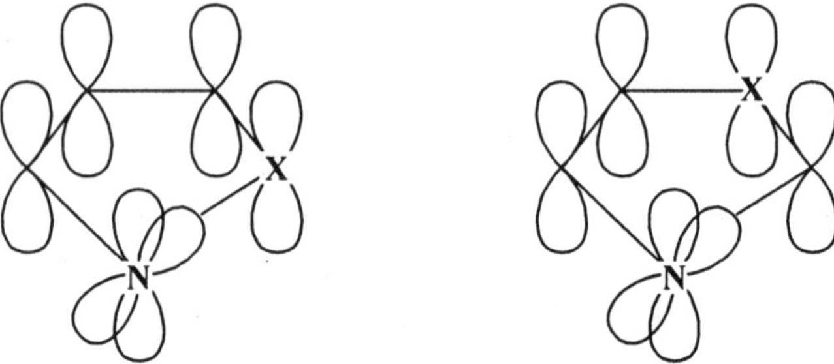

possesses an electron-pair which is situated in orthogonal fashion to the molecular π-could. It is this pair of electrons which permits the azoles to behave as basic compounds and as nucleophiles. In a manner similar to other aromatic compounds no single valence-bond structure can adequately represent these molecules which must be considered rather as a resonance hybrid of a number of contributing structures. For 1, 2-azoles these constitute the following:

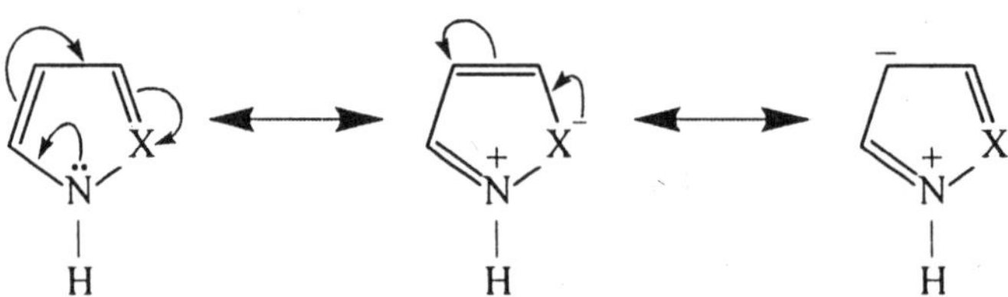

The various resonance contributing structures for the 1, 3-azoles may be written as follows:

1, 2-Azoles are much less reactive than 1, 3-azoles and do not usually undergo electrophilic substitution in acidic solutions. An electrophilic reagent generally attacks position-4 or -5 in both of these molecules. The dipole moments of some of these compounds have been determined which indicate the degree of charge separation.

Proton chemical shifts (δ ppm) in these compounds are given below:

	H-2	**H-3**	**H-4**	**H-5**
Pyrazole		7.61	7.31	7.61
Imidazole	7.86		7.25	7.25
Isoxazole		8.14	6.28	8.39
Oxazole	7.95		7.09	7.69
Izothiazole		8.54	7.26	8.72
Thiazole	8.88		7.98	7.41

A look at this data reveals that an adjacent oxygen and particularly a sulfur atom induces field shifts than either type of nitrogen atom.

10.1 PYRAZOLES AND BENZOPYRAZOLES

Pyrazole (1)[1-5] was first described by Büchner in 1889 who discovered it during the decomposition of pyrazole 3, 4, 5-tricarboxylic acid. The benzo derivative is called benzopyrazole (7). These ring system are numbered as shown.

(1)   (7)

The interest in pyrazoles stemmed from their application in drugs, dyes and as anaesthetics. Pyrazoles have also been used as antioxidants in fuels but their major applications have been in medical and agricultural fields.

Orisul is an example of a pyrazole sulfonamide.

The dihydro pyrazoles are called *pyrazolines* and three of them are possible depending on the position of the double bond. These are 1-pyrazoline (8), 2-pyrazoline (9) and 1, 3-pyrazoline (10). Out of these structure (10) is the most common.

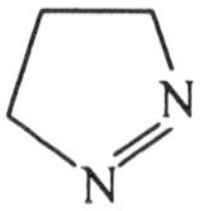

(8) 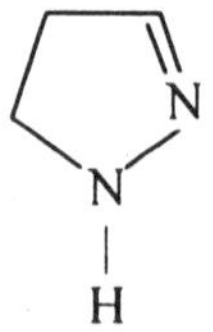   (9) 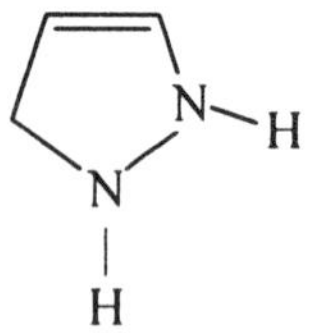   (10)

Pyrazolines are less stable than the corresponding pyrazoles but can be converted into the latter by using mild oxidizing agents such as bromine or lead tetraacetate.

10.1.1 Physical and Spectroscopic Properties

Pyrazole is a colorless solid, m.p. 70°C and is soluble in water. It possesses a penetrating pleasant smell unlike most amines. Introduction of alkyl groups in the ring causes an increase in the boiling point. Pyrazole itself has a high boiling point (187°C) but N-methylpyrazole has a lower value (127°C). The molecular weight of pyrazole is normal in the gas phase but in non-polar solvents like benzene it tends to associate through hydrogen-bonding. The hydrogen bond formation is absent in pyrrole (b.p. 130°C) and in pyridine (b.p. 115C) and as a consequence these compounds have lower boiling points than pyrazole. From these observations it is suggested that the association in pyrazole may be either linear (11) or cyclic (12) i.e. a dimer:

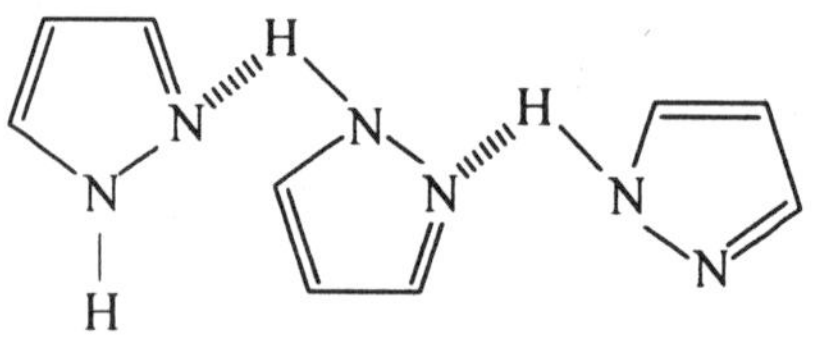

(11) 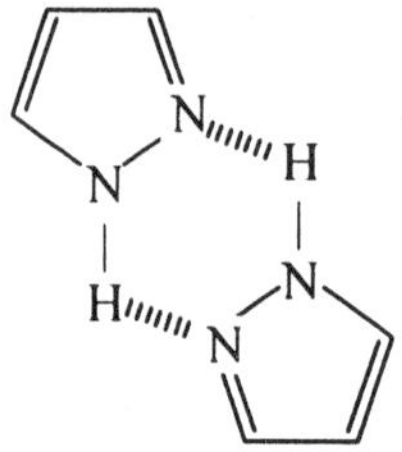   (12)

Pyrazoles which are unsubstituted at position-1 exhibit annular tautomerism. The two tautomeric froms are identical entities and are in rapid equilibrium with each other. The two nitrogen atoms are thus indistinguishable. Thus 3- methylpyrazole is the same as 5-methylpyrazole and gives rise to one signal in *n.m.r.* In case of unsymmetrical substitution, two distinct tautomers are possible and each, in fact, behaves as a mixture upon chemical reaction.

Pyrazoles with free N – H groups are amphoteric though notably more basic than acidic, the *p*Ka is 2.5. The nitrogen atom at position-1 is not basic while at position-2 is weakly basic.

The dimensions of pyrazole have been determined by X-ray crystallography[6]. Both C — C and C — N bond distances (1.35 Å) are equal within experimental error. Pyrazole exhibits aromatic character with properties resembling both pyrrole and pyridine. The resonance energy calculated from heats of combustion data has been found to be 29.3 Kcal/mole. The electron density calculations show that the electrophilic attack takes place at the 4-position. The dipole moment of pyrazole has been found to be 2.64 D (dioxane). The ultraviolet and infrared spectra of pyrazole have been well studied.[5]

10.1.2 Synthetic Methods

The Pyrazoles can be synthesized by the following general methods.

1. *From Dicarbonyl Compounds:* The most straight forward method of pyrazole synthesis involves a reaction between a 1,3-dicarbonyl compound and hydrazine or its derivative. A simple pyrazole is obtained with a 1, 3-dicarbonyl compound such as acetylacetone and hydrazine or phenyl hydrazine.

3,5-Dimethylpyrazole

3,5-Dimethyl-1-phenylpyrazole

The reaction probably proceeds *via* the formation of mono hydrazone which cyclizes under the experimental conditions. The principal drawback of this method is that unsymmetrical dicarbonyl compounds or a substituted hydrazine generally gives an isomeric mixture of two products.

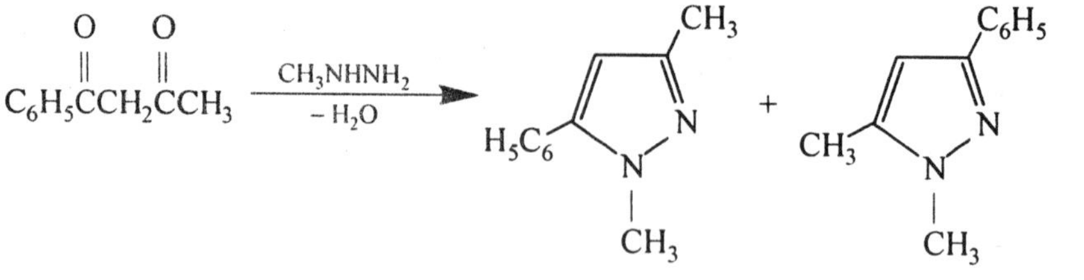

1,3-Dimethyl-5-phenylpyrazole   **1,5-Dimethyl-3-phenylpyrazole**

Analogous synthesis of indazoles can be accomplished from hydrazones of *ortho* substituted aromatic aldehydes and ketones. The cyclization is accompanied at high temperature in the presence of potassium carbonate to facilitate elimination of HCl, equation (10.1).

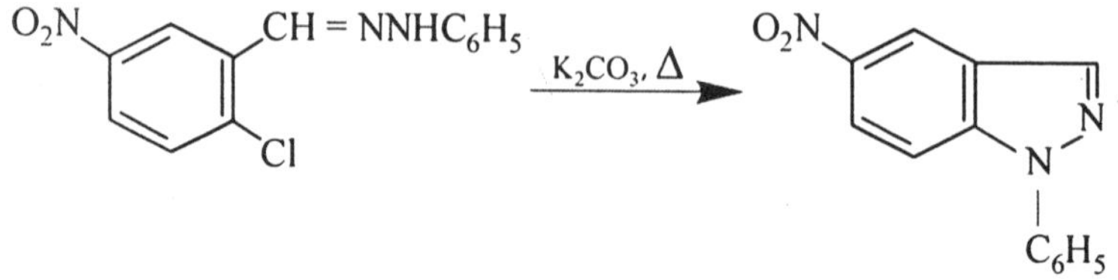

5-Nitro-1-Phenylbenzopyrazole

(10.1)

2.   *From α, β-Ethylenecarbonyl Compounds:* Another important synthesis consists of a reaction between an α, β-ethylenecarbonyl derivative and hydrazine.[12] The former must contain an easily replaceable group at the 2- or the β-position. This is illustrated in equation (10.2).

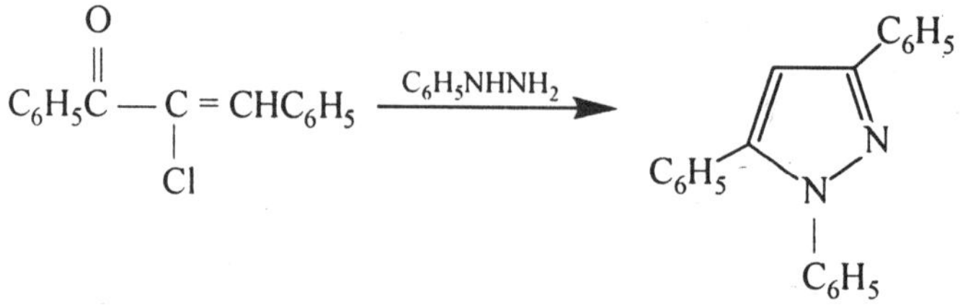

(10.2)

1,3,5-Triphenylpyrazole

α, β–Acetylene carbonyl compounds may instead be employed.[13] The hydrazine in this case may form a hydrazone or add directly to the acetylenic bond. Thus 4-phenylbut-3-yn-2-one and methylhydrazine evidently form intermediate compounds (13) and (14) which can cyclize to 1, 3-dimethyl-5-phenylpyrazole and 1, 5-dimethyl-3-phenylpyrazole. The starting acetylenic derivatives are not easily available.

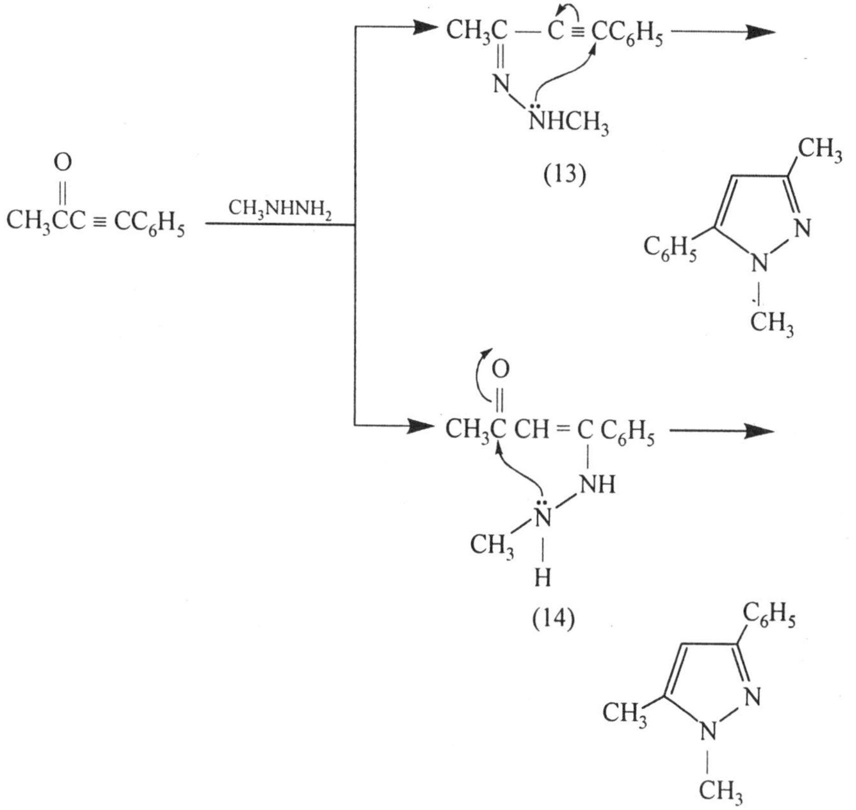

$$CH_3CC \equiv CC_6H_5 \xrightarrow{\;CH_3NHNH_2\;}$$

$$CH_3C-C \equiv CC_6H_5 \longrightarrow$$

(13)

(14)

3. From 1, 3-Dipolar Addition: A diazo compound adds to an acetylenic derivative which has its triple bond activated by an electron-withdrawing substituent. The reaction is usually carried out in a suitable solvent at room temperature. Diazomethane or methyl- or -ethyl-diazoacetate is commonly employed. Thus methyl diazoacetate and methyl phenyl propiolate yield the following isomeric pyrazoles[8] in equal amounts.

$$C_6H_5 C \equiv CCOOCH_3 \xrightarrow{N_2CH\,COOCH_3}$$

3,5-Dicarbomethoxy-4-phenylpyrazole

4,5-Dicarbomethoxy-3-phenylpyrazole

4. *From Other Ring Systems:* Various heterocyclic compounds transform to pyrazoles under appropriate conditions. Sydnone (15), for instance, and acrylonitrile result in pyrazole formation, equation (10.3). The cyanopyrazoline formed as intermediate is immediately converted to pyrazole.

$$\text{(15)} + CH_2 = CHCN \xrightarrow{70°C} \text{pyrazole} + CO_2 + HCN \quad (10.3)$$

(15)

5. The most frequently used route to indazole ring system consists of the diazotization of suitably substituted anilines, for instance *o*-toluidine under these conditions leads to the parent indazole.[9]

$$\xrightarrow[5°C]{HNO_2} \xrightarrow{(CH_3)_4N^+ OAc^-}$$

92%

Another method involves the following step[10], starting from cyclohexanone.

10.1.3 Chemical Reactions

The ring system of pyrazole is very stable and inert. In acid medium it exists as cation. Both pyrazole and indazole, are known to form coordinate complexes with several metal ions.[11] Because of the acidic character of the hydrogen at position-1 in the molecule, the ligand coordinates to the metal ions in the anionic form as, for instance, in $Ag^+(C_3H_3N_2)^-$ and $Ni^+(C_3H_3N_2)^-$

1. *Alkylation:* The alkylation of the free-NH group of pyrazoles proceeds with alkylating agents such as alkyl halides, diazomethane or dimethylsulfate. Substituted pyrazoles undergo alkylation to give a mixture of two isomeric products. However, under controlled conditions one of the isomers may be preferentially obtained. The orientation depends on the nature of the alkylating agent, and the reaction conditions, i.e., whether the free pyrazole or its anion is the main reactant. In any case, alkylation of the tautomeric mixture formed by methyl 3-methylpyrazole -5- carboxylate gives a predominance of either isomer depending on the alkylating agent employed.

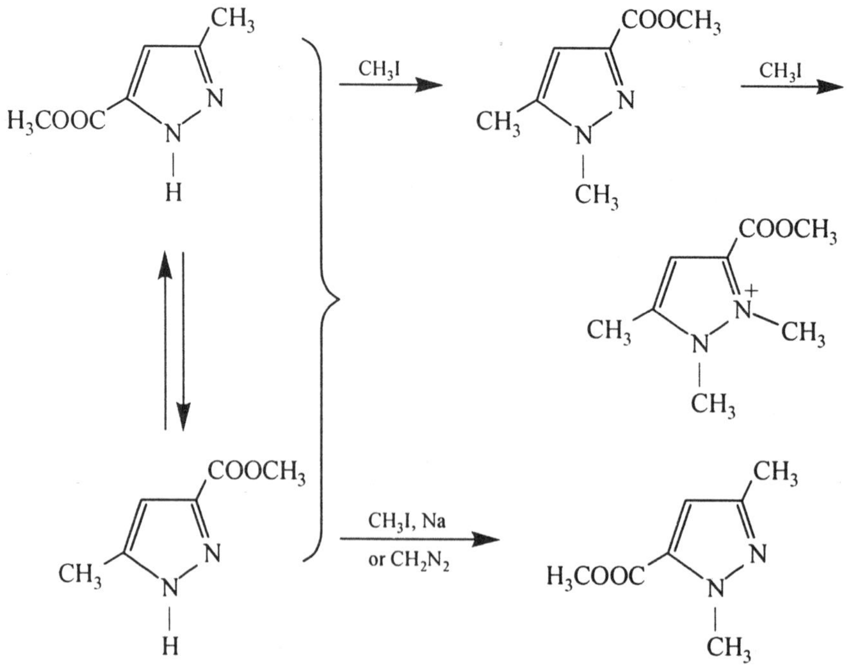

Excess of the alkylating agent causes quaternization.

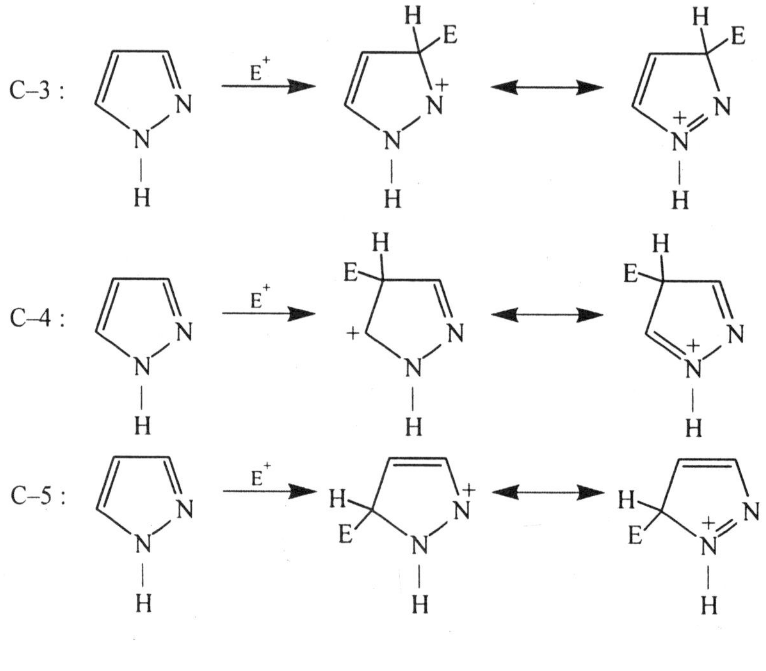

2.   *Electrophilic Substitution:* Pyrazoles are subject to electrophilic substitution and the attack takes place at position-4. This is evident from the following resonance structures of the intermediate cation. The mesomeric structures for attack of the electrophilic at position C-3 and C-5 involve forms with a highly unfavored positively charged azomethine nitrogen. Therefore, attack occurs at position-4. The electrophilic attack on pyrazole takes place readily in neutral or basic media. In acidic medium the electrophilic substitution has little use because the intermediate formed in the presence of acid catalyst is more resistant to electrophilic attack than pyrazole itself. Chlorination (SO_2Cl_2 or free Cl_2) of pyrazole yields 4-chloropyrazoles.[12] Bromination[13] (Br_2/ dioxane) similarly occurs at position-4. Nitration (HNO_3/H_2SO_4) and sulfonation (oleum) also occurs at 4-position but under more severe conditions[14] as the reaction takes place on the pyrazolium cation. Nitration of 1-phenylpyrazole can be carried out in such a manner so as to obtain any one of the three successive nitration products.[15]

The Friedel-Crafts reaction has limited application in pyrazole chemistry. Pyrazoles are 4-chloromercurated by $HgCl_2$ while with oxalyl chloride substitution takes place at 4-position in 1-phenylpyrazole[16] to give 4-carboxyl-1-phenylpyrazole.

3. ***Reaction with Oxidizing and Reducing Agents:*** The pyrazole ring is remarkably stable to the action of oxidizing agents but the side-chain may be oxidized to the carboxylic function. The oxidation proceeds well in alkaline permanaganate, equation (10.4). Pyrazole and its derivatives have been reduced under a variety of conditions. Thus with Na/C_2H_5OH, 2-pyrazoline is obtained.

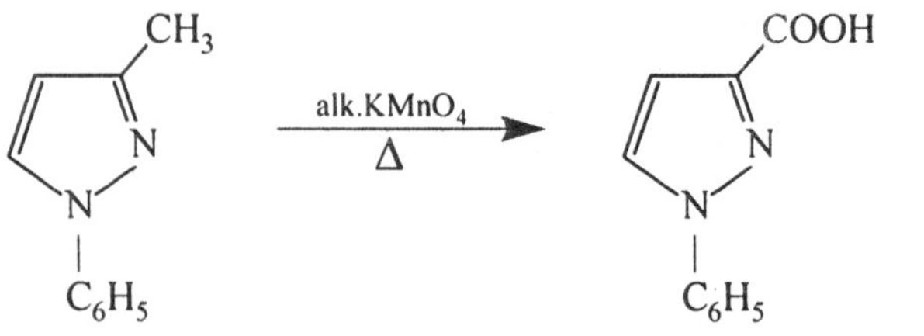

$$\text{(10.4)}$$

The catalytic reduction of 1-phenylpyrazol yields both phenylpyrazoline and 1-phenylpyrazolidine.

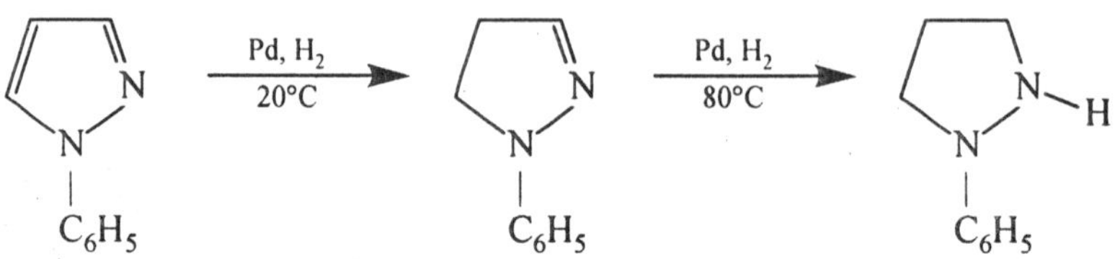

4. *Reaction with Nucleophiles:* Halogens attached to a pyrazole nucleus are exceptionally inert and thus will not undergo displacement under the usual reaction conditions. In 1-Phenylpyrazole, the halogen is most reactive in the 5-position, less so in 4-position and least in the 3-position. The presence of electron-withdrawing groups markedly assists nucleophilic displacement of halogens. Pyrazole quaternary salts are particularly reactive in nucleophilic reactions and in these compounds 3-and 5-halogen substituents can readily be replaced by a variety of nucleophiles. Direct amination of pyrazole with sodamide has not been observed, it rather causes ring opening and metallation[17] with *n*-butyllithium takes place at the 5-position. Pyrazolyl Grignard reagents have been obtained by the entrainment reaction. Pyrazoles exist partly as anions and thus react with electrophiles as phenols and undergo diazo coupling, nitrosation and Mannich reaction.

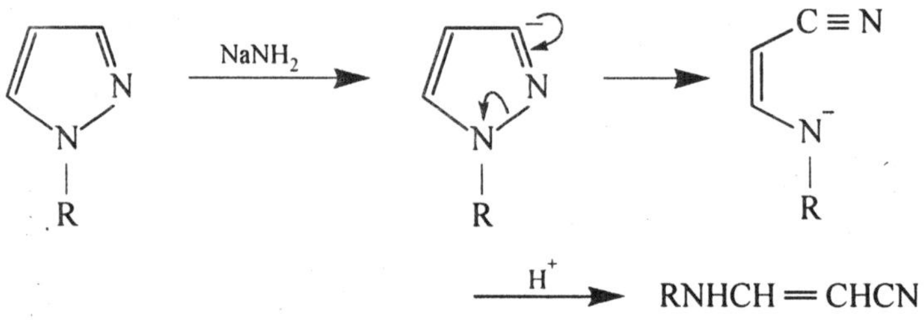

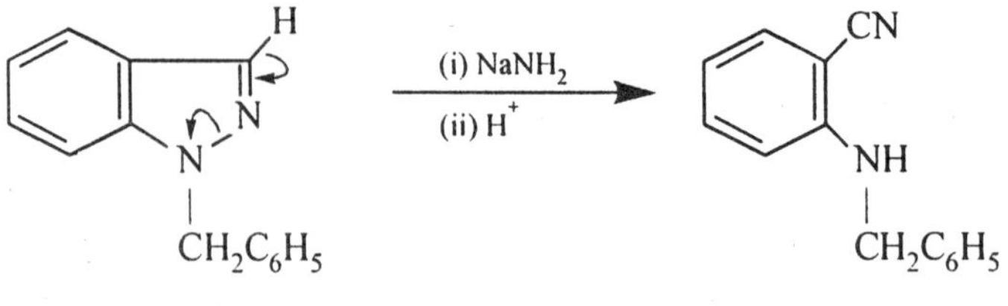

$$\xrightarrow{\text{H}^+} \quad \text{RNHCH}=\text{CHCN}$$

5.   *Photochemical Reactions:* The most important photochemical reaction of pyrazole is its conversion to imidazole.[18] The transformation can be accounted for in terms of the ring-opening to produce an azirine (16) and

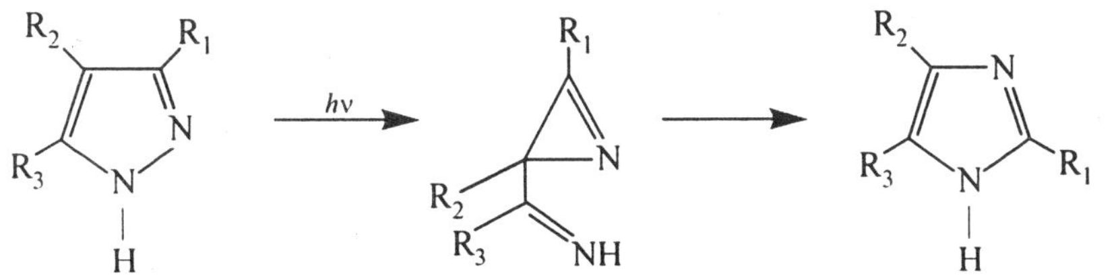

subsequent ring closure. 1-Alkyl indazoles afford 2-alkylaminobenzonitriles by photolytic ring cleavage (equation 10.5).

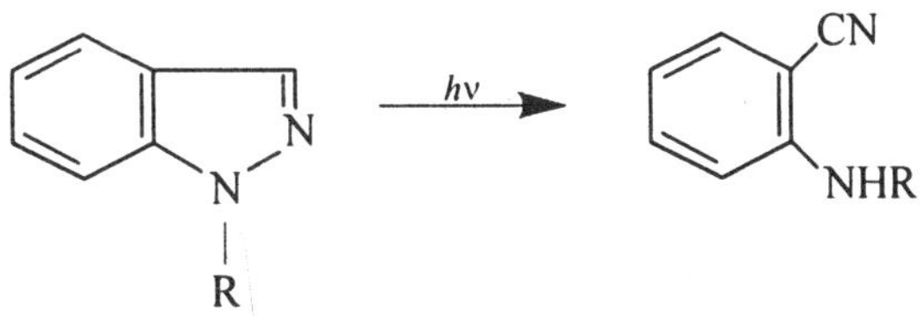

$$(10.5)$$

10.1.4 Naturally Occuring and Biologically Active Compounds

The pyrazole ring has attracted considerable attention in the last 40 years. Pyrazole ring has been shown to be the basis of a number of dyes and drugs

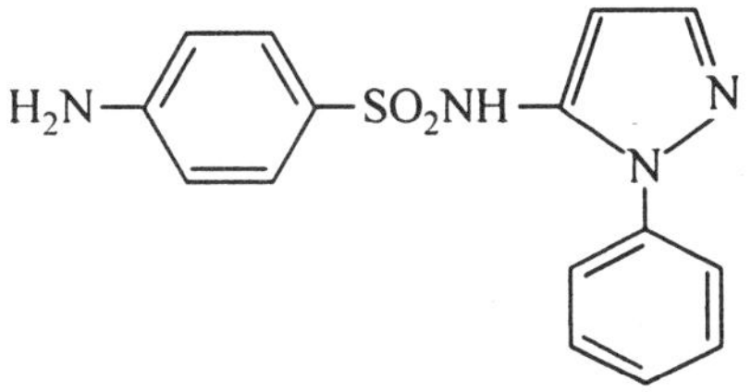

(17) 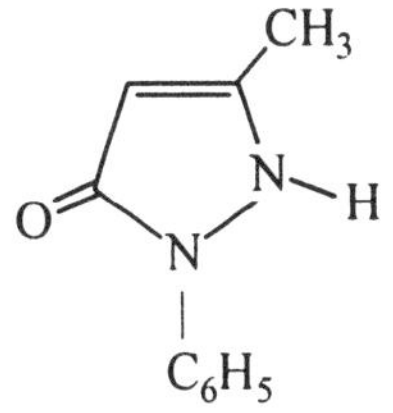   (18)

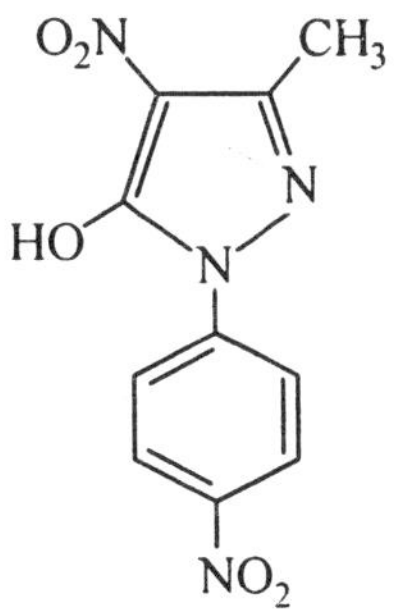

(19) 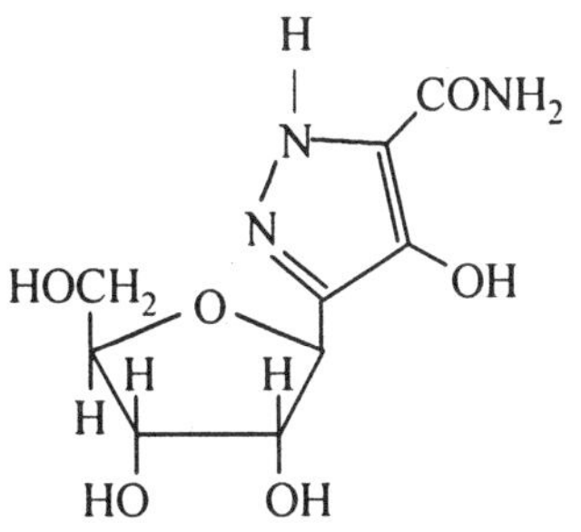   (20)

and pyrazolines are effective bleaching agents. The sulfonamides based on pyrazole are of particular interest, for instance, orisul (17) has a long bacteriostatic action *in vivo*. Antipyrine (18), a pyrazolone derivative has antipyretic properties. Similarly butazolidine (19) has proved a powerful anti-inflammatory drug used for rheumatic patients. Pyrazofurin (20) isolated from the culture of a strain of streptomyces candidus shows activity against a number of viruses.[19]

Tetrazin (21) is a yellow dye usd in dyeing wool.

(21)

10.2 IMIDAZOLES AND BENZIMIDAZOLES

Imidazole[20-22] or *iminazoline* (4) is an azapyrrole, the nitrogen atom being separated by one carbon atom. This compound was earlier also called as glyoxaline as it was first prepared in 1858 from glyoxal and ammonia.

(4)   (22)

The benzo derivative is referred to as benzimidazole (22). These ring systems are numbered as shown. The imino nitrogen is assigned position-1 while the tertiary nitrogen atom position-3. The imidazole nucleus is found in a number of naturally occurring compounds such as histamine, histidine, pilocarpine and allantoin. The benzimidazole nucleus appears in vitamin B_{12}.

A substituent present at the 2-position in pyrazole offers no problem as it is symmetrically located with respect to the two nitrogen atoms. Since imidazole also exists in tautomeric forms, either of the nitrogen atom can bear the hydrogen atom and the two nitrogens become indistinguishable and the numbering becomes rather complex for mono substituted imidazoles. As a consequence of this 4-methylimidazole (23) is identical with 5-methylimidazole (24) and depending on the position of the imino hydrogen such a compound may be designated either 4- or 5-methylimidazole. As only one substance which may be a single tautomer or an equilibrium mixture is known, it would be incorrect to call it either 4- or 5-methylimidazole. Such a compound is instead designated as 4 (5)-methylimidazole. Depending on the position of the double bond, the

(23)   (24)

dihydroimidazoles are referrd to as *imidazolines* and three of them are possible. There are 2-imidazoline (25), 3-imidazoline (26) and 4-imidazoline (27). Out of these 2-imidazolines are most common. These are mono acidic bases and form mono hydrogen halide salts.[23]

(25)   (26)   (27)

10.2.1 Physical and Spectroscopic Properties

Imidazole is a colorless liquid, b.p., 256°C as compared to pyrazole (b.p. 187) and is high boiling than all the other five-membered heterocyclic compounds. In marked contrast to imidazole, the boiling point of 198°C for 1-methyl-imidazole is comparatively low. This data demonstrates that hydrogen bonding exists in imidazole ring of the type (28) and may consist of upto 20 molecules.

(28)

Imidazole is more basic (*p*Ka 7.2) than pyrazole and even pyridine (*p*Ka 5.2). It shows amphoteric properties and behaves as an acid because it contains "pyrrole type" amino nitrogen in the ring and forms metallic salts with $NaNH_2$ and RMgX which are extensively hydrolyzed by water. The introduction of alkyl groups into the ring increases the basicity, 2-methylimidazole (*p*Ka 7.86) and 4 (5)-methylimidazole (*p*Ka 7.52). Imidazole is an aromatic compound and possesses a resonance energy of 14.2 Kcal/mole which is almost half the value for pyrazole. The dipole moment of imidazole has been measured in several solvents, in dioxane the value is 4.8 D. This large value indicates a considerable polarization in the imidazole ring although the extent of polarization is much less than that required to yield an ionic structure.

10.2.2 Synthetic Methods

Several approaches are available for the synthesis of imidazoles.

1.   *The Radiszewski Synthesis:* This is probably the most important synthesis of imidazoles. It consists of condensing a dicarbonyl compound such as glyoxal, α-keto aldehydes or α-diketones with an aldehyde in the presence of ammonia. Benzil, for instance, with benzaldehyde and two molecules of ammonia react to yield 2, 4, 5-triphenylimidazole. Formamide[24] often proves a convenient substitute for ammonia.

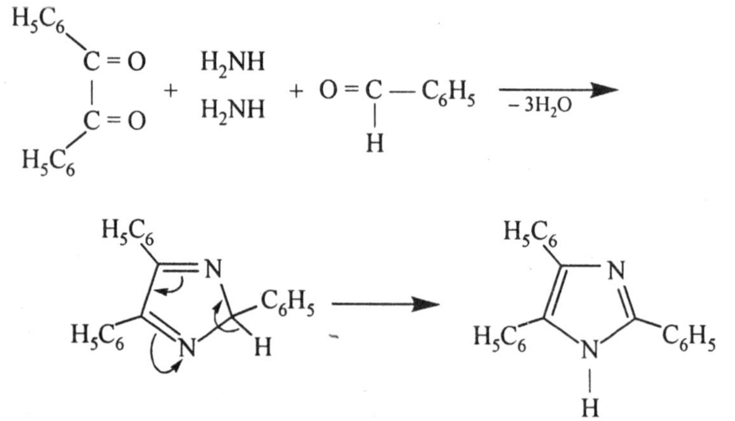

2.   *Dehydrogenation of Imidazolines:* Imidazoles have been prepared by dehydrogenation of imidazolines in the presence of sulfur.[25] However, a milder reagent, barium managanate has been recently reported by Knapp and coworkers[26] Imidazolines thus obtained from alkyl nitriles and 1, 2-ethanediamine[27] on reaction with $BaMnO_4$ yields 2-substituted imidazole and it has also been obtained as one of the by-product in other reactions.[28]

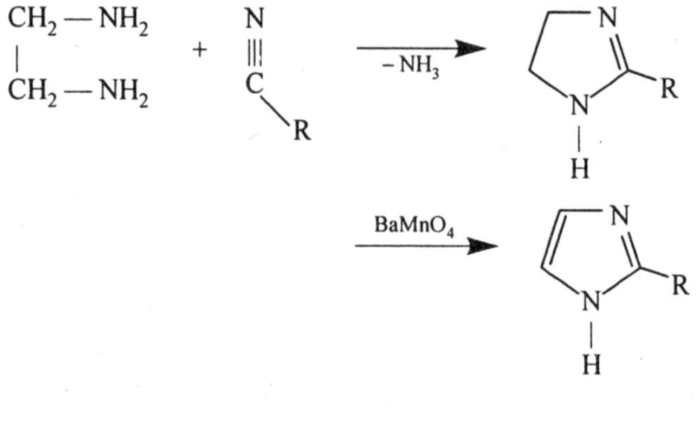

3.   *From α-Halo Ketones:* This reaction involves an interaction between an amidine and an α-halo ketone. This method[29,30] has been applied successfully for the synthesis of 2, 4- (or 2, 5-) diphenylimidazoles. Phenacyl bromide

and benzimidine according to this method afford 2, 4-diphenylimidazole similarly amidine reacts with acyloins or α-halo ketone to yield imidazoles.[31]

4.   Benzimidazole is more important than imidazole as the former occurs in Vitamin B_{12}, and has been prepared by a number of methods. 1, 2-Diaminobenzene condenses with a carboxylic acid on heating in an acidic medium to give benzimidazole,[32] equation (10.6).

$$(10.6)$$

The cyclization of N-haloamidines with sodium ethoxide forms benzimidazole *via* a nitrene intermediate.

10.2.3 Chemical Reactions

1.   *Reaction with Acids:* Imidazole is a mono acidic base and forms crystalline salts with acids. It also possesses weakly acidic properties *(pseudo acidic)*

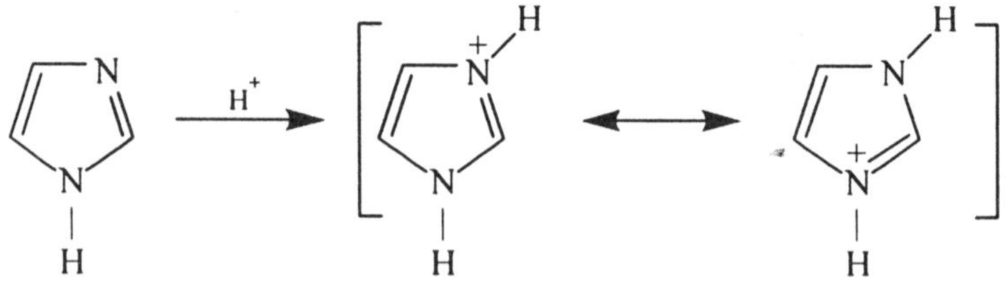

and is even more acidic than pyrrole and thus forms salts of the following type with Grignard reagents or metal ions.

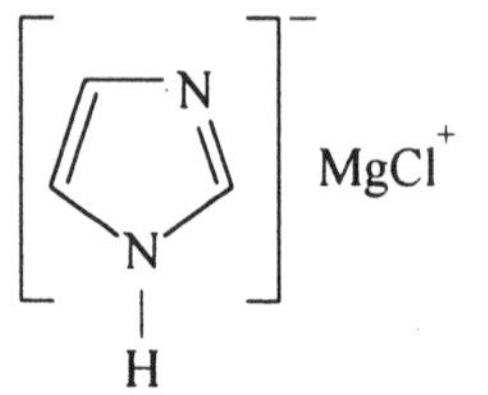

Imidazole magnesium chloride

With ammoniacal silver nitrate imidazole forms a silver salt which is sparingly soluble in water.

2.   *Quaternization:* Quaternization of imidazoles at the nitrogen atom is normally achieved by the reaction of alkyl halides or dialkyl sulfates under strongly basic conditions in an organic solvent while in water the yields are usually low owing to the strong water solubility of the reaction product. Alkylation of imidazoles has been achieved satisfactorily by heating 1-carboethoxyimidazoles obtained by reacting imidazoles and ethyl chloroformate.[33]

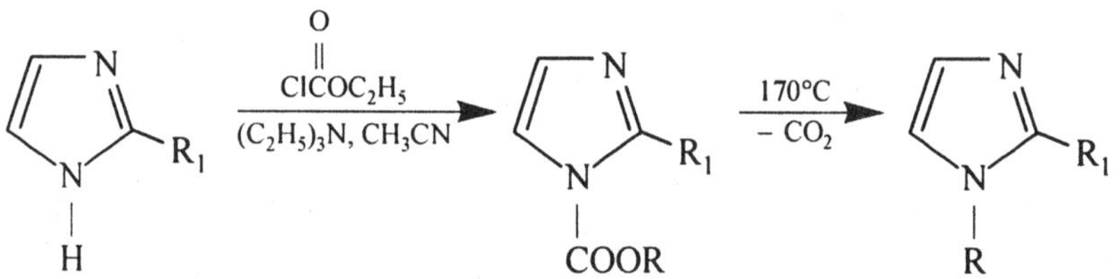

3.   *Electrophilic Substitution:* The contribution of charged structures of imidazole are more important than those of benzene. For this reason imidazole possesses increased reactivity toward electrophilic attack. It is more susceptible to electrophilic attack than pyrazole or thiazole and more so than even furan and thiophene. From the following resonance structures of the intermediate ion, it is evident that the attack takes place at the 4 (5) position in imidazole. It may be noticed that the attack at C-2 involves a canonical form with a highly unfavored positive nitrogen at position-3.

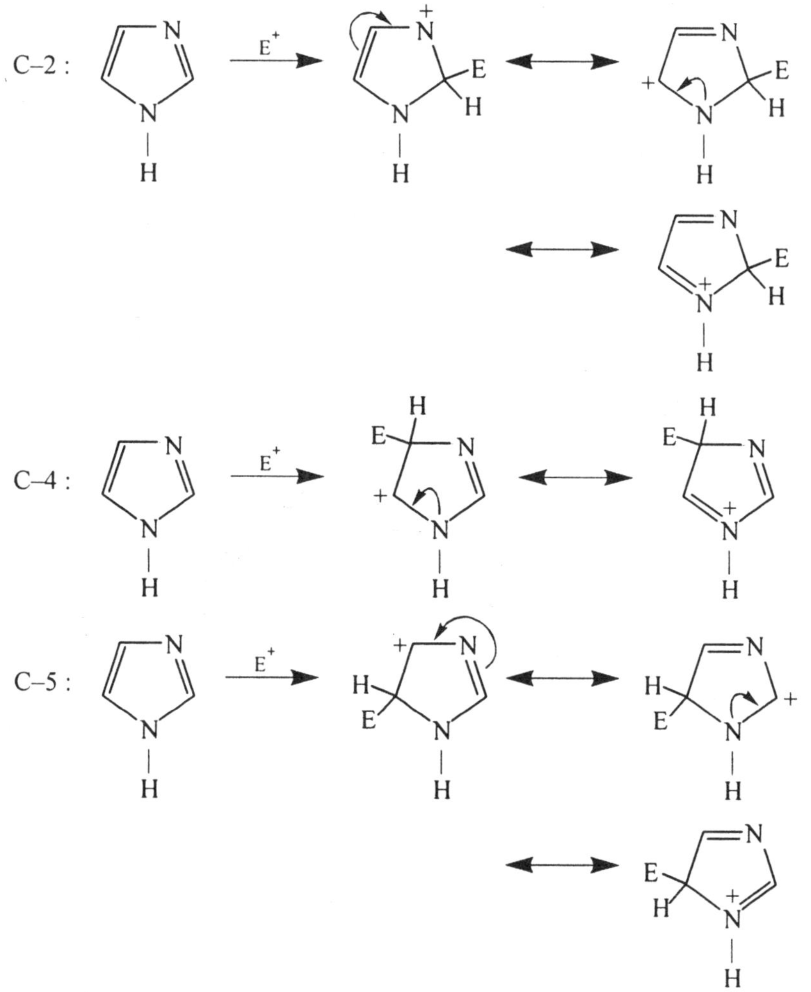

C–2 :

C–4 :

C–5 :

Halogenation of imidazole is very complex and varies considerably depending on the substrate, reagents and reaction conditions. Direct chlorination gives undefined products. Bromnination (Br_2/$CHCl_3$) yields 2, 4, 5-tribromo derivative, equation (10.7). Iodination of imidazole takes place in alkaline

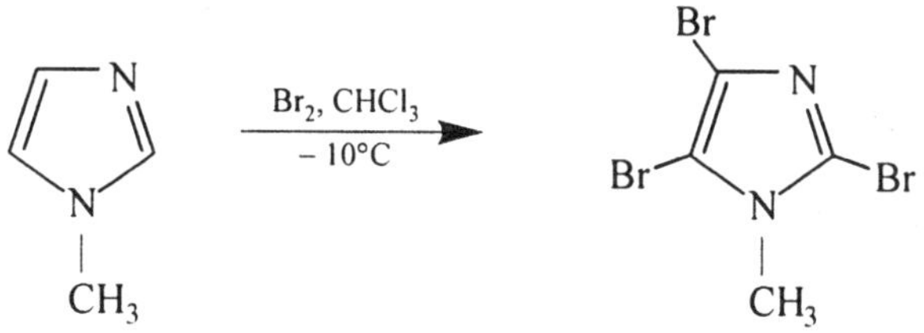

$$Br_2,\ CHCl_3 \quad -10°C \qquad (10.7)$$

2,4,5-Tribromo-1-methylimidazole

conditions to give 2, 4, 5-triiodoimidazole. That 3-substituted imidazole fails to iodinate lends support to the theory that it is the imidazole anion which reacts. The anion is actually ten times more reactive than the neutral molecule. Iodination of 2, 4, 5-trideuterioimidazole shows a large kinetic isotope effect ($K_H/K_D \sim 4.5$) which indicates that the loss of a proton from C-4 is rate-controlling

in mono iodination. This observation complies with the above mechanism. The kinetics of the iodination of 2- and 4-methylimidazole has also been studied.[34] Bromoimidazoles can be converted into chloroimidazoles by refluxing with hydrochloric acid.[35] Both nitration (fuming HNO_3/H_2SO_4) and sulfonation (oleum/100°C) afford the 4 (5)-imidazole derivatives. If these positions are occupied than imidazole fails to react. Diazo coupling, on the other hand, takes place at position-2. If this position is occupied, then it occurs at 4 (5)-position. Mercuration[36] takes place also at 4 (5)-position. In imidazole, the electrophilic substitution takes place at 4 (5)-position but the reverse is true for deuteration.[37,38]

4. ***Reaction with Oxidizing and Reducing Agents:*** Imidazole itself is stable to autoxidation and to the action of chromic acid but is attacked by potassium permanganate. Oxygen in the presence of a sensitizer (singlet oxygen) reacts to give an imidazolidine derivative. Imidazolium dichromate, a mild oxidizing agent has been employed for the oxidation of allylic and benzylic alcohols to the corresponding carbonyl compounds[39].

5. ***Reaction with Nucleophilic Reagents:*** 1, 3-Azoles, in general, do not react by nucleophilic substitution unless electron-withdrawing groups are present. A halogen atom in the 2-position can be replaced by nucleophiles such as alkoxy, thiol or aminoalkyl, etc. Often a nucleophilic attack results in ring fission. In particular, amino imidazoles are unstable owing to hydrolytic cleavage and quaternary midazolium salts decompose to primary amines in the presence of alkali when the classical Schotten Baumann method is employed to prepare benzoylimidazoles, the ring cleaves to yield 1, 2-dibenzamidoethene.[40]

6.   *Imidazole Catalyzed Ester Hydrolysis:* The imidazolyl group of a histidine residue may be implicated in the mode of action of such hydrolytic enzymes as trypsin and chymotrypsin, a great deal of work has been done on the study of imidazoles as catalysts of ester hydrolysis. Two pathways—a general base catalysis and nucleophilic catalysis have been envisaged.

In general base catalysis, the imidazole molecule functions as a base and activates a water molecule for attack at the carbonyl carbon of an acyl compound.[41]

The nucleophilic catalysis properties of imidazole are due to the intermediate formation of 1-acylimidazole.

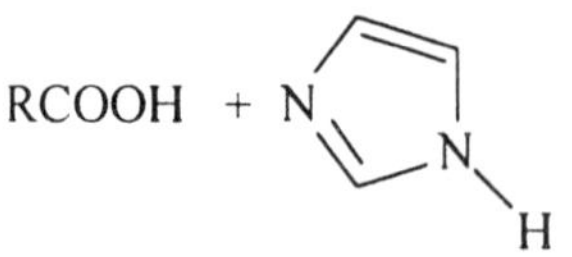

RCOOH + (imidazole)

7. *Cycloaddition Reactions:* Imidazoles undergo addition across the carbon – carbon double bond under photochemical conditions. The reactions of imidazoles with benzophenone and acrylonitrile are representative examples, equation (10.8) and (10.9).

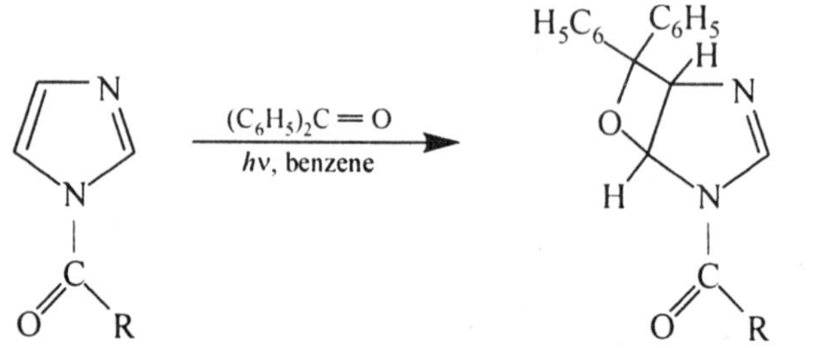

$$(10.8)$$

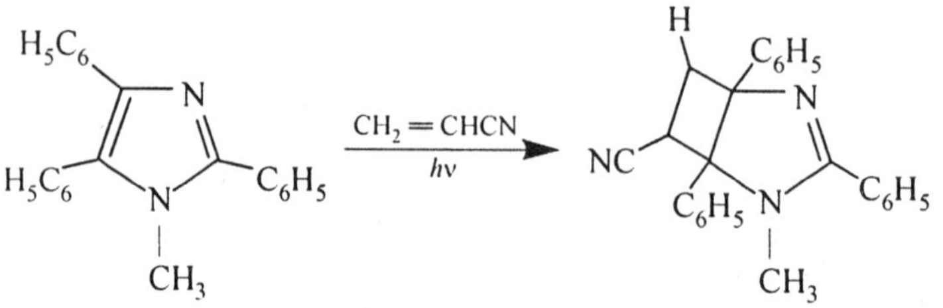

$$(10.9)$$

10.2.4 Naturally and Biologically Active Compounds

As stated earlier, compounds containing the imidazole ring are very important in living systems, such as vitamin B_{12} as discussed under pyrrole, biofin discussed under thiophene and several pilocarpine alkaloids. Imidazole also forms a part of some very important compounds such as purine, adenine, xanthine, guanine, coenzyme-a, etc. It is also widely distributed in essential amino acid 1-histidine (29) and its derivative carnosine (30) which is a dipeptide and is a constituent of mamalian muscle. A number of synthetic imidazoles

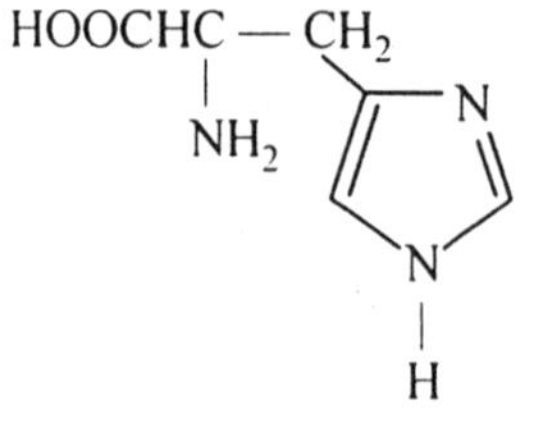

(29) 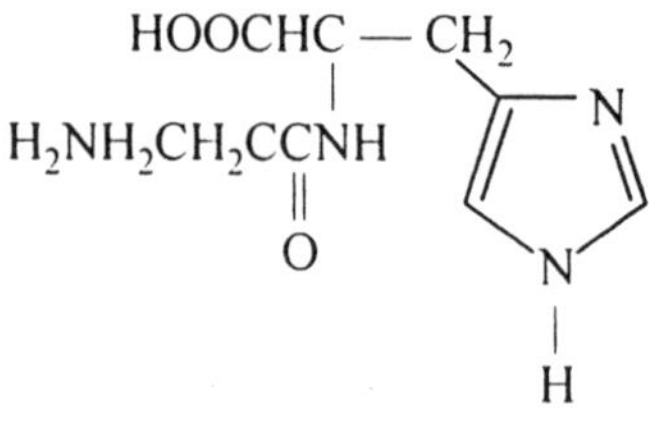   (30)

such as priscol (31), privine (32) and antihistamine (33) have physiological properties. The antihistamine, azamycin (34) is 2-nitroimidazole.

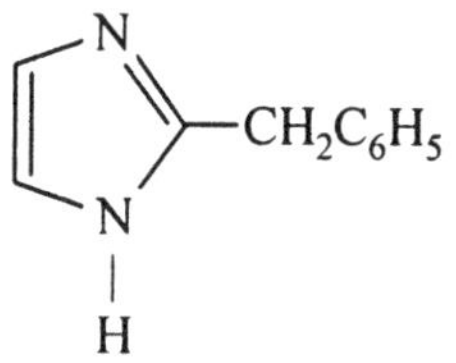

(31) 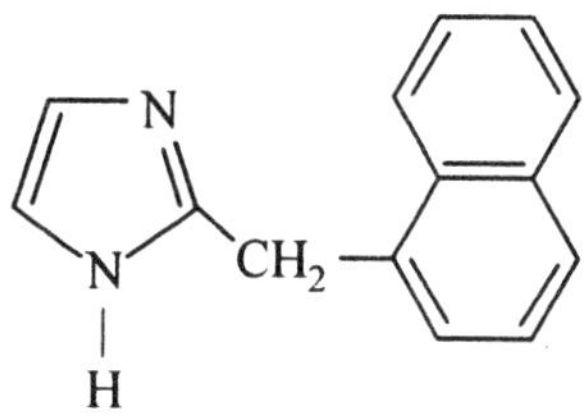   (32)

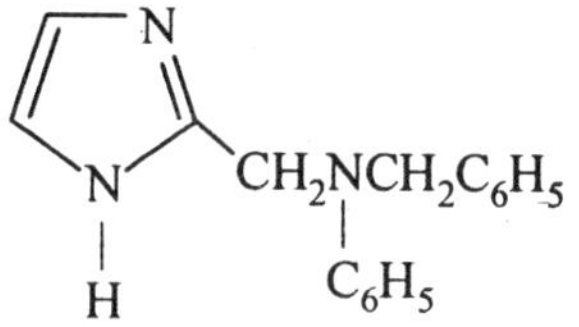

(33) 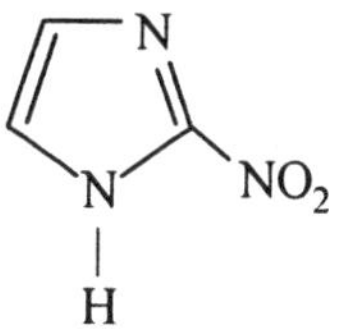   (34)

10.3 ISOXAZOLES AND BENZOISOXAZOLES

In isoxazole (2) the oxygen and nitrogen atoms are present in 1, 2-relationship. This compound was first obtained in 1903. Two classes of benzofused analogues exist, *benzoisoxazole* or 1, 2-*benzoisoxazole* or indoxazene (35) and the 2, 1-benzoisoxazole or *anthranil* or *benz [C] isoxazole* (36).

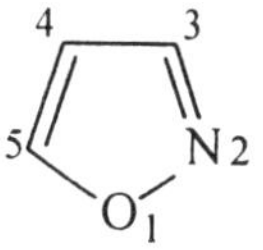

(2) 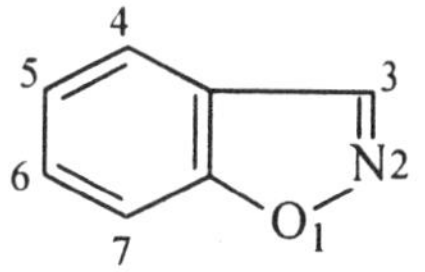   (35) 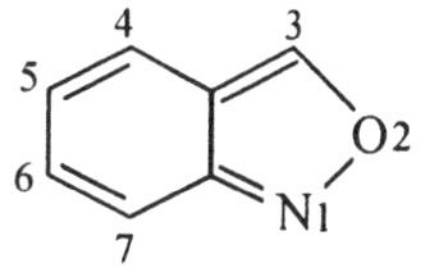   (36)

The numbering of these ring systems is shown in the structures. Certain isoxazole derivatives are known to possess antituberclosis, analgesic and local anaesthetic activities. The dihydro derivatives are called isoxazoline such as 2-isoxazoline (37) and isoxazolidine (38).

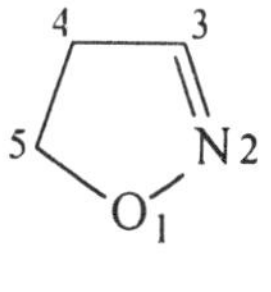

(37) 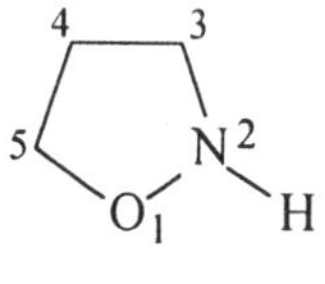   (38)

10.3.1 Physical Properties

Isoxazole is a liquid, b.p. 95°C and resembles pyridine in odor. It is very weak base (*p*Ka 1.3) like pyrazole and is not dissolved in weak acids. Isoxazole nucleus is considerably less aromatic than other five-memebred heterocycles, including oxazole and furan. Chemical reactions suggest that isoxazole is best represented as a resonance hybrid of several resonance structures. Spectroscopic properties of isoxazoles have been reviewed.[42,43]

Indoxazole is a high boiling liquid. It possesses a dipole moment of 3.3 D (benzene). Anthranil is a colorless oily liquid and possesses a characteristic smell. It possesses a dipole moment of 3.06 D. The chemistry of these compounds has been reviewed.[44]
Prototropic tautomerism of isoxazole derivatives has been known. Isoxazole - 5-ones exist in three tautomeric forms (39) to (41), where R is methyl.

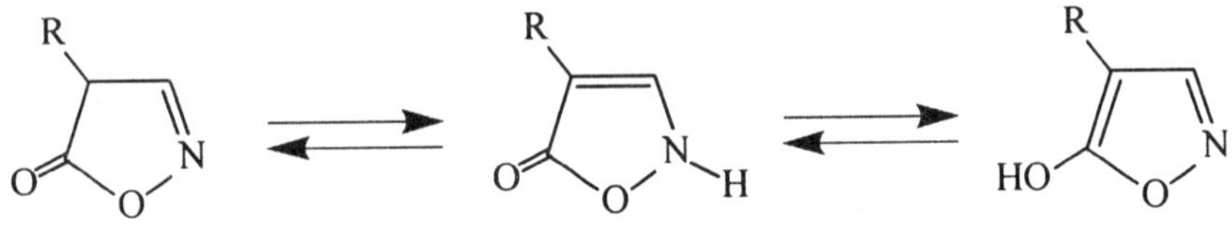

(39)   (40)   (41)

Evidence obtained from *u.v.* and *n.m.r.* indicates that in solvents (chloroform or cyclohexane) the form $(39, R = H)$ predominates. While in aqueous solution an equilibrium between form (39) and the NH form (40) takes place. In dioxane, on the other hand, all three tautomers are present. For the 3-substituted $(R = Ph)$ isoxazole-5-ones the tautomeric forms normally present are the OH form (41) and the CH_3 form (39) in all solvents appears to influence the equilibrium between the OH and the NH forms, for instance, in dioxane the OH and NH forms exist in equal amounts. However, in basic solvents the OH form predominates.

10.3.2 Synthetic Methods

Several general procedures are available for isoxazole synthesis.

1.   *From Diketones:* Probably the most general method for the synthesis of isoxazoles involves the condensation-cyclization of an α, β-diketone with hydroxylamine,[45,46] equation (10.10). The reaction involves an initial attack

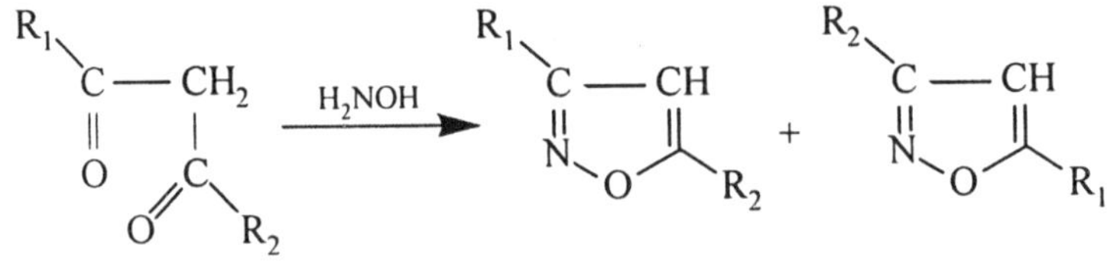

$$(10.10)$$

of hydroxylamine on the carbonyl carbon which on subsequent cyclization yields a mixture of isomeric isoxazoles. The use of acetal derivatives of 1, 3-dicarbonyls is important.[56] 1, 1, 3, 3-Tetraalkoxyalkanes are readily available and react with hydroxylamine. For instance, 1, 1, 3, 3-tetraethoxypropane on reaction with hydroxylamine yields the parent compound.[47]

$$(C_2H_5O)_2CH\ CH_2\ CH\ (OC_2H_5)_2 \xrightarrow{H_2NOH}$$

Quilico and coworkers[48] described an unequivocal synthesis of substituted isoxazoles from ketone oximes having an α-hydrogen and an aromatic ester. Thus from acetophenone oxime and phenylacetate, isoxazole (42) is obtained.

(42)

Isoxazoles have also been obtained from enamines.[49]

2. *From γ-Pyrones:* Nucleophilic attack on the γ-pyrone ring system usually proceeds through ring opening followed by recyclization to afford various heterocyclic compounds. However, scattered examples of oxime formation have been reported because of the difficuly in differentiating γ-pyrone oxime from the isomeric isoxazole derivatives. For example, the reaction of flavone (43) with hydroxylamine was earlier reported to be the oxime (44) but a reinvestigation[50] of its structure showed it to be the isoxazole (45).

(44)

(43)

(45)

Chromone (46) also gives isoxazole (47) and not the oxime on treatment with hydroxylamine,[50-53] equation (10.11). It may be stated that certain α, β-unsaturated ketoximes also cyclize to isoxazoles in the presence of palladium complex.[54]

$$(10.11)$$

$$(46) \qquad\qquad (47)$$

Thus (48) on treatment with sodium phenoxide and dichloro *bis* (triphenylphosphine) palladium in dry benzene on refluxing gives (49).

$$(48)$$

$$(49)$$

3. *From Nitrile Oxides:* Isoxazoles are generally prepared by cycloaddition of alkenes or acetylenes to nitrile oxides. The treatment of chloroximes (50) with bases particularly with triethylamine is the most commonly employed base for the preparation of nitrile oxides (51).

$$(50) \qquad\qquad\qquad\qquad (51)$$

The nitrile oxides react with a monosubstituted, alkyl- or aryl- alkyne regioselectively *via* 1, 3-cycloaddition to yield 5-substituted isoxazoles.[55]

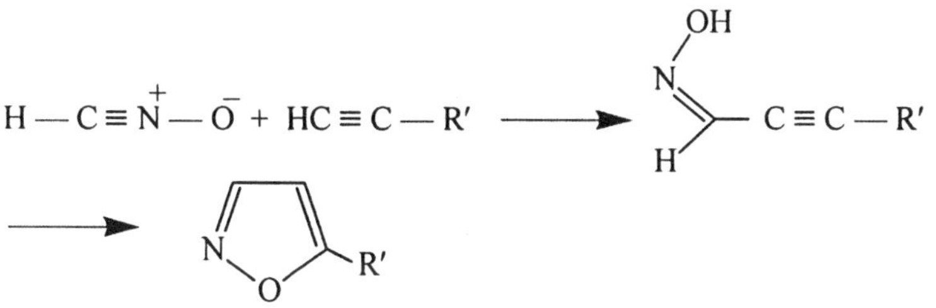

$$H-C\equiv \overset{+}{N}-\overset{-}{O} + HC\equiv C-R' \longrightarrow$$

The synthesis proceeds in two stages. The reaction of nitrile oxides with alkenes containing a leaving group which leads to elimination is rather facile, and the reaction is stereospecific with respect to the alkene whose geometry in preserved in the product. The addition is also regioselective but not always so. vinyl halide reacts with nitrile oxide to afford the following 3-substituted isoxazole.

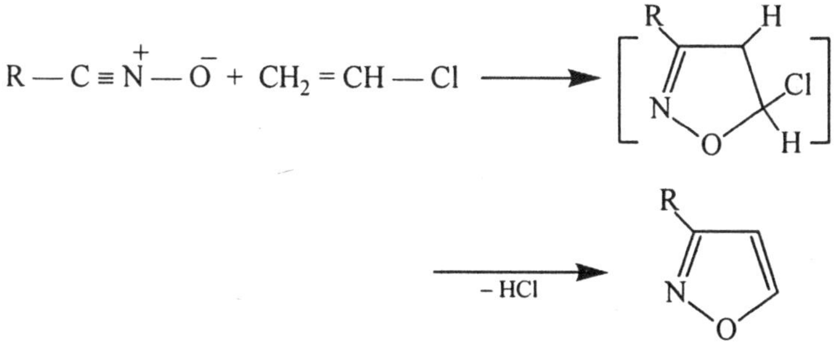

$$R-C\equiv \overset{+}{N}-\overset{-}{O} + CH_2=CH-Cl \longrightarrow$$

Fulminic acid with vinyl acetate gives 95% yield of the parent compound, isoxazole.

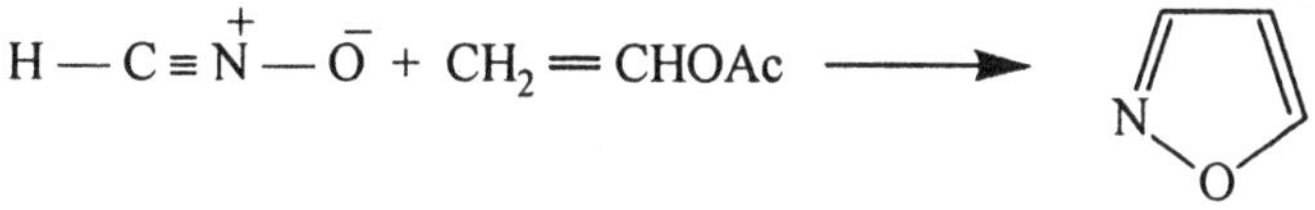

$$H-C\equiv \overset{+}{N}-\overset{-}{O} + CH_2=CHOAc \longrightarrow$$

Recently O - (trimethylsilyl) acetohydroxyimoyl chloride[56] and O-ethoxy-carbonyl hydroxymoyl chloride[57] have been used which lead to a nitrile oxide on thermolysis. The resultant nitrile oxide can be trapped with an alkyne to the corresponding isoxazole.

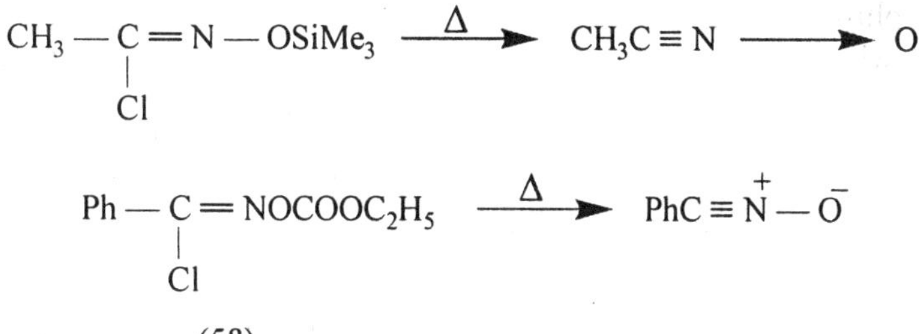

$$CH_3-\underset{\underset{Cl}{|}}{C}=N-OSiMe_3 \xrightarrow{\Delta} CH_3C\equiv N \longrightarrow O$$

$$Ph-\underset{\underset{Cl}{|}}{C}=NOCOOC_2H_5 \xrightarrow{\Delta} PhC\equiv \overset{+}{N}-\overset{-}{O}$$

(58)

Alternatively nitrile oxide can be generated *in situ* using a complex catalyst and under milder conditions, using microwave irradiation, followed by reaction with an acetylene. This method gives better yields than other methods.

4. Indoxazin may be obtained from *o*-halogenobenzoyl compounds and hydroxylamine in the presence of a base.

$$(10.12)$$

3-Phenylbenzoisoxazole

Thermal decomposition of salicylaldehyde-O-acetyloxime produces the unsaturated indoxazin, equation (10.13).

$$(10.13)$$

Reduction (Zn/CH$_3$COOH) of *o*-nitrobenzoyl compounds furnishes the corresponding anthranils.[59,60]

10.3.3 Chemical Reactions

Isoxazoles and its benzo-derivatives form crystalline salts with alkyl iodides or dialkyl sulfate.

1. *Electrophilic Substitution:* The 1,2-azoles are generally much less reactive than 1,3-azoles. Isoxazoles are more reactive than are isothiazoles but less than pyrazoles. The electrophilic attack occurs readily at C-4 and frequently fails if this position is occupied. Both the hetero atoms influence the rate of electrophilic substitution in the isoxazole ring. Because of the electron-withdrawing nature of the nitrogen atom, the electrophilic attack is retarded as in pyridine. The electron-donating property of the oxygen atom, on the other hand facilitates such a reaction.

Isoxazole derivatives are chlorinated or brominated with free halogen on heating. N-bromosuccinimide also brominates to give 4-bromoisoxazole.[61] Nitration (HNO$_3$, H$_2$SO$_4$) of 5-methylisoxazole gives 5-methyl-4-nitroisoxazole according to the following mechanism.[62] Sulfonation of isoxazole has been relatively less investigated than nitration. Aryl substituted isoxazoles on chlorosulfonation tend to give product of homocyclic ring sulfonation rather than the isoxazole ring.[63] Isoxazoles display no reactivity in the Friedel-Crafts reaction or the Vilsmeier formylation.

Benzoisoxazoles undergo electrophilic substitution exclusively in the aryl ring.

2. *Reaction with Nucleophilic Reagents:* Isoxazole is very labile towards the action of nucleophilic reagents or strong bases. This makes the chemical behavior of isoxazole fundamentally different from that of pyridine. Chichibabin reaction fails in the case of isoxazole as it opens the isoxazole ring. Reaction involving displacement of a halo atom at C-3 or C-5 position in isoxazole are known.[84] Strong bases result in the cleavage of the isoxazole ring, yielding β-keto nitriles as the end-products.

3. *Reaction with Oxidizing and Reducing Agents:* Reduction (Na/C_2H_5OH or Raney nickel) of isoxazole normally leads to ring[65] cleavage. The cleavage of the N – O bond takes place. If appropriate functional groups are present than new rings are formed by intramolecular cyclization. This is illustrated by the following example:

In contrast, isoxazole ring system is stable to many oxidizing agents though not in alkali.

4. *Condensation Involving Methyl Group:* A unique feature of picolines and many azoles is the familiar reaction of the methyl group to undergo condensation of the aldol type. Alkyl isoxazoles respond to such a reaction and the methyl group at C-5 and not at C-3 is more reactive.[66] Thus 3, 5-dimethylisoxazole condenses with benzaldehyde to give 3-methyl-4-(benzoylmethyl) isoxazole (equation 10.14).

$$\text{(10.14)}$$

5. *Photochemical Reactions:* The photochemical reactions of isoxazoles have been thoroughly investigated.[67] In many cases azirine intermediates have been identified in the conversion of isoxazole to oxazole. Thus 3, 5-diphenylisoxazole rearranges to 2, 5-diphenyloxazole.[68]

Isoxazoles also rearrange thermally to oxazoles as exemplified below:[69]

In addition, ring contraction and ring expansion products have also been observed by Ullman and Singh.[70]

Polycyclic isoxazoles yield oxazoles on irradiation *via* azirine and isonitrile as intermediates, 4, 5, 6, 7-tetrahydro-1, 2-benzisoxazole isomerizes to 4, 5, 6, 7-tetrahydrobenzoxazole in quantitative yield.[71]

99%

6. *Cycloaddition Reactions:* Isoxazoles behave as dienophiles in the Diels-Alder reaction in contrast to oxazole which acts as a diene. The $(4 + 2)$ cycloaddition between ethyl-4-nitro-3-phenylisoxazole carboxylate (54) and cyclohexa - 1, 3-diene (55), yields the adduct[72] (56), equation (10.15).

(54)   (55)   (56)

(10.15)

2, 1-Benzisoxazole, in contrast, behaves as a diene.

10.4 OXAZOLES AND BENZOXAZOLES

Oxazole (5) is a 1, 3-azole having an oxygen atom and a pyridine-type nitrogen atom at the 3-position in a five-membered ring. This compound was first introduced by Hantzsch in 1887 but was not synthesized until 1947.[73-76] The benzo derivative is known as *benzoxazole* (57). Oxazole does not occur in nature and thus does not play any part in fundamental metabolism as do imidazole and thiazole.

(5)   (57)

Partially reduced oxazoles are called oxazolines and three types are possible depending on the position of the double bond. These are 2-oxazoline (58),

3-oxazoline (59) and 4-oxazoline (60). The fully saturated system is called oxazolidine (61). Oxazolidines are solids and behave as bases and are readily hydrolyzed by water.

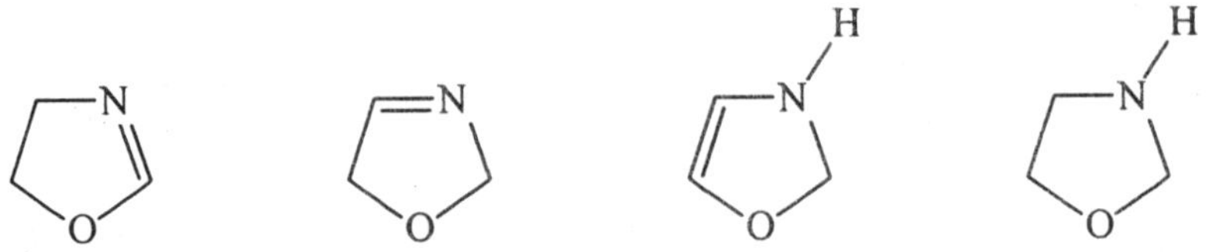

(58)   (59)   (60)   (61)

10.4.1 Physical Properties

Oxazole is a liquid, b.p. 69° C, and has an odor resembling that of pyridine. It is miscible with water and many organic solvents. It is weakly basic (pKa 0.8) but more so than isoxazole. Its reactions suggest it to be an aromatic compound. Its spectroscopic properties have been summarized.[73] The dipole moment of oxazole has been measured to be 1.5 D.

Although oxazole possesses a sextet of π-electrons, all its properties demonstrate that the delocalization is incomplete, as a result it has but little aromatic character. Chemically oxazoles function as dienes in the Diels-Alder reaction and electrophilic substitution is rare.

10.4.2 Synthetic Methods

The synthesis of the title compound has been achieved by several methods.

1.   *From Ethyl α-Hydroxy Keto Succinate:* Oxazole was first prepared by Cornforth and Cornforth[76] by a rather long reaction sequence. A simple method has been reported by Bredereck and Bangert[77] which involves a reaction between ethyl α-hydroxy keto succinate and formamide to give diethyloxazole-4, 5-dicarboxylate which is subsequently hydrolyzed and decarboxylated to isoxazole.

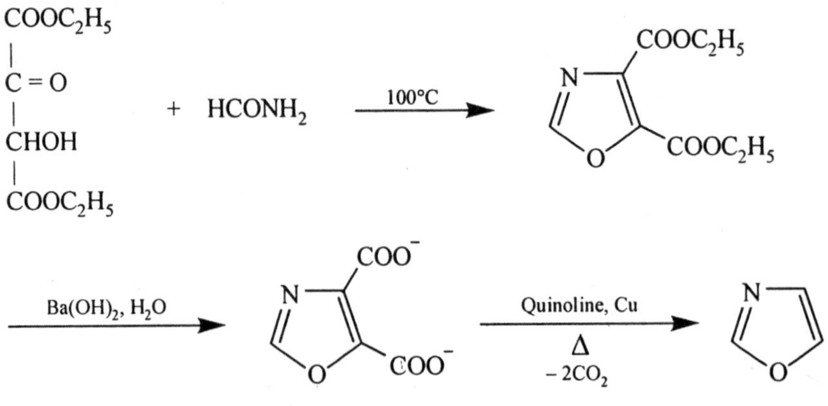

2. *Robinson-Gabriel Synthesis:* This is perhaps the most common method used for the preparation of oxazoles. This method involves an α-acylamino ketone which undergoes cyclization and dehydration in the presence of phosphorus pentoxide or a strong mineral acid. This synthesis is especially applicable for the formation of 2, 5-diaryloxazoles. The mechanism outlined above has been confirmed[78] using ketone labelled with[18]O.

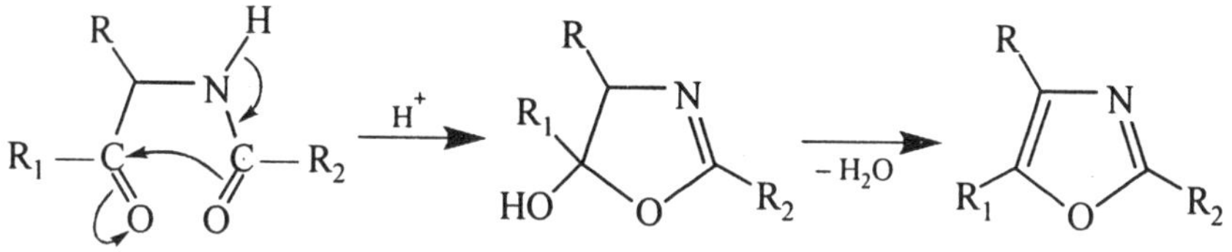

3. *From Isocyanides:* Reaction of α-metallated isocyanides with acid chlorides or an hydrides yields substituted oxazoles.[79] A number of variations of this method have beeen developed by Van Leusen.[80]

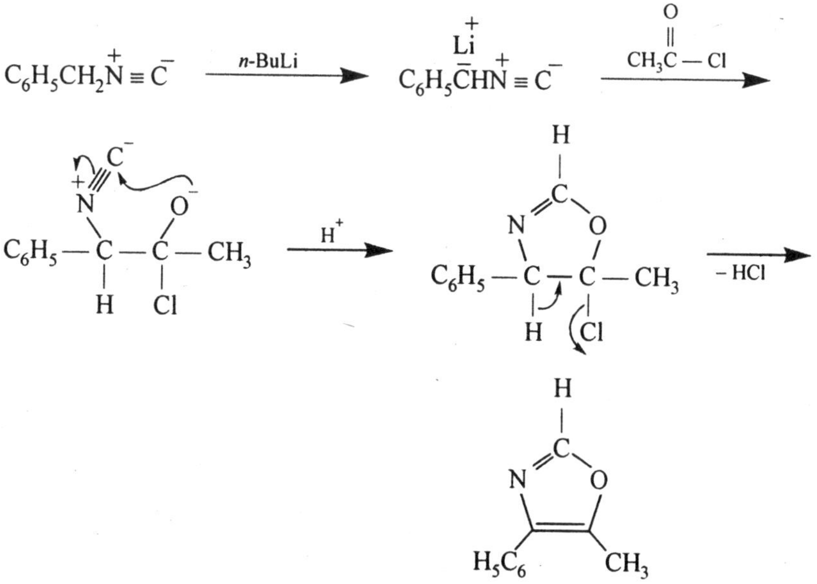

4. *From α-Aminocarbonyl Compounds:* This method[81] consists of refluxing an α-amino carbonyl compound with an imino ester to give an oxazole.

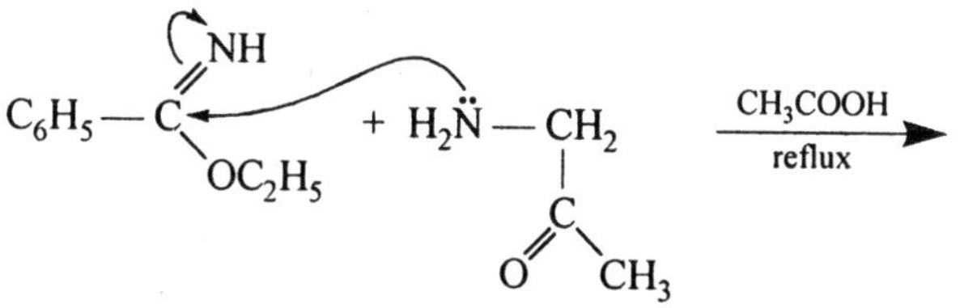

5-Methyl-2-phenyloxazole

Mechanistically the imino ester first adds to the carbonyl compound in the presence of acetic acid, then ring closure results in oxazole formation.

5. *Acylaziridines:* Thermolysis of 1-substituted-2-aroyl-3-phenylaziridine (62) gives rise to 2-aryl-5-phenyloxazole in high yields.[82] The reaction probably proceeds *via* the azomethine ylid (63) obtained by the heterolytic cleavage of the C – C bond in (62). The resulting ylid cyclizes to oxazoline which on oxidation leads to a derivative of oxazole.

(62)

(63)

96%

6. *Pomeranz-Fritsch Synthesis:* This procedure has been employed for the synthesis of a number of isoquinolines with varying yields.[83] It involves the cyclization of benzalaminoacetals in the presence of phosphorus pentoxide. Brown[84], however, found that if a nitro substituent is present on the ring than oxazoles are formed exclusively, equation (10.16).

$$(10.16)$$

7. Benzoxazoles are usualy prepared by heating 2-aminophenol derivatives of carboxylic acids with various leaving groups present on the acid itself. Condensation of these two substances under milder conditions using boric acid has been recently reported.[86]

The reaction of N-ethoxycarbonylthioamide with *o*-aminophenol similarly affords 2-substituted benzoxazole.[87]

Treatment of N-(2-hydroxyphenyl) thioamides with superoxide ($O_2^{=}$) at room temperature in dry acetonitrile results in the formation of 2-substituted benzoxazoles in excellent yields.[88]

93%

Superoxides ($O_2^{=}$) are electron reduction products of molecular oxygen. They have the potential of acting on redox agent, a nucleophile or a base.[89] Indoxazene photochemically[90] rearranges to benzoxazole.

Certain additional methods also have recently been reported[91-94] in the literature.

10.4.3 Chemical Reactions

Oxazoles and benzoxazoles can be quaternized using methyl iodide, dimethyl sulfate or triethyloxonium fluoroborate.[95]

1.   *Electrophilic Substitution:* Oxazole is more reactive towards electrophilic substitution than thiazole but less so than imidazole. The preferred attack as in the case of other 1, 3-azoles takes place at position-5. Electrophilic attack, however, occurs readily when the ring is activated by electron-donating substituents. Bromination (NBS) of 2-phenyloxazole results in 5-bromo-2-phenyloxazole. Nitration of this compound furnishes (*p*-nitrophenyl) oxazole.[96] Nitration and sulfonation of oxazole itself are difficult because of the presence of pyridine type nitrogen. Oxazoles are mercurated in acetic acid and mercuric acetate, the ring positions react in the order 5 > 4 >2.

2.   *The Diels-Alder Reaction:* The introduction of a second hetero atom does not affect the diene nature of oxazole and thus not unexpectedly oxazoles behave similar to furans in the Diels-Alder reaction. The adducts so obtained on reaction with dienophiles are important precursors for the preparation of pyridine or furan derivatives[97,98] by loss of appropriate fragments.

The orientation of oxazole dienophile system is correctly predicted through the localization energy calculations. A general rule is that the more electronegative group on the dienophile will occupy the 4-position in the final adduct. 4-Phenyloxazole similarly on treatment with anthranilic acid and isoamyl nitrite in refluxing dioxane leads to *bis* benzyne adduct 9, 10-ethoxydihydroanthracene,[99,100] equation (10.17).

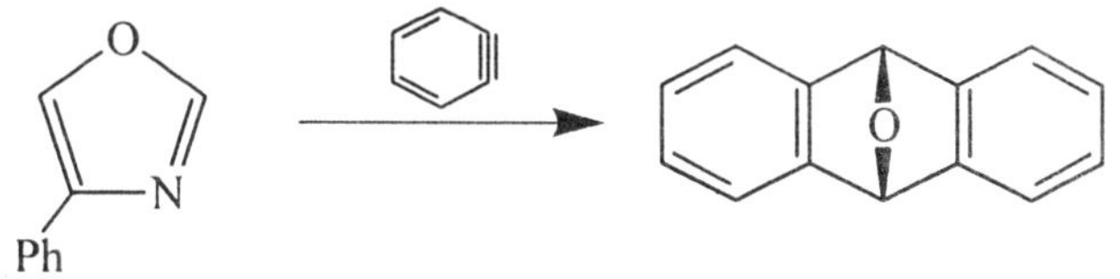

$$(10.17)$$

The oxazole adducts offer an entry into substituted furans.[101]

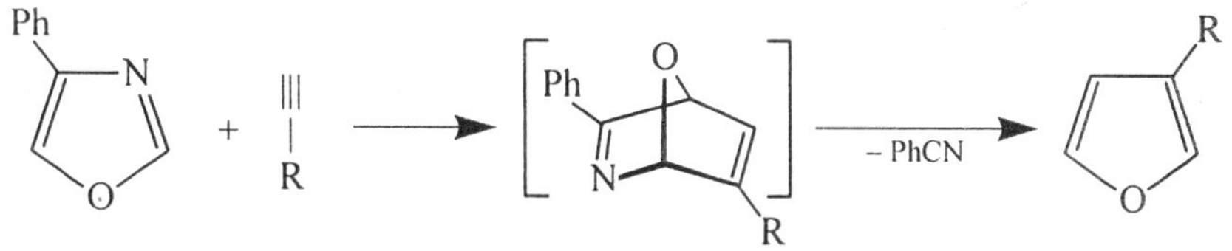

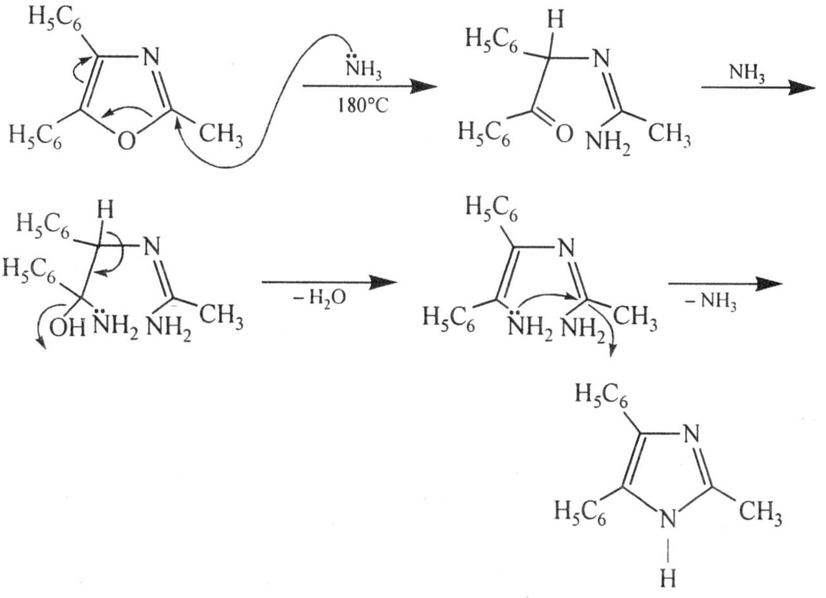

3.   *Reaction with Nucleophilic Reagents:* Nucleophilic substitution reactions on the oxazole ring are uncommon. The ease of displacement of halogens on the oxazole ring is C-2 >> C-4 > C-5.[73] The cleavage of the oxazole ring by nucleophiles is rather a more frequent reaction than nucleophilic substitution. In some cases a cyclic intermediate may be isolated while in others it may cyclize to afford another ring system. The reaction of ammonia or amine with alkyl-or acyl-oxazoles at a high temperature yields imidazoles, for instance, 2-methyl-4, 5-diphenyloxazole which rearranges to 2-methyl-4, 5-diphenylimidazole.

4.   *Reaction with Oxidizing and Reducing Agents:* The oxazole ring is not stable to oxidative conditions and thus is opened by the action of oxidizing agents such as cold potassium permanganate, chromic acid and ozone.[73] 2, 4, 5-Triphenyloxazole affords a quantitative yield of benzoic acid with chromic acid. Oxazole is normally stable to the action of hydrogen peroxide, but 4-(2-oxazolyl) pyridine with H_2O_2/HOAc gives the pyridine N-oxide derivative,[102] equation (10.18).

$$\text{(10.18)}$$

Oxazoles are, for the most part, remarkably stable towards a variety of reducing agents but reduction of the ring to oxazolidines can be affected with sodium in ethanol. A range of reductive ring cleavages is known. 2, 5-Diphenyloxazole is reduced[103] finally to 2-benzylamino-1-phenylethanol by lithium aluminum hydride ($LiAlH_4$), equation (10.19).

$$\text{(10.19)}$$

Catalytic hydrogenation has been achieved with platinum oxide or platinum in acetic acid to yield acyclic products.[73]

5.   *Photochemical Reaction:* The light induced reaction of oxazole is more complex than that of isoxazole. This is demonstrated[104] by the irradiation of 2, 5-diphenyloxazole which gives a mixture of products including 3, 5-diphenyloxazole, 2, 4-diphenyloxazole and benzoic acid. Oxazole ring is thermally stable.

10.4.4 Naturally Occurring and Biologically Active Compounds

Occurrence of oxazole in nature is rare. Naturally occurring oxazole include the alkaloid pimprinine (64) isolated from *streptomyces pimprina.*

Oxazole derivatives (65) are known to have anti-inflammatory, analgesic, antibacterial and antiviral properties. 2-Aminoazoles are useful as hypertensive agents. The oxazole derivatives (66) and (67) possess antibiotic activity.

(64)

(65)

(66)

Calcinomycin

(67)

Griseoviridin

10.5 ISOTHIAZOLES AND BENZOISOTHIAZOLES

Isothiazole or 1, 2-thiazole (3) is the sulfur analog of isoxazole. This compound was first described by Adams and Slack in 1965. The only isothiazole of any importance is saccharin (68) which is used as a substitute for sugar. There are two isomeric fused bicyclic compounds incorporating both an isothiazole ring and a benzene ring. Therse ring systems are known as 1, 2-benzisothiazole or *benzo[d]isothiazole* (69) and 2, 1-benzisothiazole or *benzo[d]isothiazole* (70). Comprehensive reviews of these compounds are available.[105-108]

(3)   (68)   (69)

(70)

10.5.1 Physical and Spectroscopic Properties

Isothiazole is a pale yellow liquid, b.p. 112°C which is closer to pyridine (115°C) and thiazole (117°C). It has a typical odor resembling that of pyridine. It is miscible with most organic solvents. Isothiazole is weakly basic (pKa –0.51) because the lone-pair on the azomethine nitrogen is not part of the aromatic sextet. The presence of amino group at 3-, 4- or 5-position increases the basicity (pKa 2.49 to 2. 70). 2, 1-Benzisothiazole is also a very weak base (pKa 0.05). Saccharin is a strong acid with $p$ka of 1.30. 3-Hydroxyisothiazoles are also acidic, the-OH group gives certain phenolic reaction.

Isothiazole exhibits *i.r.* spectrum, similar to many other five membered heterocycles, three bands of medium intensity between 1600 and 300 cm^{-1}.Proton chemical shift[109] data demonstrates that isothiazole is aromatic in character, i.e., it has an aromatic delocalized π-electron system.

10.5.2 Synthetic Method

Isothiazole and its derivatives have been prepared by a number of methods.

1.   *From Nitrile Sulfides:* Nitrile sulfides generated by the thermolysis of 1, 3, 4-oxathiazol-2-one add to alkenic or alkynic esters such as dimethylacetylenedicarboxylate (DMAD) to give 3-substituted-4, 5-isothiazole dicarboxylate.[110]

2. *From Oxazoles:* Oxazoles are reduced to enamino-ketones which are subsequently treated with phosphorus pentasulfide and chloranil to effect cyclization to isothiazoles.[111]

3. *From Acetylenes:* Several routes from acetylene exist including the reaction of acylacetylene with sodium thiosulfate to give the intermediate (71) which is cyclized in the presence of ammonia to give 3-substituted isothiazoles[112], equation (10.20).

$$\text{(71)} \qquad \xrightarrow{NH_3} \qquad (10.20)$$

Certain alkenes react with sulfur dioxide and ammonia in the presence of activated alumina as catalyst to give high yields of isothiazoles. Propylene under these conditions gives the parent compound isothiazole.[113] Isobutylene yields 4-methylisothiazole while α-methylstyrene gives 4-phenylisothiazole.

4. 1, 2-Benzoisothiazoles are often prepared from *o*-mercaptobenzaldoximes on cyclization with polyphosphoric acid,[106] equation (10.21).

$$\xrightarrow[- H_2O]{PPA, 130°C} \qquad (10.21)$$

2, 1-Benzoisothiazoles may be prepared by reductive-cyclization of *o*-nitrotoluene thiol with stannous chloride and hydrochloric acid.

10.5.3 Chemical Reactions

1.   *Quaternization:* Isothiazoles form salts on treatment with strong acids or Lewis acids. It however, decomposes on attempted alkylation with alkyl halide and heat. The more powerful reagents like dimethyl sulfate and triethyloxonium fluoroborate do give quaternary salts.[105-114] 1, 2-Benzisothiazole does not readily quaternize although 2, 1-benzisothiazole forms stable salts.

2.   *Electrophilic Substitution:* Isothiazole is the least reactive towards electrophilic attack as compared to pyrazole and isoxazole. The C-4 is susceptible to electrophilic attack. Isothiazole is almost completely protonated in strong acids, but conc. nitric acid yields 4-nitroisothiazole.

4-Nitroisothiazole

Halogen substituted at the 5-position in isothiazole undergoes facile displacement by uncleophiles. Isothiazoles react with hexacarbonyls $M(CO)_6$ to yield N-coordinated M(CO) derivatives.

3.   *Reaction with oxidizing and Reducing Agents:* The isothiazole ring is generally stable to a range of oxidizing and reducing agents including potassium permanganate or chromic acid.

4.   *Photochemical Reactions:* The principal photoreaction of isothiazoles is isomerism. Alkyl and phenylisothiazoles on irradiation yield the corresponding thiazoles, *via* a zwitterionic type intermediate.[114]

10.5.4 Naturally Occurring and Biologically Active Compounds

Sulfasomizole (72) is an antibiotic and has a medium duration of action against organisms. The growing understanding of the isothiazole chemistry has enabled the chemists to incorporate this ring system into a large number of compounds with potential biological activity. Isothiazole analogs of histamine, histidine and thiamine have been prepared.[111] Certain penicillins have been prepared from isothiazole 3-, 4-, and 5-acetic acids.

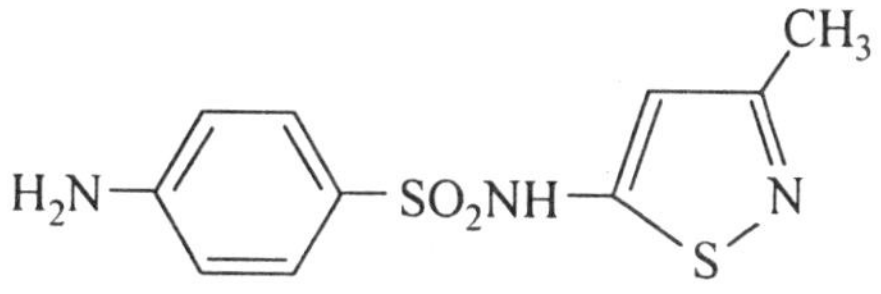

(72)

Various isothiazole derivatives have been employed as chemicals in agriculture. Compounds (73) to (75) for instance, have been employed as herbicides and fungicides.

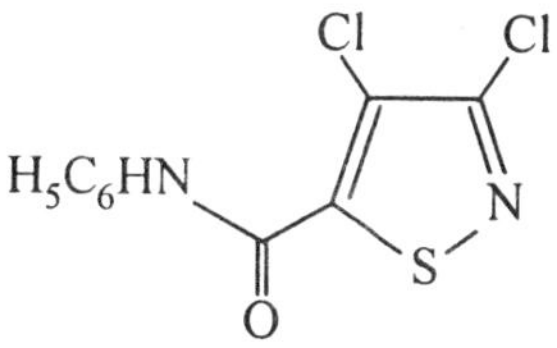

(73) 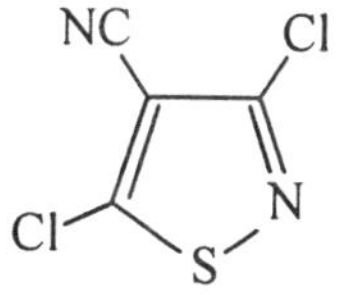   (74) 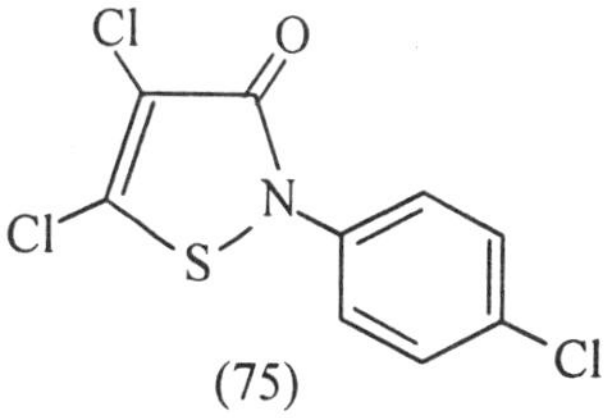   (75)

10.6 THIAZOLES AND BENZOTHIAZOLES

Thiazole (6) is structurally related to thiophene and pyridine but in most of its properties it resembles the latter.[115,116] Structure (76) is benzothiazole. Thiazole was first described by Hantzsch and Weber in 1887. Popp confirmed

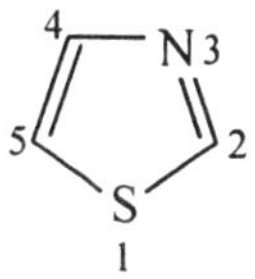

(6) 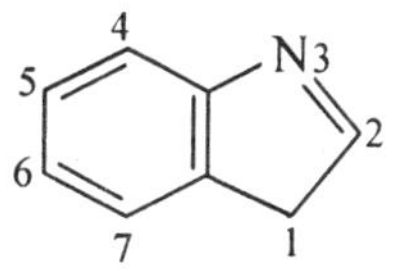   (76)

its structure in 1889. The numbering in thiazole starts from the sulfur atom. The thiazole ring has been extensively studied and it forms a part of vitamin B, penicillins and the antibacterial thiazoles. Reduced thiazoles serve in the study of polypeptides and proteins and occur as structural units in compounds of biological importance.

The partially reduced thiazoles are called *thiazolines* such as 2-thiazoline (77), 3-thiazoline (78) and 4-thiazoline (79). Out of these only 2-thiazolines have been studied.

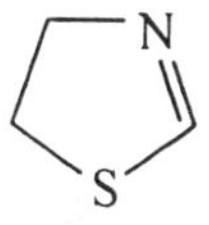

(77) 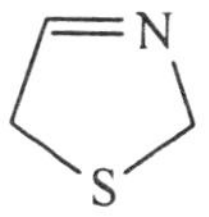   (78) 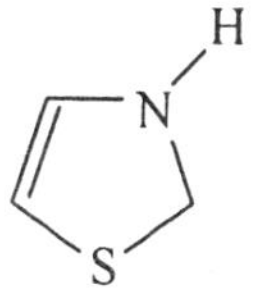   (79)

10.6.1 Physical and Spectroscopic Properties

Thiazole is a colorless liquid, b.p. 177°C which is close to pyridine (115°C) than thiophene (84°C). It resembles pyridine in odor and is miscible with water. It is weakly basic (pKa 2.5) and behaves as a tertiary base. The alkyl derivatives of thiazole are more basic, for instance, 2-methylthiazole (pKa 3.40) and 4-methylthiazole (pKa 3.16). Its aromatic character is reflected in its chemical properties though it has been questioned.[117,118] The chemical shift in the *n.m.r.* are correlated to the occurrence of a ring current connected with the *pseudo* aromatic character of the thiazole ring. Its spectroscopic properties have been summarized.[119,120]

10.6.2 Synthetic Methods

1.   *From α-Halocarbonyl Compounds:* This is an old but most important method for the preparation of thiazoles. In a general case an α-haloketone is reacted with an appropriate thiamide, or thiourea, equation (10.22).[121] Chloroacetaldehyde, for instance, on condensation with thioformamide affords thiazole.

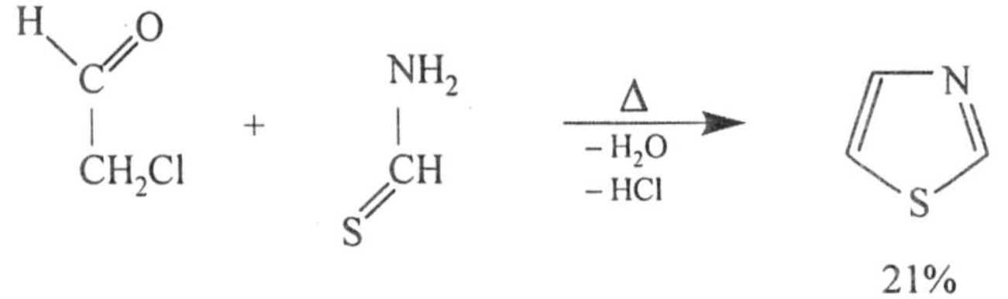

$$\text{(10.22)}$$

21%

The yields in this reaction are low because of side product formation. Thioamide for the purpose can be obtained by reacting phosphorus pentasulfide and formamide at room temperature,[122] equation (10.23).

$$\underset{\text{HCNH}_2}{\overset{\text{O}}{\|}} + P_2S_5 \longrightarrow \underset{\text{HCNH}_2}{\overset{\text{S}}{\|}} \qquad \text{(10.23)}$$

A plausible mechanism for this reaction could involve the stepwise elimination of water and hydrogen chloride as depicted below:

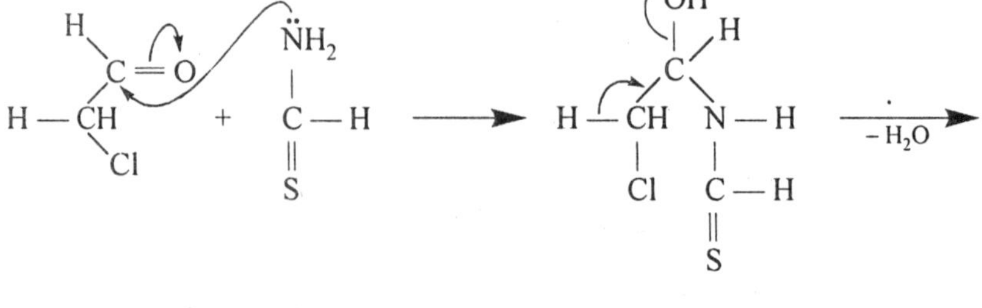

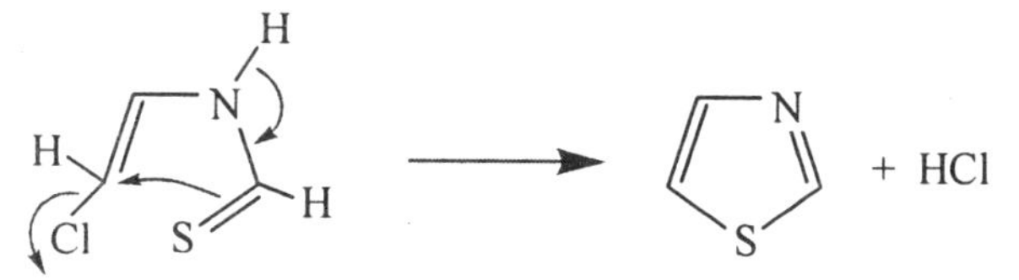

Substituted thiazoles can be prepared by selecting a suitable chlorocarbonyl compound and the thioamide. Thus α-halo derivatives of the higher aldehydes give 5-substituted thiazoles.[133]

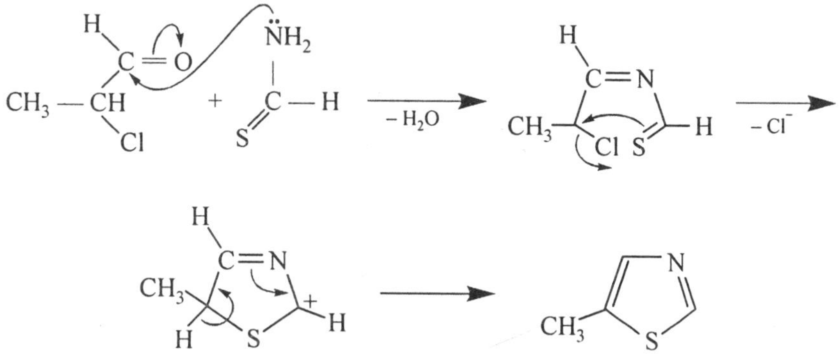

A method closely related to the above involves the reaction of α-halocarbonyl compound with ammonium dithiocarbamate. This gives rise to 2-mercaptothiazoles.[124]

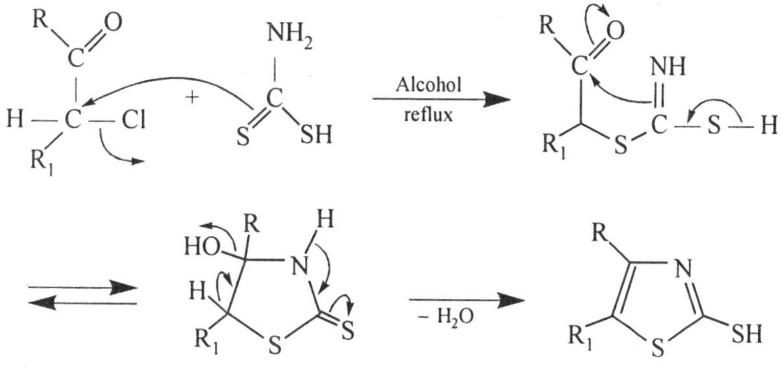

By varying R and R_1 substituted thiazoles can be obtained.[125]

2.   *From α-Thiocyanato Ketones:* α-Halo-carbonyl compounds and metal thiocyanate react to give α-thiocyanato ketone which cyclizes on treatment with acid or alkali.[137] 2-Hydroxy-4-methylthiazole has been obtained in about 60% yield from chloroacetone and sodium thiocyanate.

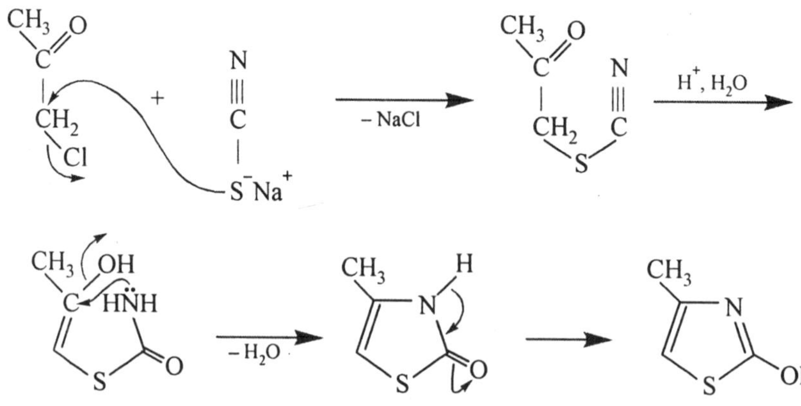

3. *From Thioamides:* Thiazole derivatives may also be obtained[126] by reacting a thioamide with substituted 2-chloroxiranes.

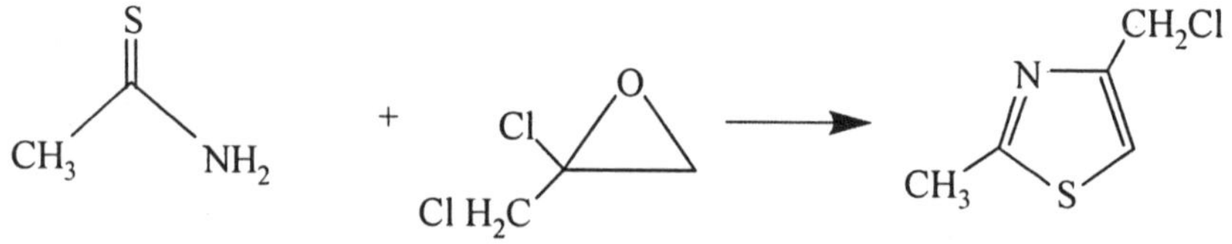

4. *The Gabriel Synthesis:* This synthesis involves the heating of acylamino compounds with phosphorus pentasulfide as shown below to give thiazole derivatives.

The carbonyl group is probably attacked first by P_2S_5, the oxygen atom being replaced by sulfur and the intermediate cyclizes to thiazole. This method is similar to the Paul-Knorr synthesis for thiophene. The driving force in this process results from the stabilization obtained in the formation of the aromatic heterocycle.[127]

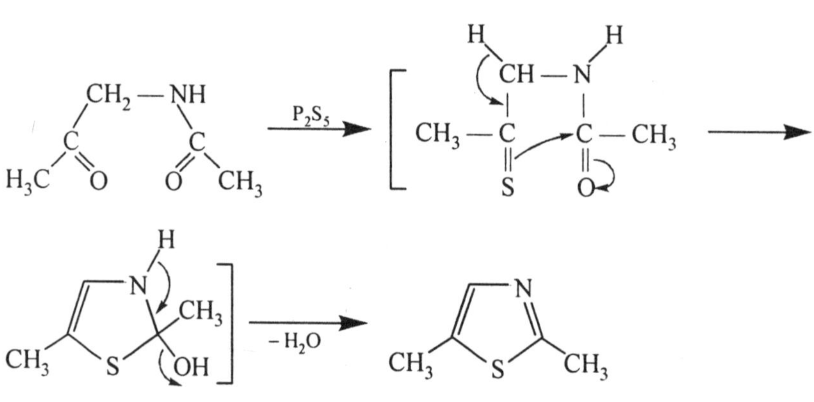

2,5-Dimethylthiazole

5. Benzothiazoles may be obtained from 2-aminothiophenol and a carboxylic acid or an anhydride. They can also be obtained by the direct

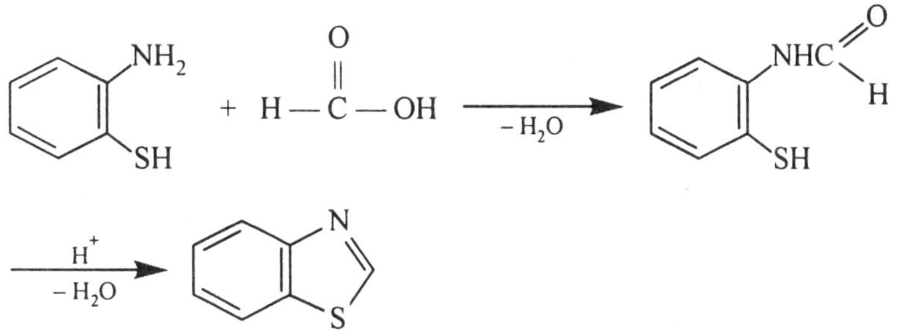

thiocyanation ($KSCN/CuSO_4$) of substituted anilines.[128] Condensation of

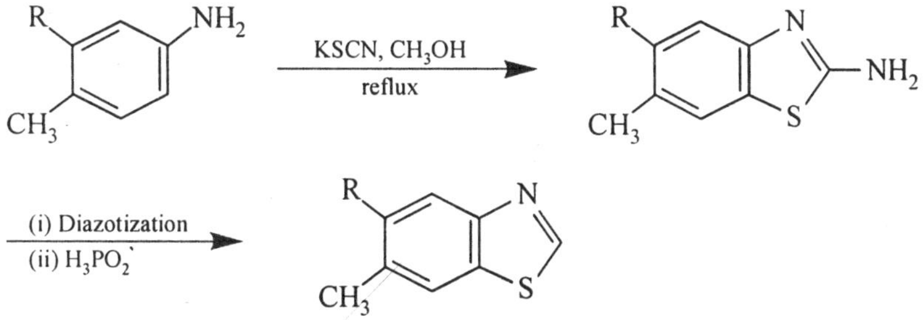

N-carboethoxythiacetamide with *o*-aminothiophenol yields 2-substituted benzothiazoles,[129] equation (10.23). 2-Arylbenzothiazoles can be obtained

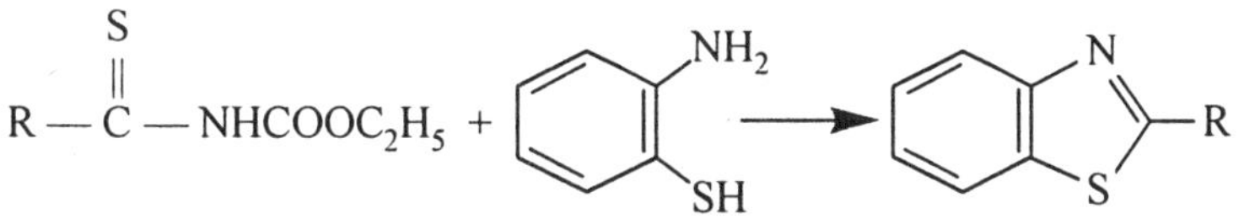

$$R - \overset{\overset{\displaystyle S}{\|}}{C} - NHCOOC_2H_5 + \quad\longrightarrow\quad \qquad (10.23)$$

directly from copper thiobenzoate and 2-iodoaniline on heating in hexamethyl-phosphoric triamide (HMPT).[130]

10.6.3 Chemical Reaction

Thiazoles and benzothiazoles can be quaternized.

1. *Reaction with Acids:* The lone-pair of electrons on the azomethine nitrogen is not involved in aromatic sextet and is thus available for protonation. Thiazole as a result forms stable crystalline salts with strong acids.

2. *Electrophilic Substitution:* The reactivity of thiazoles in electrophilic substitution is well documented[117,131]. The reactivity is intermediate between pyridine and thiophene, but it is less reactive than imidazole. The attack preferably takes place at position-5. In acid media, the attack usually takes place through the formation of thiazolium cation. Thiazole itself is resistant to bromination ($Br_2/CHCl_3$), however, if electron-donating groups are present then bromination is achieved rather easily. Thiazole can be brominated directly in the gas phase but in this case 2-bromothiazole is obtained. This reaction probably involves attack by bromine radicals (equation 10.24). 2-Chlorothiazole can be obtained by treating 2-thiazolone with phosphorus pentachloride, equation (10.25).

$$\text{(thiazole)} \xrightarrow[\text{250–400°C}]{Br_2} \text{(2-bromothiazole)} \qquad (10.24)$$

2-Bromothiazole

$$\text{(2-thiazolone)} \xrightarrow{PCl_5} \text{(2-chlorothiazole)} \qquad (10.25)$$

2-Chlorothiazole

Bromination with N-bromosuccinimide also yields 2-bromothiazole though in a low yield.[132] Nitration of thiazole is rather difficult and is not nitrated though even in oleum at 160°C. The mononitro derivatives of thiazoles are obtained by indirect means. Thus 2-nitrothiazole is obtained from 2-aminothiazole by diazotization and subsequent treatment with sodium nitrite[133]. Sulfonation of thiazole may be achieved under drastic conditions in the presence of mercury sulfate as catalyst to give thiazole 5-sulfonic acid. Benzothiazole on nitration gives 6-nitrobenzothiazole. Arylthiazoles are preferentially nitrated in the aromatic ring. The Gattermann and the Friedel-Crafts reactions are unsuccessful. Thiazoles are also mercurated in acetic acid similar to oxazoles and the ring positions react in the order 5 > 4.[58]

3. *Reaction with Nucleophilic Reagents:* In contrast to imidazole, the reaction of potassium amide in liq. ammonia with thiazole affords 2-aminothiazole. The C-2 position being susceptible to nucleophilic attack. This demonstrates its closeness to pyridine. The 2-halothiazoles easily react with nucleophilic reagents leading to expected products of nucleophilic substitution. The quaternization of thiazole takes place at 3-position and the thiazolium salts are preferentially susceptible to nucleophilic attack and resutls in ring-opening.

With *n*-butyllithium the 2-position is metallated to yield thiazole-2-lithium.[134] 2-Methylthiazole also condenses with benzaldehyde in the presence of $ZnCl_2$ to yield 2-styrylthiazole. The 4-isomer does not condense under the same conditions.

4.   *Reaction with Oxidizing and Reducing Agents:* Thiazole ring is resistant to oxidation by nitric acid but the ring is opened by potassium permanganate. 2, 5-Dimethylthiazole and benzothiazole may be oxidized by peracids to give the N-oxides.

Thiazole ring is also resistant to the action of many reducing agents, although Raney nickel can be used for desulfurization reactions (equation 10.26).

$$\text{(10.26)}$$

5.   *Photochemical Reactions*[135]. As in the case of oxazoles, isomerization and interconversion with isothiazoles are the predominant photochemical processes. Benzoisothiazole undergoes isomerization to the corresponding benzothiazole photolytically or thermally.[135a]

10.6.4 Naturally Occurring and Biologically Active Compounds

Thiazoles and thiazolines are structural compounds of a number of peptide derived natural products, among them are antibiotics siomycin, thiostereption and micrococin; antitumor antibiotics; phleonycin and bleomycin and Jadot's novel dicarboxylic amino acid (80) obtained from mushroom *Yerocomus subtmentosus*.

(80)

Another important naturally occurring thiazole is thiamine (Vit B$_1$) (81) in the form of its pyrophosfate in the coenzyme of a number of biological reactions involved in carbohydrate metabolism i.e., cleavage of C-C bond adjacent to the carbonyl group (as in pyruvic acid). Thiamine contains a pyrimidine and a thiazole nucleus but the latter is the chemically active center in the coenzyme. Reduction of the thiazolium ring results in complete loss of activity.[136] It occurs widely in plants and specially in eggs, yeast and rice polishing. The deficiency of thiamine in diet causes a disease called *beriberi*.

(81)

REFERENCES

1. For a review see, T. L. Jacob in, *Heterocyclic Compounds,* R. C. Elderfield (Ed.)., Vol. V, Interscience, New York (1957).
2. I. Bhatnagar and M. V. George, *Tetrahedron,* **24**, 1293 (1968).
3. A. F. Pozharskii, A. D. Garnovskii and A. M. Simonov, *Russ. Chem. Rev.,* 35, **122** (1966).
4. R. Fusco, in *The Chemistry of Heterocyclic Compounds,* R. H. Wiley, (*Ed.*), Interscience, New York (1967).
5. A. N. Kost and I. I. Grandberg, *Adv. Heterocyclic Chem.,* **6**, *347 (1966).*
6. H. Ehrlich, Acta, Cryst., **13**, 940 (1960).
7. I. L. Finar and R. J. Hurlock, *J. Chem. Soc.,* 3259 (1958).
8. W. Kirmse and L. Horner, *Ann.,* **614**, 1 (1958).
9. V. R. Vasil'eva *et al., Zh. Obshch. Khim.,* **31**, 1501 (1961).
10. C. Ainsworth, *J. Am. Chem. Soc.,* **79**, 5242 (1957).

11. A. Albert, *Physical Methods in Heterocyclic Chemistry,* A. R. Katritzky, *(Ed.),* Vol. I, Academic Press, New York (1963).

12. R. Hüttel *et al., Ann.,* **598**, 186 (1956).

13. M. W. Austin, J. R. Blackbrown, J. H. Ridd and B. V. Smith., *J. Chem. Soc.,* 1051, (1965).

14. J. D. Vaughan, D. G. Lambert and V. L. Vaughan, *J. Am. Chem. Soc.,* **86**, 2857 (1964).

15. I. L. Finar and R. J. Hurlock, *J. Chem. Soc.,* 3024 (1957).

16. C. I. Chiriac, *Synthesis,* 753 (1986).

17. H. R. Synder, F. Verbanar and O. B. Bright, *et al., J. Am. chem. Soc.,* **74**, 3243 (1952).

18. P. Beak and W. Messer, *Tetrahedron,* **25**, 3287 (1969).

19. M. R. Grimmett, *Adv. Heterocyclic. Chem.* **12**, 103 (1970).

20. A. R. Day, in *Heterocyclic Compounds,* R. C. Elderfield, (Ed.), Vol. V, Wiley, New York (1957).

21. K. Hoffmann, *Imidazole and its Derivatives,* Interscience, New York (1953).

22. G. E. Gutowski *et al., Ann. N. Y. Acad. Sci.,* **255**, 594 (1975).

23. S. W. Fox, *Chem. Ber.,* **32**, 47 (1943).

24. H. Bredereck *et al., Ber.,* **92**, 338 (1959).

25. E. Jassmann and H. Schulz, *Pharamazie,* **18**, 461 (1963); C. A., 64, 3518 (1966).

26. J. L. Hugmey, S. Knapp and H. Schugan, *Synthesis,* 489, (1980).

27. P. Oxley and W. E. Short, *J. Chem. Soc.,* 497 (1947).

28. B. B. jarvis and G. P. Stahly, *J. Org. Chem.,* **45**, 2604(1980).

29. H. Schubert, *J. Prakt. Chem.,* **3**, 146 (1956).

30. I. W. Cornforth and H. T. Huang, *J. Chem. Soc.,* 1960 (1940).

31. H. Bredereck *et al., in Newer methods of Preparative organic Chemistry,* W. Forest, *(Ed.),* Vol. III, Academic Press, New York (1964), p. 241.

32. D. W. Hein, R. J. Alheim and J. J. Leavitt, *J. Am. chem., soc.,* **79**, 427 (1957), Also see S. Rajappa and R. Sreenivasan, *Ind. J. Chem.,* **19B**, 533 (1980).

33. H. J. J. Loozen, J. J. M. Droven and O. Piepers, *J. org. Chem.,* **40**, 3279 (1975).

34. J. D. Vaughan, V. L. Vaughan, S. S. Daly and W. A. Smith, *J. Org. Chem.,* **45**, 3108 (1980).

35. A. W. Lutz and S. De Lorenzo, *J. Heterocyclic Chem.* **4**, 399 (1967).

36. A. P. Koran *et al., J. Inorg. Bio. Chem.,* **10**, 235 (1979).

37. T. M. Haris and J. C. Rendall, *Chem. and Ind.,* 1728 (1965).

38. J. D. Vaughan, A. Mughrabi and E. C. Wu, *J. Org. Chem.,* **35**, 1141 (1970). *J. Org. Chem.,* **35**, 1141 (1970).

39. S. Kim and D. C. Lhim, *Bull. Chem. Soc. Japan,* **59**, 3297 (1986).

40. P. W. Alley, *J. Org. Chem.,* **40**, 1837 (1975).

41. D. A. de Bie and H. C. van Der Plas, *Rec. Trav. Chim.,* 88, 1246 (1969).

42. N. K. Kochetkov and S. D. Sokolov, *Adv. Heterocyclic Chem.,* *2*, 365 (1963); K. C. Joshi *et al.,Pharmazie,* **34**, 2 (1979).

43. *Five-and Six-Membered Compounds with Nitrogen and Oxygen,* in Chemistry of Heterocyclic Compounds, A. Weissberger *(Ed.),* Interscience, New York (1962).

44. K. H. Wunsch and A. J. Boulton, *Adv. Heterocyclic Chem.,* **8**, 277 (1967).

45. *The Chemistry of Heterocyclic Chemistry,* R. C. Elderfield, *(Ed.),* Vol. V. Wiley, New York (1956).

46. K. M. Johnson and R. G. Schotter, *J. Chem. Soc.,* (C), 1774, (1960).

47. R. Justoni and R. pessia, *Gazz, Chim. Ital.,* **85**, 34 (1955).

48. A. Quilico, in *The Chemistry of Heterocyclic Compounds,* *(Ed.),* R. L. Wiley, Vol. XVII, John Wiley (1962), Chapter I.

49. Y. Lin and S. A. Lang, *J. Org. Chem.,* **45**, 4857(1980).

50. W. D. Ollise *et al., J. Chem. Soc.,* 1307 (1952).

51. G. witting and E. Banquest, *Chem. Ber.,* 58, 2636 (1925).

52. W. Basinski and Z. Jerzmanousky, *Rocz, Chem.,* **48**, 2217 (1974).

53. R. Beugelmans and C. Morin, *J. Org. Chem.,* **42**, 1356 (1977).

54. K. Maeda, T. Hosokawa, S. Murahasi and I. Moritani, *et al., Tet. letters,* 5075 (1973).

55. B. J. Wakefield and D. J. Wright, *Adv. Heterocycl. Chem.,* **25**, 147 (1979).

56. R. F. Cunico and L. Bedell, *J. Org. Chem.,* **48**, 2780 (1983).

57. T. shimizu, Y. Hayashi, N. Furukawa and K. Teramura, *Bull, Chem, Soc. Japan,* **64**, 318 (1991).

58. G. Giacomelli *et al, Tetrahedron* **59**, 5437 (2003).

59. C. H. Wong *et al., J. Am. Chem. Soc.,* **69**, 1909 (1947).

60. F. Korte and O. Behner, *Ann. Chem.,* **621**, 51 (1959).

61. N. K. Kochetkov *et al., Zhur. Obshchei Khim.,* **32**, 325 (1962).

62. A. Quilico and C. Musante, *Gazz. Chim. Acta.,* **71**, 317 (1941).

63. R. B. Woodard *et al., J. Am. Chem. Soc.,* **83**, 192 (1953).

64. G. Speroni and E. Giachetti, *Gazz. Chim. Ital.,* **83**, 192 (1953).

65. A. A. Akhrem. F. A. Lakhvich, V. A. Khripach and I. B. Klebanovich, *Tet. Letters,* 3983 (1976), D. N. McGregor, U. Corbin, J. E. Swigor and L. C. Chemy, *Tetrahedron,* **25**, 389 (1969).

66. C. Kashima, M. Uemori, Y. Tsuda and Y. Omote, *Bull. Chem. Soc. Japan,* **49**, 2254 (1976).

67. T. L. Gilchrist. in *Aromatic and Heteroaromatic Chemistry,* (Specialist Periodical Reports). The Chemical Society, London (1973).

68. A. Padwa, E. Chen and A. Ku., *J. Am. Chem. Soc.,* **97**, 6484 (1975).

69. J. D. Pevez *et al., J. Org. Chem.,* **46**, 3505 (1981).

70. E. F. Ullman and B. Singh, *J. Am. Chem. Soc,* **89**, 6911 (1967).

71. J. P. Ferris and R. W. Trimmer, *J. Org. Chem.,* **41**, 13 (1976); also see J. P. Ferris *et al., J. Am. Chem. Soc.,* **95**, 919 (1973).

72. R. Nesi, D. Giomi, S. Papaleo and M. Corti, *J. Org. Chem.* **55**, 1227 (1990).

73. For a review see. I. J. Turchi and M. J. S. Dewar, *Chem. Rev.,* **75**, 389 (1975).

74. J. W. Cornforth, in *The Chemistry of Penicillin,* Princeton University Press, Princeton, N. J. (1949).

75. *The Chemistry of Heterocyclic Compounds*, R. C. Elderfield, (*Ed.*), Vol. V. Wiley, New York (1956).

76. J. W. Cornforth and R. H. Cornforth, *J. Chem. Soc.,* 96 (1947).

77. H. Bredereck and R. Bangert, *Chem. Rev.,* **97**, 1414 (1964); *Angew. Chem. Int. Edin.* (Engl.), **1**, 662 (1962).

78. H. H. Wasserman and F. J. Vinick, *J. Org. Chem.,* **38**, 2467 (1973).

79. U. Scholkoff and R. Schroder, *Synthesis*, 148 (1972); U. Scholkoff and E. Blume, *Tet. letters,* 629 (1973).

80. A. M. van Leusen, B. E. Hoogenboom and H. Siderius, *Tet. letters,* 2369 (1972).

81. J. W. Cornforth and H. T. Huang, *J. Chem. Soc.,* 1960 (1948).

82. A. Padwa and W. Eisenhardt, *Chem. Comm.,* 380 (1968).

83. J. Gensler, *Org. Reactivity*, **6**, 191 (1951).

84. E. V. Brown, *J. Org. Chem.,* **42**, 3208 (1977).

85. A. Maquestian, E. Puk and R. Flammang, *Tet. letters,* 4022 (1986), also see A. Dondoni, M. Fogagolo, A. Mastellari, A. Peopini and F. Ugozzoli, *ibid.,* 3915 (1986).

86. M. Terashima and M. Ishii, *Synthesis,* 484 (1984).

87. B. George and E. P. Papadopulos, *J. Org. Chem.,* **42**, 441 (1977).

88. Y. H. Kim, Y. Kimi, H. S. Chang and D. C. Yoon, *Heterocycles,* **29**, 213 (1989).

89. D. T. Sawyer and M. J. Gibbian, *Tet. Letters,* 1471 (1979).

90. J. P. Ferris and F. R. Antonucci, *J. Am. Chem. Soc.,* **96**, 2014 (1974).

91. R. D. Clark and J. W. Caroor, *J. Org. Chem.,* **47**, 2804 (1982).

92. E. C. Taylor, A. H. Katz, S. I. Alvarddo and A. Mckillop, *ibid.,* **51**, 1607 (1986).

93. R. Garner, *J. Chem. Soc. (C),* 1988 (1966).

94. R. J. Perry, B. D. Wilon and R. J. Miller, *J. Org. Chem.,* **57**, 2883 (1992).

95. V. M. Dziomko and I. V. Ivashchenko, *Zhur. Org. Khim.,* **9**, 2191 (1973).

96. R. H. Wiley, *Chem. Rev.*, **37**, 401 (1945).

97. For a review see, M. Ya. Karpeiskii and V. L. Elorent'ev, *Russ. Chem. Rev.*, **38**, 540 (1969).

98. T. Matsno and *T. Miki, Chem. Pharma. Bull.* (Japan), **20**, 806 (1972); P. A. Jacobi *et al., J. Org. Chem.*, **46**, 2065 (1981); A. P. Kozikoswski and N. M. Hassan, *ibid.*, **42**, 1039 (9177).

99. S. E. Whitney, M. Winters and B. Rickborn, *J. Org. Chem.*, **55**, 929 (1990).

100. G. S. Reddy and M. V. Bhatt, *Tet. Letters*, 3627 (1980).

101. D. Liotta, M. Saindane and W. Ott, *Tet. Letters*, 2473 (1983).

102. J. V. Bauer *et al., J. Med. Chem.*, **12**, 945 (1969).

103. N. G. Gaylord and D. J. Kay, *J. Am. Chem. Soc.*, **78**, 2167 (1956).

104. M. Kojima and M. Maeda, *Tet. letters*, 2379 (1969).

105. R. Slack and K. R. H. Wooldridge, *Adv. Heterocyclic Chem.*, **4**, 107 (1965).

106. M. Davis, *ibid.*, **14**, 43 (1972).

107. L. L. Bambas, in *The Chemistry of Heterocyclic Compounds*, A. Weissberger, *(Ed.)*, Vol. IV. Wiley Interscience, New York (1952).

108. F. Hubenett *et al., Angew, Chem. Int. Edin.* (Engl.), **2**, 714 (1963).

109. D. Califano *et al., Spectrochem. Acta.*, **20**, 339 (1964).

110. R. K. Howe, T. A. Gruner, L. G. Carter, L. L. Black and J. E. Franz, *J. Org. Chem.*, **43**, 3736 (1971); *ibid.*, 3743 (1978), Also see, S. Rajappa *et al., Proceedings Ind. Acad. Sci.*, **91**, 451 (1982).

111. K. R. H. Wooldrige, *Adv. Heterocyclic Chem.*, **14**, 1 (1972).

112. F. Wille L. Capeller and A. Steiner, *Angew. Chem. Int. Edin.* (Engl.), **1**, 335 (1962).

113. F. Hubenett F. H. Flock and H. Hofmann, *ibid.*, **1**, 508 (1962).

114. G. Vernin *et al., Bull. Soc. Chim. France*, **5**, 1793 (1973).

115. For a review see, *"Heterocyclic Compounds"*, A. Weissberger and E. C. Taylor; *(Ed.)*, Vol. XXXIV, part I, Interscience, New York (1979).

116. B. Iddon and P. A. Lowe, *Spec. Period. Rep., Organic Compounds, Sulfur, Selenium and Telerium* (1977), Chapter 12.

117. L. Forlani, L. Lunazzi and A. Medici, *et al., Tet. letters*, 4525 (1977).

118. L. Forlani and A. Medici, *J. Chem. Soc., Perkin Trans.*, **1**, 1169 (1978).

119. A. R. Katritzky and J. M. Lagowski, *"The Principles of Heterocyclic Chemistry"*, Chapman and Hall, London (1968).

120. J. Panizzi, G. Davidovics, R. Gingliemetti, G. Mille, J. Metzzer and J. Chouteau, *Canad. J. Chem.*, **49**, 956 (1971).

121. R. Willstatter and T. Wuth, *Ber.*, **42**, 1908 (1909); R. H. Wiley, in *Organic Reactions,* Wiley New York (1951).

122. T. S. Griffin *et al., Adv. Heterocyclic Chem.*, **18**, 109 (1975).

123. H. Singh, C. S. Gandhi and M. S. Bal, *Chem. and Ind.,* 420 (1980); Also see, H. Singh, A. S. Ahuja and N. Malhotra, *Ind. J. Chem.,* **19B**, 1019 (1980).

124. M. Charon and J. Metzger, *Bull. Chem. Soc. France,* 7, 2842 (1968).

125. F. O. Stewart and R. A. Mathis, *Ind. Eng. Chem.,* **43**, 1569 (1951).

126. J. Gasteiger and J. Herzig, *Tetrahedron,* **37**, 2607 (1981). Also see, T. S. Jagodzinski, *Chem. Rev.,* **103**, 197 (2003).

127. R. M. Ansidei and G. Travagli, *Gazz. Chem. Ital.,* **31**, 680 (1941); *ibid.,* **32**, 7021 (1942).

128. I. A. Ismail, D. E. Sharp and M. R. Chedekel, *J. Org. Chem.,* **45**, 2243 (1980).

129. B. George and E. P. Papadoulos, *ibid.,* **42** 441 (1977).

130. A. Osuka, Y. Una., H. Horiuchi and H. Suzuki, *Synthesis,* 145 (1984).

131. L. Forlani, L. Lunazzi and A. Medici, *Tet. letters,* 4525 (1977).

132. B. M. Mikhailow and V. P. Brenovitskaza, *Zh. Obschei Khim.,* **27**, 726 (1957); *C. A.,* **51**, 16436 (1958).

133. J. Beraud and J. Metzger, *Bull. Soc. Chim. France,* 2072 (1962).

134. S. M. Hecht *et al., J. Org. Chem.,* **43**, 1024 (1978).

135. M. S. Kulyk and D. C. Neckers, *Tel. letters,* 525 (1981); *ibid.,* 529 (1981); D. C. Neckers *et al., J. Org. Chem.,* **35**, 1582 (1970); A. H. A. Tinnemans and D. C. Neckers, *ibid.,* **43**, 2493 (1978).

135a. T. R. Sharp *et al., Tet letters,* **44**, 1559 (2003).

136. P. N. Moorthy and E. Hayon, *J. Org. Chem.,* **42**, 879 (1977).

REVIEW PROBLEMS

10. 1 Write the major product for each of the following reactions:

$$(a) \quad \text{(bicyclic ketone with Br)} + \underset{\displaystyle RCNH_2}{\overset{\displaystyle S}{\|}} \longrightarrow$$

$$(b) \quad C_6H_5\overset{O}{\overset{\|}{C}}CH_2COCH_3 + CH_3\overset{O}{\overset{\|}{C}}NHNH_2 \longrightarrow$$

$$(c) \quad R-\overset{O}{\overset{\|}{C}}CH_2\overset{O}{\overset{\|}{C}}-R + H_2NOH \longrightarrow$$

(d)   [thiazole]   + C_6H_5Li $\xrightarrow[-80°C]{Ether}$

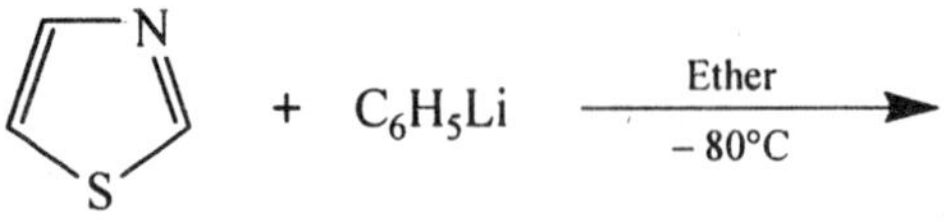

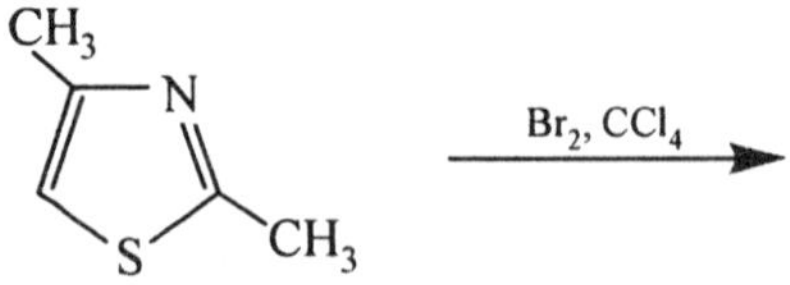

(e)   [4-methyl-2-methylthiazole] $\xrightarrow{Br_2,\ CCl_4}$

(f) $\begin{array}{c} CHO \\ | \\ CHO \end{array}$ $+ 2NH_3 + RCHO \longrightarrow$

(g)   [1-allylpyrazole] $\xrightarrow[(ii)\ \Delta]{(i)\ Br_2,\ CCl_4}$

(h)   [acenaphthenequinone] $\xrightarrow[(ii)\ 2NH_3]{(i)\ Ph\ CHO}$

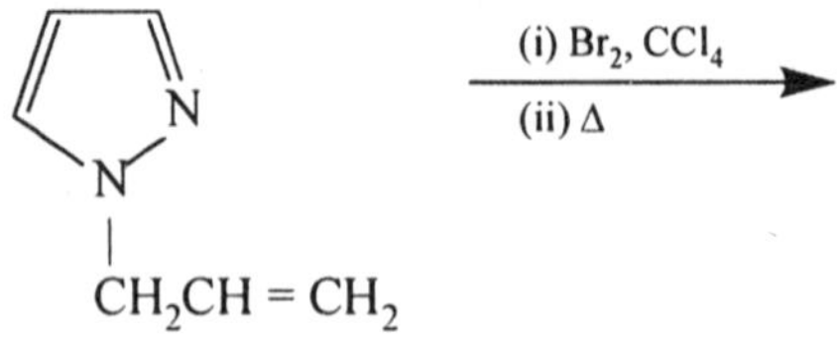

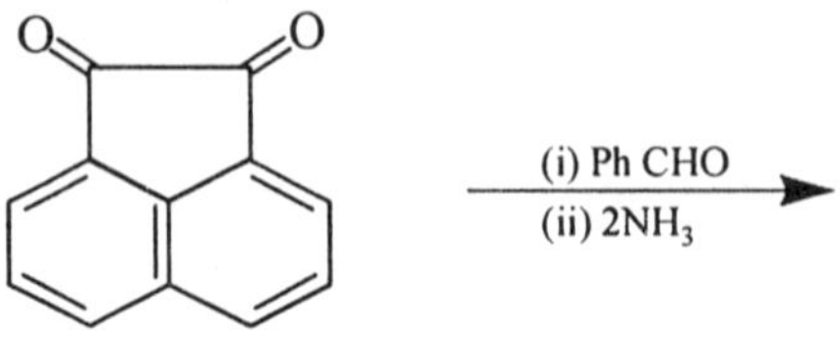

(i) $C_6H_5\overset{\overset{\displaystyle O}{\|}}{C}Cl + TOsCH_2N^+ \equiv C^-$ $\xrightarrow{Base}$

(j)   [4-phenyl-2-methyl-5-methyloxazole] $+$ [malononitrile/dicyanoethylene] $\longrightarrow$

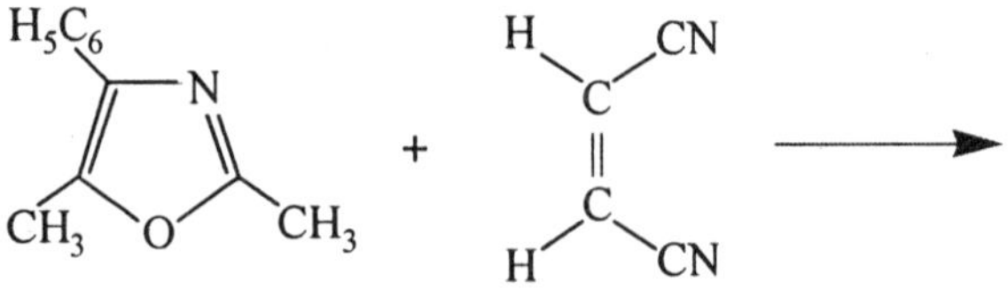

(k)   [2-aminophenyl phenyl ketone] $\xrightarrow{Sn,\ HCl}$

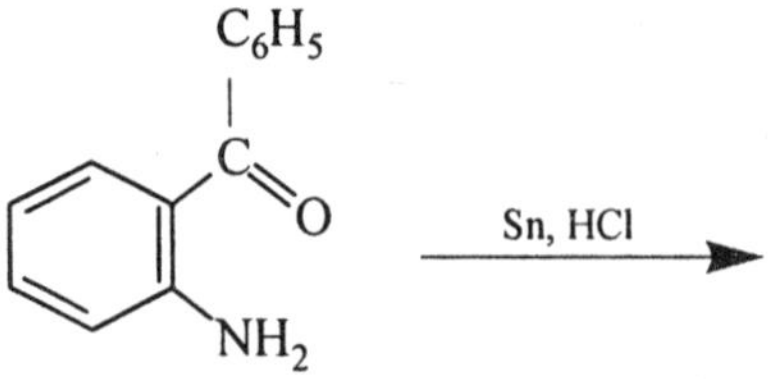

(*l*)   [structure: 2-methylchromone] $\xrightarrow{\text{H}_2\text{NOH}}$

(*m*)   [structure: 4,5-dimethyl-2-iPr-imidazole]   +   [N-phenylmaleimide] $\xrightarrow{\Delta}$

(*n*)   [structure: 2-methylimidazole] $\xrightarrow[\text{100°C}]{\text{H}_2\text{SO}_4,\ \text{SO}_3}$

(*o*)   [structure: thiazole-4-carboxamide] $\xrightarrow{\text{P}_2\text{O}_5}$

(*p*)   [structure: C_6H_5-CHCl-CHO]   +   [structure: H_2N-C($=S$)N / thiourea-type with CN and S] $\longrightarrow$

(*q*)   [structure: 2-R-benzothiazole] $\xrightarrow{C_6H_5CO_2H}$

(*r*) $R-C\equiv N^{+}-O^{-}$ +   [structure: 3-chlorocyclohex-2-enone] $\xrightarrow{\Delta}$

(s) [structure] $\xrightarrow[\text{(ii) DMAD}]{\text{(i) } \Delta}$

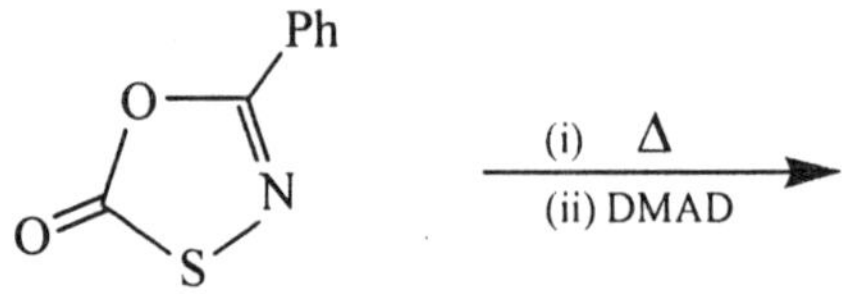

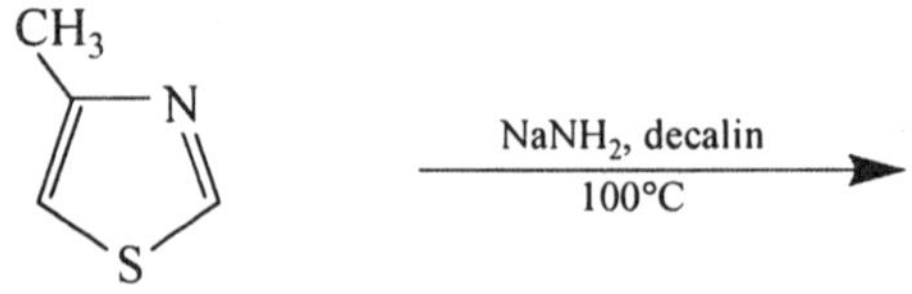

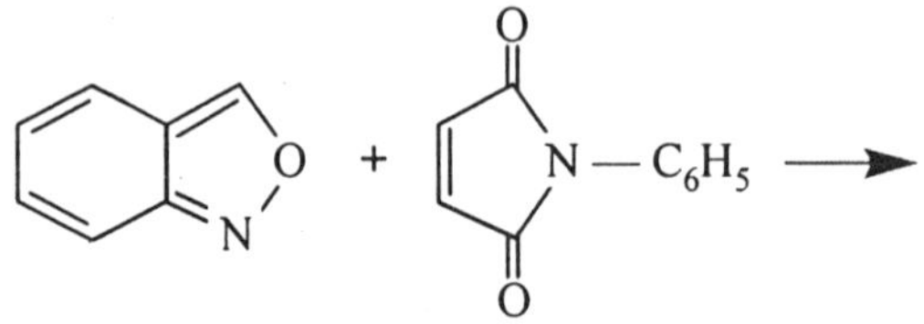

(t) [structure] $\xrightarrow[100°C]{\text{NaNH}_2\text{, decalin}}$

(u) [structure] + [structure] $\longrightarrow$

(v) [structure] + $C_6H_5C \equiv CH$ $\xrightarrow[\substack{\text{microwave}\\\text{irradiation}}]{\text{Catalyst}}$

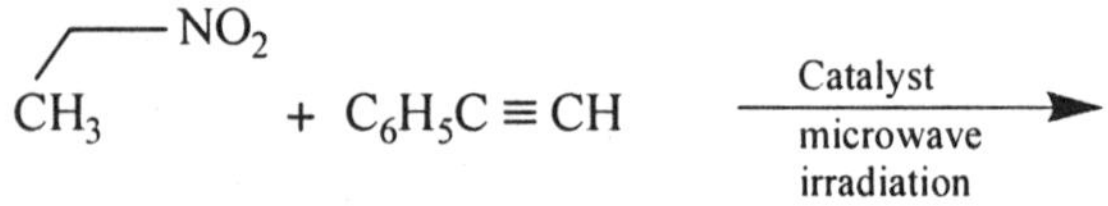

10. 2 Predict a mechanism for each of the following reactions:

(a) [structure] $\xrightarrow{\Delta}$ [structure]

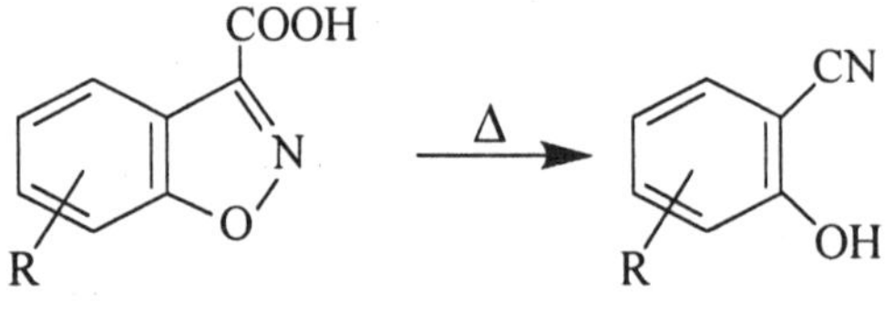

(b) [structure] $\xrightarrow{\text{H}_2\text{SO}_4,\ \Delta}$ [structure]

(c) [structure] $\xrightarrow[180°C]{\text{NH}_3}$ [structure]

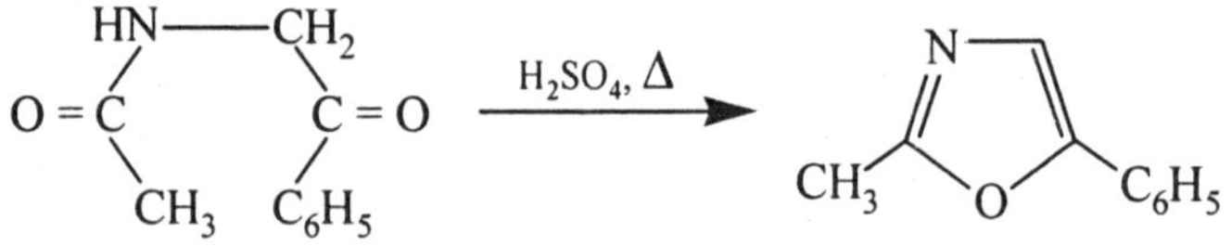

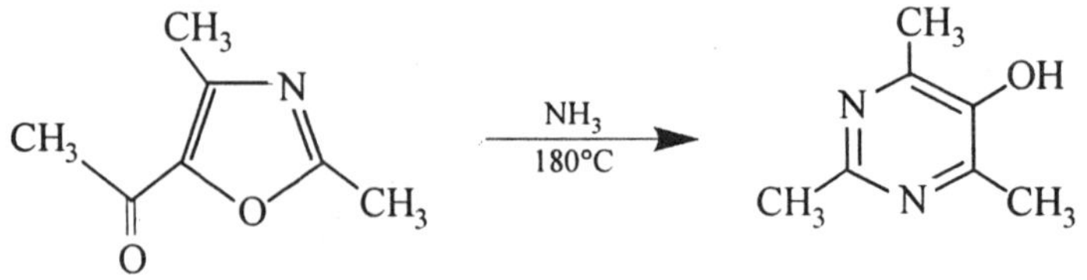

(d)

(e)

(f)

(g)

(h)   RCN + $CH_3COOC_2H_5$

(i)

10. 3 Offer an explanation for the following observations:

(a) Pyrazole has a higher boiling point than its N-alkyl derivatives.

(b) Isoxazole is less basic than oxazole.

(c) Imidazole undergoes electrophilic substitution only under vigorous conditions.

(d) Imidazole is more acidic than pyrrole.

(*e*) Electrophilic attack on pyrazole is hindered in acid medium.

(*f*) Pyrazole is resistant to the action of oxidizing agents.

10. 4 Discuss the ring opening reactions of isoxazole.

10. 5 Furan and oxazole behave as dienes in the Diels-Alder reaction. Elaborate your answer with appropriate examples.

10. 6 Give two methods for the synthesis of imidazole. Why is it more basic than pyrazole?

10. 7 Describe five synthesis of oxazoles.

10. 8 Compare the electrophilic aromatic substitution of imidazole, oxazole and thiazole.

10. 9 Discuss the preparation and properties of benzopyrazole.

Chapter 11

Six Membered Heterocyclic Compounds with Two Hetero Atoms

Two ring carbon atoms in benzene can be replaced by nitrogen, oxygen or sulfur to give six-membered heterocyclic compounds containing two hetero atoms. The diazines are group of compounds formaly derived from benzene by the replacement of two ring carbon atoms by nitrogen. Three isomeric diazines are possible with the nitrogen atoms in a 1, 2-, 1, 3-, or a 1, 4-relationship giving rise to *pyridazine* (1) *pyrimidine* (2) and *pyrazine* (3) respectively.

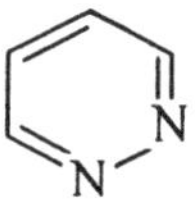

(1) 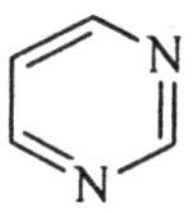 (2) 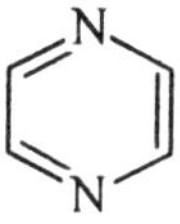 (3)

The corresponding benzo derivatives of these compounds are also known by the trivial names and are designated as cinnoline (4), phthalazine (5), quinazoline (6) and quinoxaline (7).

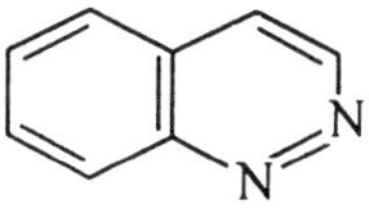

(4) 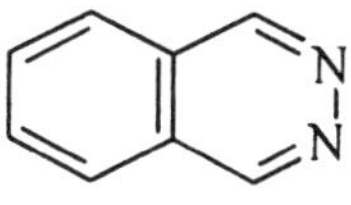 (5) 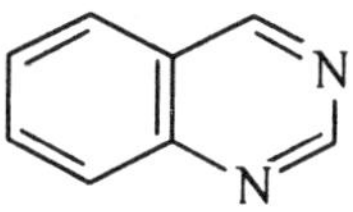 (6) 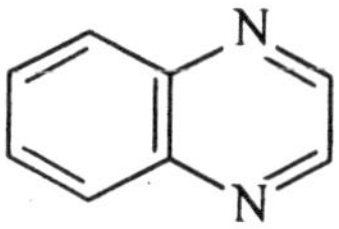 (7)

Six-membered heterocyclic rings containing one oxygen or sulfur atom cannot be aromatic unless they carry a positive charge. Diazines with two nitrogen atoms in the ring are aromatic but if nitrogen and oxygen atoms are present, the name of such a ring system given is oxazine and is apparently non-aromatic. There are three isomeric oxazines namely 1, 2-oxazine (8), 1, 3-oxazine (9) and 1, 4-oxazine (10).

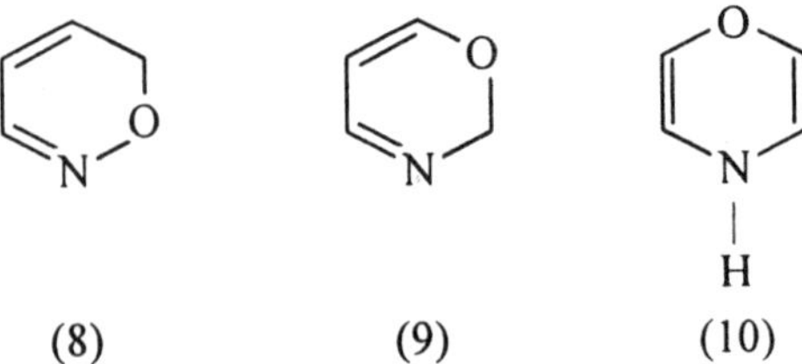

(8)   (9)   (10)

The *thiazines* are six-membered heterocyclic compounds containing nitrogen and sulfur atoms in the ring.

Pyrimidines, however, are by far the most widely studied compounds because this ring occurs in many compounds vital to living system. The diazines are stable, colorless compounds and soluble in water. They are essentially non-basic and are much weaker bases than pyridine as is evident from their pKa value of 2.3, 1.3 and 0.6 respectively. The drop in basicity is believed to be the consequence of destabilization of the N-1 protonated cations by inductive electron-withdrawal by the second nitrogen. They are all aromatic and possess high resonance energy. The unsubstituted diazines are much more resistant to electrophilic attack than is pyridine which undergoes this reaction only under drastic conditions. It is thus not surprising that induction of an additional azomethine nitrogen in pyridine will further deactivate the ring towards electrophilic attack.

The proton *n.m.r.* spectra of the azines in ppm is given below:

Position	Pyridazine	Pyrimidine	Pyrazine
1	N	N	N
2	N	9.26	8.6
3	9.17	N	8.6
4.	9.52	8.78	N
5.	7.52	7.36	8.6
6.	9.17	8.78	8.6

A large number of compounds containing more than two hetero atoms in one or more rings are known. Most of them, however contain nitrogen atoms and the important among these are the *triazines* i.e. 1, 3, 5-triazine (11), *pteridine* (12), *purine* (13) and *sydnone* (14).

(11)   (12)   (13)   (14)

11.1 PYRIDAZINES

Pyridazine is one of the three possible isomeric diazines. In pyridazine (1) the two nitrogen atoms are present adjacent to each other. This ring system does not form a part of any natural product and thus has been less extensively investigated[1-3] than other diazines. It was obtained as early as 1886 by Fischer and was first synthesized by Tauber in 1895. Because of the natural proximity of the nitrogen atoms pyridazine displays properties different from the other isomeric diazines.

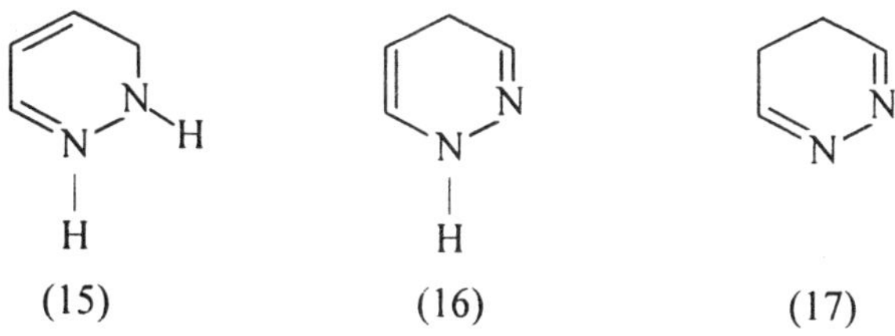

Pyridazine is a colorless liquid, b.p. 207°C and is miscible with water as well as with benzene. The lone electron pairs on nitrogen atoms are involved in H-bond formation with protic solvents and benzene. It is weakly basic (*p*Ka 2.3). Electron-donating groups augument basicity, thus 4-methylpyridazine has a *p*Ka of 2.92. It possesses a high dipole moment (4 D) value. This rather high value is attributed to the fact that the two nitrogen atoms in pyridazine are located on the same side of the ring. Therefore, there is a greater pull of electrons to that side resulting in high dipole moment. It is assumed to be planar and structures (1) and (1a) contribute to the resonance hybrid, the resonance energy is estimated to be 12.3 Kcal/mole.[4]

Of the six possible reduced pyridazines, only 1, 2-(15), 1, 4-(16) and 4, 5-dihydropyridazines (17) are known.

Appropriately substituted pyridazines exhibit tautomerism. Thus 3- and 4-hydroxypyridazines (18) and (19) exist predominantly in the oxo form. The conclusion is based on spectroscopic evidence. Though the 3- and 5- hydroxypyridazine-1-oxides exist in the hydroxy N − oxide forms (20) and (21), the 4- and 6- hydroxypyridazine 1-oxides, in

(18)   (19)

contrast, exist predominantly in the N-hydroxypyridazinone forms (22) and (23).

(20)   (21)   (22)   (23)

11.1.1 Synthetic Methods

Most syntheses of pyridazine are based on the addition of hydrazine or its derivatives to an appropriately 1, 4-disubstituted carbon chain.

1. *From Maleic Anhydride:* The condensation of hydrazine or its derivatives with maleic anhydride derivatives[5], 1, 4-dicarbonylethenes[6], butenolides[7] or dihydrofuran derivatives[8] result in the formation of pyridazine rings. The intermediate (24) is refluxed with phosphorus oxychloride to

(24)

give dichloropyridazine which on subsequent dehalogenation with hydrogen over Pd-C yields pyridazine. It may be noted that no dione intermediate formation takes place in the above reaction but a dione is initially formed when a substituted maleic anhydride is employed.[9]

4-Ethyl-5-methylpyridazine

γ-Ketoacids may instead be employed as the starting material to prepare pyridazine itself.

2. *From 1, 4-Dicarbonyl Compounds:* The most common method for the preparation of alkyl-or acyl-substituted pyridazines consist of the direct one step cyclization from an unsaturated diketone and hydrazine.[10]

(25)

3,6-Dimethylpridazine

The corresponding diketones, for this purpose, may be obtained from furans of type (25) by treating with acetic acid. A useful modification of this method is the use of hydroxy-1, 4-diketone such as 3-hexanol-2, 6-dione which is condensed directly with hydroxylamine to yield 3, 6-dimethylpyridazine, equation (11.1).

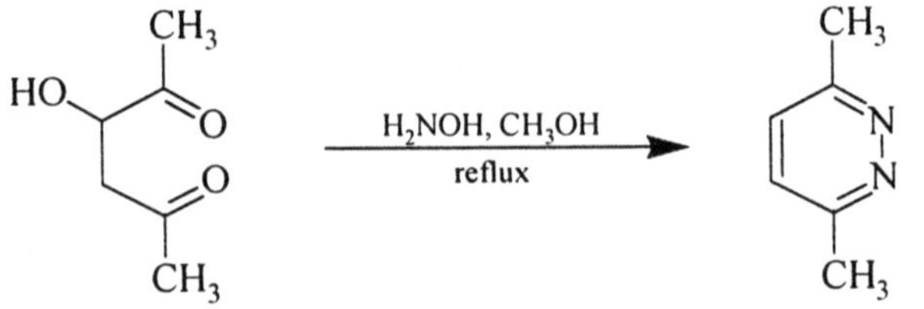

$$\text{(11.1)}$$

3.   From 1, 2, 4, 5-Tetrazines: The cycloaddition reaction of 1, 2, 4, 5-tetrazines with acetylenes or alkenes result in the formation of pyridazines. This method was first discovered by Carboni and Lindsey.[11]

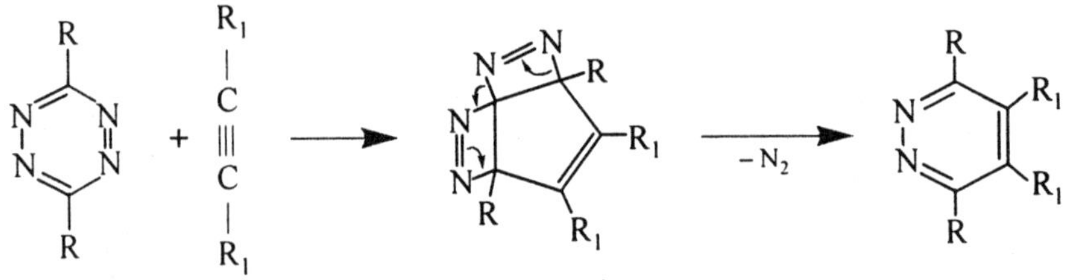

Analogous reactions of tetrazines with enol ethers, ketone acetals, enol esters and enamines was shown by Sauer and coworkers[12] to yield pyridazine derivatives. Recently it has been reported[13] that treatment of 3, 6-diphenyl-1, 2, 4, 5-tetrazine with aldehydes or ketone at room temperature gives the corresponding pyridazines.

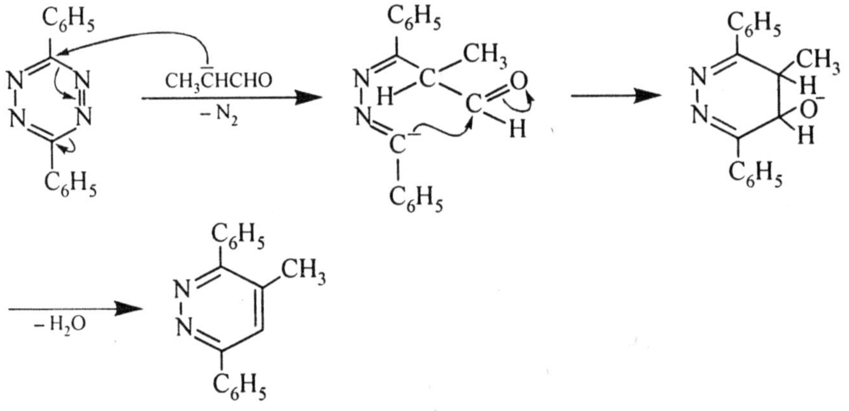

11.1.2 Chemical Reactions

1. *Reaction with Acids:* Pyridazine is a weak base and thus forms salts with mineral acids. The protonation of second nitrogen atom is difficult because of the high energy required to generate two positive charges on adjacent atoms.

2. *Quaternization:* The pyridazine ring reacts with alkyl halides or dialkyl sulfates in the presence of a base to furnish monoquaternary salts though less readily than pyridine. The position of monoalkylation is determined by the presence of alkyl groups on the ring. In the following reaction (equation 11.2) the methyl group directs the alkylation of 4-methylpyridazine to N-1 position.

$$\text{(11.2)}$$

3. *Electrophilic Substitution:* The 3-, 4-, 5- and 6- positions in pyridazine nucleus are electron deficient due to the inductive effect of nitrogen atoms. Pyridazine itself is very resistant to electrophilic substitution and can undergo reacion only under drastic conditions. No sulfonation or nitration of pyridazine has been reported. Direct halogenation of pyridazine is also not expected to be a method of wide application. Bromine in acetic acid is used for dehydrogenation of reduced pyridazines. The 3- and 4-chloropyridazines are obtained by the treatment of pyridazinones with excess phosphorus oxychloride. The 3-iodopyridazine could be obtained from the corresponding 3-chloropyridazine with sodium iodide and a small quantity of hydroiodic acid in refluxing acetone.[14] When suitable activating groups are present on the pyridazine ring, nitration becomes feasible.

4. *Reaction with Nucleophilic Reagents:* The diazine, in general, are very susceptible to the action of nucleophilic reagents. The effect of the presence of a second nitrogen atom is to make the carbon atoms of the ring even more electron-deficient than they are in pyridine. Aminopyridazines are prepared by the direct displacement of the halo group with conc. ammonia or amines.

Arylation of 3, 6-dichloropyridazine takes place though with powerful aromatic nucleophiles such as 3-aminophenol to yield 3-chloro-6-(3-hydroxyamino) pyridazine.[14a]

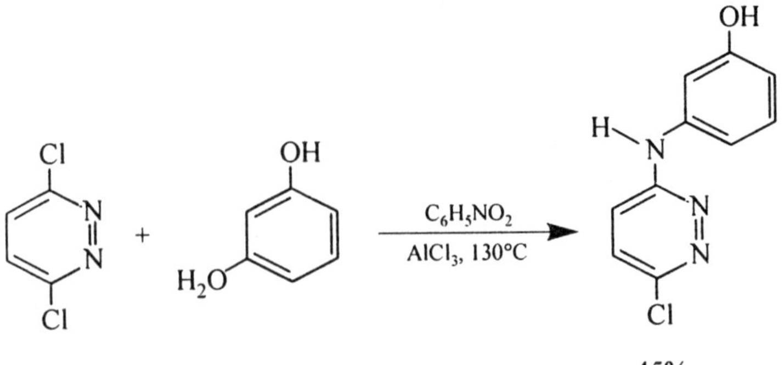

45%

Alkylpyridazines undergo condensation reaction aided by weak bases.

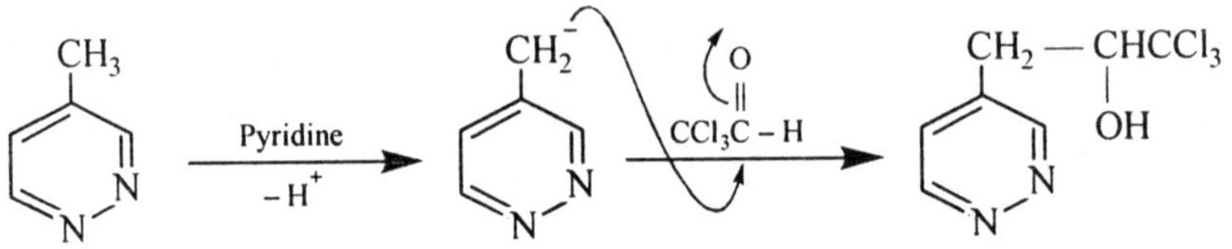

5.   *Reaction with Oxidizing and Reducing Agents:* Pyridazine is also resistant to the attack of oxidizing agents because of electron deficiency in the ring. With hydrogen peroxide, however, the N – oxide formation takes place but no di-N – oxide is obtained, equation (11.3). The pyridazine -1- oxide on treatment

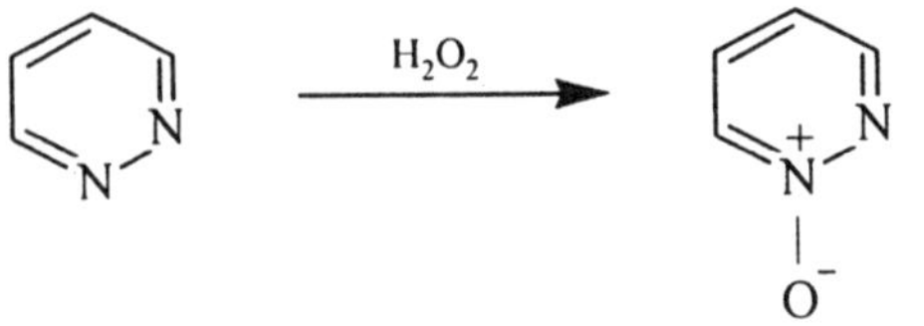

$$\qquad\qquad (11.3)$$

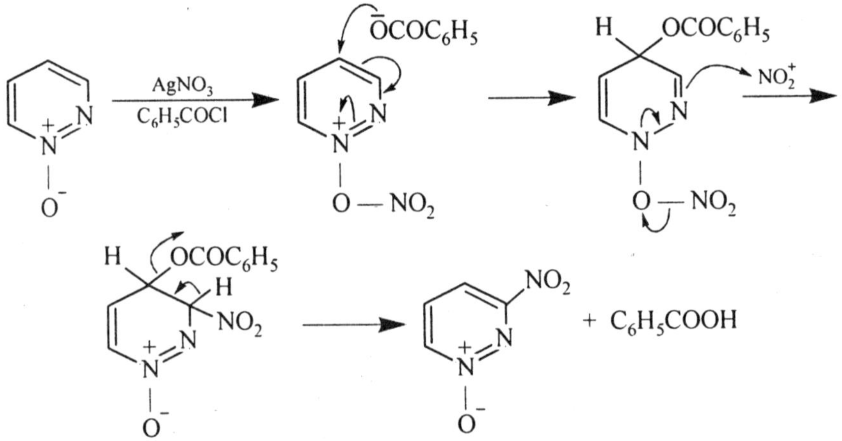

with silver nitrate and an acid chloride results in 3-nitropyridazine-1-oxide. The proposed mechanism involves the formation of acyl nitrate which decomposes to give NO_2^+ ion. Diazines are readily reduced. The ring *in situ* can be easily cleaved when the two nitrogens are adjacent. The ring can be completely reduced with Na/CH_3OH, equation (11.4).

$$(11.4)$$

6. *Photochemical Reactions:* Chambers and coworkers[15] have reported that pyridazine is rearranged to pyrazine on irradiation according to the mechanism given below:

7. *Cycloaddition Reaction:* Pyridazine with maleic anhydride forms a 1:2 adduct (27) at room temperature.

$$(27)$$

11.2 CINNOLINES AND PHTHALAZINES

The pyridazine ring can be fused into a benzene ring in two ways. Fusion at the 3, 4-position gives cinnoline (4) while fusion at the 4, 5-position gives phthalazine (5).

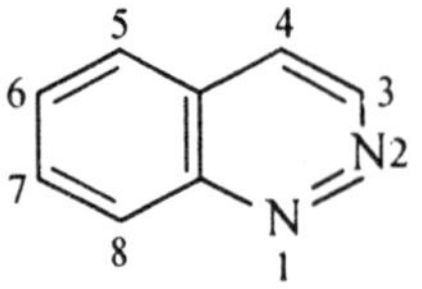

Cinnoline[16,17] is a pale yellow solid, m.p. 24-25°C and was first discovered by von Richter in 1883. He also prepared a cinnoline derivative from 2-aminophenylpropiolic acid *via* intramolecular cyclization of the following diazonium salt.

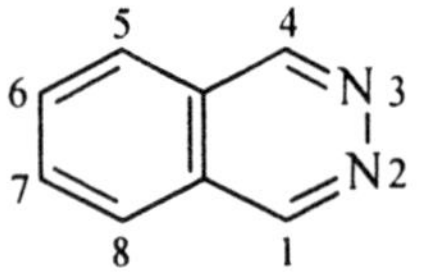

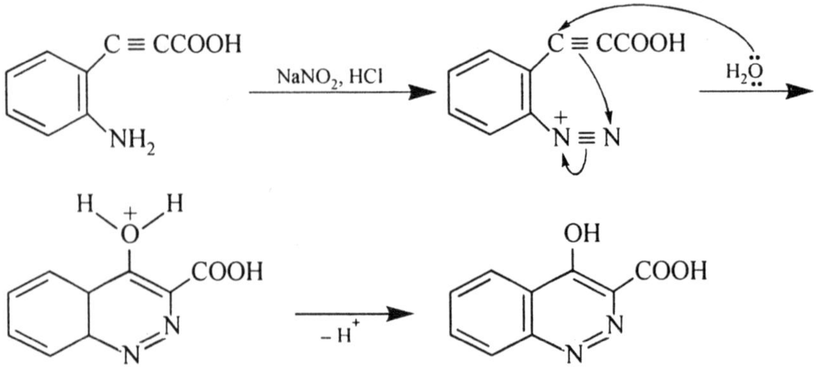

A related reaction is the Borsche[18] synthesis which was thoroughly strudied by Schofield.[19] It involves the cyclization of diazotized 2-aminoacetophenone according to the following steps:

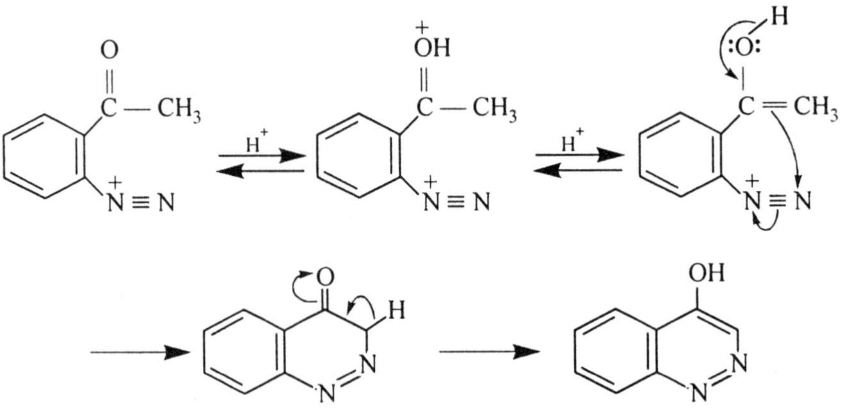

Another method which is conceptually different from the diazonium salt cyclization involves the treatment of a performed benzaldehyde phenylhydrazone with oxalyl chloride followed by the Friedel-Crafts cyclization of the N-benzylideneamino N-phenyloxamoyl chloride to N-benzylideneaminoisatin. This on reaction with hot potassium hydroxide solution yields 4-hydroxy-3-phenylcinnoline.

Cinnoline is a weak base (*p*Ka 2.51). It forms salts with acids like hydrochloric acid and picric acid. It is also quaternized with alkyl halides. Cinnoline is toxic and shows some antibacterial action against *Escherichia coli*.

In general, the electrophilic substitution in the benzodiazines series has not been extensively examined. Certain trends, however, may be predicted. The diazines do not undergo electrophilic attack unless electron-donating groups are present. The benzodiazines, on the other hand, are reactive by virtue of the presence of a benzene ring and the electrophilic attack takes place in this ring. Nitration[20] in sulfuric acid takes place to give a mixture of 5- and 8-nitrocinnoline, equation (11.5). Halide, nitrile and mercapto groups present at the 4-position in cinnoline are labile and can be replaced by a variety of nucleophiles such as alkoxide, phenoxide, ammonia, amines. With H_2O_2 in acetic acid, cinnoline gives a mixture of mono N – oxides which are formulated as cinnoline 1-oxide[21] and cinnoline 2-oxide, equation (11.6). If a bulky

$$(11.6)$$

group is present at the 3-position then formation of 1-oxide is favored, when no such substitution is present than 2-oxide formation is favored.

Cinnolines form either a dihydro derivatives or indole by ring opening and reclosure, depending on the reacion conditions. 3, 4-Benzocinnoline is also called phenazone (28).

(28)

Pathalazine (5)[22,23] is a pale yellow solid, m.p. 90-91°C. It has a characteristic odor and is readily miscible with water. It was first prepared by Gabriel and Pinkus in 1893 by heating tetrachloro o-xylene with hydrazine, equation (11.7).

$$(11.7)$$

A second method consists of condensing phthaldehyde or its derivative such as *o*-dibenzoyl ketone with hydrazine[24], equation (11.8).

$$(11.8)$$

1,4-Diphenylphthalazine

Another approach[25] to making 1, 4-disubstituted phthalazines involves the reaction of a benzyne with 1, 4-disubstituted tetrazines as shown in the following sequence of reactions:

Phthalazine is weakly basic and it forms salts with acids. It is somewhat more reactive than pyridazine and has been found to undergo direct electrophilic substitution. Since the 5- and 8-positions of phthalazine are equivalent and are expected to be the most reactive towards substitution.

Nucleophilic substitution in the phthalazine ring centers around the displacement of halogen atoms from halophthalazines. 1-Halophthalazine readily reacts with refluxing solution of alkoxides in appropriate alcohol to yield 1-alkoxyphthalazines.

Phthalazine at 0°C yields phthalazine 2-oxide on oxidation with monoperxyphthalic acid, while vigorous oxidation ($KMnO_4$) gives phthalazine-4, 5-dicarboxylic acid. Phthalazines on reduction result in 1, 2, 3, 4-tetrahydrophthalazines (Na/Hg) or the ring opened product (Zn/HCl).

11.3 PYRIMIDINES

Pyrimidine (2) is the most important member of all the diazines as this ring

(2)

system occurs widely in living organisms. Purines, uric acid, alkoxan, barbituric acid and a mixture of anti-malarial and anti-bacterials also contain the pyrimidine ring. The chemistry[26,27] of pyrimidine has been widely studied. Pyrimidine was first isolated by Gabriel and Colman in 1899. Since pyrimidine is symmetrical about the line passing C-2 and C-5, the positions C-4 and C-6 are equivalent and so are N-1 and N-3. When a hydroxy or amino group is present at the 2-, 4-, or 6-, position than they are tautomeric with oxo and imino respectively as illustrated below:

Lactin form   **Lactam form**

There are eight possible partially hydrogenated pyrimidines alhough it is not certain that all of them are stable to be isolated.

11.3.1 Physical and Spectroscopic properties

Pyrimidine is a colorless compounds, m.p. 225°C, b.p. 124°C. It is weakly basic (*p*Ka 1.3) as compared to pyridine (*p*Ka 5.2) or imidazole (*p*Ka 7.2). The decrease in its basicity is due to the electron-withdrawing effect of the second nitrogen atom present in the ring. Moreover, the addition of the proton does not increase the probability for mesomerism and hence the resonance energy. Presence of alkyl groups, however, enhances the basicity, thus 4-methylpyrimidine has *p*Ka 2.0 while 4, 6-dimethylpyrimidine has a value of 2.8 The 2-and 4-aminopyrimidines are more basic with *p*Ka 3.54 and 5.71

respectively. In these two compounds more resonance stuctures are possible in the cation than in the neutral molecule. The close relationship of pyrimidine

with benzene suggests the former is also highly aromatic and the ring is virtually plannar. The following cannonical structures contribute to the resonance hybrid. But pyrimidine ring is less aromatic compared to pyridine and benzene. This view is corroborated by the resonance energies which are, benzene

(36 Kcal/mole), pyridine (31 Kcal/mole) and pyrimidine (26 Kcal/mole). Pyrimidine has a dipole moment of 2.40 D. The molecular dimensions of pyrimidine have not been determined.

11.3.2 Synthetic Methods

Pyrimidines have been prepared by a number of methods but the most important are those in which the ring is formed from two fragments which contribute the C—C—C and N—C—N atoms respectively.

1.   *From Malonic Estes:* A simple synthesis[28,29] of pyrimidine ring involves a condensation between a malonic ester and urea in the presence of a base to yield barbituric acid. A modification of this method consists of using substituted malonic esters. Besides malonic esters, a series of other compounds such as β-keto acids or ester may be employed. Uracil, for instance, is obtained from α-formylacetic acid (produced *in situ* by decarboxylation of malic acid with

conc. sulfuric acid and reaction of the β-keto acid with urea. Uracil can be converted to pyrimidine in the following steps:

Urea may be replaced by amidine, guanidine or thiourea and condensed with a 1, 4-diketone. Thus acetylacetone on reaction with benzamidine give 4, 6-dimethyl-2-phenylpyrimidine.

A 1, 4-diketone may be condensed with an aldehyde and ammonia to furnish pyrimidine derivatives.[30]

2.   *From Ethyl Cortonate:* Another useful method involves the condensation of amidines or urea with unsaturated compounds such as ethyl crotonate in the presence of a base. A dihydropyrimidine is the initial product which is readily oxidized to the corresponding pyrimidine.

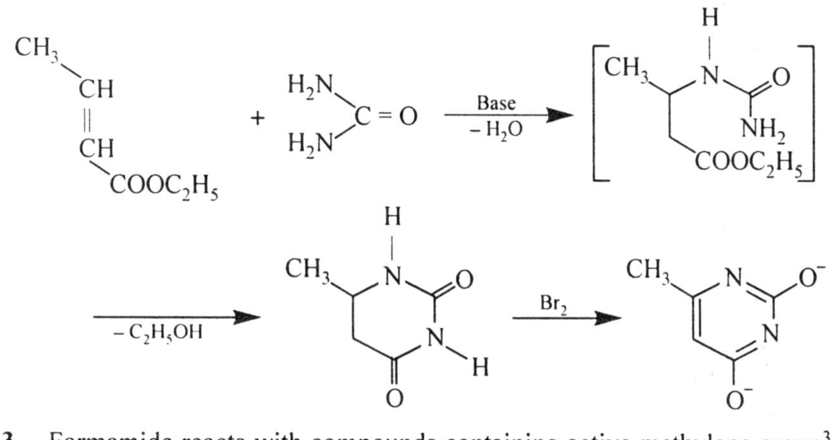

3. Formamide reacts with compounds containing active methylene group[31] in a way to form β-enaminoketones. With excess formamide, the β-enamino-ketone cyclizes to form pyrimidines. Acetophenone, for instance, reacts with formamide to yield 4-phenylpyrimidine. A closely related procedure involves the reaction of β-diketones with an excess of formamide.[32]

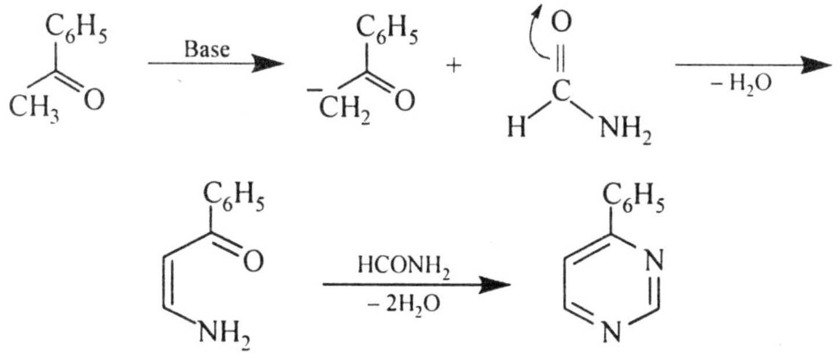

4. Pyrimidine itself can be obtained by the decarboxylation[33] of pyrimidine-4, 6-dicarboxylic acid or by the dechlorination of 2, 4-dichloropyrimidine, equation (11.9).

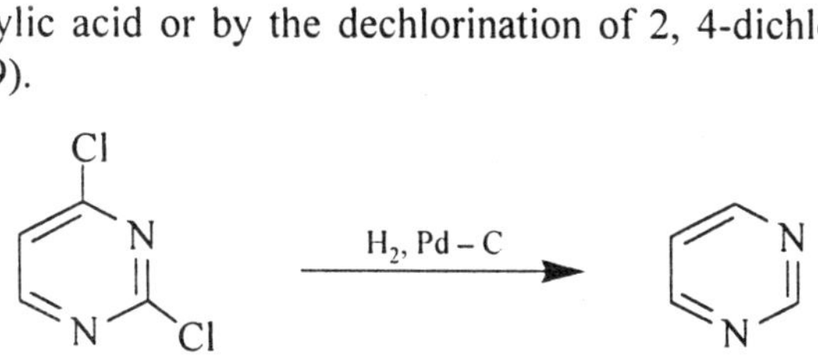

$$\text{(11.9)}$$

2. *From α, β-Unsaturated Ketones:* An interesting reaction of simple α, β–unsaturated ketone with amidines to give pyrimidines has been reported.[34] The initial product of this reaction is probably a dihydropyrimidine which is readily oxidized by a stream of air to the corresponding pyrimidine. Benzamidine and β-benzoylstyrene furnish 2, 4, 6-triphenylpyrimidine.

11.3.3 Chemical Reactions

1. *Reaction with Acids:* Pyrimidine though a weak base can be protonated in the presence of acids. Diprotonation unlike pyridine takes place in strong acids. Diprotonation is possible because the nitrogen atoms are not present in adjacent positions as in pyridazine.

2. *Electrophilic Substitution:* Pyrimidine ia also resistant to electrophilic substitution. The attack at positions 2-, 4- and 6- is particularly retarded because of the electron-deficiency at these positions. The 5-position is also difficult to attack as it is influenced by the inductive effect of the two nitrogen atoms and this resembles position-3 in pyridine. Electrophilic subsitution at position-5 is easy if one or more electron-releasing groups are present on the ring. The iodination[36] of aminopyrimidines has been investigated.[36] Bromination of pyrimidine yields 5-bromopyrimidine,[37] equation (11.10). The reaction is not simple electrophilic substitution rather proceeds *via* a perbromide intermediate.

$$\text{(11.10)}$$

If at least two electron-donating groups are present on the ring electrophilic substitution becomes facile. Thus, 2- or 4-hydroxypyrimidine does not nitrate but 2, 4-dihydroxypyrimidine (uracil) does so in boiling fuming nitric acid to give 2, 4-dihydroxy-5-nitropyrimidine.

The direct 5-sulfonation of pyrimidines which bear at least one electron-releasing group is possible but has been used sparingly.
Pyrimidines with two or more electron-donating groups undergo diazocoupling at the 5-position.

3. *Reaction with Nucleophilic Reagents:* The attack of a nucleophile takes place easily on the pyrimidine ring similar to pyridine, quinoline and isoquinolne. The positions suceptible to attack are 2, 4, and 6. Pyrimidine is stable in cold alkali but in boiling hydrazine it rearranges to pyrazole,[38] *via* a ring opened intermediate (29).

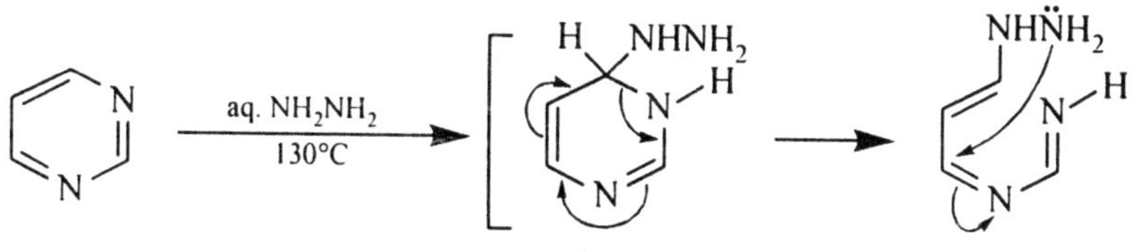

(29)

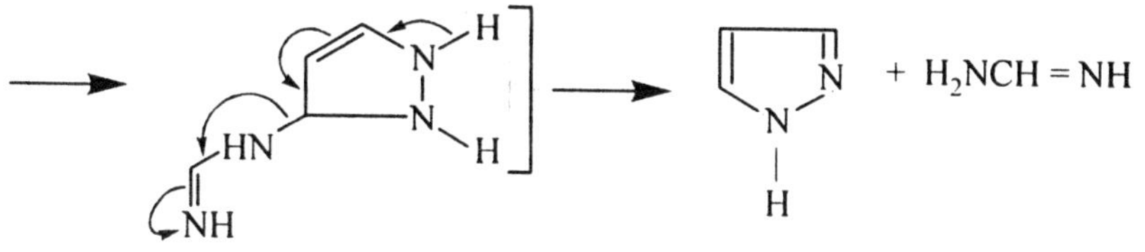

75%

Grignard and organolithium reagents readily add to the 3, 4-bond of pyrimidines at room temperatures as illustrated by the reaction of phenylmagnesium bromide on pyrimidine. The intermediate on hydrolysis and subsequent oxidation of the dihydropyrimidine with $KMnO_4$ in acetone yields 4-phenylpyrimidine.[39]

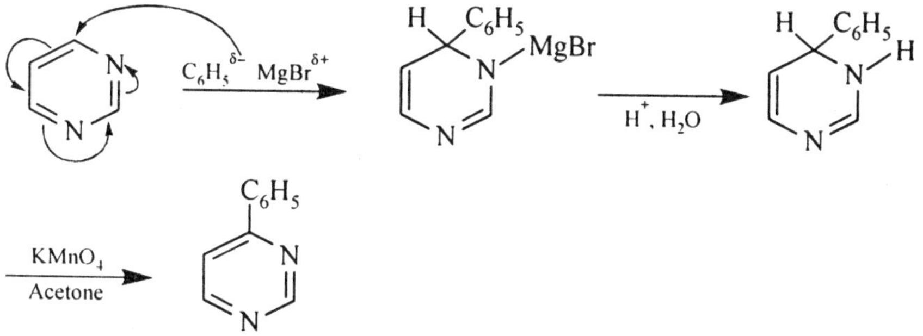

Methyl groups in the 2-, 4-, and 6-positions of pyrimidine behave as active methylene groups and can be converted to styryl derivatives by condensation with benzaldeehyde or *p*-dimethylaminobenzaldehyde. There is, however, an increased acidity of the methyl group at the 2-position because of electron-withdrawing effect of the nitrogen atom, equation (11.11).

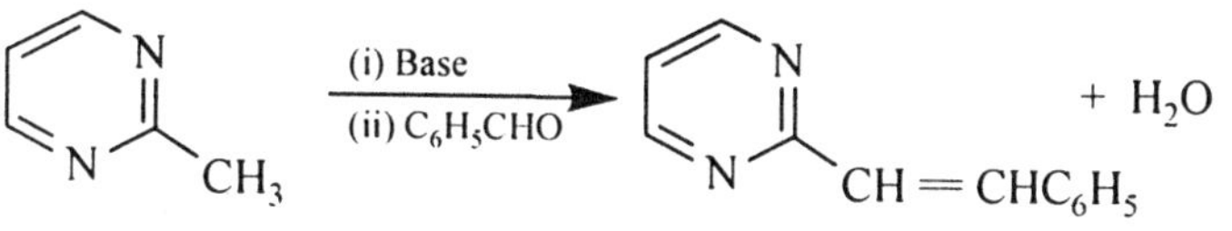

(11.11)

4.   *Reaction with Oxidizing and Reducing Agents:* Pyrimidine gives a low yield of N – oxide on oxidation with a peracid. The ring is largely destroyed during N – oxide formation. But the alkylpyrimidines give a satisfactory yield of N – oxides. 4-Methylyrimidine with H_2O_2 yields 4-methylpyrimidine N – oxide, equation (11.12). This can be converted back to pyrimidine by refluxing

$$\text{(11.12)}$$

with phosphorus oxychloride. Reissert addition reaction appears to be fairly common in pyrimidine N – oxides. 4-Methoxypyrimidine N – oxide on reaction with sodium cyanide and benzoyl chloride under alkaline conditions furnishes 2-cyano-4-methoxypyrimidine.[39]

5.   *Thio-Claisen Rearrangement:* Certain suitably substituted thiopyrimidines undergo the familiar thio-Claisen rearrangement. Thus 3-methyl-4-allylthiopyrimidin-2-one on heating yields 5-allyl-3-methyl-4-thiouracil,[40,41] equation (11.13).

$$\text{(11.13)}$$

(30)

11.3.4 Naturally Occurring and Biologically Active Compounds[42,43]

Pyrimidine itself is not found in nature but substituted pyrimidines and compounds containing the pyrimidine ring are widely distributed in nature.

Derivatives of barbituric acid (31), i.e., oxygenated pyrimidines are perhaps the most widely used in medicines, for example, Veronal (32), *Luminal* (33) are used as hypnotics while *pentothal* (34) is used as an anaesthetic. Several

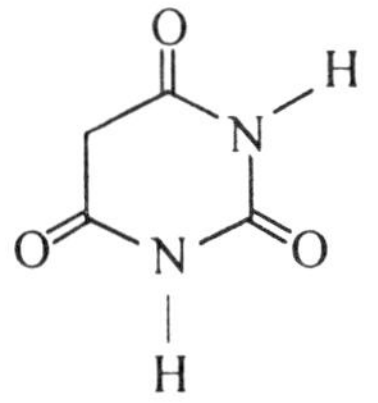

(31) 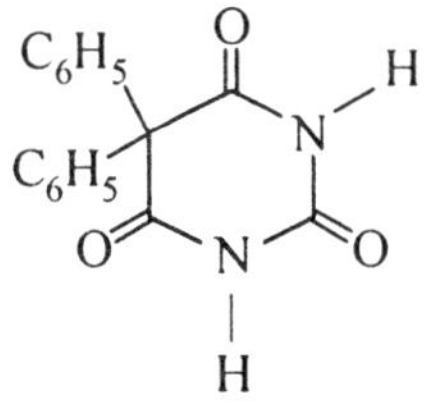 (32)

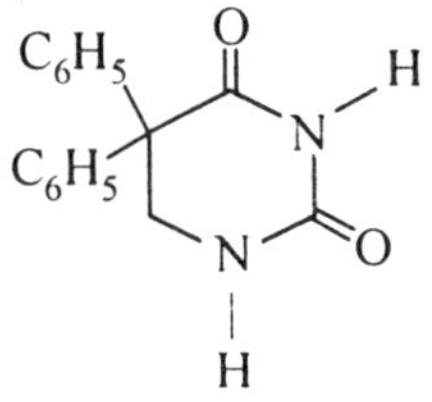

(33) 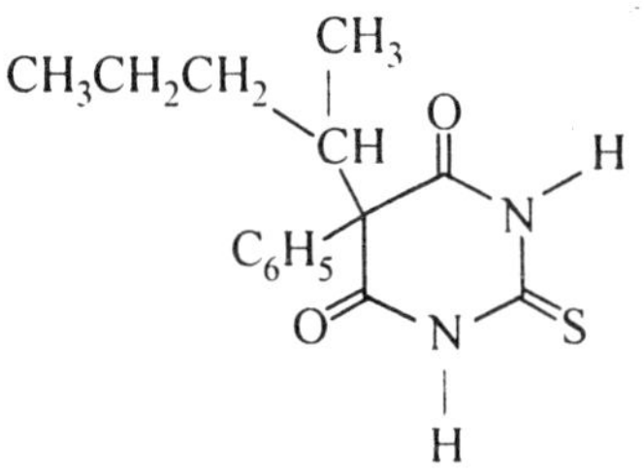 (34)

important sulfa drugs are pyrimidine derivatives namely sulfadiazine (35) sulfamerazine (36) and sulfadimidine (37).

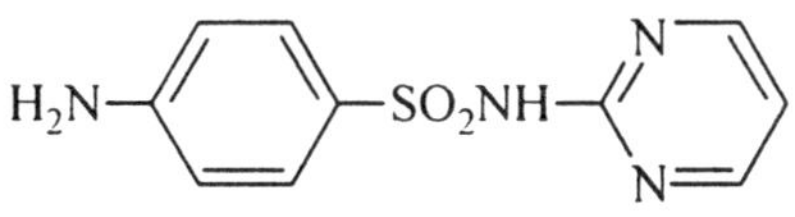

(35) 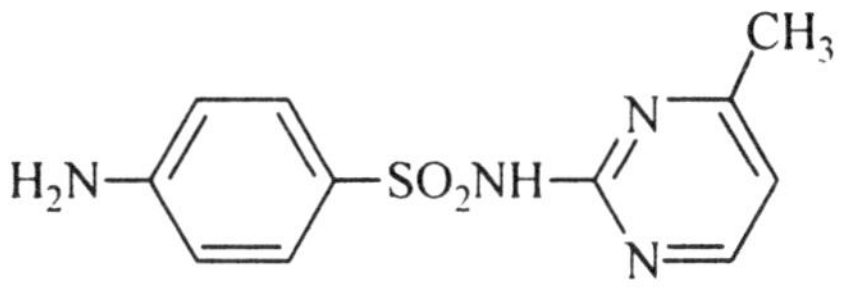 (36)

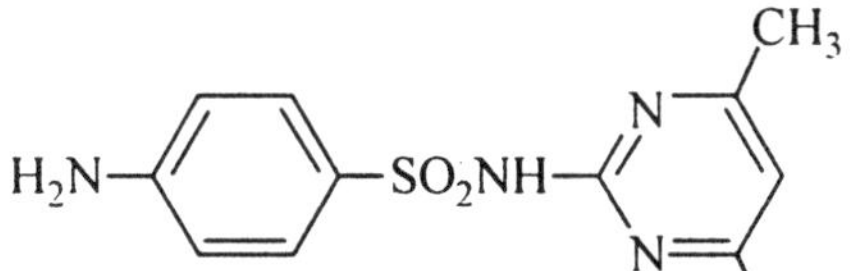

(37)

Sulfadiazine is still widely used but the last two are no longer used for chemotheraphy of infections.

Three pyrimidines are of considerable biological importance because of their relation to the nucleic acids, these are uracil (38), thymine (39) and cytosine (40). The purine ring system (41) obtained by the fusion of pyrimidine and

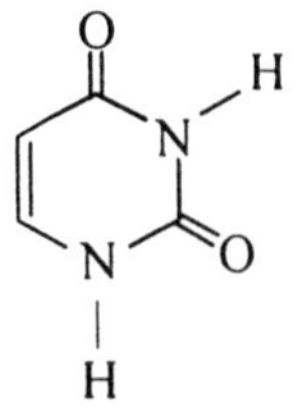

(38) 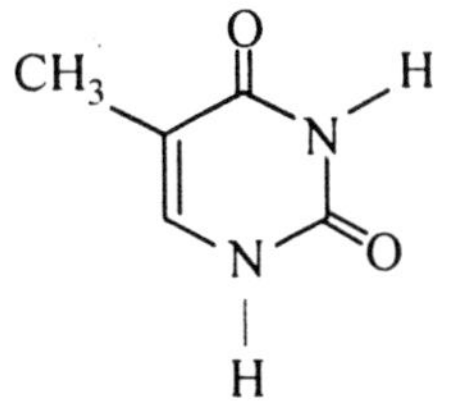   (39) 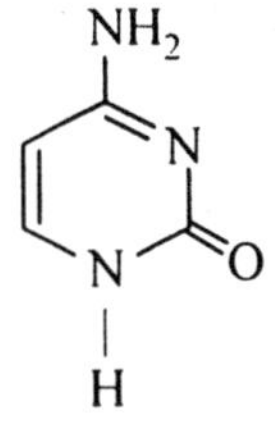   (40)

imidazole unclei also is important because certain of its derivatives, in particular adenine (42) and guanidine (43) are building blocks of RNA and DNA.

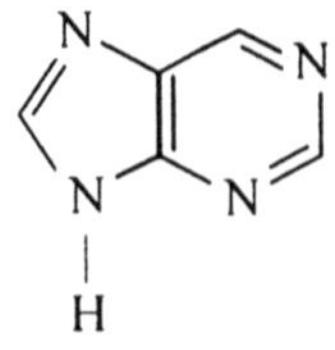

(41) 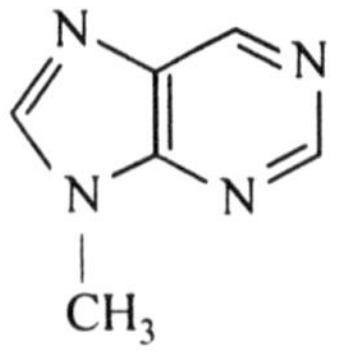   (42) 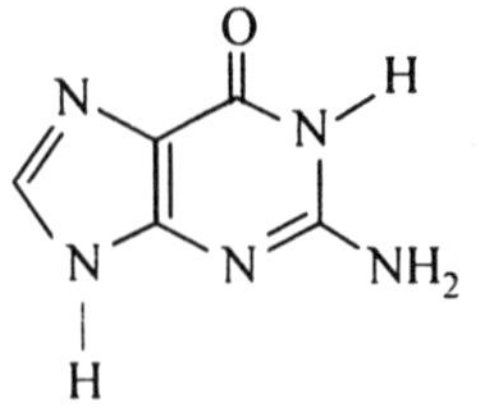   (43)

A variety of natural products such as alkaloids also contain the pyrimidine ring system, these include hypoxanthine (44), and xanthine (45) which occur in tea and caffeine (46) and theophylline (47) are the constituents of tea leaves. Theobromine (48) is found in cocoa beans.

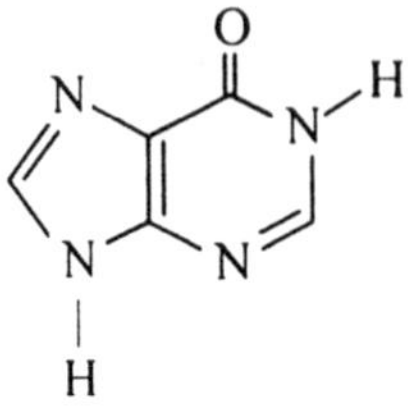

(44) 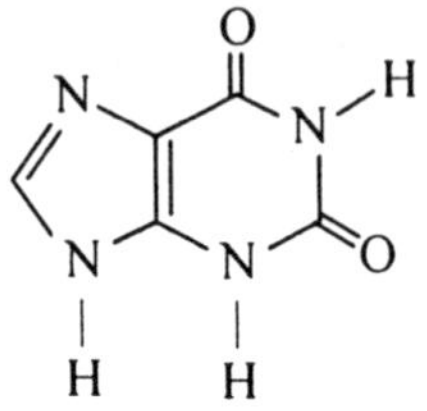   (45) 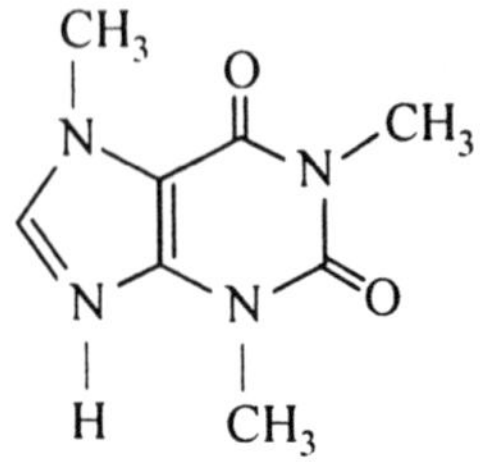   (46)

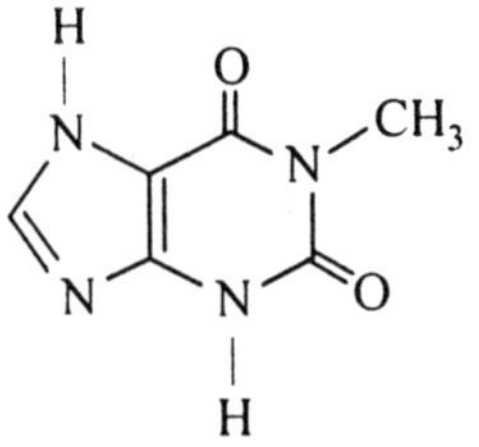

(47) 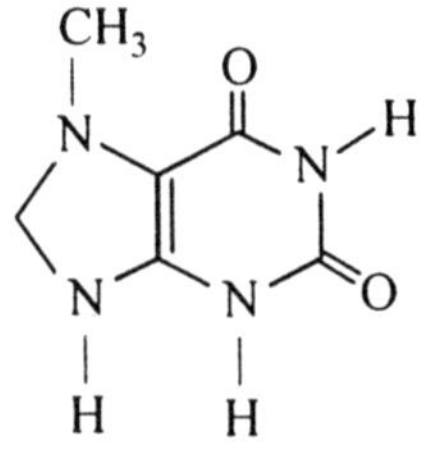   (48)

Folic Acid: The pteridine (49) ring system is also widely distributed in nature. The important growth factor folic acid, vitamin B_{10} (50) is constructed of a pteridine ring, *p*-aminobenzoic acid and glutamic acid, i.e. pteroylglutamic acid. It is widely distributed and has been isolated from liver and yeast.

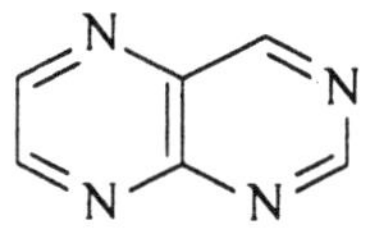

(49)

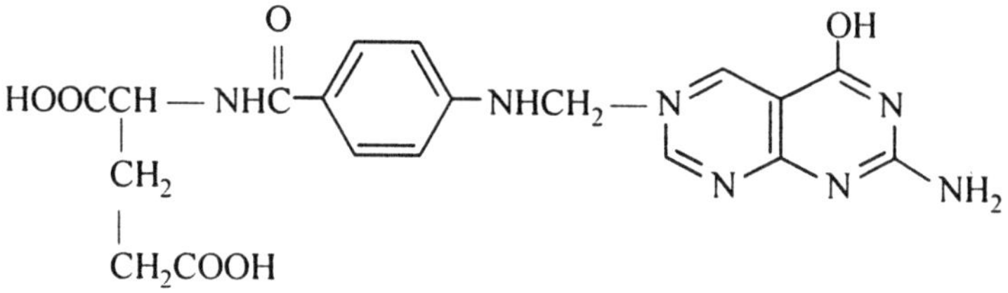

(50)

Nucleosides: The linking of sugar residues D- ribose or D-2-deoxyribose at N-3 of pyrimidine or at N-9 of purine gives rise to the formation of glycosides known as *nucleosides.* Although nucleosides occur as such in living cells they are also obtained by hydrolysis of nucleotides (the phosphoric acid

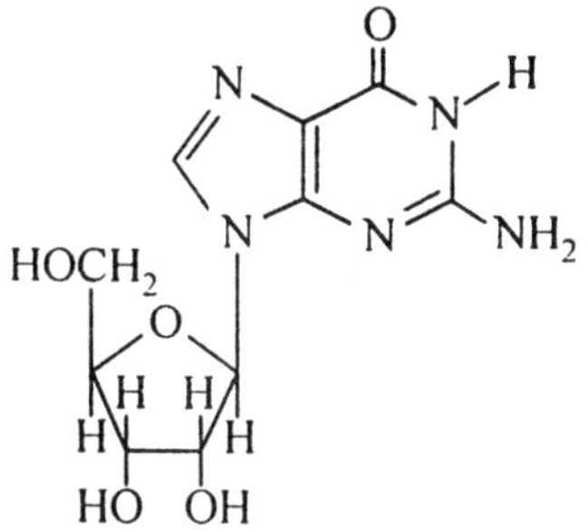

(51)

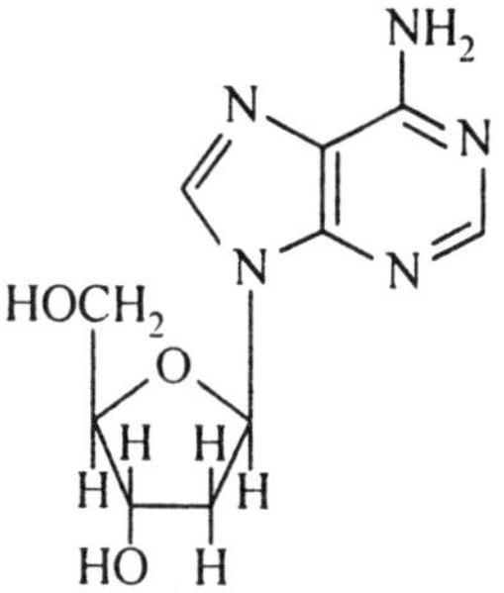

(52)

esters of nucleosides). The hydrolysis of ribonucleic acid (RNA) gives the nucleoside guanosine (51) together with pyrimidine nucleosides, adenosine (52), pteridine (53) and cytidine (54).

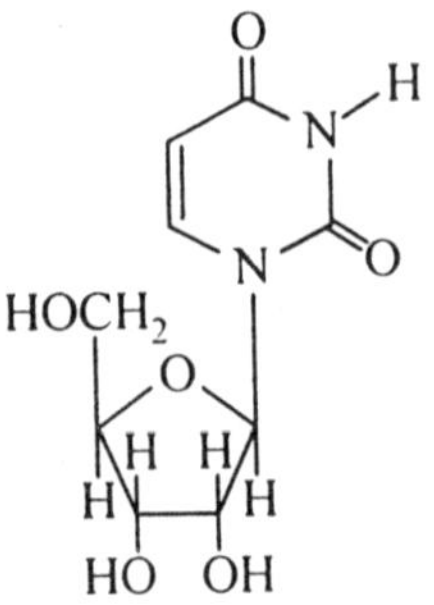

(53) 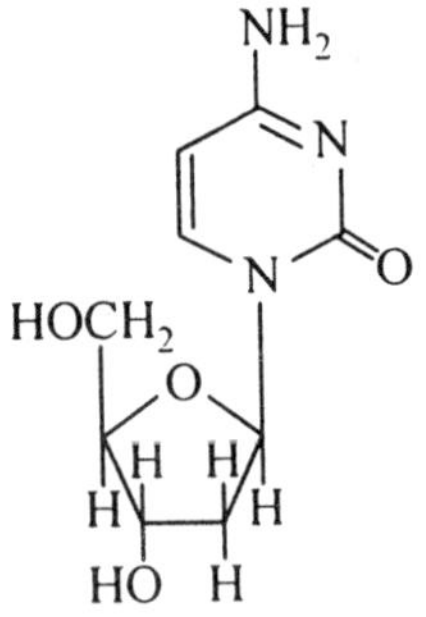   (54)

Nucleotides: The phosphoric acid ester of nucleosides are called *nucleotides.*
Nucleotides are also obtained by chemical or enzymatic hydrolysis of nucleic
acids. Ribonucleic acid, thus gives ribonucleotide and deoxyribonucleic acid
give deoxyribonucleotide. Many coenzymes are nucleotides and as such play
a vital role in the living cells. Adenosine triphosphate (ATP) (55) functions as
the key agent linking phosphate donors and phosphate acceptors.

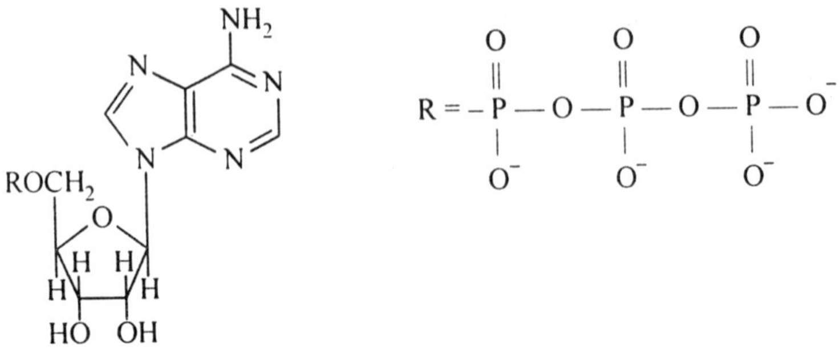

$$R = -\overset{\overset{\displaystyle O}{\|}}{\underset{\underset{\displaystyle O^-}{|}}{P}} - O - \overset{\overset{\displaystyle O}{\|}}{\underset{\underset{\displaystyle O^-}{|}}{P}} - O - \overset{\overset{\displaystyle O}{\|}}{\underset{\underset{\displaystyle O^-}{|}}{P}} - O^-$$

(55)

Nucleic Acid: Nucleic acids are molecules which are composed of chains of
mononucleotide units adapted for the storage and transcription of biological
information. Two main types of nucleic acids have been recognized, ribonucleic
acid (RNA) and deoxyribonucleic acid (DNA). Both types are biopolymers in
which the repeating monomer units are nucleotides. As discussed earlier a
nucleotide is made up of one unit each of phosphate, a pentose and a

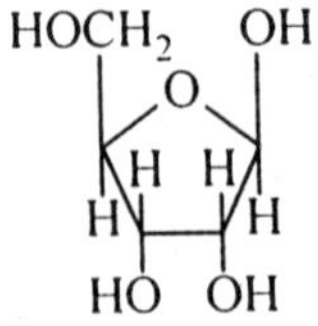

(56) 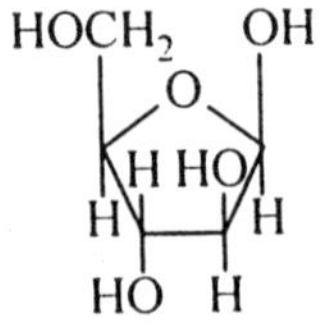   (57)

heterocyclic base. For each class of nucleic acid, there are four main nucleotidce monomers. For RNA, the sugar is ribose (56) and the heterocyclic bases are the pyrimidines, uracil or cytosine, or the purines, adenine or guanine. For DNA, the sugar is 2-deoxyribose (57) and the heterocyclic bases are the same, except that thiamine replace uracil.

A widely distributed nucleotide that plays a decisive role in many metabolic processes is called coenzyme A (58). It is known to take place in processes like transacetylation reactions, fatty acids oxidation, etc.

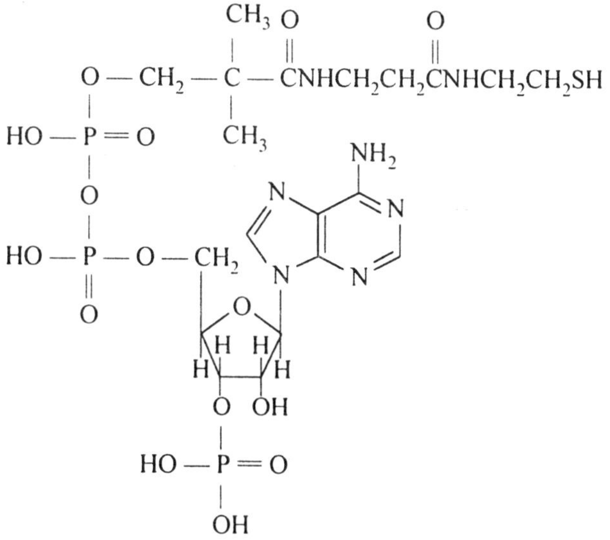

(58)

Riboflavin: This substance has been called vit. B_2, lactaflavin or riboflavin (59) and is a member of the water soluble vitamin B complex. It is widely distributed in nature both in animals and plants. It is unstable to light.

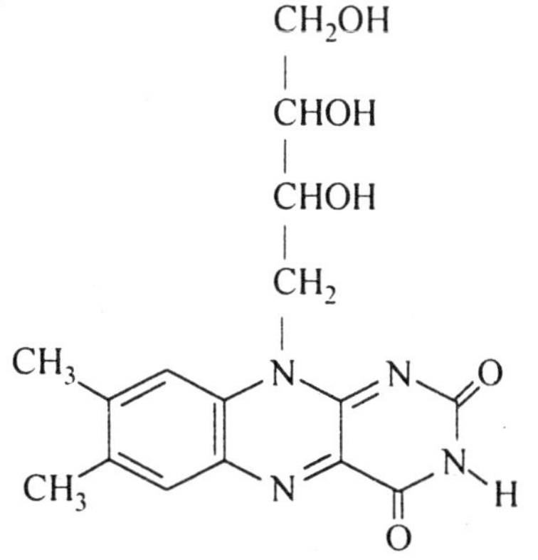

(59)

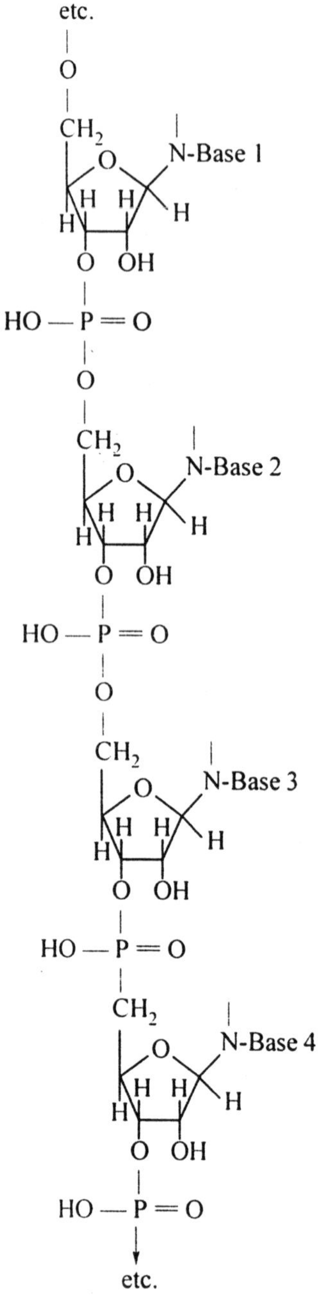

RNA backbone

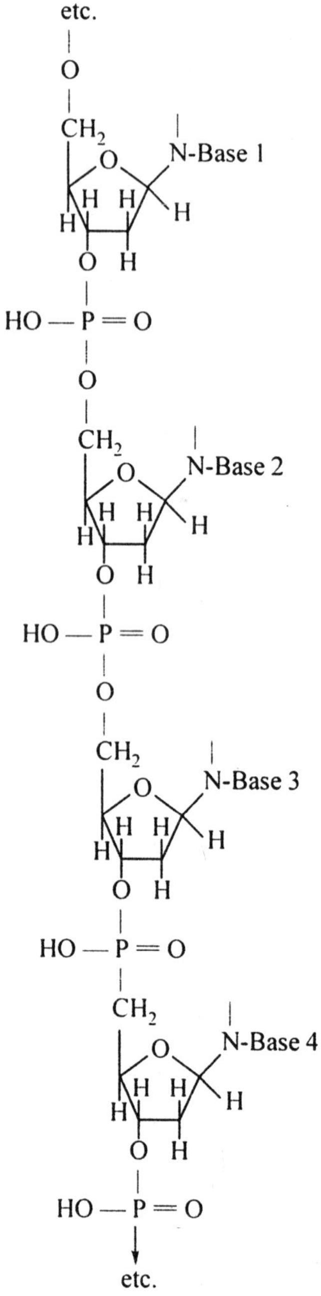

DNA backbone

11.4 QUINAZOLINE

Quinazoline[44,45] or 1, 3-diazonaphthalene represented by structure (6) has also been given names like *phenmiazine, benzo-1, 3-diazine* or 5, 6-*benzopyrimidine*. It was first prepared by Gabriel in 1903.

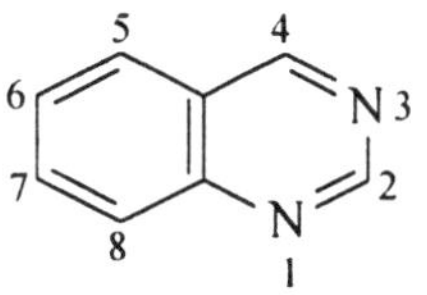

(6)

Quinazoline is a solid, m.p. 48°C and can be recrystallized from ether. It is soluble in water and most organic solvents. It is basic (*p*Ka 1.4) in nature.

Generally a suitably substituted anthranilic acid serves as precursor of quinazolines. Schofield *et al*[46] synthesized quinazoline derivatives by heating the acyl derivative of o-aminoacetophenone with ammonia. The reaction (equation 11.14) is of general applicability.[47] 2, 4-Dimethylquinazoline can be obtained staring from *o*-acetamidoacetophenone.

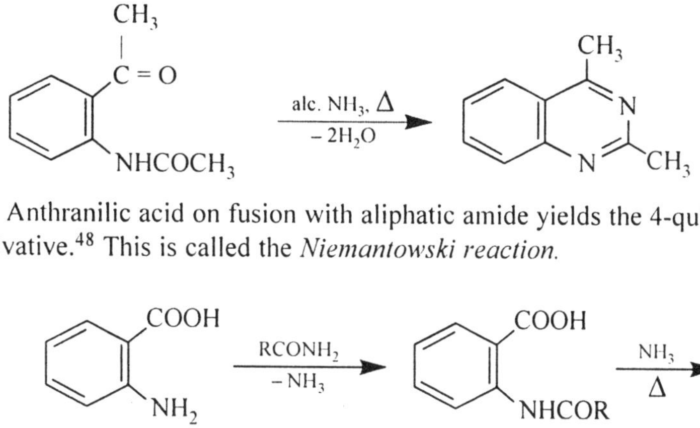

$$\text{(11.14)}$$

Anthranilic acid on fusion with aliphatic amide yields the 4-quinazolone derivative.[48] This is called the *Niemantowski reaction.*

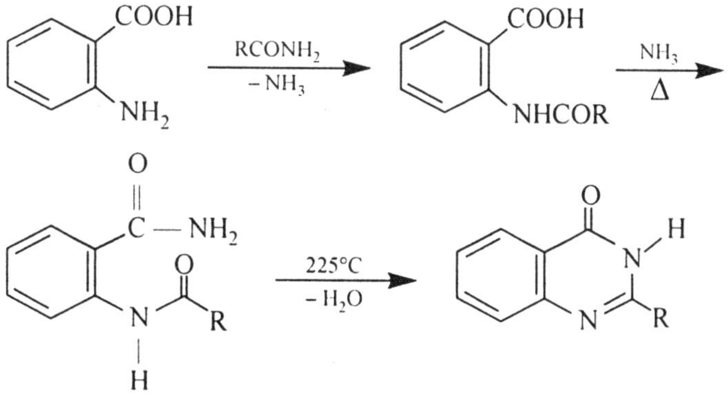

o-Aminobenzophenone condenses with urea on heating to afford 2-hydroxy-4-phenylquinazoline[49], equation (11.15).

$$(11.15)$$

Quinazoline forms stable monobasic salts. Nitration appears to be the only known electrophilic substitution reaction of quinazolines. Nitration (fuming nitric acid, conc. H_2SO_4) gives 6-nitroquinazoline though theoretically 8-position is expected to be more reactive.[50]

Quaternization of quinazoline takes place at 3- position. Grignard reagents and alkyllithium readily add to the 3, 4-bond of quinazoline analogous to pyrimidine[51], equation (11.16). The methyl derivatives at C-2 and C-4 positions of quinazoline behave in a manner similar to methyl groups that are placed at α-or γ-position to the nitrogen atom in pyridine. These groups thus react like *"active methylene"* groups. 4-Chloroquinazoline reacts readily with methanol to give 4-methoxyquinazoline.

$$(11.16)$$

Quinazoline is stable to oxidation but under drastic conditions the benzene ring is ruptured to yield 4, 5-pyrimidinedicarboxylic acid, (equation 11.17).

$$(11.17)$$

Quinazoline-1-oxide is unknown although 4-substituted derivatives have been prepared. Quinazoline 3-oxide, however, has been prepared. Quinazolines on reduction (Na/CH_3OH) yield the 1, 2, 3, 4-tetrahydro derivtive.

11.5 PYRAZINES

Pyrazine[52-54] or 1, 4-diazine (3) is a symmetrical molecule as the nitrogen atoms occupy the 1, 4-positions. Pyrazine occurs naturally though not in

(3)

appreciable amounts but several polycyclic derivatives of pyrazine ring system such as pteridine (49) and phenazine (60) occur in nature. The pyrazines have long been of interest to medicinal chemists. Their derivatives many of which are natural products have proven to be useful as antibiotics, diuretics and anti-tumor agents.

(60)

Various dihydro- and tetrahydro-pyrazines are known but the most important is hexahydropyrazine or piperazine (61).

(61)

Pyrazine is a colorless solid, m.p. 54°C. It is a weaker bse (*p*Ka 0.65) than either pyridazine (*p*Ka 2.3) or pyrimidne (*p*Ka 1.3). The reason for the decreased basicity in this series is not clear.

The pyrazine molecule may be represented as a resonance hybrid of a number of canonical structures. The resonance energy has been calculated to be 24.3 Kcal/more. It has a dipole moment of zero because of the symmetric nature of the molecule. The π-electron density data indicate and increase in π-electron density at the nitrogen atoms and a depletion at the carbon atoms.

11.5.1 Synthetic Methods

1. *From 1, 2-Diketones:* In this method 1, 2-diketones undergo condensation with 1, 2-diamines to yield an intermediate dihydropyrazine. Subsequent dehydrogenation to the corresponding pyrazine is carried out over copper chromite at 300°C.[55] This method appears to offer a general procedure for the synthesis of alkyl, aryl and aralkylpyrazines.

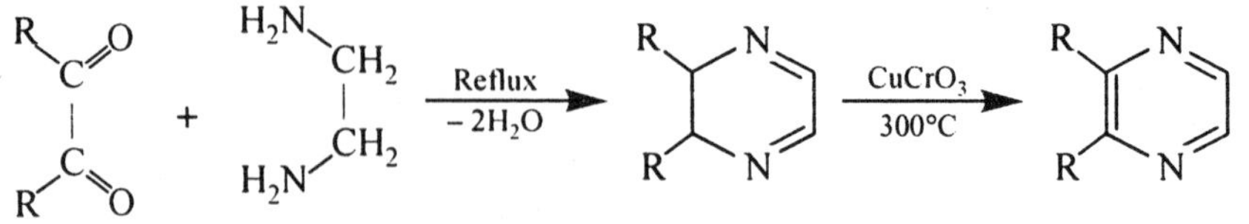

2. *From α-Aminocarbonyl Compounds:* This method involves a self condensation of two molecules of α-aminocarbonyl compounds to yield 2, 5-dihydropyrazine. Subsequent oxidation as in the earlier method affords a pyrazine derivative. The α-aminocarbonyl compound for this purpose can be obtained by nitrosation of a ketone followed by reduction of the aminoketone with stannous chloride, zinc/acetic acid or zinc/alkali.[57]

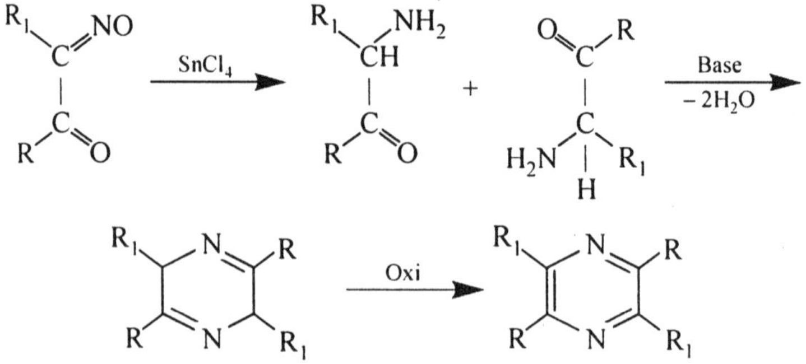

α-Amono acids and their esters normally undergo a facile dimerization to the corresponding 2, 5-dioxopiperzines which can be alkylated with triethyloxonium fluoroborate. The derivative diethoxydilyoropyrazine is oxidized by DDQ to piperazine.

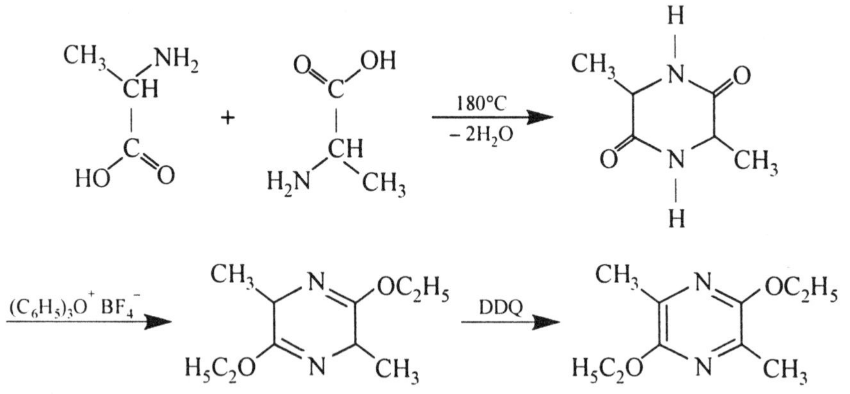

3. *From Azine Derivatives:* Padwa and coworkers[57] reported that irradiation of diphenylazirine on rearrangement yields 2, 3, 4, 5-tetraphenylpyrazine among other products.

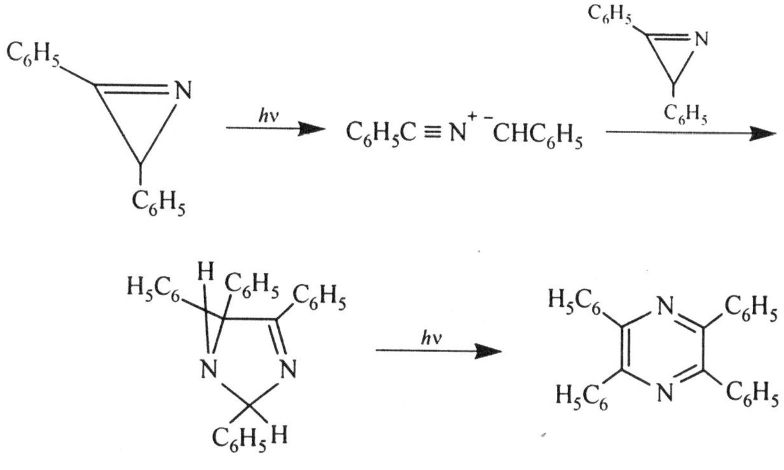

11.5.2 Chemical Reactions

Pyrazines show considerable aromatic character and the main features of their reactions may be interpreted by regarding them as pyridines into which a nitrogen atom has been introduced at the *para* position.

1. **Reaction with Acid:** Pyrazine is a monobasic compound and is protonated at N-1. Diprotonaion is possible only in the presence of strong acids.

2. **Electrophilic Substitution:** The presence of two nitrogen atoms deactivates the ring towards electrophilic attack. Presence of electron-donating groups, however, facilitates electrophilic attack. This is evident by the chlorination (Cl_2, CCl_4) of 2-methylpyrazine under milder conditions to furnish 3-chloro-2-methylpyrazine, equation (11.18). This reaction probably proceeds *via* and addition-elimination mechanism. No nitrations on any pyrazine derivatives seeem to have been reported.

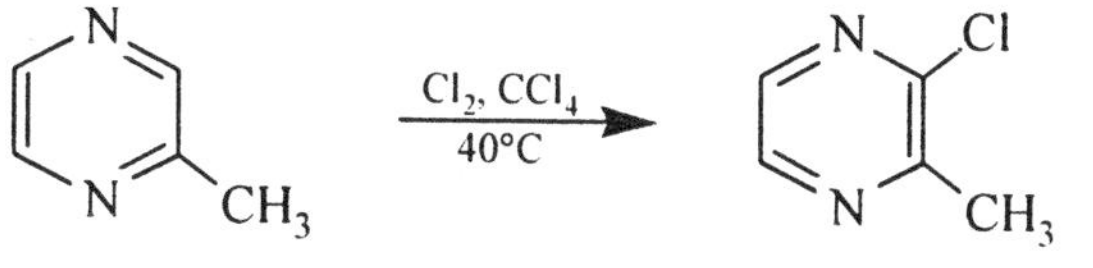

$$\text{(11.18)}$$

3-Chloro-2-methylpyrazine

3. **Reaction iwth Nucleophilic Reagents:** Pyrazine reacts in a manner similar to pyridine in the Chichibabin reaction (equation 11.19).

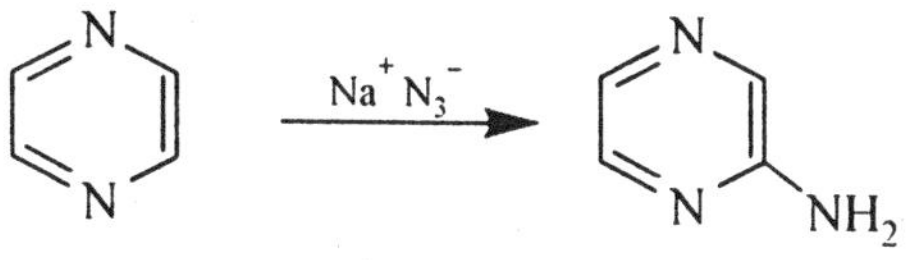

$$\text{(11.19)}$$

2-Fluoropyrazine undergoes a facile reaction with sodium azide[58] to give 2-azidopyrazine which exists in dynamic equilibrium with tetrazolo [1, 5-a] pyrazine.

Because of the inductive and resonance effects of the nitrogen atoms protons attached to the α- or β-methyl carbon atoms are acidic and are easily removed by strong bases. The resultant carbanions undergo typical condensation[59] and alkylation[60] reactions. This is illustrated by the following example, equation (11.20).

$$(11.20)$$

4.   *Reaction with Oxidizing Agents:* The pyrazine ring system is reasonably stable to the action of oxidizing agents. The alkyl group present on the ring can be oxidized to the carboxyl group. Pyrazine itself has been converted to *di*-N – oxide with CF_3CO_3H directly. N – oxide formation of substituted pyrazines with H_2O_2/CH_3COOH or peracids takes place on the most basic and least sterically hindered nitrogen.[61]

$$(11.21)$$

$$(11.22)$$

Thus 2-chloropyrazine yields exclusively the 4-oxide, equation (11.21) while a mixture of N – oxide is obtained from 2-methylpyrazine, equation (11.22). Oxidation of 2-chloropyrazine on the other hand, takes place at the nitrogen atom adjacent to the chloro group with Caro's acid (peroxysulfuric acid) in conc. sulfuric acid.[62]

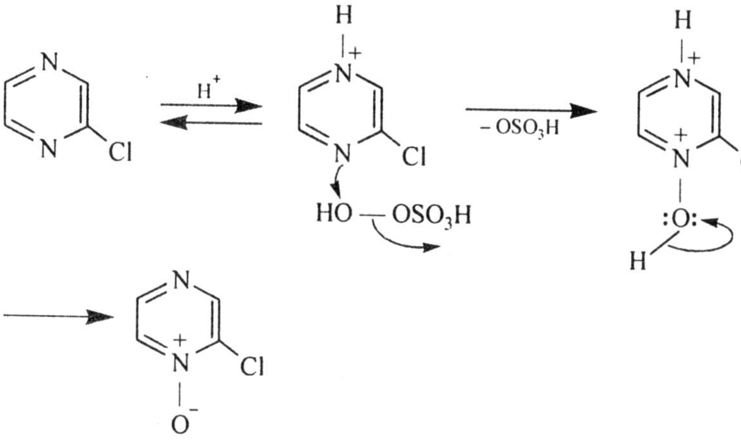

2-Chloropyrazine N-oxide

Sodium perborate yields a mixture of the oxides.[63]

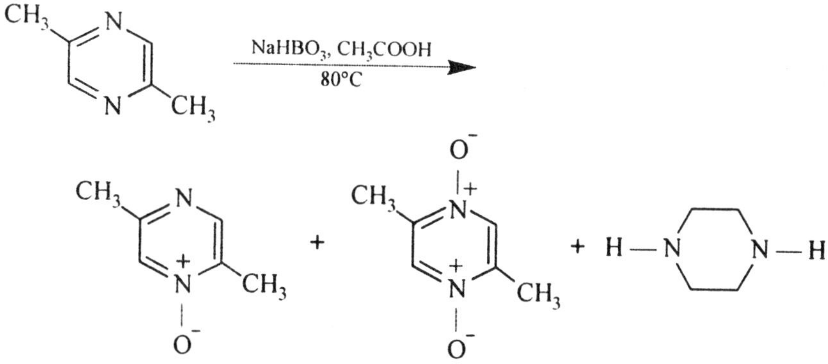

Besides chemical oxidation, enzymatic method of oxidation of methyl groups or aromatic heterocycles is a versatile technique for the preparation of heteroaromatic carboxylic acids. For this purpose *Pseudomonas pitida* ATCC 33015 has been found to be a useful biocatalyst.

2, 5-Dimethylpyrazine with this catalyst is oxidized to 5-methylpyrazine-2-carboxylic acid in 90% yield[64], equation (11.23).

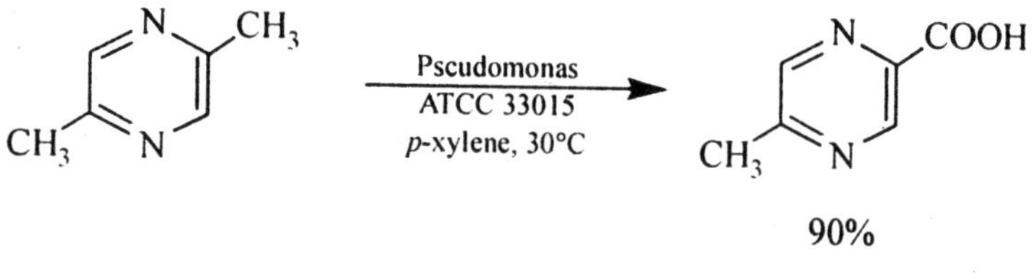

$$(11.23)$$

90%

Various other heterocycles such as 2, 3, 6-trimethylpyrazine, 2, 5-dimethylpyrrole, 2, 5-dimethylfuran, 4-methylthiazole, etc have been oxidized with this reagent to the corresponding 5, 6-dimethylpyrazine-2-carboxylic acid, 5-methyl-2-carboxylic acid, 5-methylfuran-2-carboxylic acid, thiazole 4-carboxylic acid respectively. Similar to pyridine N – oxide the pyrazine N – oxide also reacts with acetic anhydride on warming. In the case of the alkyl substituent, the nature of the product is a function of the location of the substituent[65] as shown below, equation (11.24) and (11.25).

$$(11.24)$$

4-Methyl-6-pyrazinone

$$(11.25)$$

2-Hydroxymethylpyrazine

5.   *Formation of 2, 3-Pyrazyne:* In contrast to the chemistry of pyridines, there are some scattered instances of diazines which are thought to involve dehydroxiazine intermediates.[66-68] Brown and coworkers[73] examined the flash pyrolysis of pyrazine-2, 3-dicarboxylic anhydride and identified an isomeric mixture of maleonitrile and furmaronitrile amongst the products. This led them to suggest a mechanism involving the frangmentation of an intermediate 2, 3-pyrazyne (62).

$$(62)$$

6.   *Photochemical Reactions:* It is known that pyrazine ring is isomerized to other diazine under photolytic conditions. As an example 2, 6-dimethylpyrazine is isomerized to 2, 5-dimethylpyrimidine (63) and 2, 4-dimethylpyrimidine (64).[74] Certain suitably substituted pyrazines undergo retroene type reaction on thermolysis.[71]

(63)

(64)

11.6 QUINOXALINES[72,73]

Quinoxaline (7) is commonly called 1, 4-*diazanaphthalene* or benzopyrazine. The approved numbering of the ring atoms is shown.

(7)   (65)

Quinoxaline and its derivatives are mostly of synthetic origin. Some quinoxaline derivatives are known to possess antibacterial activities. The quinoxaline antibiotic are agents of bicyclic desipeptide antibiotic that have been reported activity against gram positive bacteria and certain tumors and to inhibit RNA synthesis.[74] Quinoxaline has also been used in reactive dyes and pigments, azo dyes, fluorescein dyes and it also forms a part of certain antibiotics.

Quinoxaline is a low melting solid, m.p. 29-30°C and is miscible with water. It is weekly basic (*p*Ka 0.56) and is thus considerably weaker base than the isomeric diazonaphthalenes namely cinnoline (*p*Ka 2.42), phthalizine (*p*Ka 3.47) or quinazoline (*p*Ka 1.95). 2-Hydroxy-but not 2-amino-quinoxaline exist in tautomeric forms.

The fusion of one or two benzene rings in quinoxaline and phenazine (65) increases the number of resonance structures which are available to these systems. It possesses a dipole moment of zero.

Quinoxaline itself is prepared by the reaction of *o*-phenylenediamine and glyoxal.[75] Similarly 2-methylquinoxaline has been prepared from *o*-phenylenediamine and pyruvaldehyde, equation (11.26).

$$(11.26)$$

Taylor[80] recently reported that 1-(*p*-tolylsulfonyl)-2-phenyloxirane, obtained from the condensation of chloromethyl *p*-tolylsulfone with benzaldehyde, on reaction with *o*-phenylenediamine yields 2-phenylquinoxaline in good yield, equation (11.27)

$$(11.27)$$

60%

A number of simple variations of the dialdehyde/diamine reaction appear to work well. Thus replacement of the dialdehyde with an α-halogenoketone results in the formation of 2-substituted quinoxaline. 2-Phenylquinoxaline has been prepared in this manner from phenacyl chloride and *o*-phenylenediamine.[77]

2-Phenylquinoxaline

Compounds containing 1, 2, 3- tricarbonyl functionality have been used in the synthesis of a variety of heterocyclic derivatives.[78,79] The tricarbonyl group containing compounds can be prepared by treating 3-keto ester with p-nitro sulfonyl peroxide,[80] to give 2-(((p-nitrophenyl)-sulfonyl) oxy)- 3-keto esters (66).

(66)   (67)

Treatment of the resulting 2- (nosyloxy)-3-keto ester (66) with triethylamine (TEA) in benzene at room temperature results in *vic* tricarbonyl compound (67). The tricarbonyl compound can be trapped *in situ* with *o*-phenyldiamine to give quinoxaline derivative.[81]

Additional methods for the preparation of quinoxaline and its derivatives have been listed.[83,84]

Quinoxaline forms salts with acids. Nitration occurs only under forcing conditions (conc. HNO_3, oleum, 90°C) to give 5-nitroquinoxaline (1.5%) and 5, 7-dinitroquinoxaline (24%),[85] equation (11.28). 2-Chloroquinoxaline has been prepared by the action of phosphorus oxychloride on quinoxaline-2-one[86] or quinoxaline 1-oxide.[87]

$$(11.28)$$

Oxidation of quinoxaline results in the formation of the product depending on the nature of the oxidizing agent employed. With alkaline potassium permangnate pyrazine 2, 3-dicarboxylic acid is formed, while with peracid quinoxaline *di*-N – oxide resutls. 2-Methylquinoxaline on selenium dioxide oxidation affords quinoxaline 2-carboxaldehyde.

Alkyl radicals produced from acyl peroxides or alkyl hydroperoxides give high yields of 2-substituted alkyl derivatives. Reduction (Na, C_2H_5OH) of quinoxaline gives a 1, 2, 3, 4-tetrahydro derivative.

REFERENCES

1. M. Tisler and B. Stanovanik, *Adv. Heterocyclic Chem.,* 9, 211 (1968)
2. T. L. Jacob, in *"Heterocyclic Compounds"* R. C. Elderfield, (Ed.), Vol. VI, Wiley, New York (1957), p. 101.
3. J. D. Manson and D. L. Aldous, in, *"Heterocyclic Compounds"* (Ed.), R. N. Castle, Vol. XXVIII wiley, New York (1973).
4. T. Tjebbes, *Acta. Chem. Scand.,* 16, 916 (1962).
5. A. R. Katritzky and A. J. Warning, *J. Chem. Soc.,* 1523 (1964).
6. J. Levisailles and P. Baranger, *Compt. rend.,* 24, 444 (1955).
7. J. A. Giles and J. N. Schumacher, *Tetrahedron,* 14, 246 (1961).

8. G. W. H. Cheeseman, *J. Chem. Soc.,* 242 (1960).

9. J. levisalles, *Bull. Chem. Soc. France,* 1004 (1957).

10. T. harada, E. Akiba and A. Oku, *Tet. letters,* 655 (1985).

11. R. A. Carboni and R. V. lindsey, *J. Am. Chem. Soc.,* 82, 1342 (1959).

12. J. Sauer, *Angew. Chem. Int. Edin.,* 6, 16 (1967).

13. J. A. Deyrup and H. L. Gingrich, *Tet. letters,* 3115 (1977).

14. G. B. Barlin and C. Y. Yap, *Aust. J. Chem.,* 30, 2319 (1977).

14a. W. J. Coats and A. Mckillop. *J. Org. Chem.,* 55, 5418 (1990).

15. R. D. Chambers, J. R. Maslakiewicz and K. C. Srivastava, *Chem. Soc. perkin Trans.,* 1, 1130 (1975).

16. J. C. E. Simpson, in *"The Chemistry of Heterocyclic Compounds"* (Ed.), A. Weissberger, Vol. V. Interscience, New York (1953).

17. G. M. Singerman, in *"The Chemistry of Heterocyclic Compounds"* (Ed.), R. N. Castle, Vol. XXVI, Interscience, New York (1972).

18. L. Borsche and H. Herbert, *Ann.,* 546, 293 (1941).

19. K. Schofield and J. C. F. Simpson, *J. Chem. Soc.,* 512 (1945); K. Schofield and T. Swain, *ibid.,* 293 (1949).

20. M. J. S. Dewar and P. M. Maitlis, *J. Chem. Soc.,* 2521 (1957).

21. C. M. Atkinsion and J. C. E. Simpson, *J. Chem. Soc.,* 1469 (1947).

22. R. C. Elderfield and G. L. Wythe, in *"Heterocyclic Compounds",* (Ed.), R. C. Elderfield, Vol. 6, Wiley, New York (1947); p. 186.

23. W. R. Vaughan, *Chem. Revs.,* 43, 447 (1948).

24. G. W. Kerner and A. Todd, in *"HBeterocyclic Compounds",* (Ed.), R. C. Vol. 6, Wiley New York (1957); p. 234.

25. J. Sauer and G. Heinrichs, *Tet. letters.,* 4979 (1966).

26. D. J. Brown, *Revs. Pure Appl. Chem.,* **3,** 115 (1953).

27. D. J. Brown, *"The pyridines"* (*Ed.*), A. Weissberger, Interscience, New York (1962); pp. 183-210.

28. J. B. Dickey and A. R. Gray, *Org. Syn. Coll. Vol.,* 2, 60 (1943).

29. S. R. James and C. B. Reese, *Tet. letters,* 1453 (1981).

30. A. L. Weis and V. Rosenbach, *Tet. letters,* 1453 (1981).

31. G. W. Kenner and A. Todd, in, *"Heterocyclic Compounds",* (*Ed.*), R. C. Elderfield, Vol. 6. Wiley, New York (1957).

32. H. Bredereck *et al., Angew. Chem.,* **71,** 573 (1959); *Chem. Ber.,* **92,** 1314 (1957).

33. R. R. Hunt, J. F. W. McOmie and E. R. Sayer *J. Chem. Soc.,* 525 (1959).

34. R. M. Dodson and J. K. Seyler, *J. Org. Chem.,* **16,** 461 (1952). .

35. E. K. Ryn and M. MacCross, *J. Org. Chem.,* **46,** 2819 (1981).

36. H. Bredereck, *et al., Angew. Chem.,* **70,** 571 (1958).

37. H. Yamanaka, T. Sakamoto and Y. Kondo, *Synthesis,* 252 (1984).

38. J. H. Chesterfield, J. F. W. McOmie and M. S. Tute *J. Chem. Soc.,* 4590 (1960).

39. R. Huisgen, *Angew. Chem.,* 75, 628 (1965).

40. G. W. Rice and R. S. Tobia, *Chem. Comm.,* 994 (1975).

41. B. A. Otter, A. Tamble and J. J. Fox, *J. Org. Chem.,* **36,** 1261 (1971).

42. V. M. Ingram, *"The Biosynthesis of Macromolecules",* W. A. Benjamin, Inc., New York (1965).

43. R. K. Robins, in *"Hetrerocyclic Compounds",* R. C. Elderfield, (*Ed.*), Vol. 8, Wiley, New York (1967).

44. W. L. F. Armarego, *Adv. Heterocyclic Chem.,* **1,** 253 (1963).

45. T. A. Williams, in *"Heterocyclic Compounds",* R. C. Elderfield, (*Ed.*), Wiley, New York (1957), Chapter 8.

46. K. Schofield, T. Swain and R. S. Theobald., *J. Chem. Soc.,* 1924 (1952).

47. A. Albert and A. Hampton, *J. Chem. Sco.,* 505 (1954), also see H. A. El-Sherief, *Bull. Chem. Soc. Japan,* **57,** 1138 (1984).

48. W. L. F. Armergo, *J. Appl. Chem.,* **11,** 70 (1961).

49. K. Schofield, *J. Chem. Soc.,* 1927 (1952).

50. R. C. Elderfield, T. A. Williamson, W. J. Gensler and C. B. Kremer, *J. Org. Chem.,* **12,** 405 (1947).

51. A. Albert, W. L. F. Armarego and F. Spinner, *J. Chem. Soc.,* 2689 (1961).

52. G. W. H. Cheeseman and E. S. G. Werstink, *Adv. Heterocyclic Chem.,* **14,** 99 (1972).

53. Y. T. Pratt, in *"Heterocyclic Compounds",* R. C. Elderfield, (*Ed.*) Vol. 6, Wiley, New York (1957), p. 377.

54. G. R. Ramage and J. K. Landquist, in *"Chemistry of Carbon Compounds",* (*Ed.*), E. H. Rodd, vol. 4B, Elsevier, Amsterdam (1959), p. 1318.

55. I. Flament and M. Stoll, *Helv. Chem. Acta.,* **50,** 1754 (1967).

56. P. G. Sammes *et al., J. Chem. Soc. Perkin Trans.,* **1,** 2494 (1972).

57. A. Padwa, M. Dharan, J. Smolanoff and S. I. Wetmove, Jr., *J. Am. Chem. Soc.* **95,** 1954 (1973).

58. H. Rutner and P. E. Spoerri, *J. Heterocyclic Chem.,* **3,** 435 (1966).

59. M. R. Kamal and R. Levine, *J. Org. Chem.,* **27,** 1355 (1962).

60. J. D. Behun and R. Levine., *ibid.,* **26,** 3379 (1961).

61. (*a*) E. Ochiai, *"Aromatic amine Oxides",* Elsevier, Amsterdam (1967). (*b*) N. Sato, *J. Org. Chem.,* **43,** 3307 (1978).

62. G. E. Mixan and R. G. Pews, *J. Org. Chem.,* **42,** 1809 (1977).

63. A. Oliata and M. Oliata, *Synthesis,* 216 (1985), A. Oliata, *et al., J. Heterocyclic Chem.,* **20,** 311 (1983).

64. A. Kienev, *Angew. Chem. Int. Edin.,* (Engl.), **31,** 774 (1992).

65. C. F. Kolesch and W. H. Gumpercht, *ibid.,* **23,** 1603 (1958).

66. H. C. van dern Plas and G. Geurtsen, *Tet. letters,* 2093 (1964).

67. T. Kauffman, J. Hanesen, K. Udluff and R. Wirthhweir, *Angew. Chem.,* **76,** 590 (1964).

68. T. Kauffmann and A. Risberg, *Tet. letters,* 3913 (1967).

69. R. F. C. Brown *et al., Chem. and Ind.,* 343 (1966).

70. F. Lahamani and B. Ivanoff, *Tet. letters,* 3913 (1967).

71. Y. Houminer, *J. Org. Chem.,* **45,** 999 (1980).

72. G. W. H. Cheesman and R. F. Cookson, in *"The Chemistry of Heterocyclic Compounds",* A. Weissbergerr and E. C. Taylor, (*Eds.*) Vol. 35, John Wiley, New York (1979).

73. J. C. E. Simpson, *"Condensed Pyradazine and Pyrazine Rings",* Interscience, New York (1979).

74. H. Otsuka and J. Shoji, *Tetrahedron,* **23,** 1536 (1967).

75. R. G. Jones and K. C. McLaughlin, *Org. Synth.,* **30,** 86 (1950).

76. E. C. Taylor, C. A. Maryanoff and J. S. Skotnicki, *J. Org. Chem.,* **45,** 2512 (1980).

77. G. W. H. Cheeseman, *Adv. Heterocyclic Chem.,* **2,** 203 (1963).

78. H. H. Wasserman, R. Amici, R. Frechette and J. H. van Duzer, *Tet. Letters,* 869 (1989).

79. H. H. Wasserman, *Aldrichimica Acta.,* **63,** 20 (1987).

80. K. Schank and C. Lick, *Synthesis,* 392 (1983).

81. R. V. Hoffman, H. O. Kim and A. L. Wilson, *J. Org. Chem.,* **55,** 2820 (1990).

82. R. V. Hoffman and D. J. Huizenga, *ibid,* **56,** 6435 (1991).

83. F. Sannicolo, *J. Org. Chem.,* **48,** 2924 (1983).

84. M. Z. Aminbadr, G. M. El-Naggar, H. A. H. El Sherief and S. A. Mahgoub, *Bull. Chem. Soc. Japan,* **57,** 1653 (1984).

85. M. J. S. Dewar and P. M. Mailtis, *J. Chem. Soc.,* 2518 (1957).

86. G. W. H. Cheeseman, *J. Chem. Soc.,* 3236 (1957).

87. Y. Ahmad, M. S. Habib, and B. Bakhtiari, *J. Org. Chem.,* **31,** 2613 (1966).

REVIEW PROBLEMS

11.1 Predict the products of the following reactions :

(a) A 3,6-diphenyl-1,2,4,5-tetrazine $+ CH_3CH_2CH_2\overset{\text{O}}{\overset{\|}{C}}CH_2CH_3 \longrightarrow$

(b) 4-chloro-1,2-diaminobenzene $+ PhCH\overset{\text{O}}{\diagup\!\!\diagdown}CH\text{-}SO_2\text{-}Tol\text{-}p \xrightarrow{C_2H_5OH,\ \Delta}$

(c) $\xrightarrow{\text{NH}_3}$

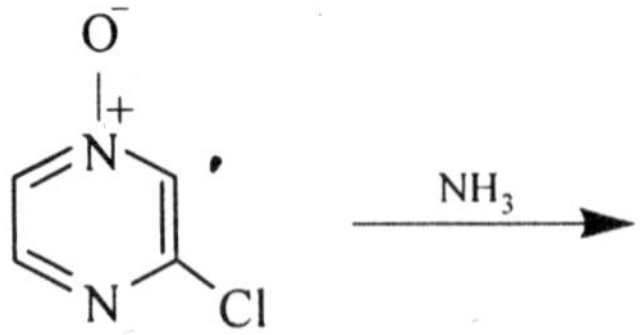

(d) $\xrightarrow[\text{[2H]}]{\text{HI, P}}$

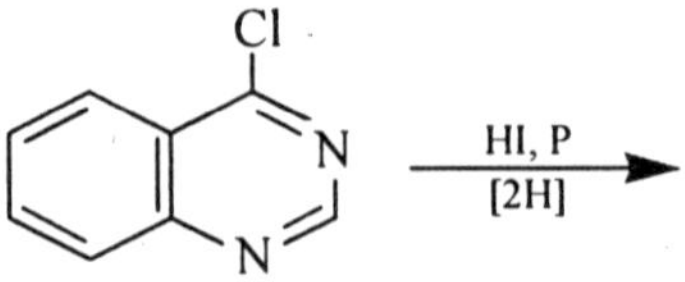

(e) $\xrightarrow{\text{H}^+}$

(f) $\xrightarrow{\text{NaOH}}$

(g) $(C_2H_5)_2C\begin{smallmatrix}CN\\COOH\end{smallmatrix} \quad + \quad \begin{smallmatrix}H_2N\\H_2N\end{smallmatrix}C=O \longrightarrow$

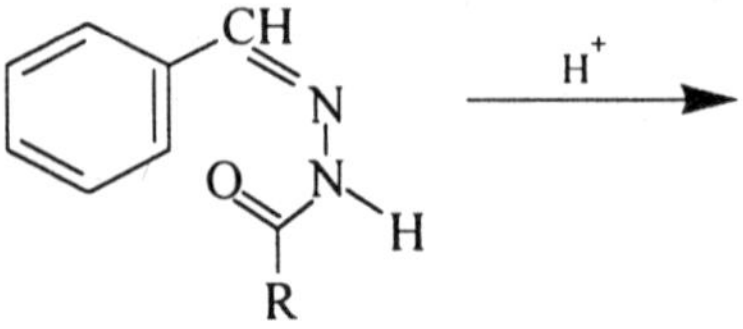

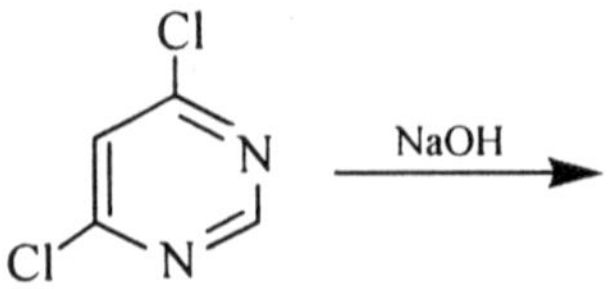

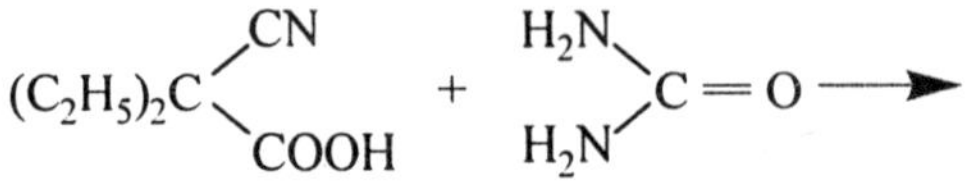

(h) 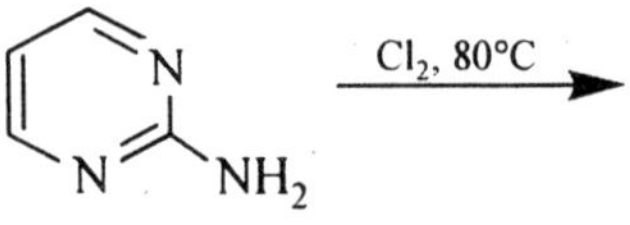 $\xrightarrow{\text{Cl}_2,\ 80°C}$

(i) $\xrightarrow{\text{OCH}_3^-\ (\text{excess})}$

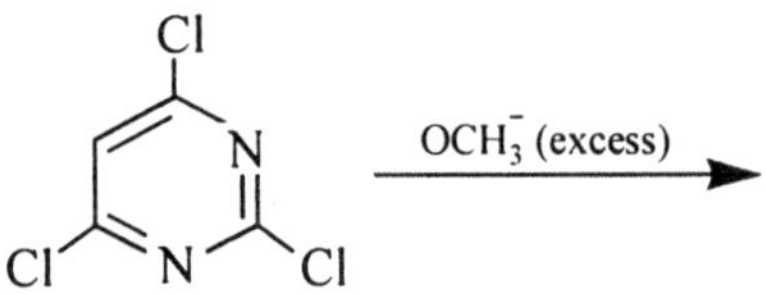

(j) $\longrightarrow$

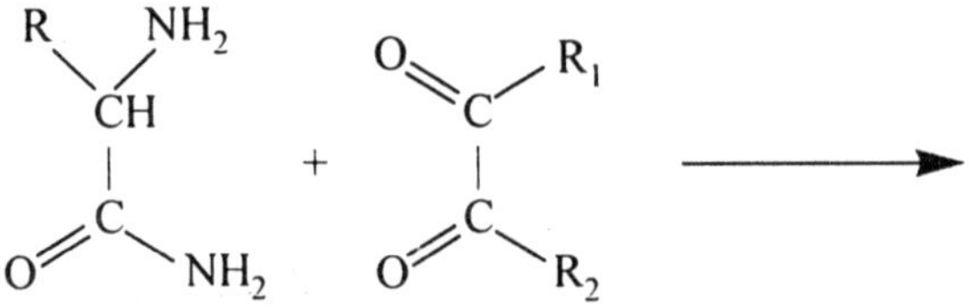

(*r*) 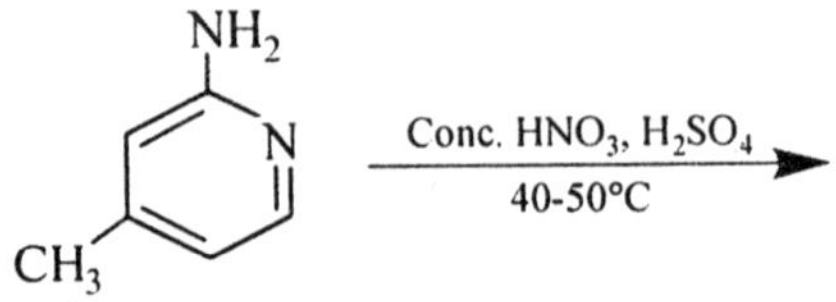
Conc. HNO_3, H_2SO_4
40-50°C

(*s*) 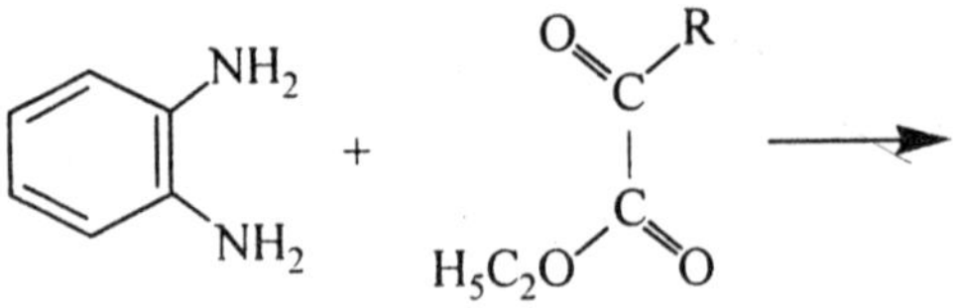

(*t*) 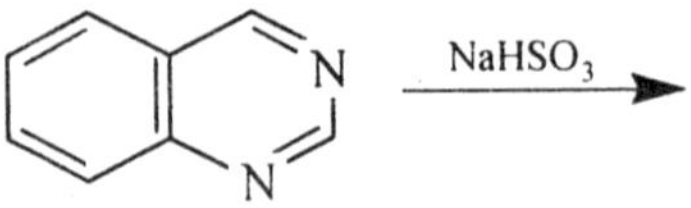
$NaHSO_3$

11.2 Postulate a suitable mechanism for each of the following reactions :

(*a*) 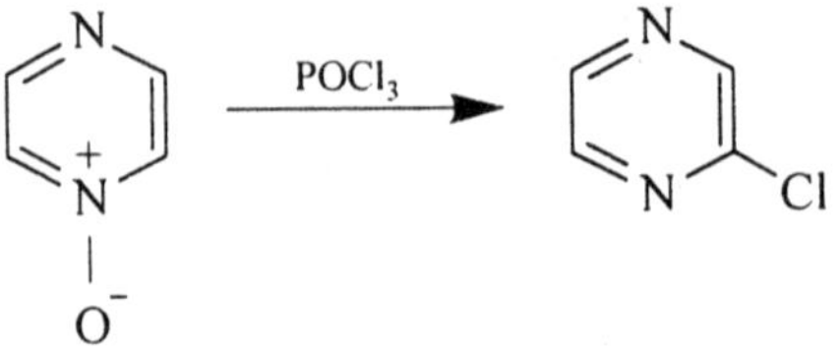
$POCl_3$

(*b*) 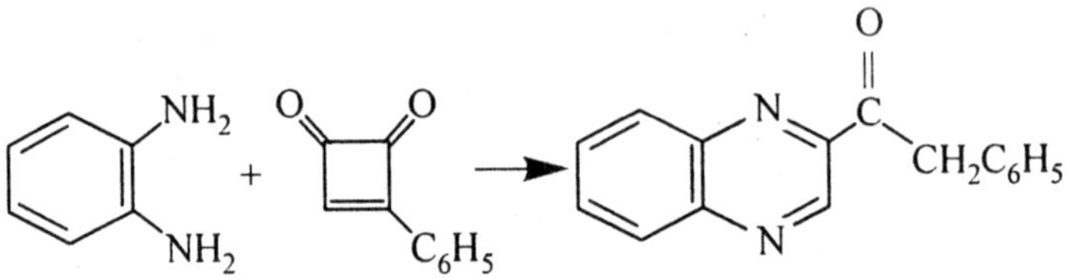

(*c*) 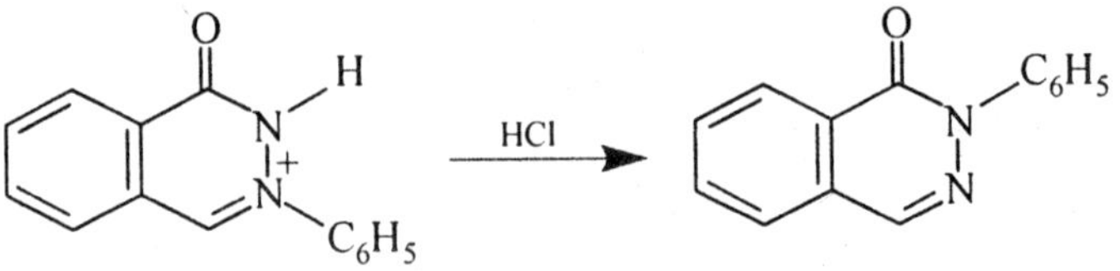
HCl

(*d*) 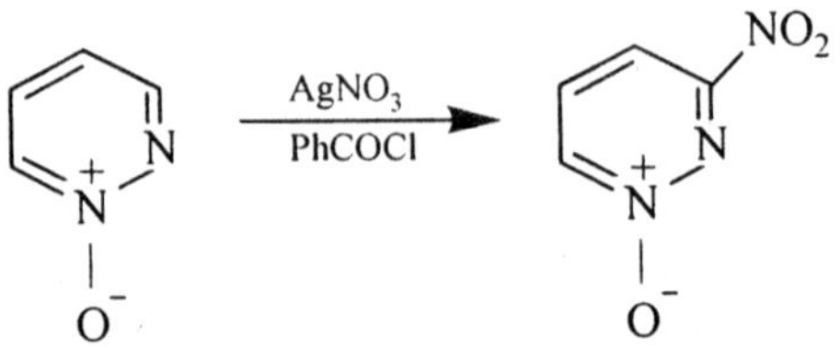
$AgNO_3$
PhCOCl

(e)   (i) CH$_3$MgI, ether (ii) H$_2$O

(f)   HNO$_2$, HCl

11.3 Offer explanation for the following observations:

(*a*) The benzo derivatives of diazines undergo electrophhilic substitution in the benzene ring of the molecule.

(*b*) A hydroxy derivative of pyrimidine is less soluble in water than pyrimidine itself.

(*c*) Pyridazine has a high dipole moment whereas pyrazine has a value of zero.

(*d*) Pyrimidine (*p*Ka 1.30) is much less basic than pyridine (*p*Ka 5.2).

(*e*) The diazines are more resistant to electrophilic attack than pyridine.

(*f*) Pyrimidine and pyrazine but not pyridazine can be diprotonated.

11.4 Give the preparation and properties of quinoxalines.

11.5 Discuss the synthesis of phthalazine.

11.6 Describe the nucleophilic reactions of pyrimidines and quinazolines.

11.7 Compare the electrophilic and nucleophilic properties of pyrazines and pyrimidines.

11.8 Give an account of some naturally occurring pyrimidine containing compounds.

Index

Hemin, 175
Hetarynes, 253
    reactions, 254
Heterocyclic compounds, 1
Hexadecyltriethylammonium chloride,
    113
Hexahydropyridine, 231
Hexamethylenetriamine, 89
Histidine, 462
Hoch Campbell aziridine synthesis, 17
Hiickle's rule, 1
3-Hydroxyazitidine, 101
4-Hydroxy-2-butenolide, 192
4-Hydroxycarbonyl-2-quinolone, 343
β-Hydroxyethylamine, 20
3-(2-Hydroxyimino) propyliden-
    2-oxindole, 320
6-Hydroxyimino-7,7-dimethyl-2-
    azobicyclo [2.2.1] heptan-1-one, 7
N-Hydroxymethylamino azetidine, 101
O-Hydroxymethyldiphenylamine, 407
2-Hydroxy-4-methylthiazole, 485
2-Hydroxymethylpyrazine, 534
2-Hydroxy-4-phenylquinazoline, 527
4-Hydroxy-3-phenylcinnoline, 511
3-Hydroxypyridazine, 503
2-Hydroxy-2*H*-pyran, 278
N-(2-hydroxyphenyl) thioamide, 475
8-Hydroxyquinoline, 385
    metal complex, 385
2-Hydroxyquinazolinium salt, 403
Hydroxylammino-O-sulfonic acid, 76
Hypoxanthine, 522

I

Imidazoles, 454
    basicity, 455
    dipole moment, 456
    H-bond formation, 455
    reactions:
        with acids, 458
        cycloaddition, 462
        electrophilic, 458
        nucleophilic, 460
        oxidizing agents, 460
        reducing agents, 460
    synthetic methods:
        Radiszeowski synthesis, 456
Imidazolines, 455

8,9-Indan oxide, 430
4-Indanol, 432
Indazole, 448
Indene, 394
Indigo, 328
3-Indolesulfonic acid, 315
Indoles, 299
    basicity, 300
    derivatives, 321
    reactions:
        alkylation, 313
        electrophilic substitution, 313
        nucleophilic substitution, 319
        oxidizing agents, 316
        photochemical, 320
        reducing agents, 287
    synthetic methods:
        Bischler, 308
        Fischer, 300
        Gassman, 311
        Japp-Klingemann, 303
        Madelung, 304
        Nenitzescu, 305
        Reissert, 308
Indolizines, 334
    basicity, 335
    resonance, 335
    reactions, 336
    synthetic methods:
        from 2-picoline, 335
Indoline, 318
Indophenine, 327
Indoxazene, 463
Indoxazole, 419
Indoxyl, 325
2-Iodo-2-phyenylethanol, 39
(+)-Ipomeamarone, 199
Isatin, 325
Isobenzofuran, 345
    reactions:
        Diels-Alder, 347
    synthetic methods, 345
Isobenzothiophene, 353
Isocarbostyril, 397
Isoindoles, 331
    reactions:
        Diels-Alder, 334
    synthetic methods, 332
1-N-Isopropylidene-endene-3-amine, 61
Isoquinolines, 390